Analytical Methods and Concepts in Chemistry

Analytical Methods and Concepts in Chemistry

Editor: Taylor Spencer

STATES
ACADEMIC PRESS
www.statesacademicpress.com

Published by States Academic Press
109 South 5th Street,
Brooklyn, NY 11249, USA
www.statesacademicpress.com

Analytical Methods and Concepts in Chemistry
Edited by Taylor Spencer

International Standard Book Number: 978-1-63989-047-7 (Hardback)

Cataloging-in-Publication Data

Analytical methods and concepts in chemistry / edited by Taylor Spencer.
p. cm.
Includes bibliographical references and index.
ISBN 978-1-63989-047-7
1. Chemistry. 2. Chemistry, Analytic. I. Spencer, Taylor.
QD31.3 .A53 2022
540--dc23

Contents

Preface

Chemistry is a natural science, which is involved with the study of elements and compounds, their structures, properties, composition, behavior and reactions with other substances. The branch of chemistry that deals with the separation, identification and quantification of matter is known as analytical chemistry. Such analyses of material samples help in understanding of chemical structure and composition. This can be performed through various classical or wet chemical methods as well as modern or instrumental methods. Separation of materials is possible through processes like distillation, extraction and precipitation, or through electrophoresis, chromatography or field flow fractionation. Material analysis usually employs the interaction of the material with light, heat, and electric and magnetic fields. Such analytical methods have major applications in the fields of science, engineering, forensics and medicine. This book contains some path-breaking studies in modern analytical chemistry. Also included is a detailed explanation of the various analytical methods of chemistry and their applications. It is an essential guide for both academicians and those who wish to pursue this discipline further.

This book is the end result of constructive efforts and intensive research done by experts in this field. The aim of this book is to enlighten the readers with recent information in this area of research. The information provided in this profound book would serve as a valuable reference to students and researchers in this field.

At the end, I would like to thank all the authors for devoting their precious time and providing their valuable contribution to this book. I would also like to express my gratitude to my fellow colleagues who encouraged me throughout the process.

Editor

1

High-Precision In Situ $^{87}Sr/^{86}Sr$ Analyses through Microsampling on Solid Samples: Applications to Earth and Life Sciences

Sara Di Salvo,[1] Eleonora Braschi,[2] Martina Casalini,[1] Sara Marchionni,[1] Teresa Adani,[1] Maurizio Ulivi,[1] Andrea Orlando,[2] Simone Tommasini,[1] Riccardo Avanzinelli,[1,2] Paul P. A. Mazza,[1] Sandro Conticelli,[1,2] and Lorella Francalanci[1,2]

[1]*Dipartimento di Scienze della Terra, Università degli Studi di Firenze, via Giorgio La Pira 4, 50121 Firenze, Italy*
[2]*C.N.R., Istituto Geoscienze e Georisorse, U.O. di Firenze, via Giorgio La Pira 4, 50121 Firenze, Italy*

Correspondence should be addressed to Eleonora Braschi; eleonora.braschi@igg.cnr.it

Academic Editor: Veronica Termopoli

An analytical protocol for high-precision, in situ microscale isotopic investigations is presented here, which combines the use of a high-performing mechanical microsampling device and high-precision TIMS measurements on micro-Sr samples, allowing for excellent results both in accuracy and precision. The present paper is a detailed methodological description of the whole analytical procedure from sampling to elemental purification and Sr-isotope measurements. The method offers the potential to attain isotope data at the microscale on a wide range of solid materials with the use of minimally invasive sampling. In addition, we present three significant case studies for geological and life sciences, as examples of the various applications of microscale $^{87}Sr/^{86}Sr$ isotope ratios, concerning (i) the pre-eruptive mechanisms triggering recent eruptions at Nisyros volcano (Greece), (ii) the dynamics involved with the initial magma ascent during Eyjafjallajökull volcano's (Iceland) 2010 eruption, which are usually related to the precursory signals of the eruption, and (iii) the environmental context of a MIS 3 cave bear, *Ursus spelaeus*. The studied cases show the robustness of the methods, which can be also be applied in other areas, such as cultural heritage, archaeology, petrology, and forensic sciences.

1. Introduction

In situ radiogenic isotope determinations with microscale resolution, especially of Sr, can represent a powerful tool in different fields of geological and life sciences. In particular, this technique is nowadays one of the most important methods for the investigation and interpretation of magmatic processes, as well as of environmentally-induced responses of terrestrial mammals; it has the potential to greatly enhance our understanding of not only volcanic systems and the related magma genesis and evolution, but also of the physiological mechanisms behind specific organic adaptations.

In situ $^{87}Sr/^{86}Sr$ provides significant data on the (i) source heterogeneities of magmas, (ii) crystallization histories within shallow level magmatic reservoirs, and (iii) magma residence times prior to eruptions (e.g., [1–10]). Since crystals record changes occurring in the environment in which they grow (e.g., [7, 11–15]), isotopic investigations at grain and subgrain scales on rock-forming minerals provide information on mineral-whole rock equilibria that constrain the magmatic processes occurring during magma evolution (e.g., mixing, mingling, crystals recycling, crustal contamination, or metasomatism). In addition, variation of Sr-isotope composition from core to rim within the same crystal can shed light on the complex pre-eruptive history of active volcanoes. Therefore, combining micro-(small-scale) isotope data with textural and petrographic data provides significant information on crystals residence time, magma production rates, and recharge dynamics (e.g., [5–8, 16–18]).

Sr-isotopic investigation has recently gained popularity in other fields, such as archaeology, anthropology, biology, cultural heritage, environmental and food studies, forensics, life and medical sciences, and palaeontology. (e.g., [19–38]). $^{87}Sr/^{86}Sr$ on tooth enamel and bone tissues gained particular attention over the last years in life studies, archaeology, palaeontology, and forensic sciences (e.g., [39–48]). In situ analyses of $^{87}Sr/^{86}Sr$ using laser ablation and multicollector-inductively coupled plasma mass spectrometry (LA-MC-ICPMS) for tooth enamel, bones, and rocks were developed in the early new millennium but have met variable success (e.g., [49–53]). At the same time, microscale sampling through drilling and micro-Sr isotope analyses by thermal ionisation mass spectrometry (TIMS) was also developed on geological materials (e.g., [6, 18, 54]). As compared with mechanical microdrilling plus TIMS procedures, LA-MC-ICPMS has the advantage of significantly reducing the time of analyses, but at the expense of precision of the measured isotope ratio; this is generally due to smaller Sr signals and the need to correct for isobaric interferences (e.g., [16, 53]).

We present a detailed protocol for in situ sampling through microdrilling, Sr purification, and thermal ionisation mass spectrometer (TIMS) high-precision determinations of small amounts of Sr (<10 ng) in both biological and geological materials at the same error levels. In situ $^{87}Sr/^{86}Sr$ analysis is presented in three case studies which deal with the following: (i) plagioclase crystals from Nisyros volcano (Greece), (ii) glassy matrix of single glassy clasts from the 2010 eruption of the Eyjafjallajökull volcano (Iceland), and (iii) bones and teeth from fossil remains of the *Ursus spelaeus*. In these three examples, the in situ Sr-isotope approach permits to constrain petrological and volcanological processes and to effectively outline the life habits and habitat exploitation of extinct living species. The three presented cases aim to show the high potential of the methodology, validating its wide-scale applicability in many other areas, such as cultural heritage, archaeology, and forensic sciences.

2. Materials and Methods

Technological improvements on mass spectrometer and microsampling devices allow researchers to collect and analyse small amounts of sample (few micrograms of sample, containing nanograms of Sr) with no loss of precision in the isotopic determination. In situ analyses have many important advantages over more typical ones on bulk samples. In geological applications, it allows to preserve the textural information and thus to combine it with the isotopic and geochemical composition of specific portions of the samples. In archaeology and palaeontology, this method has the advantage of minimising the damage and/or destruction of samples, thereby leaving significant amounts available for further applications.

The procedure consists in three main stages: (i) in situ sampling through microdrilling, (ii) sample digestion and purification of the element of interest, in our case Sr, and (iii) measurement of the isotope ratios (i.e., $^{87}Sr/^{86}Sr$) through thermal ionisation mass spectrometry (TIMS). Our Radiogenic Isotope Laboratory at the Department of Earth Sciences of the Florence University is equipped with a modern MicroMill™ grinder device, an ultraclean laboratory ("Class 1000") for microsamples digestion and elemental separation and a thermal ionisation mass spectrometer (ThermoFinningan™ Triton-Ti®) for isotopic measurements.

Compared to classic, bulk sample analyses (generally measured on 100–150 ng of Sr [55]), small-sample analysis (typically 5 to 10 ng of Sr) has the drawback of being more exposed to contamination from Sr alien to the sample. In situ micro-Sr measurements therefore require continued testing of laboratory blanks during the whole analytical procedure.

2.1. In Situ Sampling

2.1.1. The Microdrilling Device. Microsampling on minerals, glasses, and tooth enamel reported in this paper was performed using a microdrilling device capable of high resolution milling (New Wave-Merchantek MicroMill™, https://www.esi.com/products/laser-processing/milling/micromill/). The MicroMill (Figure 1(a)) combines a binocular microscope (with 6.7x–40x magnifying power) with remotely controlled submicron stage resolution and positional accuracy and a real-time video observation (at 3x digital magnification). It includes a low-eccentricity high-torque milling chuck, with variable speed (1,200–35,000 rpm), wherein a tungsten carbide or diamond-tipped bit is fixed, and an automated high-precision sliding stage on which the sample is loaded (Figure 1). The open stage architecture can accommodate thin sections or mounts and also larger solid samples such as bones, shells, ceramics, and plastics. The plug where the bit is hosted moves with adjustable speed for both spin and vertical movement, along the Z direction. The stage moves along the X-Y direction with a precision of 1 μm and maximum shift of 5 cm, allowing a high spatial resolution to the micron scale. This yields high spatial resolution, to the micron scale, and allows small-size sampling (i.e., a few μg of powder). The digital camera, placed next to the milling chunk (Figure 1(c)), shows a live image of the stage and of the bit position (Figure 2(a)). Using the microscope position mode (scope position), the optical image of the sample can be monitored on the PC screen prior to and after the sampling (Figure 2(b)).

The device allows in situ microsampling on several types of solid materials such as rocks, minerals, glasses, plastics, bone tissues, ceramics, metals, and alloys. The designed software package also allows performing different milling patterns such as holes arranged randomly, lines, or rasters (Figure 2(b)). Fine adjustment of milling velocity helps the microsampling of solid materials with different hardness.

Microsampling on geological (e.g., rock, minerals, and glasses) and archaeological materials (glasses, ceramics, teeth, and bones) is performed using either thick polished sections (some 100 μm thick) or sample mounts (e.g., small chunks, chips, fragments of bones, and teeth). In our case studies, we used thick, polished sections for mineral and volcanic glasses and mounts for the teeth; both were fixed on the stage using either a double-sided adhesive tape or hot glue, to avoid sample displacement during milling. Thick, polished sections are preferable for geological samples because they permit to

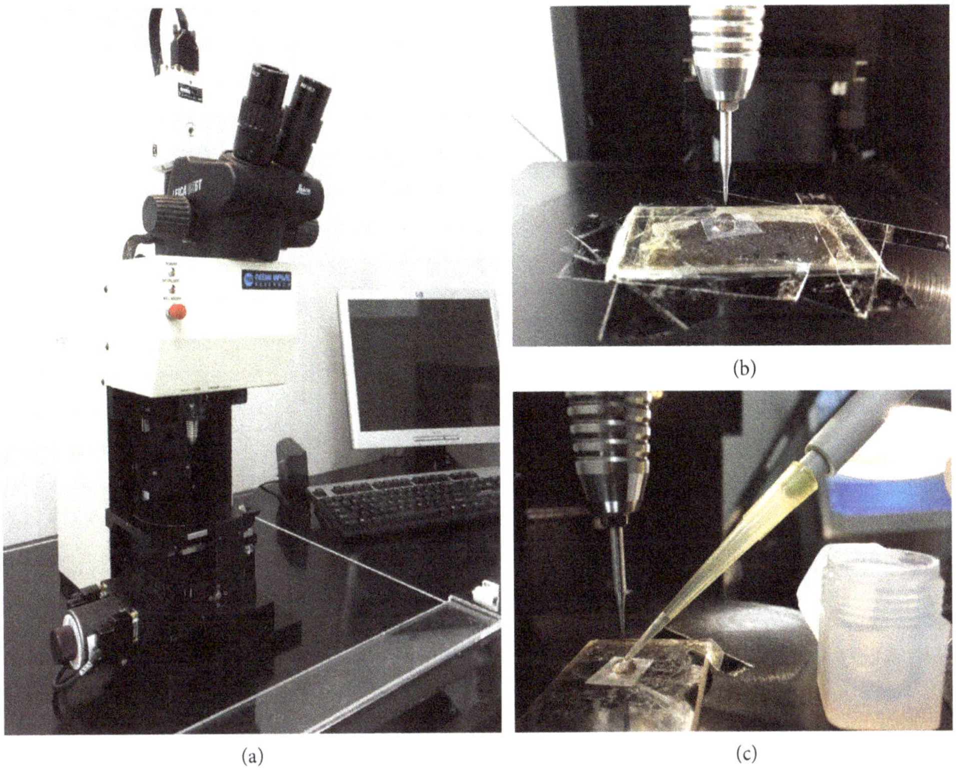

Figure 1: (a) The New Wave-Merchantek Micromill device operating at the Department of Earth Sciences–University of Firenze; (b) image of a petrographic polish thick section fixed on the sample stage under the tungsten carbide drilling bit that is locked into the milling chuck. A Milli-Q droplet constrained by the Parafilm is placed on the section in order to collect the powder during the drill; and (c) sample slurry recovery from the drilled surface into the digestion beaker.

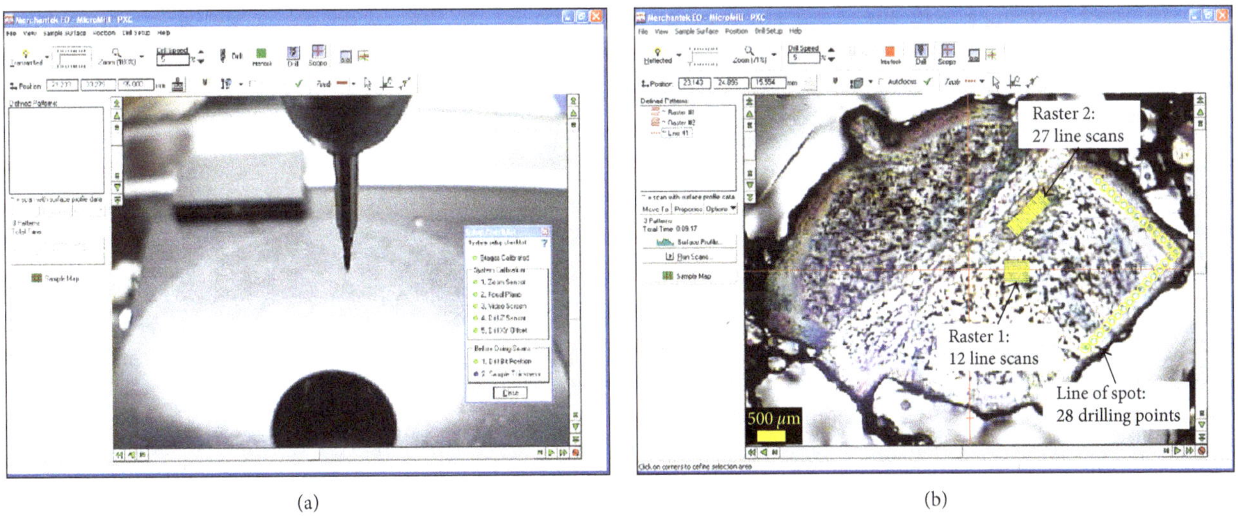

Figure 2: (a) Live image of the sample stage and drill bit shown by the digital camera placed next to the milling chuck; (b) microscope view image showing a zoned plagioclase crystal with drilling pattern. Two rasters (in yellow) are set up to drill the plagioclase core, whereas a line of spot is set up for the rim microdrilling.

characterise the petrographic features of the samples and thus to perform the microsampling according to their textural properties.

2.1.2. Milling Procedure and Sample Collection. A droplet of Milli-Q® water is placed with a micrometric pipette on the selected area prior to milling; this is performed by sticking a small punched square of warmed-up Parafilm™ on the sample surface (Figure 1(c)). The water droplet retains the powder produced by the milling, which can then be easily collected by pipetting; it has also the effect of cooling the microdrill bit while milling. Before each drilling session, the drill bit is ultrasonically cleaned with pure ethanol and then rinsed with Milli-Q water.

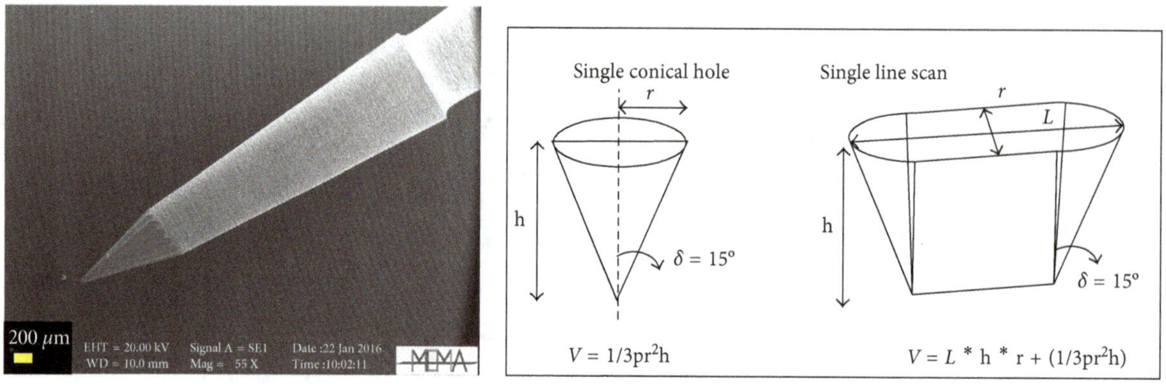

(a) (b)

FIGURE 3: (a) SEM imaging of a tungsten carbide drill bit used for milling: the tip angle is 30° and geometrical shape of (b) a single conical hole and single line scan.

The instrument software provides different drilling patterns and depths: single or multiple independent holes, spot lines, grids, line scans, or rasters (Figure 2(b)). Milling spot lines (or grids) are more accurate, but more time-consuming; they were used for the geological material (crystals and volcanic ashes) which requires more precise spatial resolution between the different zones of the same crystal or between the thin films of glass. Milling failure, such as crystal breaking, was prevented by setting a slow scan speed and splitting the milling into two or more steps. Line scans, which are performed faster but less accurately, were used for drilling the teeth. The number of points, lines, or rasters to be milled (which accounts for the amount of Sr to be collected) need to obtain a sufficient quantity of Sr for the TIMS measurements and can be calculated based on (i) the Sr content of the sample (independently determined by LA-ICPMS), (ii) the geometry of the drill bit, and (iii) the drilling pattern and the depth. Tips of different size and shape can be used for the drilling; therefore, the volume of material actually removed from different depths of a single hole or at different depths and lengths of a single line needs to be carefully calculated. The tungsten carbide mill bits supplied with the microdrill device (Komet–Brassler), have conical shape with an angle of 30° (Figure 3). The volume removed during each drilling is equivalent to that of the conical tip and dependent on the geometry of the drilling pattern, as well as on the specific depth (Figure 3). The minimum amount of sample that needs to be drilled depends also on the total procedural blank, which should be at least two orders of magnitude lower than the total amount of Sr collected from the sample.

After milling was completed, the sample slurry was collected with a micropipette in a PFA beaker and then transferred in the clean lab for sample digestion and elemental purification. The blanks of the milling procedure were determined by keeping the drill bit tip into a Milli-Q water droplet on the sample surface (accurately cleaned before use) for as long as the average sampling time; the droplet was then processed as an ordinary sample. The amount of Sr in the blank was then determined through isotope dilution, by adding a single-spike solution (enriched in ^{84}Sr).

2.2. Sample Dissolution and Sr Purification. The purification of the element of interest, in our case Sr, is crucial to obtain high-precision isotopic measurements for at least two reasons. First, it avoids isobaric interferences on the masses that will be analysed by mass spectrometry; in the specific case of Sr isotope measurements, even a small amount of ^{87}Rb will add to ^{87}Sr, yielding an overestimate of the ^{87}Sr/^{86}Sr ratio. Secondly, the presence of other elements of the matrix will compete with Sr during the thermal ionisation process, reducing the Sr signal and thus yielding less accurate measurements. The possibility of collecting and processing the sample for the purification of the element of interest is one of the major advantages of the method presented here over other methodologies, which do not achieve the same degree of accuracy and precision. The LA-MC-ICPMS methods allow faster data acquisition and higher sample throughput than mechanical microdrilling plus TIMS procedures, thanks to the possibility of introducing the samples directly into the mass spectrometer without chemical separation. On the other hand, LA-MC-ICPMS measurements require careful monitoring and corrections to minimize isobaric interferences in order to achieve suitable analytical accuracy and precision (e.g., [49, 53, 54, 56]).

Powder digestion and Sr purification were carried out in our ultraclean laboratory ("Class 1000") aiming at the following: (i) optimising the separation of Sr and Rb to avoid interference of ^{87}Rb with ^{87}Sr, (ii) purifying the Sr collection form all the matrix analytes, (iii) maximising the yield of the columns during the chromatographic purification, and (iv) preserving low procedural blanks. Sample digestion was performed by sequential HF-HNO_3-HCl as described in [55]. Chromatographic Sr purification was performed using Eichrom® Sr-Spec™ resins (100–150 μm) in quartz micro-columns (0.14 ml volume; Figure 4). Matrix elements were flushed out through elution with 14 column volumes of 3 N HNO_3. Sr was then collected in Milli-Q (13 column volumes). The collected Sr fractions were further treated with concentrated HNO_3 and H_2O_2 (fluxing at 150°C on a hot-plate) to remove any organic residue. After this final step, samples were diluted in HNO_3 (10 vol.%) and were finally ready for loading on filaments for mass measurements. The

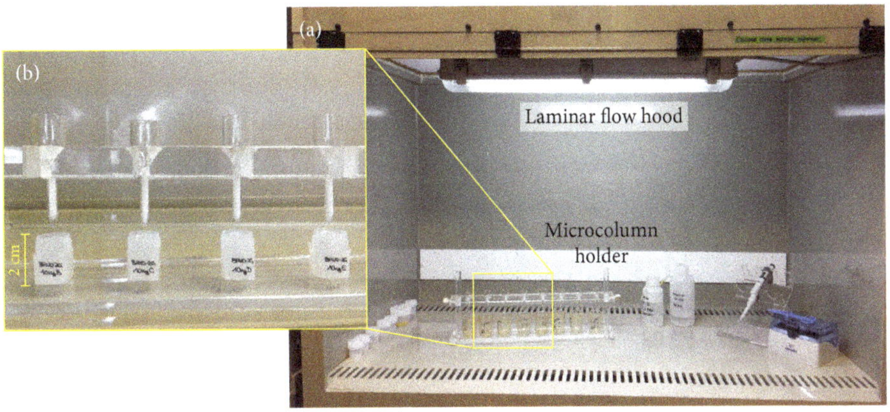

FIGURE 4: (a) Image of the laminar flow hood used for chemical processing and Sr separation of microdrilled samples; (b) image of quartz microcloumns filled with approximately 140 μl of specific chromatographic resin (Eichrom Sr-Spec, 100–150 μm) with high Sr-recovery efficiency, for Sr element extraction.

TABLE 1: Blank contamination level.

	Laboratory blanks			Isotope ratios on blanks		
	Sr pg	1 SD	*n*	$^{87}Sr/^{86}Sr$	1 SD	*n*
Standard procedure on large size samples	127	60	32	0.707497	0.000060	5
Procedure on small size samples						
Total procedure (drilling, digestion, and elemental selection)	38	19	16			
Chemical digestion and separation	17	6	12			

1 SD, standard deviation (external precision); *n*, numbers of blank measurements; the average Sr-isotope value obtained from 5 unspiked blanks is reported to fully characterize the potential contamination component.

whole analytical procedure was performed with acids of ultra-pure quality.

In order to thoroughly assess the contamination levels, we measured two types of blanks, one considering only the amount of Sr deriving from the chemical digestion and Sr separation, the other accounting for the whole procedure, including the drilling process, as described in the previous section. The results were 17 ± 6 (1 SD, $n = 12$) and 38 ± 19 pg (1 SD, $n = 16$), respectively, over a 14-month period (Table 1), thus allowing sampling as low as 4 ng of Sr for isotope analysis.

2.3. $^{87}Sr/^{86}Sr$ Measurements on TIMS. Sr isotope ratios were determined using a multicollector, thermal ionization mass spectrometer (TIMS: ThermoFinningan Triton-Ti™) (Figure 5), equipped with nine moveable collectors, which allow to simultaneously detect all the natural masses of Sr (^{84}Sr, ^{86}Sr, ^{87}Sr, and ^{88}Sr). The mass of ^{85}Rb was also measured to monitor possible ^{87}Rb interference, but it was always lower than the detection limit of the instrument, confirming the quality of the separation procedure described above. A detailed description of instrumental characteristics and performances are given in [55], along with standardised routine, measuring conditions, and setting for normal-sized samples (100–150 ng of Sr). Instrumental mass bias (e.g., [57–59]) was corrected to the natural value of $^{86}Sr/^{88}Sr = 0.1194$ using an exponential law (e.g., [55, 59]).

The most critical aspects of measuring small-size samples are related to (i) the procedure of sample loading onto the filaments and (ii) the measurement mode (i.e., static versus multidynamic, e.g., [55, 59]). Both are very important to maximise the Sr signal during the measurements, to balance the analysis time and the analytical errors that are a function of sample size.

The measurement protocol was tested by replicate analyses of an international certified standard (NIST-SRM987), properly diluted to attain sample sizes (5 to 10 ng Sr) comparable to those of the microsamples. Then, we tested the whole procedure, from in situ sampling (<10 ng of Sr) to isotope measurement, on the international glass reference sample BHVO-2G. BHVO-2G reference sample is a synthetic basaltic glass (provided by USGS) obtained by melting the BHVO-2 powder collected from a Hawaiian lava flow. The glassy slices are supplied in epoxy resin mounts (https://crustal.usgs.gov/geochemical_reference_standards/microanalytical_RM.html).

2.3.1. Sample Loading onto Filament. The Sr fraction collected from the columns was dissolved into 1 μl HNO_3 10 vol.% and loaded on single Re filaments under a horizontal laminar flow hood. Due to the small amount of Sr available, it is important to confine the sample on the smallest possible area on the filament, so that the whole loaded sample can be ionised at the same time from a single spot. To attain this, a thin layer of Parafilm™ was melted at both sides of the filament surface, to prevent any spreading of the solution, leaving a small gap (1 mm) for the droplet at the center on the filament (Figure 5(b)). The loading was performed by sandwiching the sample between two 0.5 μl

(a) (b) (c) (d)

FIGURE 5: (a) The ThermoFinningan Triton-Ti multicollector, thermal ionization mass spectrometer (TIMS) at the University of Firenze; (b) single-rhenium filament with a sample droplet on top; (c) detail of a single filament, holding a dried sample loaded on a sample holder turret that can host 21 filament positions; (d) the turret is ready for the installation into the thermal ionization mass spectrometer.

drops of $TaCl_5$ activator solution [55] and 0.5 μl of H_3PO_4 solution (6 vol.%), respectively. $TaCl_5$ activator was added for enhancing Sr ionisation efficiency and H_3PO_4 for stabilising Sr isotope fractionation during measurement. All the solutions (i.e., activator, sample, and H_3PO_3) were slowly dried by passing a current on the filament, which was increased at the end of the procedure until the filament starts glowing. The loaded filaments were then placed on the filament-holder turret and then inserted into the mass spectrometer (Figures 5(c) and 5(d)).

2.3.2. Measurement Procedure Reproducibility and Accuracy. The measurement routine was established to obtain the best internal and external precisions, and the accuracy, on $^{87}Sr/^{86}Sr$, was achieved by experimentally comparing runs performed in *static* versus *dynamic* conditions at a variable number of cycles and integration times. A detailed description of the *static* and *dynamic* methods is provided in [55, 59]. In brief, the static mode consists of simultaneous measurements of all isotopes in a single "jump," so that the magnetic field remains static and the masses always hit the same detectors (Table 2). *Static* measurements have the advantage of considerably reducing the acquisition time in comparison with the *dynamic* mode, which becomes important when little amount of Sr is available, as dealing with small samples. The main limitation of this method is related to the uncertainty on the Faraday cup efficiency and on the drift of the electronics (i.e., the amplifiers). The Triton-Ti is equipped with a virtual amplifier, which enables a variable connection between amplifiers and Faraday cups and allows a complete switching between amplifiers and cups during a single measurement. However, the virtual amplifier is not able to correct for the different Faraday cup efficiency and its variation with time.

In contrast, the *dynamic* (or *multidynamic*) mode is a peak-jumping procedure where a number of different cup configurations are employed for determining a single isotopic ratio (Table 2). This means that each isotope beam is measured sequentially in different Faraday cups, so that two $^{87}Sr/^{86}Sr_{double}$ values (Table 2) can be calculated without cup efficiency biases and drifts of the electronics (Table 2). The two $^{87}Sr/^{86}Sr_{double}$ values are then geometrically averaged to obtain a single $^{87}Sr/^{86}Sr_{triple}$ value.

The best configuration for *static* mode measurements was found by measuring 300 cycles with an integration time of 8 s, which corresponds to a total measuring time of about 35 minutes for each sample. For *dynamic* mode measurements, we performed 120 cycles (each including 3 magnetic jumps), with 8 s of integration time and an idle time of 3 s between the different jumps, for a total of 70 minutes for each sample.

In both *static* and *dynamic* methods, the filament was slowly warmed up, for a total of about 45 minutes, to stabilise the ion emission until the suitable intensity is achieved. During the heating, the beam was accurately optimised by peak-centering and focusing. The optimal beam intensity for

Table 2: Cup configuration schemes of static (a) and dynamic (b) mode measurements.

Cup	L4	L3	L2	L1	C (Far)	H1	H2	H3	H4
(a) Static collection mode		^{84}Sr	^{85}Rb	^{86}Sr	^{87}Sr	^{88}Sr			
(b) Dynamic collection mode									
Jump 1			^{85}Rb	^{85}Rb	^{86}Sr	^{87}Sr	^{88}Sr		
Jump 2 (main)		^{84}Sr	^{86}Sr	^{86}Sr	^{87}Sr	^{88}Sr			
Jump 3				^{87}Sr	^{88}Sr				

Sr double

Combining the measurements from two different magnetic field position (jump 1-2 and jump 2-3 above), it is possible to evaluate two independent $^{87}Sr/^{86}Sr_{double}$ defined as follows: $[^{87}Sr/^{86}Sr]_{1\text{-}2} = \sqrt{[^{87}Sr_{H1}/^{86}Sr_{C}]_1 \cdot [^{87}Sr_{C}/^{88}Sr_{H1}]_2 \cdot [^{88}Sr/^{86}Sr]_N}$ and $[^{87}Sr/^{86}Sr]_{2\text{-}3} = \sqrt{[^{87}Sr_{C}/^{86}Sr_{L1}]_2 \cdot [^{87}Sr_{L1}/^{88}Sr_{C}]_3 \cdot [^{88}Sr/^{86}Sr]_N}$. The numbers outside the parentheses are relative to the three different magnetic field positions, the subscript of each isotope refers to the cup on which it is measured, and $[^{88}Sr/^{86}Sr]_N$ is the natural ratio (i.e., 8.375209)

the measurement varied from *static* to *dynamic* mode, with higher intensity allowed by the shorter *static* (3–3.5 V on ^{88}Sr) mode with respect to *dynamic* (1.5–2 V on ^{88}Sr) mode, which instead requires maintaining a stable beam, owing to the longer duration of the measurement.

The results are shown in Figure 6 and Table 3. *Static* and *dynamic* mode measurements on NIST-SRM987 reference samples (10 ng of Sr measured) yielded $^{87}Sr/^{86}Sr$ average values of 0.710247 ± 0.000026 (2 SD, $n = 30$) and 0.710251 ± 0.000018 (2 SD, $n = 51$), respectively, with internal precisions of 13 ppm (2 SE) and 16 ppm (2 SE), respectively. Both values are within the recommended reference value for NIST-SRM987 ($^{87}Sr/^{86}Sr = 0.710248 \pm 0.000011$; Figures 6(a) and 6(b) and Table 3 [59]). *Static* measurement reduced the experimental time but showed a worse external reproducibility than that obtained in *dynamic* mode (Figures 6(a) and 6(b)), yet maintaining similar internal precision. Therefore, the *dynamic* mode was chosen for the experimental work both on the international glass standard BHVO-2G and on the unknown samples. Further measurements of SRM987 in the *dynamic* mode were performed after the initial testing, along with the studied samples, yielding consistent results ($^{87}Sr/^{86}Sr = 0.710252 \pm 0.000018$, 2 SD, $n = 47$; Figure 6(c)).

Results on the BHVO-2G are reported in Table 3. The results were also compared to standard measurements (150 ng of Sr) on the BHVO-2 powder reference sample. BHVO-2G versus BHVO-2 results found for micro- and normal-size samples, respectively, are well within the internal analytical error (Table 3) and in agreement with the reference values reported in [60, 61] for bulk powder (i.e., standard BHVO-2). The significantly larger standard deviation of the micro-drilled BHVO-2G measurements, with respect to both micro-Sr SRM987 and BHVO-2 powder data, is likely partly related to small isotopic heterogeneities of the glass standard. The few available micro-Sr data on the same sample provide similar averages and reproducibility of our data [54, 62] (Table 3).

Comparing our results with LA-MC-ICPMS data is more difficult; in fact, the latter vary largely depending on the material used for the analyses. External reproducibility obtained with LA-MC-ICPMS on material with high Sr contents and low Rb/Sr (e.g., apatite [49, 53], marine shells, and synthetic plagioclase [16]) is comparable or slightly worse than that attained with our method; yet, small but significant differences in accuracy have been reported [53]. On the other hand, the internal and external reproducibility worsen significantly (e.g., by a factor 5 to 10 in [16]) in materials with low Sr and high Rb/Sr.

In summary, the method presented here generally provides more accurate and precise results than LA-MC-ICPMS, independently on the nature of the analysed material, despite being more time-consuming. It is therefore suitable for a wider range of applications.

3. Applications

In this section, we report three case studies, two of which were previously published [8, 10], as examples of possible applications of the presented methodology in different fields of science. Indeed, in the last decade, the in situ isotope microsampling approach has been used and applied in many pilot studies in a wide range of research fields, including, among others, palaeoenvironmental and palaeoecologic reconstructions (i.e., [53, 63]) and climate changes (i.e., [64]).

In the three presented case studies, the samples were thoroughly characterised both texturally (optical microscope and SEM) and chemically (electron micro probe analyses) before drilling. The strontium element concentrations (in ppm) in all the samples was then determined through LA-ICPMS.

3.1. Micro-Sr Isotope in Minerals. Rock-forming minerals in igneous rocks display variable chemical composition depending on several processes and parameters such as (i) the physicochemical conditions of the magmas, (ii) open system processes (e.g., magma mixing and mingling), and (iii) recycling of cumulated crystals triggered by new arrivals of magma within crustal reservoirs. Radiogenic isotope ratios in minerals, or portions of them, can be used as a petrogenetic “DNA” to record the history of the magma reservoir (*crystal isotope stratigraphy*, e.g., [7]) and their evolution within the crust (e.g., [1, 5, 7, 8, 65–67]). Combining in situ Sr-isotope fingerprints with other approaches, such as textural evidences and crystal size distribution, offers the opportunity to understand the processes and timescales through which magmas are stored, differentiated, and delivered prior to eruption (e.g., [5–7, 13, 14, 16, 18, 54, 68, 69]).

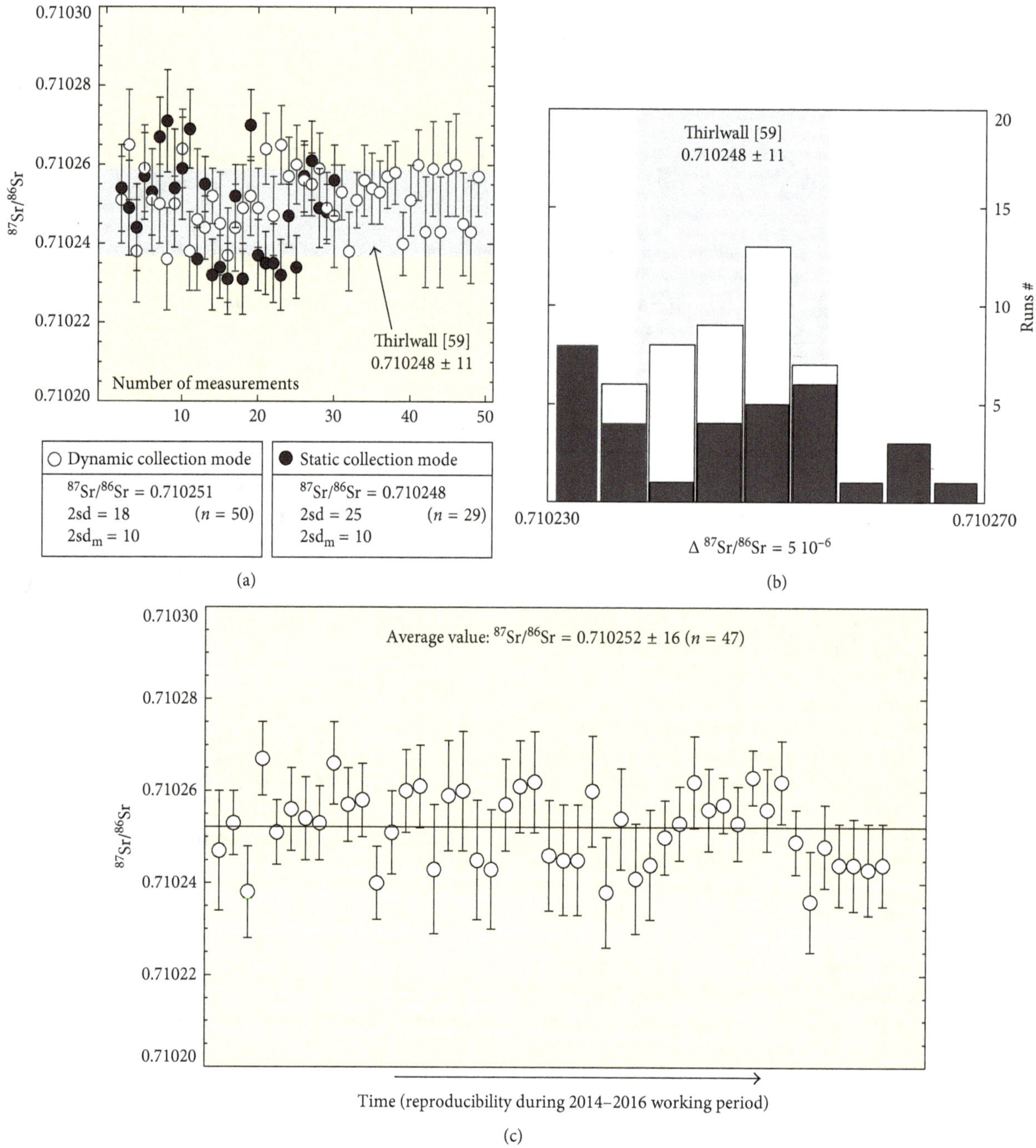

FIGURE 6: (a) Reproducibility and accuracy for repeated measurements of $^{87}Sr/^{86}Sr$ on 12 ng load size of NSIT-SRM987 standard material for static versus dynamic collection mode over a period of 10 months. Each single measurement is plotted with the relative error bars. The grey-shaded field shows the [59] recommended value range. (b) Distribution of the $^{87}Sr/^{86}Sr$ values measured all over the period of analysis with the static and dynamic collection mode: the static collection mode (black filled columns) shows a worst external reproducibility than that obtained in the dynamic mode (open columns); indeed the dynamic mode gives measurements that reasonably fit a Gaussian distribution pattern with the more representative $^{87}Sr/^{86}Sr$ values centered within the [59] recommended interval (grey shade fields); (c) reproducibility and accuracy trend of $^{87}Sr/^{86}Sr$ 12 ng load size of NSIT-SRM987 measured in the dynamic collection mode throughout the setup period (from 2014 to 2016).

The case study presented here is related to the active Nisyros volcano, the easternmost volcanic island of the South Aegean Active Volcanic Arc (Figure 7) [5, 70–76]. Nisyros volcanic products are typically porphyritic rocks, with clear petrographic evidence of recurrent mixing and mingling of different magmas during the whole volcano's history, which is likely interpreted as the triggering mechanism for its eruptions (e.g., [74, 77, 78]). Sr-isotope determinations at the subcrystal scale, along with detailed petrographic microscopic textural evidence, provided significant data for better defining the interaction of different magmas, concerning pre-eruptive mechanisms. The

Table 3: Accuracy and reproducibility on reference standard material (NIST-SRM987 and BHVO-2).

Standard	Sr content	$^{87}Sr/^{86}Sr$	2 SD	n	Reference
		Within run measurement on international standard NIST-SRM987			
Large size loading (ng)					
SRM987	150	0.710253	0.000016	59	Long-term reproducibility (from 2013)
SRM987	150	0.710248	0.000011	427	Thirlwall [59]
Small size loading					
SRM987	10	0.710252	0.000016	47	This study
SRM987	12	0.710259	0.000018	92	Charlier et al. [54]
		Drilling procedure on BHVO-2 glass			
BHVO-2 glass	<10	0.703490	0.000092	9	This study
BHVO-2 glass	10	0.703492	0.000094	3	Charlier et al. [54]
		$^{87}Sr/^{86}Sr$	2 SE		
BHVO-2 powder	150	0.703469	0.000004	1	This study
		$^{87}Sr/^{86}Sr$	2 SD		
BHVO-2 powder	—	0.703479	0.000020	12	Weis et al. [60]

2 SD, two standard deviation (external precision); 2 SE, two standard error of the mean; n, numbers of measurements; literature data are reported in italics below our mean values for references.

study focused, in particular, on postcaldera, rhyodacitic dome magmas of the final Nisyros activity emplaced after the caldera-forming rhyolitic explosive eruption of upper pumice. These rhyodacitic lavas contain magmatic enclaves (Figures 7(a) and 7(c)), with basaltic andesite to andesite compositions interpreted, based on their textural features, as quenched portions of mafic magmas included in the cooler, more evolved rhyodacitic host melt (Figures 7(b) and 7(c)) (e.g., [8, 78]).

In situ Sr-isotope ratios were determined on plagioclase phenocrysts (Figures 7(c)–7(e)), from both domes and enclaves, which preserve evidence of the complex history of interaction between the mafic (i.e., enclaves) and felsic (i.e., rhyodacitic domes) magmas in their growing zones [8]. The $^{87}Sr/^{86}Sr$ values determined on micromilled samples from the different growth zones of the plagioclase phenocrysts show clear Sr-isotope disequilibria between (i) cores and rims of single crystals (Figures 7(c)–7(e)) and (ii) the crystals and the host magmas (Figure 7(e)). This suggests that some of the phenocrysts that had formed in the rhyolitic magmas were later enclosed (as xenocrysts) within the more mafic one (i.e., as enclaves). Whereas the rim of the phenocrysts is isotopically intermediate between the rhyolitic and mafic magmas (Figure 7(f)), their cores show higher $^{87}Sr/^{86}Sr$ values, quite close to that of the previously erupted upper pumice magma. This clearly indicates that the phenocrysts originated in a different, older system. In this light, the large plagioclase phenocrysts found inside the dome lavas and enclaves can be interpreted as recycled from previously cumulated crystals (called "antecrysts" by Davidson et al. [7]).

These results have also demonstrated that the dome lavas are multicomponent magmas formed by progressive mingling/mixing processes between (i) a rhyolitic, and more Sr-radiogenic melt derived from the original upper pumice magmas, and (ii) the enclave-forming mafic, and less Sr-radiogenic, melts refilling the felsic magma chamber.

The constraints involved in interpreting in situ isotope data have further implications for the timing and style of eruption. The inferred delay between the mafic input (i.e., enclaves) and the relative dome eruption allows time for reheating and consequent drop in magma viscosity, thus favouring dome extrusion rather than explosive activity [8, 78].

3.2. Micro-Sr Isotope in Natural Glasses. Glasses are found in nature generated by rapid quenching of molten material. They represent a volumetrically small component of crustal rocks and can have different genesis (i.e., volcanic, lightning strikes, meteorite impact, and anthropogenic). In this light, the radiogenic isotopic compositions (i.e., Sr, Nd, and Pb) can provide fundamental information to discriminate among the processes involved in their formation. Glasses constitute the main component of ash and pyroclastic deposits, and their composition and Sr-isotope signature can provide important information in defining the triggering mechanisms of explosive eruptions (e.g., [10, 69, 79–84]). Glasses may also be found in ceramics, as well as in other artefacts. Sr-isotope data are therefore also important to define the possible source of raw materials for pottery, which is particularly relevant for cultural heritage, or to track trade routes in archaeology (e.g., [85–88]). Microscopic scale Sr-isotopic measurements on glasses would help in minimising the amount of sample that needs to be milled, which is of crucial importance both dealing with small-sized volcanic ashes and ejecta and with human artefact of archaeological interest.

The present study is a case study of submicroscopic scale Sr-isotopic determination on ashes of the 2010 Eyjafjallajökull (Iceland) volcano's explosive eruption, which caused enormous disruption to air travel across the northern hemisphere (Figure 8(a)) (e.g., [89–94]). This eruption represents a unique opportunity to test the potential of microscale Sr-isotope determinations on ash glasses from tephra deposits that were well preserved within the ice/snow pack [10]. The population of tephra is composed by four different types of ash fragments: (i) fluidal, (ii) coarsely vesicular, (iii) spongy fine vesicular, and (iv) blocky (Figure 8(b); [93]), generated by different fragmentation processes during the eruption. A detailed microanalytical geochemical and in situ Sr-isotope study performed on glassy groundmass of single ash clast showed unusually high $^{87}Sr/^{86}Sr$ values (up

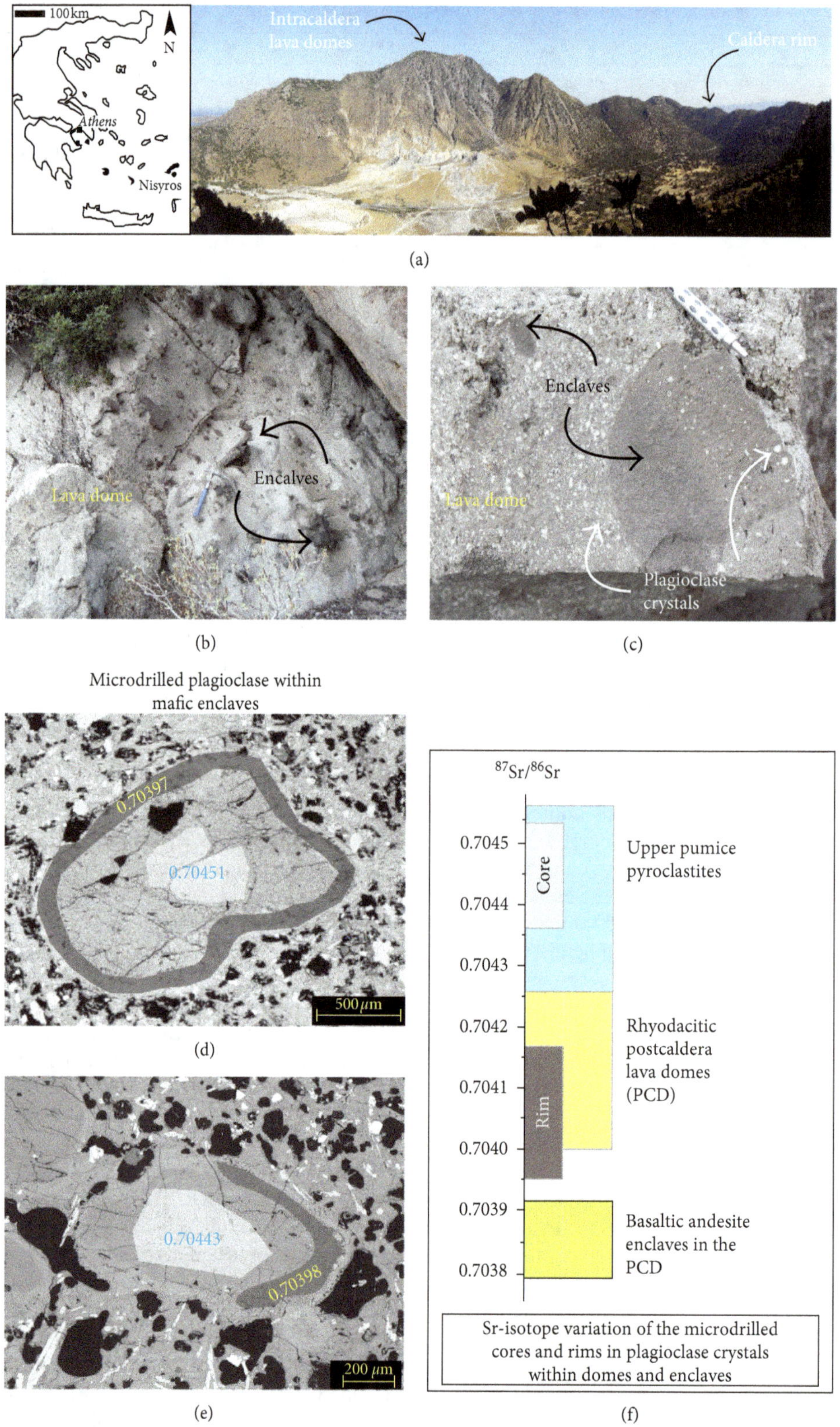

FIGURE 7: Results of micro-Sr isotope studies on plagioclase crystals from the last magmatic activity of Nisyros volcano (Greece). (a) Landscape view of the Nisyros caldera and its lava domes outpoured during the final magmatic activity of the volcano; (b) image of a lava dome outcrop rich in magmatic enclaves. Notably, the enclaves occurs as well-defined body with rounded and smooth surfaces; (c) specific image of a lava dome and enclave, both rich in large plagioclase crystals; (d and e) back-scattered electron microscope images of two representative plagioclase crystals selected for micro-Sr investigation with microdrill. The shaded areas represent the crystal zones drilled for Sr-isotope analyses, both on cores and rims; (f) $^{87}Sr/^{86}Sr$ variation of the drilled cores and rims compared to the range of $^{87}Sr/^{86}Sr$ of host whole rock (domes and enclaves) and upper pumice whole rock.

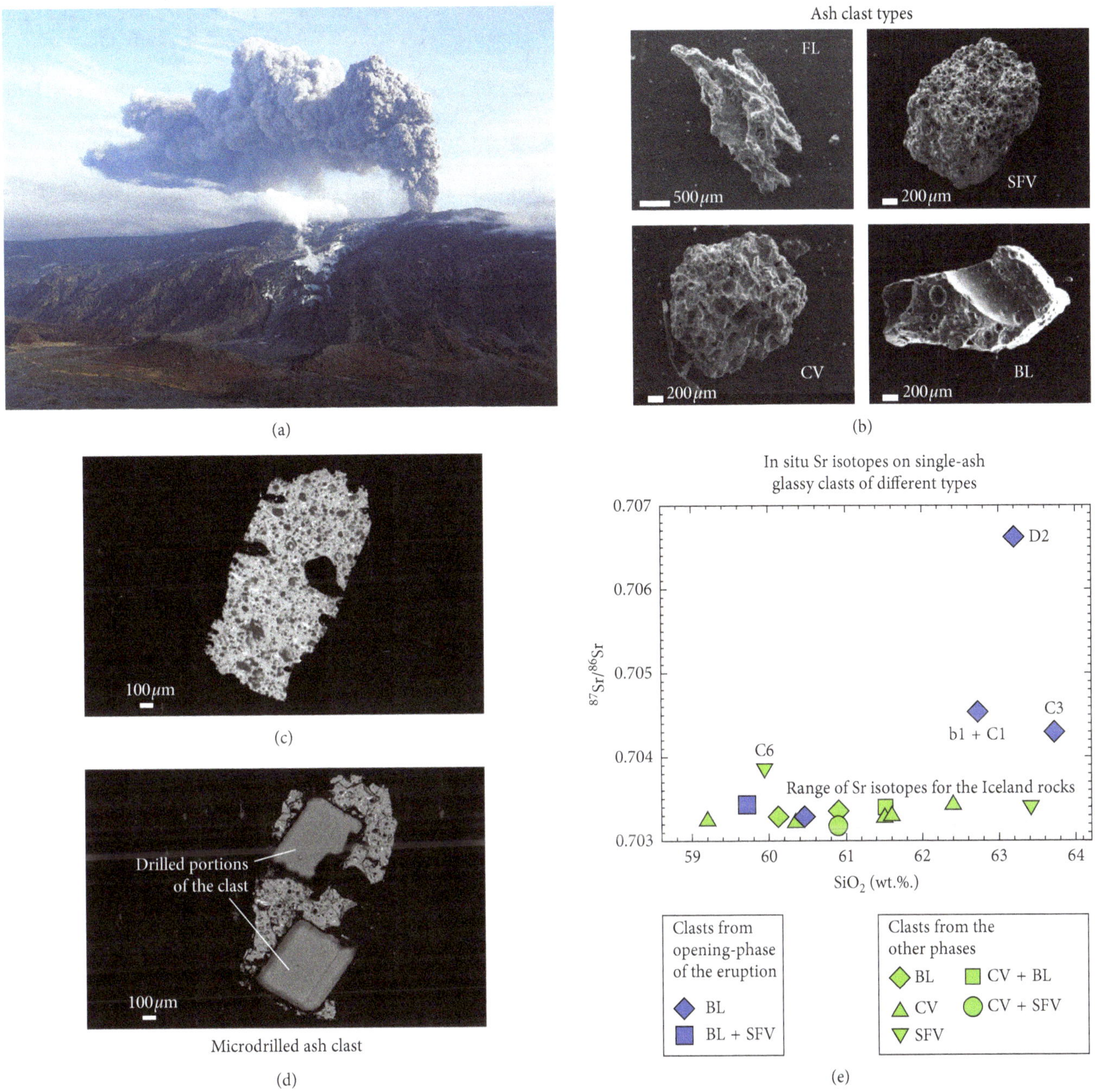

FIGURE 8: Results of micro-Sr isotopes in volcanic ash glasses of the 2010 explosive eruption of Eyjafjallajökull volcano (Iceland). (a) Spectacular image of the ash cloud erupted by the volcano during the early stage of the 2010 activity (https://hiticeland.com/places_and_photos_from_iceland/eyjafjallajokull causing); (b) back-scattered electron microscope images of different clast types sampled in the basal level of the fallout deposit related to the open phase of the 2010 eruption. FL = fluidal; SFV = spongy finely vesicular; CV = coarse vesicular; and BL = blocky. Scale bars are in micron; comparison between back-scattered electron microscope images of clasts before (c) and after (d) in situ microdrilling. Scale bars are in micron; (e) $^{87}Sr/^{86}Sr$ versus SiO_2 (wt.%) diagram of matrix glasses sampled from single, glassy ash clasts. In situ microsampling for Sr-isotope determinations on glass and plagioclase have been obtained by microdrilling technique. Error bars are inside the symbols.

to 0.70668) for Icelandic volcanism (0.7026–0.7037 [94]) (Figure 8(c)); these high isotopic values were also associated to atypical elemental compositions compared to most of the juvenile ash fragments of the eruption. The anomalous, high Sr-radiogenic clasts belong to the blocky type (Figure 8 (b)) and are concentrated in the first, thin ash level emplaced during the initial phase of eruptive activity. These clasts originated from the magma quenched from the contact with the ice cap filling the summit caldera of the volcano [10]. These anomalous findings in the Icelandic magmatic environment can be explained supposing that during its rise and before intruding into the ice cover, the erupting magma selectively assimilated hydrothermal minerals (i.e., zeolites, silica phases, and anhydrite) with seawater-related high-Sr isotopic ratios, hosted in altered volcanic/epiclastic rocks. This selective assimilation took place at the tip edge of the first rising magma body, resulting in a high degree of contamination restricted to the rather small amount of melt directly in contact with the hydrothermal veins. Indeed, evidence for this process is recorded only by the very first erupted juveniles (i.e., the blocky clasts). The results obtained through submicroscopic-scale micromilling and relative Sr-isotope determination revealed the dynamics of the processes involved in the initial stages of magma ascent to the surface; this provides significant insights into the interpretation of the precursory signals of the eruption (mostly consisting of ground deformation or increased seismicity) [10]. These transient processes, which interested only a small, well-confined part of the magma, cannot be detected using traditional, Sr-isotope determination on whole-rock samples but can be revealed only analysing single-glassy clasts separately.

3.3. Micro-Sr Isotope in Teeth and Bones. Due to their similar chemical properties, Sr can substitute for Ca in the bioapatite [$Ca_5(PO_4,CO_3)_3(OH,F)$] of mammalian bones and teeth, reaching contents of few hundreds of ppm that allows the isotope analysis by microdrilling. The Sr isotope composition of human and animal hard tissues is a function of their dietary habits (e.g., [22]) and depends on the isotopic composition of the food and water ingested during life, which in turn are related to the geological substrate [95–97].

Sr isotopes have been successfully used, along with other stable and radiogenic isotope systematics (i.e., $\delta^{13}C$ and $\delta^{18}O$ and Pb isotopes), not only to track the source regions of migrants and migration pathways, as well as the hunting and trading areas of human populations, but also to study and define the dietary habits of humans and animals (e.g., [23, 44, 95–106]).

A pioneering study [107] demonstrated that migrant individuals who moved between different geologic regions might be traced by comparing $^{87}Sr/^{86}Sr$ in adult tooth enamel, formed between four and twelve years of age, and in the bones, which remodel throughout life and therefore representative of adulthood [30]. Unlike bones and dentine, dental enamel formed during childhood [108] remains unaltered throughout the years. Different $^{87}Sr/^{86}Sr$ in the teeth and bones of an individual may thus reflect the fact that it moved around the landscape passing through different isotopic environments during its youth and maturity [109, 110]. Teeth enamel is generally preferred to dentine and bones in the analysis of Sr concentrations and isotope ratios because it is virtually unaffected by postmortem diagenesis (e.g., [111–113]).

The micro-scale Sr-isotope measurements of samples obtained using the submicroscopic-scale micromill technique is perfectly able to discriminate between enamel and dentine in single-tooth samples. In addition, this technique increases the accuracy of sampling and also reduces the amount of specimen to be destroyed for high-precision Sr-isotope analysis.

Ursus spelaeus was an endemic, widespread European Late Pleistocene species. In contrast to many other taxa, it has fairly rich fossil records, especially thanks to its recurrent use of caves or shelters for winter hibernation [114–116]. Caves were frequently used by bears for many generations, and numerous individuals eventually died in them, so that significant quantities of their remains accumulated over considerable periods of time.

Many studies used stable isotopes to determine the dietary habits of living and extinct bears (e.g., [117–121]), but until now only a small number of papers have considered employing Sr isotopes to possibly elucidate the factors influencing habitat use and gain insights into the foraging behaviour of cave bears [122]. We report here the first $^{87}Sr/^{86}Sr$ data obtained through in situ microsampling on teeth and bones of *Ursus spelaeus* found in Grotta all'Onda cave. The study was aimed at defining the lifestyle and feeding behaviour of one of the most prominent European Late Pleistocene mammals [123–125].

Grotta all'Onda cave is located 708 m above sea level (a.s.l.) in the Apuan Alps nearby the village of Camaiore (Tuscany, Italy) (Figure 9), in a sub-Mediterranean habitat [126]. The cave opens at the base of the Tuscan Nappe, at the contact between the "Calcare a Rhaetavicula Contorta Formation" (i.e., Upper Triassic dolomitic-limestone of the Tuscan Nappe) and the "Argilliti Varicolori Formation" (i.e., Lower Cretaceous shales) (Figure 9). The Rhaetavicula Contorta Formation is a polygenic breccia, mainly including metamorphic clasts, known as "Brecce di Grotta all'Onda" (Figure 9) (http://www502.regione.toscana.it/geoscopio/geologia.html). The fossil remains of *Ursus spelaeus* were recovered during a 1999 excavation. Radiocarbon dating of bone yielded ages ranging from 38.22 to 38.28 ky (BP) [127]. Six different specimens were selected for Sr isotope analysis; these include three lower molars (SCT4, SCT5, and SCT6) from layers 7J4, 7J3, and 7J5, respectively and three metapodial bones, SCT 1 (third metacarpal), SCT 2 (fourth metatarsal), and SCT3 (fourth metatarsal), from layers 7J4, 7J5, and 7J3, respectively (Figures 9(c) and 10(a)–10(b)–10(c)–10(d)). Two whole soil (i.e., cave earth) samples, SCT7 and SCT8, were also collected from the representative layers 7J4 and 7J5 (Figure 9(c)).

All the three teeth had well-preserved dentin and very thin enamel layer (Figure 10). In contrast, the three bones were rather differently preserved and had different porosity. In particular, SCT3 was the most heavily mineralised and best preserved, whereas SCT1 was densely vacuolated and preserved higher amounts of organic components.

Major element analyses of these specimens revealed that the enamel bioapatite was more mineralised than that of dentin and bones. Moreover, the dentin bioapatite was found enriched in Sr and in other trace elements [128].

For the purpose of this study, dentin and enamel of the teeth and cortical sections of the bones from Grotta all'Onda were micromilled and analysed for micro-Sr isotope determination. The two soil samples were also processed for Sr-isotope determination using the traditional, large sample method [37, 55]. The $^{87}Sr/^{86}Sr$ data are reported in Table 4. Tooth enamel shows higher $^{87}Sr/^{86}Sr$ and lower Sr/Ca than

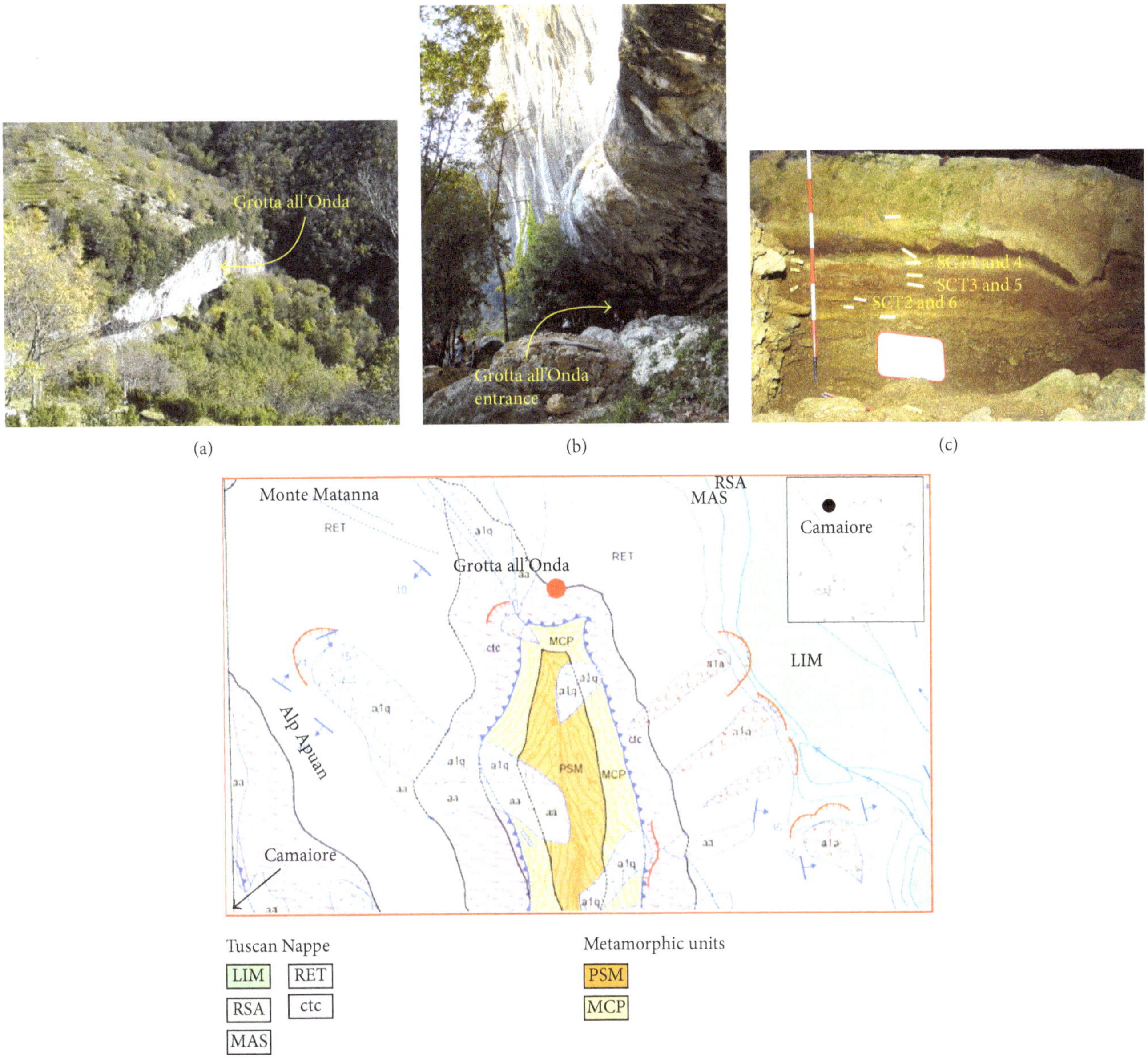

(d)

FIGURE 9: The *Ursus spelaeus* finding environment. (a) Image of the Grotta all'Onda cave from the above, showing the well-defined cut of the Mt. Matanna flank where the cave is located; (b) image of the entrance of the Grotta all'Onda cave; (c) image of the soil stratigraphy from which the *Ursus spelaeus* fossil remains have been sampled during the 1999 field campaign; (d) geological sketch map of the area around the Grotta all'Onda cave (http://www502.regione.toscana.it/geoscopio/geologia.html#). Formation of the Tuscan Nappe unit: LIM: "*Calcare Secifero di Limano*" formation; RSA: "*Rosso Ammonitico*" formation; MAS: "Calcare Massiccio" formation; RET: "*Raethavicola Contorta*" formation, Late Triassic dolomitic-limestones; ctc: cataclastic formation made by a polygenic breccias of mainly metamorphic limestone clasts. Tuscan metamorphic units: PSM: "Pseudomacigno" formation; MCP: "Cipollini" formation.

microsamples of dentin and bones do (Figures 10(e) and 10(f)). This strongly suggests that $^{87}Sr/^{86}Sr$ of tooth enamel is unaffected by diagenetic alteration, in contrast to the other organic-rich samples (i.e., bones and dentin). The $^{87}Sr/^{86}Sr$ composition of the dentin samples is close to that of the local soil samples (Figure 10(e)); even closer to the latter is that of the bone samples, with the only exception of SCT3 due to its high degree of mineralisation.

These results, despite the relatively recent age of the fossil specimens (ca. 40 ka [127]), indicate that bone tissues have been more exposed to *postmortem* diagenetic exchange processes compared to more heavily mineralized enamel. This speaks for a possible isotope reequilibration between dentin and bones (but not enamel) and soil, due to Sr exchange with percolating fluids. The soil samples have $^{87}Sr/^{86}Sr$ values comparable to those of the "Calcare a Rhaetavicula Contorta" formation, which forms the cave's bedrock [129], through which fluids filter into the cave (Figure 10(e)).

In summary, our study shows that the Grotta all'Onda bones and dentin are unsuitable to determine the

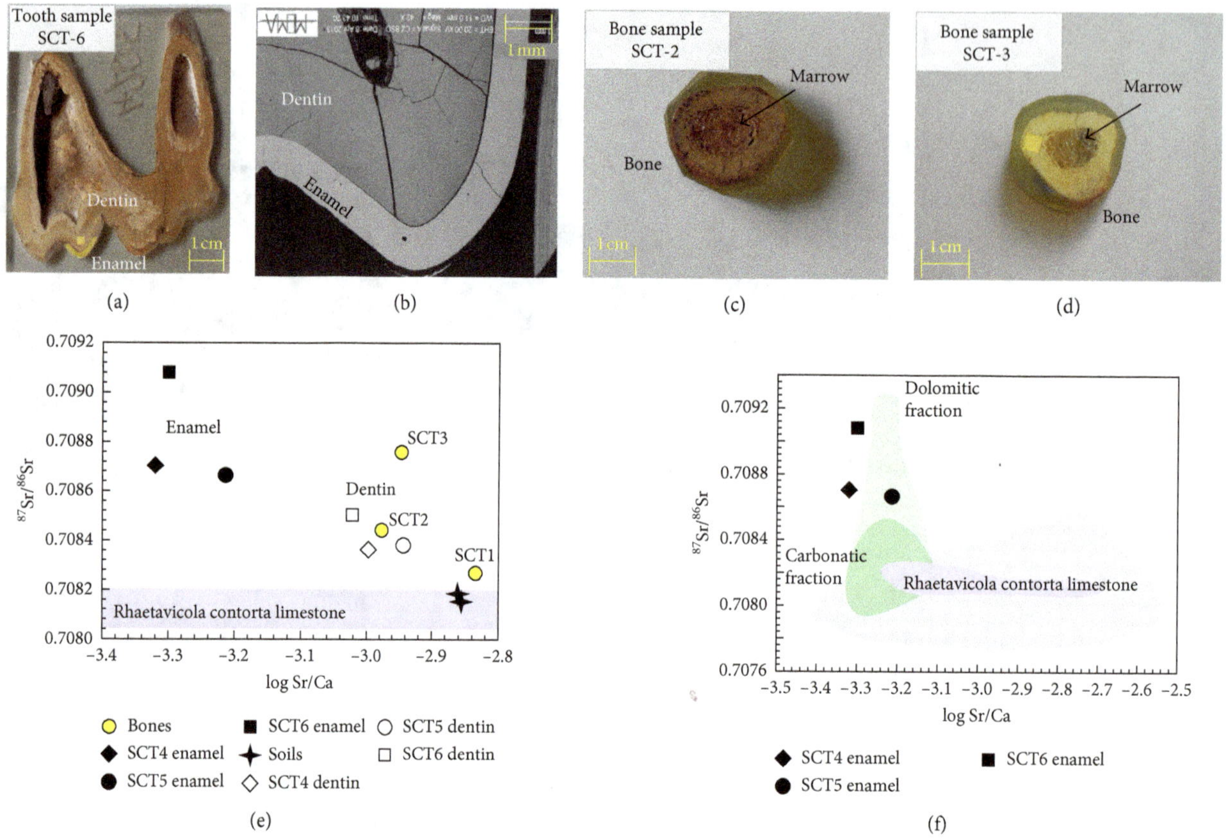

Figure 10: Results of micro-Sr isotopes in teeth and bones of the *Ursus spelaeus*. (a) Representative image of one tooth sample, sectioned for microdrilling. The thin layer of enamel is well evident with respect to the lighter inner dentin. In yellow are reported the two drilling sites on dentin and enamel; (b) particular back-scattered electron microscopy image of the edge of the tooth showing dentin (dark grey) and enamel (light grey) portion; (c and d) images of two of the analysed bones properly prepared in epoxy resin mounts for microdrilling sampling. In yellow is reported the drilling site; (e) $^{87}Sr/^{86}Sr$ versus Sr/Ca (in log scale) diagram showing the results obtained from the analyses of enamel and dentin in the three teeth, compared with bones of the *Ursus spelaeus* specimen. The $^{87}Sr/^{86}Sr$ of soils in which the fossils have been sampled are also reported together with the Sr-isotope field of the *Raetavicola Contorta* limestone for comparison. (See text for detail) (f) $^{87}Sr/^{86}Sr$ versus Sr/Ca (in log scale) diagram comparing the Sr-isotope values of dentin from the three teeth samples with the Sr-isotope range fields of the different geologic formations forming the bedrock outcropping in the area of the Grotta all'Onda cave.

Table 4: (a) $^{87}Sr/^{86}Sr$ results obtained on microsamples of tooth dentine and enamel and bones of the *Ursus spelaeus* specimen. (b) $^{87}Sr/^{86}Sr$ soils sampled in the Grotta all'Onda cave from which the fossil remains were collected.

Sample	Type	Sr content (ppm)	$^{87}Sr/^{86}Sr$	2 SE	Log (Sr/Ca) Average
(a) Teeth and bones					
STC1	Bone	510	0.708268	0.000006	−2.84
STC2	Bone	352	0.708441	0.000006	−2.98
STC3	Bone	359	0.708758	0.000006	−2.95
STC4-1	Enamel	187	0.708703	0.000005	−3.32
STC4-2	Dentin	360	0.708362	0.000006	−3.00
STC5-1	Enamel	211	0.708663	0.000005	−3.21
STC5-2	Dentin	384	0.708380	0.000006	−2.94
STC6-1	Enamel	179	0.709081	0.000005	−3.30
STC6-2	Dentin	339	0.708504	0.000006	−3.02
(b) Soils					
STC7	Soil	204	0.708184	0.000006	−2.86
STC8	Soil	228	0.708155	0.000006	−2.86

2 SE: two standard error of the mean.

characteristics of the habitat where the cave bear lived, due to their interaction with percolating water and their consequent contamination by the soils in which they had been preserved.

In contrast, the $^{87}Sr/^{86}Sr$ ratios of the tooth enamel results unaltered and realistically reflects the original values achieved during the cave bear's life. The isotopic composition of the enamel samples is consistent with that of the "Calcare Massiccio Formation" and in particular with the dolomite fraction [129] (Figure 10(f)). The mismatch between the $^{87}Sr/^{86}Sr$ values of the enamel samples and those of the cave soils (Figure 10(e)) (i.e., local substrata of the cave) indicate that the cave bear died away from its customary habitat. Bears cannot indeed find food in caves, where they find refuge as shelter for winter hibernation or for night resting. The enamel isotopic values obtained during our study indicate that *Ursus spelaeus* from Grotta all'Onda roamed in search for food within a confined area not far from the cave, where the "Calcare Massiccio" is largely exposed and did not move too far from the area during its whole life.

4. Summary and Conclusions

The present study show the potential of $^{87}Sr/^{86}Sr$ determination by TIMS on micro-scale samples, based on micromilling solid specimens, not only for geological applications, but also for other fields, such as archaeology, forensics, medical, and life sciences, where it has hardly, if ever, been used. Reported here is a detailed description of all the analytical protocols, including results on replicate analyses of international standards (SRM 987 and BHVO-2G), which yield good accuracy and precision. In addition, three case studies are presented, performed in our laboratory, where in situ microdrilled Sr isotopes have been used in different fields of application.

The first case study on micro-Sr isotope determination at subgrain microscopic scale regards the petrogenetic processes relevant to the understanding of the plumbing system dynamics under active volcanoes. This example revealed the role played by the interaction of different magmas, which are normally characterised by distinct $^{87}Sr/^{86}Sr$ signatures, comingled in the plumbing system of the Nisyros volcano, which was capable of triggering the eruption.

The second case regards volcanic glasses with extremely low total Sr content (i.e., tholeiitic). The micromilling determination of $^{87}Sr/^{86}Sr$ ratio was performed on ashes, with different shapes and nature, erupted by different phases, during the 2010 eruption of Eyjafjallajökull volcano (Iceland). The $^{87}Sr/^{86}Sr$ data provide information on the eruptive mechanism involved during the eruption, as well as on the interaction between the magma and the hydrothermally-derived minerals attained before the thawing of the ice cap; it also provides significant insights into the interpretation of the precursory signals of the eruption.

The third case displays the use of $^{87}Sr/^{86}Sr$ microdrilled in enamel, dentin, and bones, to show that only enamel has more chances to preserve the original Sr-isotope signatures than bone and dentin. The analysis also revealed the close relationship existing between the radiogenic-Sr of the organic materials and that of the geologic Cretaceous substratum of the Apuan Alps, which provides valuable insights into the palaeoenvironment of the local cave bears. In this case, the Sr isotopes proved particularly useful for determining the foraging habits of extinct mammals, which substantiates the well-known statement "YOU ARE WHAT YOU EAT" (cit. Anthelme Brillat-Savarin).

Conflicts of Interest

The authors declare that there are no conflicts of interest regarding the publication of this paper.

Authors' Contributions

Lorella Francalanci provided the input to the set-up and the development of the in situ micro-Sr isotope procedure at the Radiogenic Isotope Laboratory of Firenze, providing also continuous scientific stirring and encouraging to the other research fellows, coauthors of the present paper.

Acknowledgments

The research on volcanic rocks was financially supported by PRIN 2010-2011 and 2015 with Grants 2010TT22SC_001 and 20158A9CBM, respectively. Regione Toscana government supported the salary of Sara Di Salvo through a "Pegaso" fellowship. The studies on the *Ursus spelaeus* had no financial support, and the analytical work expenses were covered by the "Radiogenic Lab" of the University of Florence. The authors are grateful for the suggestions of an anonymous reviewer, which improved the paper.

References

[1] F. J. Tepley and J. P. Davidson, "Mineral-scale Sr-isotope constraints on magma evolution and chamber dynamics in the Rum layered intrusion, Scotland," *Contributions to Mineralogy and Petrology*, vol. 145, no. 5, pp. 628–641, 2003.

[2] G. Perini III, F. J. Tepley, J. P. Davidson, and S. Conticelli, "The origin of K-feldspar megacrysts hosted in alkaline potassic rocks from central Italy: a track for low-pressure processes in mafic magmas," *Lithos*, vol. 66, no. 3-4, pp. 223–240, 2003.

[3] D. Morgan, S. Blake, N. Rogers et al., "Time scales of crystal residence and magma chamber volume from modelling of diffusion profiles in phenocrysts: vesuvius 1944," *Earth and Planetary Science Letters*, vol. 222, no. 3-4, pp. 933–946, 2004.

[4] W. Siebel, E. Reitter, T. Wenzel, and U. Blaha, "Sr isotope systematics of K-feldspars in plutonic rocks revealed by the Rb-Sr microdrilling technique," *Chemical Geology*, vol. 222, no. 3-4, pp. 183–199, 2005.

[5] L. Francalanci, G. R. Davies, W. Lustenhouwer, S. Tommasini, P. R. Mason, and S. Conticelli, "Intra-grain Sr isotope evidence for crystal recycling and multiple magma reservoirs in the recent activity of Stromboli volcano, southern Italy," *Journal of Petrology*, vol. 46, no. 10, pp. 1997–2021, 2005.

[6] L. Francalanci, R. Avanzinelli, I. Nardini, M. Tiepolo, J. P. Davidson, and R. Vannucci, "Crystal recycling in the steady-state system of the active Stromboli volcano: a 2.5-ka story inferred from in situ Sr-isotope and trace element data," *Contributions to Mineralogy and Petrology*, vol. 163, no. 1, pp. 109–131, 2012.

[7] J. P. Davidson, D. J. Morgan, B. L. A. Charlier, R. Harlou, and J. M. Hora, "Microsampling and isotopic analysis of igneous rocks: implications for the study of magmatic systems," *Annual Review of Earth and Planetary Sciences*, vol. 35, no. 1, pp. 273–311, 2007.

[8] E. Braschi, L. Francalanci, S. Tommasini, and G. E. Vougioukalakis, "Unraveling the hidden origin and migration of plagioclase phenocrysts by in situ Sr isotopes: the case of final dome activity at Nisyros volcano, Greece," *Contributions to Mineralogy and Petrology*, vol. 167, no. 3, p. 988, 2014.

[9] S. Conticelli, E. Boari, L. Burlamacchi et al., "Geochemistry and Sr-Nd-Pb isotopes of Monte Amiata Volcano, Central Italy: evidence for magma mixing between high-K calc-alkaline and leucititic mantle-derived magmas," *Italian Journal of Geosciences*, vol. 134, no. 2, pp. 266–290, 2015.

[10] M. Pistolesi, R. Cioni, L. Francalanci et al., "The onset of an eruption: selective assimilation of hydrothermal minerals during pre-eruptive magma ascent of the 2010 summit eruption of Eyjafjallajökull volcano, Iceland," *Journal of Volcanology and Geothermal Research*, vol. 327, pp. 449–458, 2016.

[11] J. N. Christensen, A. N. Halliday, D. C. Lee, and C. M. Hall, "In situ Sr isotopic analysis by laser ablation," *Earth and Planetary Science Letters*, vol. 136, no. 1-2, pp. 79–85, 1995.

[12] J. P. Davidson and F. J. Tepley, "Recharge in volcanic systems: evidence from isotope profiles of phenocrysts," *Science*, vol. 275, no. 5301, pp. 826–829, 1997.

[13] J. P. Davidson, L. Font, B. L. Charlier, and F. J. Tepley, "Mineral-scale Sr isotope variation in plutonic rocks–a tool for unravelling the evolution of magma systems," *Earth and Environmental Science Transactions of the Royal Society of Edinburgh*, vol. 97, no. 4, pp. 357–367, 2006.

[14] B. L. A. Charlier, O. Bachmann, J. P. Davidson, M. A. Dungan, and D. J. Morgan, "The upper crustal evolution of a large silicic magma body: evidence from crystal-scale Rb–Sr isotopic heterogeneities in the Fish Canyon magmatic system, Colorado," *Journal of Petrology*, vol. 48, no. 10, pp. 1875–1894, 2007.

[15] B. L. Charlier, C. J. Wilson, and J. P. Davidson, "Rapid open-system assembly of a large silicic magma body: time-resolved evidence from cored plagioclase crystals in the Oruanui eruption deposits, New Zealand," *Contributions to Mineralogy and Petrology*, vol. 156, no. 6, pp. 799–813, 2008.

[16] J. P. Davidson, F. J. Tepley, Z. Palacz, and S. Meffan-Main, "Magma recharge, contamination, and residence times revealed by in situ laser ablation isotopic analysis of feldspar in volcanic rocks," *Earth and Planetary Science Letters*, vol. 184, no. 2, pp. 427–442, 2001.

[17] F. J. Tepley, J. P. Davidson, R. I. Tilling, and J. G. Arth, "Magma mixing, recharge and eruption histories recorded in plagioclase phenocrysts from El Chichon Volcano, Mexico," *Journal of Petrology*, vol. 41, no. 9, pp. 1397–1411, 2000.

[18] D. J. Morgan, D. A. Jerram, D. G. Chertkoff et al., "Combining CSD and isotopic microanalysis: magma supply and mixing processes at Stromboli Volcano, Aeolian Islands, Italy," *Earth and Planetary Science Letters*, vol. 260, no. 3-4, pp. 419–431, 2007.

[19] P. Horn, P. Schaaf, B. Holbach, S. Hölzl, and H. Eschnauer, "$^{87}Sr/^{86}Sr$ from rock and soil into vine and wine," *Zeitschrift für Lebensmitteluntersuchung und-Forschung A*, vol. 196, no. 5, pp. 407–409, 1993.

[20] J. L. Banner and J. Kaufman, "The isotopic record of ocean chemistry and diagenesis preserved in non-luminescent brachiopods from Mississippian carbonate rocks, Illinois and Missouri," *Geological Society of America Bulletin*, vol. 106, no. 8, pp. 1074–1082, 1994.

[21] G. Åberg, G. Fosse, and H. Stray, "Man, nutrition and mobility: a comparison of teeth and bone from the Medieval era and the present from Pb and Sr isotopes," *Science of the Total Environment*, vol. 224, no. 1, pp. 109–119, 1998.

[22] R. C. Capo, B. W. Stewart, and O. A. Chadwick, "Strontium isotopes as tracers of ecosystem processes: theory and methods," *Geoderma*, vol. 82, no. 1–3, pp. 197–225, 1998.

[23] B. L. Beard and C. M. Johnson, "Strontium isotope composition of skeletal material can determine the birth place and geographic mobility of humans and animals," *Journal of Forensic Science*, vol. 45, no. 5, pp. 1049–1061, 2000.

[24] S. Tommasini, G. R. Davies, and T. Elliott, "Lead isotope composition of tree rings as bio-geochemical tracers of heavy metal pollution: a reconnaissance study from Firenze, Italy," *Applied Geochemistry*, vol. 15, no. 7, pp. 891–900, 2000.

[25] M. Vanhaeren, F. d'Errico, I. Billy, and F. Grousset, "Tracing the source of Upper Palaeolithic shell beads by strontium isotope dating," *Journal of Archaeological Science*, vol. 31, no. 10, pp. 1481–1488, 2004.

[26] C. M. R. Almeida and M. T. S. Vasconcelos, "Lead contamination in Portuguese red wines from the Douro region: from the vineyard to the final product," *Journal of Agricultural and Food Chemistry*, vol. 51, no. 10, pp. 3012–3023, 2003.

[27] J. A. Evans and S. Tatham, "Defining "local signature" in terms of Sr isotope composition using a tenth-to twelfth-century Anglo-Saxon population living on a Jurassic clay-carbonate terrain, Rutland, UK," *Geological Society, London, Special Publications*, vol. 232, no. 1, pp. 237–248, 2004.

[28] S. Hölzl, P. Horn, A. Rossmann, and S. Rummel, "Isotope-abundance ratios of light (bio) and heavy (geo) elements in biogenic tissues: methods and applications," *Analytical and Bioanalytical Chemistry*, vol. 378, no. 2, pp. 270–272, 2004.

[29] S. Kelly, K. Heaton, and J. Hoogewerff, "Tracing the geographical origin of food: the application of multi-element and multi-isotope analysis," *Trends in Food Science & Technology*, vol. 16, no. 12, pp. 555–567, 2005.

[30] R. A. Bentley and C. Knipper, "Geographical patterns in biologically available strontium, carbon and oxygen isotope signatures in prehistoric SW Germany," *Archaeometry*, vol. 47, no. 3, pp. 629–644, 2005.

[31] P. Degryse, J. Schneider, U. Haack et al., "Evidence for glass "recycling" using Pb and Sr isotopic ratios and Sr-mixing lines: the case of early Byzantine Sagalassos," *Journal of Archaeological Science*, vol. 33, no. 4, pp. 494–501, 2006.

[32] S. Rummel, S. Hoelzl, P. Horn, A. Rossmann, and C. Schlicht, "The combination of stable isotope abundance ratios of H, C, N and S with $^{87}Sr/^{86}Sr$ for geographical origin assignment of orange juices," *Food Chemistry*, vol. 118, no. 4, pp. 890–900, 2010.

[33] S. Vorkelius, G. D. Lorenz, S. Rummel et al., "Strontium isotopic signatures of natural mineral waters, the reference to a simple geological map and its potential for authentication of food," *Food Chemistry*, vol. 118, no. 4, pp. 933–940, 2010.

[34] B. P. Kennedy, A. Klaue, J. D. Blum, C. L. Folt, and K. H. Nislow, "Reconstructing the lives of fish using Sr isotopes in otoliths," *Canadian Journal of Fisheries and Aquatic Sciences*, vol. 59, no. 6, pp. 925–929, 2011.

[35] C. Durante, C. Baschieri, L. Bertacchini et al., "Geographical traceability based on $^{87}Sr/^{86}Sr$ indicator: a first approach for PDO Lambrusco wines from Modena," *Food Chemistry*, vol. 141, no. 3, pp. 2779–2787, 2013.

[36] S. Marchionni, E. Braschi, S. Tommasini et al., "High-precision $^{87}Sr/^{86}Sr$ analyses in wines and their use as a geological fingerprint for tracing geographic provenance," *Journal of

Agricultural and Food Chemistry, vol. 61, no. 28, pp. 6822–6831, 2013.

[37] S. Marchionni, A. Buccianti, A. Bollati et al., "Conservation of $^{87}Sr/^{86}Sr$ isotopic ratios during the winemaking processes of "Red" wines to validate their use as geographic tracer," *Food Chemistry*, vol. 190, pp. 777–785, 2016.

[38] V. Vinciguerra, R. Stevenson, K. Pedneault, A. Poirier, J.-F. Hélie, and D. Widory, "Strontium isotope characterization of wines from Quebec, Canada," *Food Chemistry*, vol. 210, pp. 121–128, 2016.

[39] P. L. Koch, J. Heisinger, C. Moss, R. W. Carlson, M. L. Fogel, and A. K. Behrensmeyer, "Isotopic tracking of change in diet and habitat use in African elephants," *Science*, vol. 267, no. 5202, p. 1340, 1995.

[40] A. Sillen, G. Hall, and R. Armstrong, "Strontium calcium ratios (Sr/Ca) and strontium isotopic ratios ($^{87}Sr/^{86}Sr$) of *Australopithecus robustus* and *Homo* sp. From Swartkrans," *Journal of Human Evolution*, vol. 28, no. 3, pp. 277–285, 1995.

[41] G. Grupe, "Preservation of collagen in bone from dry, Sandy soil," *Journal of Archaeological Science*, vol. 22, no. 2, pp. 193–199, 1995.

[42] G. Grupe, T. D. Price, and F. Söllner, "Mobility of Bell Beaker people revealed by strontium isotope ratios of tooth and bone: a study of southern Bavarian skeletal remains. A reply to the comment by Peter Horn and Dieter Müller-Sohnius," *Applied Geochemistry*, vol. 14, no. 2, pp. 271–275, 1999.

[43] P. Willey, A. Galloway, and L. Snyder, "Bone mineral density and survival of elements and element portions in the bones of the Crow Creek massacre victims," *American Journal of Physical Anthropology*, vol. 104, no. 4, pp. 513–528, 1997.

[44] K. A. Hoppe, P. L. Koch, R. W. Carlson, and S. D. Webb, "Tracking mammoths and mastodons: reconstruction of migratory behavior using strontium isotope ratios," *Geology*, vol. 27, no. 5, pp. 439–442, 1999.

[45] T. D. Price, J. H. Burton, and R. A. Bentley, "The characterization of biologically available strontium isotope ratios for the study of prehistoric migration," *Archaeometry*, vol. 44, no. 1, pp. 117–135, 2002.

[46] R. A. Bentley, T. D. Price, and E. Stephan, "Determining the "local" $^{87}Sr/^{86}Sr$ range for archaeological skeletons: a case study from Neolithic Europe," *Journal of Archaeological Science*, vol. 31, no. 4, pp. 365–375, 2004.

[47] J. A. Evans, J. Montgomery, G. Wildman, and N. Boulton, "Spatial variations in biosphere $^{87}Sr/^{86}Sr$ in Britain," *Journal of the Geological Society*, vol. 167, no. 1, pp. 1–4, 2010.

[48] J. H. Burton and T. D. Price, "Seeking the local $^{87}Sr/^{86}Sr$ ratio to determine geographic origins of humans," in *ACS Symposium Series*, pp. 309–320, ACS Publication, Washington, DC, USA, 2013.

[49] F. Lugli, A. Cipriani, C. Peretto, M. Mazzucchelli, and D. Brunelli, "In situ high spatial resolution $^{87}Sr/^{86}Sr$ ratio determination of two Middle Pleistocene (ca 580 ka) *Stephanorhinus hundsheimensis* teeth by LA–MC–ICP–MS," *International Journal of Mass Spectrometry*, vol. 412, pp. 38–48, 2017.

[50] T. Prohaska, C. Latkoczy, G. Schultheis, M. Teschler-Nicola, and G. Stingeder, "Investigation of Sr isotope ratios in prehistoric human bones and teeth using laser ablation ICP-MS and ICP-MS after Rb/Sr separation," *Journal of Analytical Atomic Spectrometry*, vol. 17, no. 8, pp. 887–891, 2002.

[51] S. R. Copeland, M. Sponheimer, P. J. le Roux et al., "Strontium isotope ratios ($^{87}Sr/^{86}Sr$) of tooth enamel: a comparison of solution and laser ablation multicollector inductively coupled plasma mass spectrometry methods," *Rapid Communications in Mass Spectrometry*, vol. 22, no. 20, pp. 3187–3194, 2008.

[52] A. Simonetti, M. R. Buzon, and R. A. Creaser, "In situ elemental and Sr isotope investigation of human tooth enamel by Laser Ablation-(MC)-ICP-MS: success and pitfalls," *Archaeometry*, vol. 50, no. 2, pp. 371–385, 2008.

[53] J. Lewis, C. D. Coath, and A. W. G. Pike, "An improved protocol for $^{87}Sr/^{86}Sr$ by laser ablation multi-collector inductively coupled plasma mass spectrometry using oxide reduction and a customised plasma interface," *Chemical Geology*, vol. 390, pp. 173–181, 2014.

[54] B. L. A. Charlier, C. Gimbre, D. Morgan et al., "Methods of microsampling and high-precision analysis of strontium and rubidium isotopes at single crystal scale for petrological and geochronological applications," *Chemical Geology*, vol. 232, no. 3-4, pp. 144–133, 2006.

[55] R. Avanzinelli, E. Boari, S. Conticelli et al., "High precision Sr, Nd, and Pb isotopic analyses using the new generation Thermal Ionisation Mass Spectrometer ThermoFinningan Triton-Ti®," *Periodico di Mineralogia*, vol. 74, no. 3, pp. 147–166, 2005.

[56] D. L. Hoffmann, C. Spötl, and A. Mangini, "Micromill and in situ laser ablation sampling techniques for high spatial resolution MC-ICPMS U-Th dating of carbonates," *Chemical Geology*, vol. 259, no. 3-4, pp. 253–261, 2009.

[57] A. Eberhart, R. Delwiche, and Z. Geiss, "Isotopic effects in single filament thermal ion sources," *Zeitschrift für Naturforschung A*, vol. 19, no. 6, pp. 736–740, 1964.

[58] W. A. Russel, D. A. Papanastassiou, and T. A. Tombrello, "Ca isotope fractionation on the Earth and other solar system materials," *Geochimica et Cosmochimica Acta*, vol. 42, no. 8, pp. 1075–1090, 1978.

[59] M. F. Thirlwall, "Long-term reproducibility of multicollector Sr and Nd isotope ratio analysis," *Chemical Geology*, vol. 94, no. 2, pp. 85–104, 1991.

[60] D. Weis, B. Kieffer, C. Maerschalk et al., "High-precision isotopic characterization of USGS reference materials by TIMS and MC-ICP-MS," *Geochemistry, Geophysics, Geosystems*, vol. 7, no. 8, 2006.

[61] J. B. Mahoney, D. Weiss, B. Keiffer et al., "Ongoing isotopic characterization of USGS standards: MC-ICPMS and TIMS data from the Pacific Centre for isotopic and geochemical research, University of British Columbia," in *Proceedings of Geological Society of America, Seattle Annual Meeting, Paper 117-118*, Seattle, WA, USA, October 2003.

[62] M. Elburg, P. Vroon, B. van der Wagt, and A. Tchalikian, "Sr and Pb isotopic composition of five USGS glasses (BHVO-2G, BIR-1G, BCR-2G, TB-1G, NKT-1G)," *Chemical Geology*, vol. 223, no. 4, pp. 196–207, 2005.

[63] C. Spötl and D. Mattey, "Stable isotope microsampling of speleothems for palaeoenvironmental studies: a comparison of microdrill, micromill and laser ablation techniques," *Chemical Geology*, vol. 235, no. 1-2, pp. 48–58, 2006.

[64] C. Saenger, R. I. Gabitov, J. Farmer, J. M. Watkins, and R. Stone, "Linear correlations in bamboo coral $\delta^{13}C$ and $\delta^{18}O$ sampled by SIMS and micromill: evaluating paleoceanographic potential and biomineralization mechanisms using δ^{11} B and Δ_{47} composition," *Chemical Geology*, vol. 454, pp. 1–14, 2017.

[65] K. M. Knesel, J. P. Davidson, and W. A. Duffield, "Evolution of silicic magma through assimilation and subsequent recharge: evidence from Sr isotopes in sanidine phenocrysts, Taylor Creek Rhyolite, NM," *Journal of Petrology*, vol. 40, no. 5, pp. 773–786, 1999.

[66] G. S. Wallace and G. W. Bergantz, "Reconciling heterogeneity in crystal zoning data: an application of shared characteristic diagrams at Chaos Crags, Lassen Volcanic Center, California," *Contributions to Mineralogy and Petrology*, vol. 149, no. 1, pp. 98–112, 2005.

[67] O. Bachmann and G. W. Bergantz, "Gas percolation in upper-crustal silicic crystal mushes as a mechanism for upward heat advection and rejuvenation of near-solidus magma bodies," *Journal of Volcanology and Geothermal Research*, vol. 149, no. 1-2, pp. 85–102, 2006.

[68] L. Font, J. P. Davidson, D. G. Pearson, G. M. Nowell, D. A. Jerram, and C. J. Ottley, "Sr and Pb isotope micro-analysis of plagioclase crystals from Skye lavas: an insight into open-system processes in a flood basalt province," *Journal of Petrology*, vol. 49, no. 8, pp. 1449–1471, 2008.

[69] V. M. Martin, J. Davidson, D. Morgan, and D. A. Jerram, "Using the Sr isotope compositions of feldspars and glass to distinguish magma system components and dynamics," *Geology*, vol. 38, no. 6, pp. 539–542, 2010.

[70] G. M. Di Paola, "Volcanology and petrology of Nisyros island (Dodecanese Greece)," *Bulletin Volcanologique*, vol. 38, no. 3, pp. 944–987, 1974.

[71] G. P. Wyers and M. Barton, "Polybaric evolution of calc-alkaline magmas from Nisyros, southeastern Hellenic Arc, Greece," *Journal of Petrology*, vol. 30, no. 1, pp. 1–37, 1989.

[72] K. S. Seymour and D. Vlassopoulos, "Magma mixing at Nisyros volcano, as inferred from incompatible trace-element systematics," *Journal of Volcanology and Geothermal Research*, vol. 50, no. 3, pp. 273–299, 1992.

[73] G. Vougioukalakis, "Volcanic stratigraphy and evolution of Nisyros island," *Bulletin of the Geological Society of Greece*, vol. 28, no. 2, pp. 239–258, 1993.

[74] L. Francalanci, J. C. Varekamp, E. G. Vougioukalakis, M. J. Defant, F. Innocenti, and P. Manetti, "Crystal retention, fractionation and crustal assimilation in a convecting magma chamber, Nisyros Voclano, Greece," *Bulletin of Volcanology*, vol. 56, no. 8, pp. 601–620, 1995.

[75] L. Francalanci, J. C. Varekamp, E. G. Vougioukalakis, F. Innocenti, and P. Manetti, "Is there a compositional gap at Nisyros volcano? A comment on: magma generation at the easternmost section of the Hellenic arc: Hf, Nd, Pb and Sr isotope geochemistry of Nisyros and Yali volcanoes (Greece)," *Lithos*, vol. 95, no. 3-4, pp. 458–461, 2007.

[76] C. Longchamp, A. Skopelitis, C. Bonadonna, and O. Bachmann, "Characterization of tephra deposits with limited exposure: the example of the two largest explosive eruptions at Nisyros volcano (Greece)," *Bulletin of Volcanology*, vol. 73, no. 9, pp. 1337–1352, 2011.

[77] E. M. Limburg and J. C. Varekamp, "Young pumice deposits on Nisyros, Greece," *Bulletin of Volcanology*, vol. 54, no. 1, pp. 68–77, 1991.

[78] E. Braschi, L. Francalanci, and G. E. Vougioukalakis, "Inverse differentiation pathway by multiple mafic magma refilling in the last magmatic activity of Nisyros Volcano, Greece," *Bulletin of Volcanology*, vol. 74, no. 5, pp. 1083–1100, 2012.

[79] M. Rautenschlein, G. A. Jenner, J. Hertogen et al., "Isotopic and trace element composition of volcanic glasses from the Akaki Canyon, Cyprus: implications for the origin of the Troodos ophiolite," *Earth and Planetary Science Letters*, vol. 75, no. 4, pp. 369–383, 1985.

[80] N. Blum and J. H. Crocket, "Repetitive cyclical volcanism in the Late Archean Larder Lake Group near Kirkland Lake, Ontario: implications of geochemistry on magma genesis," *Precambrian Research*, vol. 54, no. 2-4, pp. 173–194, 1992.

[81] M. F. Roden, T. Trull, S. R. Hart, and F. A. Frey, "New He, Nd, Pb, and Sr isotopic constraints on the constitution of the Hawaiian plume: results from Koolau Volcano, Oahu, Hawaii, USA," *Geochimica et Cosmochimica Acta*, vol. 58, no. 5, pp. 1431–1440, 1994.

[82] P. R. Castillo, E. Klein, J. Bender et al., "Petrology and Sr, Nd, and Pb isotope geochemistry of mid-ocean ridge basalt glasses from the 11 45′ N to 15 00′ N segment of the East Pacific Rise," *Geochemistry, Geophysics, Geosystems*, vol. 1, no. 11, 2000.

[83] L. Melluso, V. Morra, P. Brotzu et al., "Geochronology and petrogenesis of the Cretaceous Antampombato-Ambatovy complex and associated dyke swarm, Madagascar," *Journal of Petrology*, vol. 46, no. 10, pp. 1963–1996, 2005.

[84] G. Conde, P. D. Ihinger, and E. E. Frahm, "Water speciation in Anatolian Obsidian: quenched magmatic water vs low temperature hydration," *Geochimica et Cosmochimica Acta*, vol. 73, p. A239, 2009.

[85] P. Degryse, A. Boyce, N. E. Satullo et al., "Isotopic discriminants between late Bronze Age glasses from Egypt and the Near East," *Archaeometry*, vol. 52, no. 3, pp. 380–388, 2010.

[86] J. Henderson, J. Evans, and K. Nikita, "Isotopic evidence for the primary production, provenance and trade of Late Bronze Age glass in the Mediterranean," *Mediterranean Archaeology and Archaeometry*, vol. 10, no. 1, pp. 1–24, 2010.

[87] M. Ganio, K. Latruwe, D. Brems, P. Muchez, F. Vanhaecke, and P. Degryse, "The Sr-Nd isolation procedure for subsequent isotopic analysis using multi-collector ICP-mass spectrometry in the context of provenance studies on archaeological glass," *Journal of Analytical Atomic Spectrometry*, vol. 27, no. 8, pp. 1335–1341, 2012.

[88] E. Gliozzo, E. Braschi, F. Giannetti, A. Langone, and M. Turchiano, "New geochemical and isotopic insights into the Late Antique Apulian glass and the HIMT1 and HIMT2 glass productions—the glass vessels from San Giusto (Foggia, Italy) and the diagrams for provenance studies," *Archaeological and Anthropological Sciences*, pp. 1-30, 2017.

[89] F. Sigmundsson, S. Hreinsdóttir, A. Hooper et al., "Intrusion triggering of the 2010 Eyjafjallajokull explosive eruption," *Nature*, vol. 468, no. 7322, pp. 426–430, 2010.

[90] E. Kaminski, S. Tait, F. Ferrucci, M. Martet, B. Hirn, and P. Husson, "Estimation of ash injection in the atmosphere by basaltic volcanic plumes: the case of the Eyjafjallajökull 2010 eruption," *Journal of Geophysical Research: Solid Earth*, vol. 116, no. 9, 2011.

[91] M. T. Gudmundsson, T. Thordarson, A. Höskuldsson et al., "Ash generation and distribution from the April-May 2010 eruption of Eyjafjallajökull, Iceland," *Scientific Reports*, vol. 2, no. 1, p. 572, 2012.

[92] P. W. Webley, T. Steensen, M. Stuefer, G. Grell, S. Freita, and M. Pavolonis, "Analyzing the Eyjafjallajökull 2010 eruption using satellite remote sensing, lidar and WRF-Chem dispersion and tracking model," *Journal of Geophysical Research: Atmospheres*, vol. 117, no. 20, 2012.

[93] R. Cioni, M. Pistolesi, A. Bertagnini, C. Bonadonna, A. Hoskuldsson, and B. Scateni, "Insights into the dynamics and evolution of the 2010 Eyjafjallajökull summit eruption (Iceland) provided by volcanic ash textures," *Earth and Planetary Science Letters*, vol. 394, pp. 111-123, 2014.

[94] O. Sigmarsson, J. Maclennan, and M. Carpentier, "Geochemistry of igneous rocks in Iceland: a review," *Jökull*, vol. 58, pp. 139–160, 2008.

[95] T. D. Price, S. Nakamura, S. Suzuki, J. H. Burton, and V. Tiesler, "New isotope data on Maya mobility and enclaves at Classic Copan, Honduras," *Journal of Anthropological Archaeology*, vol. 36, pp. 32–47, 2014.

[96] P. A. Slater, K. M. Hedman, and T. E. Emerson, "Immigrants at the Mississippian polity of Cahokia: strontium isotope evidence for population movement," *Journal of Archaeological Science*, vol. 44, pp. 117-127, 2014.

[97] A. J. Waterman, D. W. Peate, A. M. Silva, and J. T. Thomas, "In search of homelands: using strontium isotopes to identify biological markers of mobility in late prehistoric Portugal," *Journal of Archaeological Science*, vol. 42, pp. 119–127, 2014.

[98] A. L. Rheingold, S. Hues, and M. N. Cohen, "Strontium and zinc content in bone as an indication of diet," *Journal of Chemical Education*, vol. 60, no. 3, pp. 233-234, 1983.

[99] C. Gilbert, J. Sealy, and A. Sillen, "An investigation of barium, calcium and strontium as palaeodietary indicators in the Southwestern Cape, South Africa," *Journal of Archaeological Science*, vol. 21, no. 2, pp. 173–184, 1994.

[100] T. D. Price, C. M. Johnson, J. A. Ezzo, J. Ericson, and J. H. Burton, "Residential mobility in the prehistoric southwest United States: a preliminary study using strontium isotope analysis," *Journal of Archaeological Science*, vol. 21, no. 3, pp. 315–330, 1994.

[101] T. D. Price, L. Manzanilla, and W. D. Middleton, "Immigration and the ancient city of Teotihuacan in Mexico: a study using strontium isotope ratios in human bone and teeth," *Journal of Archaeological Science*, vol. 27, no. 10, pp. 903–913, 2000.

[102] J. H. Burton, T. D. Price, L. Cahue, and L. E. Wright, "The use of barium and strontium abundances in human skeletal tissues to determine their geographic origin," *International Journal of Osteoarchaeology*, vol. 13, no. 1-2, pp. 88–95, 2003.

[103] D. Mueller and E. Heinzle, "Stable isotope-assisted metabolomics to detect metabolic flux changes in mammalian cell cultures," *Current Opinion in Biotechnology*, vol. 24, no. 1, pp. 54–59, 2013.

[104] T. Tütken and T. Vennemann, "Stable isotope ecology of Miocene large mammals from Sandelzhausen, southern Germany," *Paläontologische Zeitschrift*, vol. 83, no. 1, pp. 207–226, 2009.

[105] L. A. Gregoricka, "Residential mobility and social identity in the periphery: strontium isotope analysis of archaeological tooth enamel from southeastern Arabia," *Journal of Archaeological Science*, vol. 40, no. 1, pp. 452–464, 2013.

[106] E. McManus, J. Montgomery, J. Evans, A. Lamb, R. Brettell, and J. Jelsma, "To the land or to the sea: diet and mobility in Early Medieval Frisia," *Journal of Island and Coastal Archaeology*, vol. 8, no. 2, pp. 255–277, 2013.

[107] J. E. Ericson, "Strontium isotope characterization in the study of prehistoric human ecology," *Journal of Human Evolution*, vol. 14, no. 5, pp. 503–514, 1985.

[108] S. Hillson, *Dental Anthropology*, Cambridge University Press, Cambridge, UK, 1996.

[109] J. E. Ericson, "Some problems and potentials of strontium isotope analysis for human and animal ecology," in *Stable Isotopes in Ecological Research*, pp. 252–259, Springer, New York, NY, USA, 1989.

[110] J. C. Sealy, N. J. van der Merwe, A. Sillen, F. J. Kruger, and H. W. Krueger, "87Sr86Sr as a dietary indicator in modern and archaeological bone," *Journal of Archaeological Science*, vol. 18, no. 3, pp. 399–416, 1991.

[111] B. K. Nelson, M. J. Deniro, M. Schoeninger, D. J. De Paolo, and P. E. Hare, "Effects of diagenesis on strontium, carbon, nitrogen and oxygen concentration and isotopic composition of bone," *Geochimica et Cosmochimica Acta*, vol. 50, no. 9, pp. 1941–1949, 1986.

[112] P. Budda, B. L. J. Montgomery, P. Rainbird, R. G. Thomas, and S. M. Young, "Pb and Sr isotope composition of human dental enamel: an indicator of Pacific Islander population dynamics," in *Le Pacifique de 5000 à 2000 avant le présent: suppléments à l'histoire d'une colonisation = The Pacific from 5000 to 2000 BP: Colonisation and Transformations*, G. Jean-Christophe and I. Lilley, Eds., Institut de recherche pour le développement, Marseille, France, 1999.

[113] M. Sponheimer and J. A. Lee-Thorp, "Enamel diagenesis at South African Australopith sites: implications for paleoecological reconstruction with trace elements," *Geochimica et Cosmochimica Acta*, vol. 70, no. 7, pp. 1644–1654, 2006.

[114] C. Hänni, V. Laudet, D. Stehelin, and P. Taberlet, "Tracking the origins of the cave bear (*Ursus spelaeus*) by mitochondrial DNA sequencing," *Proceedings of the National Academy of Sciences*, vol. 91, no. 25, pp. 12336–12340, 1994.

[115] P. Argenti and P. P. Mazza, "Mortality analysis of the Late Pleistocene bears from Grotta Lattaia, central Italy," *Journal of Archaeological Science*, vol. 33, no. 11, pp. 1552–1558, 2006.

[116] I. Martini, M. Coltorti, P. P. Mazza, M. Rustioni, and F. Sandrelli, "The latest *Ursus spelaeus* in Italy, a new contribution to the extinction chronology of the cave bear," *Quaternary Research*, vol. 81, no. 1, pp. 117–124, 2014.

[117] H. Bocherens, M. Fizet, and A. Mariotti, "Diet, physiology and ecology of fossil mammals as inferred from stable carbon and nitrogen isotope biogeochemistry: implications for Pleistocene bears," *Palaeogeography, Palaeoclimatology, Palaeoecology*, vol. 107, no. 3-4, pp. 213–225, 1994.

[118] H. Bocherens, D. Billiou, M. Patou-Mathis, D. Bonjean, M. Otte, and A. Mariotti, "Paleobiological implications of the isotopic signatures (^{13}C, ^{15}N) of fossil mammal collagen in Scladina Cave (Sclayn, Belgium)," *Quaternary Research*, vol. 48, no. 3, pp. 370–380, 1997.

[119] H. Bocherens, M. Stiller, K. A. Hobson et al., "Niche partitioning between two sympatric genetically distinct cave bears (*Ursus spelaeus* and *Ursus ingressus*) and brown bear (*Ursus arctos*) from Austria: isotopic evidence from fossil bones," *Quaternary International*, vol. 245, no. 2, pp. 238–248, 2011.

[120] D. Fernández-Mosquera, M. Vila-Taboada, and A. Grandal-d'Anglade, "Stable isotopes data ($\delta^{13}C$, $\delta^{15}N$) from the cave bear (*Ursus spelaeus*): a new approach to its palaeoenvironment and dormancy," *Proceedings of the Royal Society of London B: Biological Sciences*, vol. 268, no. 1472, pp. 1159–1164, 2001.

[121] M. P. Richards, M. Pacher, M. Stiller et al., "Isotopic evidence for omnivory among European cave bears: late Pleistocene *Ursus spelaeus* from the Peştera cu Oase, Romania," *Proceedings of the National Academy of Sciences*, vol. 105, no. 2, pp. 600–604, 2008.

[122] Z. Nerudová, M. Nývltová Fišáková, and J. Míková, "Palaeoenvironmental analyses of animal remains from the Kůlna Cave (Moravian Karst, Czech Republic)," *Quartär*, vol. 61, 2014.

[123] O. Loreille, L. Orlando, M. Patou-Mathis, M. Philippe, P. Taberlet, and C. Hänni, "Ancient DNA analysis reveals divergence of the cave bear, *Ursus* spelaeus, and brown bear, Ursus arctos, lineages," *Current Biology*, vol. 11, no. 3, pp. 200–203, 2001.

[124] L. Orlando, D. Bonjean, H. Bocherens et al., "Ancient DNA and the population genetics of cave bears (*Ursus spelaeus*) through space and time," *Molecular Biology and Evolution*, vol. 19, no. 11, pp. 1920–1933, 2002.

[125] M. Pacher and A. J. Stuart, "Extinction chronology and palaeobiology of the cave bear (*Ursus spelaeus*)," *Boreas*, vol. 38, no. 2, pp. 189–206, 2009.

[126] A. Berton, M. Bonato, A. Borsato et al., "Nuove datazioni radiometriche con il metodo U/Th sulle formazioni stalagmitiche di Grotta all'Onda," *Rivista di Scienze Preistoriche*, vol. 53, pp. 241–256, 2003.

[127] G. Molara, "Resti faunistici provenienti dai livelli del Pleistocene superior di Grotta all'Onda (Camaiore, Lucca)," in *Atti del 6° Convegno Nazionale di Archeozoologia*, pp. 57–62, Centro visitatori del Parco dell'Orecchiella, Lucca, Italy, 2009.

[128] T. Adani, *Applicazioni delle sistematiche geochimiche ed isotopiche a reperti fossili Quaternari*, M.S. thesis, Earth Sciences Department, Università degli Studi di Firenze, 2013.

[129] G. Cortecci and L. Lupi, "Carbon, oxygen and strontium isotope geochemistry of carbonates rocks from the Tuscan Nappe, Italy," *Mineral Petrol Acta*, vol. 37, pp. 63–80, 1994.

2

A PCR Method that can be Further Developed into PCR-RFLP Assay for Eight Animal Species Identification

Feng Guan, Yu-Ting Jin, Jin Zhao, Ai-Chun Xu, and Yuan-Yuan Luo

College of Life Sciences, China Jiliang University, Hangzhou 310018, China

Correspondence should be addressed to Yuan-Yuan Luo; yyluo@cjlu.edu.cn

Academic Editor: Antony C. Calokerinos

There are many PCR-based methods for animal species identification; however, their detection numbers are limited or could not identify unknown species. We set out to solve this problem by developing a universal primer PCR assay for simultaneous identification of eight animal species, including goat, sheep, deer, buffalo, cattle, yak, pig, and camel. In this assay, the variable lengths of mitochondrial DNA were amplified using a pair of universal primers. PCR amplifications yielded 760 bp, 737 bp, 537 bp, 486 bp, 481 bp, 464 bp, 429 bp, and 359 bp length fragments for goat, sheep, deer, buffalo, cattle, yak, pig, and camel, respectively. This primer pair had no cross-reaction with other common domestic animals and fish. The limit of detection varied from 0.01 to 0.05 ng of genomic DNA for eight animal species in a 20 μl PCR mixture. Each PCR product could be further digested into fragments with variable sizes and qualitative analysis by *Ssp*I restriction enzyme. This developed PCR-RFLP assay was sufficient to distinguish all targeted species. Compared with the previous published related methods, this approach is simple, with high throughput, fast processing rates, and more cost-effective for routine identification of meat in foodstuffs.

1. Introduction

The consumption of meat and meat products is increasing each year in the world. On the other hand, food authenticity issues in the form of adulteration and improper description have existed for as long as food has been offered for sale. In China, "hang a sheep head, sell vinegar" is a widely known proverb. However, meat adulteration affects food safety, quality, and many other respects, becoming a public focus recently. Since the 1980s, many immunological and molecular methods for species identification in food products have been developed [1–8]. Concomitant with advances in large-scale integrated molecular technology, DNA analysis methods have been in rapid development. Many assays based on DNA analysis, especially PCR-based methods, have been developed for species identification in foodstuffs. Though each of these methods has its own advantages and disadvantages, the number of species identified in a single PCR was limited to 4-5 [9]. Compared with other PCR methods, the combination of universal primer PCR and restriction fragment length polymorphism (RFLP) assays has advantages of increased simplicity, specificity, and sensitivity and has been used in many meat identification practices [10–17]. On the other hand, a few modified PCR assays combining universal primer with specific primer or RFLP had been developed [18, 19]; these methods increased efficiency and versatility. Furthermore, PCR-RFLP had additional advantages over multiplex PCR and species-specific PCR, such as the ability to detect a larger number of animal species and differentiate between closed animal species.

With the development of Western China, more and more local meat products of mutton, beef, yak, deer, and camel have been sold to all parts of the country. There have been a few analytical methods for the identification of yak [20, 21], deer [17], buffalo [22, 23], and camel [24] meat in food products, but to our knowledge, there is no one method for simultaneously identifying two and more animal species, for goat, sheep, buffalo, cattle, yak, deer, and camel. Furthermore, pork is popular with Chinese people, but Muslim populations are widely distributed in China; all the meat products from pork source are forbidden in halal foods. In this regard, precise identification of meat origin has become a vital element in food quality control procedures. Thus, the development of reliable, simple, and sensitive analytical methods for evaluating meat authenticity is of critical importance.

In recent years, universal primer PCRs have been used and further developed for animal species identification combination with other DNA analysis methods [19, 25–29], which had many advantages over common and even real-time PCR assays. In this study, a new pair of universal primers was designed and a species identification assay has been developed using the variable size fragments, and this assay can be further developed into PCR-RFLP assay, which showed potential as a tool for cost-effective, rapid, specific, and sensitive detection of goat, sheep, buffalo, cattle, yak, deer, pig, and camel simultaneously.

2. Materials and Methods

2.1. Samples Collection and DNA Extraction. The samples of known species origins including goat *(Capra aegagrus hircus)*, sheep *(Ovis aries)*, deer *(Cervus axis)*, buffalo *(Bovinae)*, cattle *(Bos primigenius)*, yak *(Bos mutus grunniens)*, pig *(Sus scrofa)*, camel *(Pilus cameli)*, horse *(Equus caballus)*, donkey *(Equus asinus)*, rabbit *(Oryctolagus cuniculus)*, chicken *(Gallus gallus)*, duck *(Anas platyrhynchos)*, rat *(Rattus norvegicus)*, dog *(Canis lupus familiaris)*, frog *(Rana catesbiana)*, and fish *(Carassius auratus)* were collected from a slaughterhouse and experimental animal center in Hangzhou, China, and were stored at −20°C in our laboratory. In addition, fifteen commercial meat product samples were purchased to assess this developed method in practice; the ingredients were labeled as deer, yak, camel, beef, and mutton, respectively.

Genomic DNA was extracted from meat samples using an Animal Tissue DNA Extraction Kit (Takara) according to supplied instructions. Following DNA extraction, the purity and concentration of all the DNA samples were confirmed using a UV-Vis spectrophotometer (NanoDrop 2000, Thermo). DNA samples were diluted to a final concentration of 10 ng/μl and stored at 4°C for next use.

2.2. Universal Primers and PCR Amplification. The mitochondrial DNA (mtDNA) sequences of all the above animals were retrieved from the NCBI database and were aligned using ClustalW sequence alignment tool to select the congenerous conserved regions. Two regions were selected as primer design areas, and a pair of universal primers was designed to amplify variable length mtDNA sequences from genomic DNA of yak, deer, goat, sheep, pig, camel, cattle, and buffalo but no other animals. There were several insertion-deletion polymorphisms in the amplified sequences, which, in theory, could yield different length fragments for each animal. The primer sequences are as follows: forward primer (5′-CCTCCCTAAGACTCAGGG-AA-3′) and reverse primer (5′-AGCGGGTTGCTGGTTT-CACG-3′). The designed primers were also screened for unique specificity to check cross-species binding with other animal or plant species using the online BLAST local alignment tool in the NCBI database.

PCR amplifications were carried out using a MJ-200 thermal cycler in a total of 20 μl mixture containing 2.0 μl 10x PCR buffer, 1.2 μl of dNTP mixture (25 mM), 1.6 μl $MgCL_2$ (25 mM), 1 U of *Taq* polymerase, 2 μl each of universal primers (10 mM), and 3 μl DNA template (about 30 ng). The PCR conditions consisted of preheating at 94°C for 5 min, followed by 30 cycles of denaturation at 94°C for 30 s, annealing at 61°C for 30 s, and extension at 72°C for 30 s. Then final extension was done at 72°C for 5 min. The 5.0 μl PCR product obtained was analyzed on 2% agarose gel in 1x TAE buffer stained with 4S Red as a visualizing agent and run for 40 min at 90 V. A known DNA ladder Marker C (Shanghai Sangon Biotech) was electrophoresed simultaneously in order to assess the size of the amplification product.

2.3. Restriction Fragment Length Polymorphism (RFLP) Analysis. To further discriminate between targeted animals, expected nucleotide sequences were restriction mapped with the Mapdraw program of DNASTAR (NY, USA). After testing, one restriction enzyme *Ssp*I (TakaRa) was selected to digest the PCR products in separate reactions. Briefly, 5 μl of the PCR product and 2 μl of the restriction enzyme were prepared by mixing 2 μl of 10x buffer, and sterile free water was added for a total volume of 20 μl. The tubes were incubated at 37°C for approximately 4 h. The digested PCR products were then separated on 2.5% agarose gel.

2.4. PCR Products Sequencing and Commercial Samples Detection. To further validate this assay, each of the PCR products was purified and then sequenced in Hangzhou Qingke Biotechnology Company and then were analyzed using the BLAST local alignment tool. Fifteen commercial samples were screened using this developed assay, and each of the PCR products obtained was extracted and purified separately and then digested by the restriction enzyme *Ssp*I.

3. Results and Discussions

3.1. DNA Extraction and PCR Amplifications. In the present study, all the genomic DNA samples were measured. The spectrophotometric assessment results showed that concentrations varied between 20 and 200 ng/μl, and the purity ($A_{260}/A_{280} = 1.72$–1.96) was suitable for PCR amplification. The universal primer pair generated 760 bp, 737 bp, 537 bp, 486 bp, 481 bp, 464 bp, 429 bp, and 359 bp length fragments from goat, sheep, deer, buffalo, cattle, yak, pig, and camel DNA templates, respectively (Figure 1).

3.2. PCR Specificity and Sensitivity. The specificity of primers was checked against the extracted DNA samples from common animals, which included dog, chicken, horse, rat, donkey, rabbit, duck, frog, and fish. The specificity test results showed that no cross amplification was detected. The sensitivity was tested using serially diluted DNA samples from goat, sheep, deer, buffalo, cattle, yak, pig, and camel, and until the PCR products could not be visualized on 2.0% agarose gel. DNA band patterns (Figure 2) indicated that the detection limit of DNA from pigs was even at 0.001 ng, and the minimum detection levels of goat, sheep, deer, buffalo,

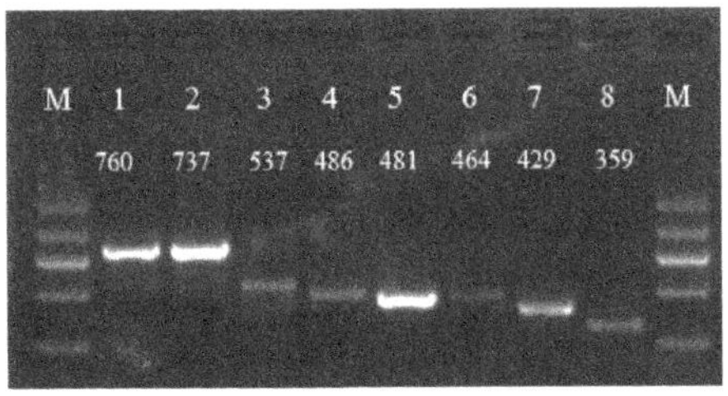

Figure 1: PCR amplification results for eight animal species. Lanes 1–8 represent goat, sheep, deer, buffalo, cattle, yak, pig, and camel, respectively. Lane M represents DNA marker; the lengths of different PCR products were noted above the band.

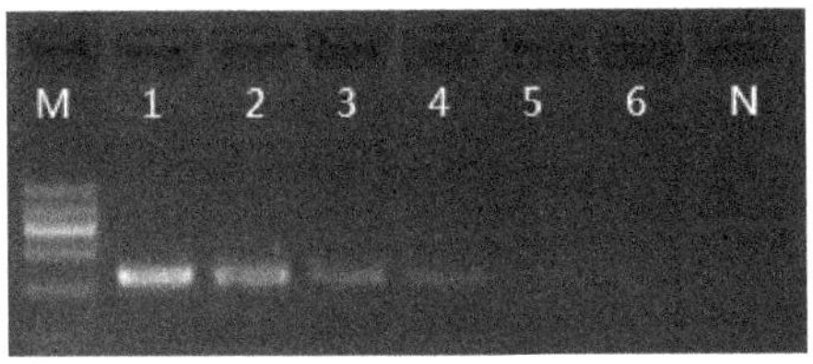

Figure 2: Sensitivity test for pig DNA. Lanes 1–6 represent the PCR products with serial dilution of pig DNA template. The content of pig DNA was 30.0, 10.0, 1.0, 0.1, 0.01, and 0.001 ng. Lane N was negative control.

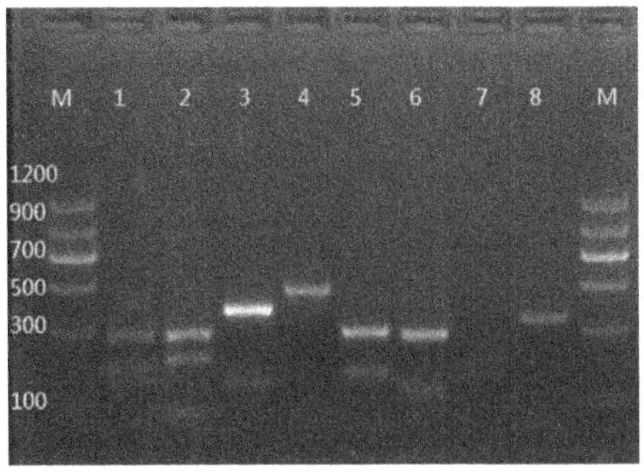

Figure 3: The PCR-RFLP result for eight target animals. Lane M represents DNA marker; lanes 1–8 represent goat, sheep, deer, buffalo, cattle, yak, pig, and camel, respectively.

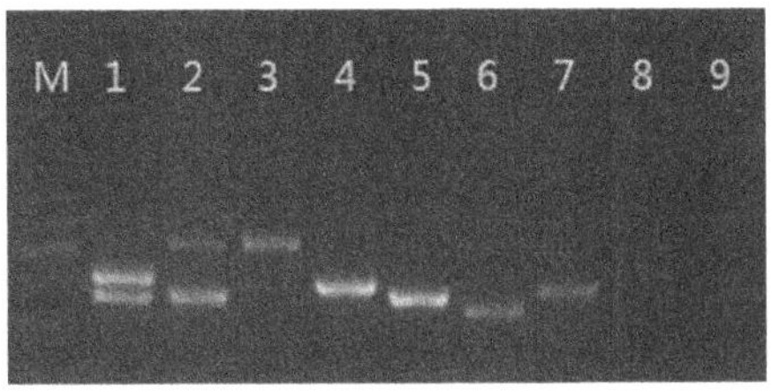

Figure 4: Identification results of commercial meat samples. Samples in lanes 1–7 were labeled as deer, mutton, mutton, beef, pork, camel, and yak meat. Lane 8 was soybean DNA, and lane 9 was negative control. But, the samples in lanes 1 and 2 contained pork.

cattle, yak, and camel DNA templates were between 0.01 and 0.05 ng.

3.3. PCR-RFLP Analysis and Sequencing. Each of the PCR products yielded different patterns for eight animal species after *Ssp*I restriction enzyme digestion. The PCR products yielded 291 bp, 192 bp, 160 bp, and 118 bp for goat; 300 bp, 213 bp, and 75 bp for sheep; 391 bp and 146 bp for deer; 303 bp and 178 bp for cattle; 214 bp, 178 bp, and 73 bp for yak; and 300 bp and 129 bp for pig. However, the PCR products from buffalo and camel DNA could not be cleaved by *Ssp*I; thus, the PCR products were always 486 bp and 359 bp, respectively. The PCR product digestion patterns are shown in Figure 3. Unique restriction pattern of *Ssp*I for each species was found to satisfactorily differentiate between all eight animal species.

To confirm the developed method, each of the PCR products was purified and sequenced. The sequencing results indicated that the sizes and sequences of all the PCR products corresponded exactly to that of the expected amplicons. The similarity parameters were as high as 100% in accordance with the GenBank database. One exception was buffalo (98.2%). This difference was caused by two single nucleotide polymorphic loci detected in the referenced sequence of buffalo compared to GenBank Accession no. AY702618. The sequencing results were fully in accordance with the expected amplicons, which indicates that this developed assay had a higher specificity.

3.4. Commercial Samples Detection. The fifteen commercial meat products were detected using this developed assay, two animal ingredients were found in two of the fifteen samples, while other samples had only one ingredient (Figure 4). The two samples were labeled as deer meat and mutton, respectively. Judging from the electrophoretic profiles, these two meat products were adulterated with pork. To confirm this finding, all the electrophoretic bands were extracted and sequenced. Sequencing results confirmed that the two samples contained pork ingredients. The other thirteen samples contained only one meat ingredient and were consistent with the labeling meat species.

With the development of economy, meat products become a daily food for consumption. Furthermore, yak, camel, and deer meats are new popular meat products in the Chinese market, especially as leisure foods. The price of these meat products is higher than that of common meat products such as beef and pork. On the other hand, the rising price and decreasing availability of high-quality meat drive some meat producers to misrepresent and/or adulterate meat and meat products. In fact, food adulteration has been around for a long time. The most common economic fraudulence widely spread in meat industry is adulteration or substitution of costlier meat with cheaper or inferior meat. Preventing meat adulteration in foods is important for economic, religious, and health reasons. In addition, identification of the species origin in processed meat products is an important task in food hygiene, food codex, food control, and veterinary forensic medicine.

In order to protect consumers from fraud and adulteration, many meat and/or animal species identification methods have been developed to date. DNA assays have become common methods in practice because they have many advantages over other analyses. The stability of mtDNA is higher than that of genomic DNA and it is distributed in all the tissues. Therefore, mtDNA sequences are preferential for DNA barcoding in species identification [9, 30–37]. It is clear that PCR assay is a preferred technique based on DNA analysis for meat species identification to date. PCR and its derived techniques have been widely used for meat authentication, which included DNA hybridization, species-specific PCR, multiplex PCR, PCR-RFLP, PCR-SSCP, and PCR sequencing [2, 31, 38, 39]. Among these techniques, PCR-RFLP is considered as a highly discriminatory, reliable, and reproducibility method. Furthermore, the advantage of mitochondrial DNA-based PCR-RFLP analysis derives from the fact that there are many mitochondria per cell and many mitochondrial DNA molecules within each mitochondrion, making mtDNA a naturally amplified source of genetic variation [10, 14, 24, 32, 40, 41].

In the past few years, PCR-RFLP assays had been used to identify food animal species targeting CytB gene [14, 17, 42] and 12S rRNA gene [16, 43]. Among these assays, the same length fragments were amplified from two or more animals in the PCR procedure, then PCR products were digested using one or more restriction enzymes to yield different patterns [14, 15]. Partis et al. [34] developed a PCR-RFLP method for the detection of 22 animal species, but this method was unsuitable for analyzing meat mixtures. Several similar multiplex species identification assays had been reported. For example, Wang et al. [32] reported a terminal restriction fragment length polymorphism (T-RFLP) method for the identification of 12 animal species by targeting the 12S rRNA gene, but it required two restriction enzymes, double-fluorescently labeled primers and capillary electrophoresis. Hanapi et al. [44] reported a common primer multiplex PCR (CP-M-PCR) for the identification of pig, ruminant, avian, and rabbit meat, but this method could not distinguish closely related species, such as buffalo and cattle, and sheep and goat. There have been other several reported related methods based on PCR analysis, but these methods were often not suitable for mixture of food products, as the process was time consuming, inadequate, and expensive.

In the present study, a modified PCR assay was developed using universal primers for simultaneous identification of goat, sheep, deer, buffalo, cattle, yak, pig, and camel. Also, this assay could be further developed into RFLP method; the fragments of different lengths produced from PCR products could be identifying the animal species. This method can detect any targeting individual species within a meat mixture, which had better discrimination and specificity than many other multiple PCR methods.

This developed PCR-RFLP assay was a simple and rapid technology for identifying multiple species. This method did not require expensive equipment, and the optimization procedure was simple in comparison with multiplex PCR, real-time PCR, and other technologies that are used for the detection of two or more animal species. Interestingly, this method could identify individual species within a meat mixture by separating different size PCR products. This method was developed based on conventional PCR platform and required a common PCR thermal cycler, so it could be carried out in most laboratories. Furthermore, the developed PCR had no cross-reaction with DNA samples from common meats, such as duck, chicken, rat, frog, fish, donkey, and horse, and the PCR system had a high sensitivity at a minimum level of 0.001 ng of DNA. In general, the detection sensitivity for animal species such as poultry, ruminant, and pig using common PCR and real-time PCR was up to 0.1–1% of the ingredients and was also able to detect common meat species down to 0.15–0.01 ng of DNA [19, 45, 46]. So, it seems that the sensitivity is even higher than that of real-time PCR targeting for genomic DNA. This high sensitivity was probably because mtDNA was a multiple copy gene in each cell, and there was an increased probability of survival under different processing conditions. These advantages ensured that amplification was successful even in certain samples containing small amounts of DNA [47]. Although this developed PCR-RFLP had advantages over conventional PCR and multiplex PCR in some respects, it could not well satisfy the needs for meat products supervision in practice. Instead, this process could be combined with other species identification methods to get a better result in practice. For example, this method could combine with other species-specific primers to establish a multiplex PCR or combined with direct PCR to improve the speed [48, 49]. In fact, the choice of an appropriate method for species identification is decided by multiple considerations, such as sensitivity, specificity, accuracy, discriminatory ability, reproducibility, cost, speed, availability of equipment, and availability of suitably trained staff. The developed method provides more choices for animal identification, especially for meat food supervision.

4. Conclusion

In conclusion, a pair of universal primers for amplifying variable length fragments in eight animal species was designed and a PCR assay was developed, which could be further developed into PCR-RFLP. This developed method can be used to identify eight animal species. This method was specific, sensitive, and reliable in the simultaneous identification of goat, sheep, deer, buffalo, cattle, yak, pig, and camel. This proposed method was relatively simple and rapid, does not require expensive equipment, and could be performed in most laboratories. It is a practical approach for routine analysis to determine fraudulent and/or mislabeled substitution in meat products.

Conflicts of Interest

The authors have declared that no conflicts of interest exist.

Acknowledgments

The authors would like to thank all the people for collecting the samples and they also gratefully acknowledge financial support from the National Natural Science Foundation of China (Nos. 31672394 and 31360540), the Science and

Technology Department of Zhejiang for public welfare Technology Application Research (2017C32081), and the Major State Basic Research Development Program of China (2017YFD0501904).

References

[1] M. Safdar and Y. Junejo, "The development of a hexaplex-conventional PCR for identification of six animal and plant species in foodstuffs," *Food Chemistry*, vol. 192, pp. 745–749, 2016.

[2] A. Kumar, R. R. Kumar, B. D. Sharma, P. Gokulakrishnan, S. K. Mendiratta, and D. Sharma, "Identification of species origin of meat and meat products on the DNA basis: a review," *Critical Reviews in Food Science and Nutrition*, vol. 55, no. 10, pp. 1340–1351, 2015.

[3] B. Hou, X. Meng, L. Zhang, J. Guo, S. Li, and H. Jin, "Development of a sensitive and specific multiplex PCR method for the simultaneous detection of chicken, duck and goose DNA in meat products," *Meat Science*, vol. 101, pp. 90–94, 2015.

[4] Z. M. Li, L. J. Zhao, Y. F. Wang, and H.-W. Zhang, "Progress of molecular identification techniques used in meat and meat products," *Journal of Food Safety and Quality*, vol. 6, no. 2, pp. 405–409, 2015.

[5] A. Doosti, P Ghasemi Dehkordi, and E. Rahimi, "Molecular assay to fraud identification of meat products," *Journal of Food Science and Technology*, vol. 51, no. 1, pp. 148–152, 2014.

[6] A. Vlachos, I. S. Arvanitoyannis, and P. Tserkezou, "An updated review of meat authenticity methods and applications," *Critical Reviews in Food Science and Nutrition*, vol. 56, no. 7, pp. 1061–1096, 2013.

[7] A. Iwobi, I. Huber, and U. Busch, "The application of PCR-based methods in food control agencies—a review," in *Polymerase Chain Reaction*, pp. 173–194, Bavarian Health and Food Safety Authority, Oberschleissheim, Germany, 2012.

[8] R. Rodríguez-Ramírez, A. F. González-Córdova, and B. Vallejo-Cordoba, "Review: authentication and traceability of foods from animal origin by polymerase chain reaction-based capillary electrophoresis," *Analytica Chimica Acta*, vol. 685, no. 2, pp. 120–126, 2011.

[9] M. T. Bottero and A. Dalmasso, "Animal species identification in food products: evolution of biomolecular methods," *Veterinary Journal*, vol. 190, no. 1, pp. 34–38, 2011.

[10] Y. Erwanto, M. Z. Abidin, E. Y. Sugiyono, and A. Rohman, "Identification of pork contamination in meatballs of Indonesia local market using polymerase chain reaction-restriction fragment length polymorphism (PCR-RFLP) analysis," *Asian-Australasian Journal of Animal Sciences*, vol. 27, no. 10, pp. 1487–1492, 2014.

[11] X. Huang, X. Zhou, Q. Lin, W. Fang, and X. Chen, "PCR-RFLP technique for species identification of molted feathers in six species of co-occurring Ardeids," *Conservation Genetics Resources*, vol. 5, no. 3, pp. 817–819, 2013.

[12] D. Maede, "A strategy for molecular species detection in meat and meat products by PCR-RFLP and DNA sequencing using mitochondrial and chromosomal genetic sequences," *European Food Research and Technology*, vol. 224, no. 2, pp. 209–217, 2006.

[13] Y.-L. Sun and C. S. Lin, "Establishment and application of a fluorescent polymerase chain reaction-restriction fragment length polymorphism (PCR-RFLP) method for identifying porcine, caprine, and bovine meats," *Journal of Agricultural and Food Chemistry*, vol. 51, no. 7, pp. 1771–1776, 2003.

[14] D. Kumar, S. P. Singh, N. S. Karabasanavar, R. Singh, and V. Umapathi, "Authentication of beef, carabeef, chevon, mutton and pork by a PCR-RFLP assay of mitochondrial cytb gene," *Journal of Food Science and Technology*, vol. 51, no. 11, pp. 3458–3463, 2014.

[15] Z. Tian, J. Du, J. Yang et al., "A PCR-RFLP Assay targeting RPS8 gene for the discrimination between bovine *Babesia* and *Theileria* species in China," *Parasites and Vectors*, vol. 8, no. 1, p. 475, 2015.

[16] P. S. Girish, A. S. Anjaneyulu, K. N. Viswas et al., "Polymerase chain reaction-restriction fragment length polymorphism of mitochondrial 12S rRNA gene: a simple method for identification of poultry meat species," *Veterinary Research Communications*, vol. 31, no. 4, pp. 447–455, 2007.

[17] I. Pfeiffer, J. Burger, and B. Brenig, "Diagnostic polymorphisms in the mitochondrial cytochrome b gene allow discrimination between cattle, sheep, goat, roe buck and deer by PCR-RFLP," *BMC Genetics*, vol. 5, no. 1, p. 30, 2004.

[18] P. Sharma and T. Kobayashi, "Are "universal" DNA primers really universal?," *Journal of Applied Genetics*, vol. 55, no. 4, pp. 485–496, 2014.

[19] J. Xu, W. Zhao, M. Zhu et al., "Molecular identification of adulteration in mutton based on mitochondrial 16S rRNA gene," *Mitochondrial DNA*, vol. 27, no. 1, pp. 628–632, 2016.

[20] P. Wang, Y. Hu, H. Yang, J. Han, Y. Zhao, and Y. Chen, "DNA-based authentication method for detection of yak *(Bos grunniens)* in meat products," *Journal of AOAC International*, vol. 96, no. 1, pp. 142–146, 2013.

[21] R. H. Yin, W. L. Bai, J. M. Wang et al., "Development of an assay for rapid identification of meat from yak and cattle using polymerase chain reaction technique," *Meat Science*, vol. 83, no. 1, pp. 38–44, 2009.

[22] P. S. Girish, S. Haunshi, S. Vaithiyanathan, R. Rajitha, and C. Ramakrishna, "A rapid method for authentication of Buffalo *(Bubalus bubalis)* meat by Alkaline Lysis method of DNA extraction and species specific polymerase chain reaction," *Journal of Food Science and Technology*, vol. 50, no. 1, pp. 141–146, 2013.

[23] B. G. Mane, S. K. Mendiratta, A. K. Tiwari, and K. N. Bhilegaokar, "Detection of adulteration of meat and meat products with buffalo meat employing polymerase chain reaction assay," *Food Analytical Methods*, vol. 5, no. 2, pp. 296–300, 2011.

[24] M. V. Mahajan, Y. P. Gadekar, V. D. Dighe, R. D. Kokane, and A. S. Bannalikar, "Molecular detection of meat animal species targeting MT 12S rRNA gene," *Meat Science*, vol. 88, no. 1, pp. 23–27, 2011.

[25] M. T. Bottero, T. Civera, D. Nucera, and R. M. Turi, "Design of universal primers for the detection of animal tissues in feedstuff," *Veterinary Research Communications*, vol. 27, no. 1, pp. 667–669, 2003.

[26] T. Kitano, K. Umetsu, W. Tian, and M. Osawa, "Two universal primer sets for species identification among vertebrates," *International Journal of Legal Medicine*, vol. 121, no. 5, pp. 423–427, 2007.

[27] A. O. Tillmar, B. Dell'Amico, J. Welander, and G. Holmlund, "A universal method for species identification of mammals utilizing next generation sequencing for the analysis of DNA mixtures," *PLoS One*, vol. 8, no. 12, article e83761, 2013.

[28] F. Ripp, C. F. Krombholz, Y. Liu et al., "All-Food-Seq (AFS): a quantifiable screen for species in biological samples by deep DNA sequencing," *BMC Genomics*, vol. 15, no. 1, p. 639, 2014.

[29] T. J. Fernandes, J. Costa, M. B. Oliveira, and I. Mafra, "DNA barcoding coupled to HRM analysis as a new and simple tool

for the authentication of Gadidae fish species," *Food Chemistry*, vol. 230, pp. 49–57, 2017.

[30] I. R. Maine, R. Atterbury, and K. C. Chang, "Investigation into the animal species contents of popular wet pet foods," *Acta Veterinaria Scandinavica*, vol. 57, no. 1, pp. 1–4, 2015.

[31] H. Shabani, M. Mehdizadeh, S. M. Mousavi et al., "Halal authenticity of gelatin using species-specific PCR," *Food Chemistry*, vol. 184, pp. 203–206, 2015.

[32] Q. Wang, X. Zhang, H. Y. Zhang et al., "Identification of 12 animal species meat by T-RFLP on the 12S rRNA gene," *Meat Science*, vol. 85, no. 2, pp. 265–269, 2010.

[33] G. Rastogi, M. S. Dharne, S. Walujkar, A. Kumar, M. S. Patole, and Y. S. Shouche, "Species identification and authentication of tissues of animal origin using mitochondrial and nuclear markers," *Meat Science*, vol. 76, no. 4, pp. 666–674, 2007.

[34] L. Partis, D. Croan, Z. Guo, R. Clark, T. Coldham, and J. Murby, "Evaluation of a DNA fingerprinting method for determining the species origin of meats," *Meat Science*, vol. 54, no. 4, pp. 369–376, 2000.

[35] J. F. Montiel-Sosa, E. Ruiz-Pesini, J. Montoya, P. Roncalés, M. J. López-Pérez, and A. Pérez-Martos, "Direct and highly species-specific detection of pork meat and fat in meat products by PCR amplification of mitochondrial DNA," *Journal of Agricultural and Food Chemistry*, vol. 48, no. 7, pp. 2829–2832, 2000.

[36] I. Martin, T. Garcia, V. Fajardo et al., "Species-specific PCR for the identification of ruminant species in feedstuffs," *Meat Science*, vol. 75, no. 1, pp. 120–127, 2007.

[37] I. Mafra, I. M. P. L. V. O. Ferreira, and M. B. P. P. Oliveira, "Food authentication by PCR-based methods," *European Food Research and Technology*, vol. 227, no. 3, pp. 649–665, 2007.

[38] R. L. Patterson and S. J. Jones, "Review of current techniques for the verification of the species origin of meat," *Analyst*, vol. 115, no. 5, pp. 501–506, 1990.

[39] V. Fajardo, I. González, M. Rojas, T. García, and R. Martín, "A review of current PCR-based methodologies for the authentication of meats from game animal species," *Trends in Food Science and Technology*, vol. 21, no. 8, pp. 408–421, 2010.

[40] R. Q. Geng, C. Yuan, and Y. L. Chen, "Identification of goat cashmere and sheep wool by PCR-RFLP analysis of mitochondrial 12S rRNA gene," *Mitochondrial DNA*, vol. 23, no. 6, pp. 466–470, 2012.

[41] S.-Y. Chen, Y.-G. Yao, and Y.-P. Liu, "Species identification of ten common farm animals based on mitochondrial 12S rRNA gene polymorphisms," *Animal Biotechnology*, vol. 23, no. 3, pp. 213–220, 2012.

[42] H. Y. Feng, C. S. Liu, J. W. He et al., "Identification of mutton from duck meat by PCR-RFLP analysis of mitochondrial DNA cyt b gene," *Science and Technology of Food Industry*, vol. 33, no. 13, pp. 319–321, 2012.

[43] P. S. Girish, A. S. Anjaneyulu, K. N. Viswas et al., "Meat species identification by polymerase chain reaction-restriction fragment length polymorphism (PCR-RFLP) of mitochondrial 12S rRNA gene," *Meat Science*, vol. 70, no. 1, pp. 107–112, 2005.

[44] U. K. Hanapi, M. N. Desa, A. Ismail, and S. Mustafa, "A higher sensitivity and efficiency of common primer multiplex PCR assay in identification of meat origin using NADH dehydrogenase subunit 4 gene," *Journal of Food Science and Technology*, vol. 52, no. 7, pp. 4166–4175, 2015.

[45] M. E. Ali, M. A. Razzak, S. B. Hamid et al., "Multiplex PCR assay for the detection of five meat species forbidden in Islamic foods," *Food Chemistry*, vol. 177, pp. 214–224, 2015.

[46] N. S. Karabasanavar, S. P. Singh, D. Kumar, and S. N. Shebannavar, "Detection of pork adulteration by highly-specific PCR assay of mitochondrial D-loop," *Food Chemistry*, vol. 145, pp. 530–534, 2014.

[47] M. A. Rodriguez, T. Garcia, I. Gonzalez, L. Asensio, P. E. Hernández, and R. Martín, "PCR identification of beef, sheep, goat, and pork in raw and heat-treated meat mixtures," *Journal of Food Protection*, vol. 67, no. 1, pp. 172–177, 2004.

[48] T. Kitpipit, K. Sittichan, and P. Thanakiatkrai, "Direct-multiplex PCR assay for meat species identification in food products," *Food Chemistry*, vol. 163, pp. 77–82, 2014.

[49] M. Miura, C. Tanigawa, Y. Fujii, and S. Kaneko, "Comparison of six commercially-available DNA polymerases for direct PCR," *Revista do Instituto de Medicina Tropical de São Paulo*, vol. 55, no. 6, pp. 401–406, 2013.

3

Development of an Amperometric Glucose Biosensor based on the Immobilization of Glucose Oxidase on the Se-MCM-41 Mesoporous Composite

Sabriye Yusan,[1,2] Mokhlesur M. Rahman,[3] Nasir Mohamad,[3] Tengku M. Arrif,[3] Ahmad Zubaidi A. Latif,[4] Mohd Aznan M. A.,[5] and Wan Sani B. Wan Nik[6]

[1]*Department of Chemistry, Faculty of Science, Universiti Teknologi Malaysia, 81310 Johor Bahru, Johor, Malaysia*
[2]*Institute of Nuclear Science, Ege University, Bornova, 35100 Izmir, Turkey*
[3]*Institute for Community Development & Quality of Life (i-CODE), Universiti Sultan Zainal Abidin, 21300 Kuala Nerus, Terengganu, Malaysia*
[4]*Faculty of Medicine, Universiti Sultan Zainal Abidin, 21300 Kuala Nerus, Terengganu, Malaysia*
[5]*Faculty of Medicine, International Islamic University Malaysia, 25200 Kuantan, Malaysia*
[6]*School of Ocean Engineering, Universiti Malaysia Terengganu, 21300 Kuala Nerus, Terengganu, Malaysia*

Correspondence should be addressed to Mokhlesur M. Rahman; mokrahman@unisza.edu.my

Academic Editor: Larisa Lvova

A new bioenzymatic glucose biosensor for selective and sensitive detection of glucose was developed by the immobilization of glucose oxidase (GOD) onto selenium nanoparticle-mesoporous silica composite (MCM-41) matrix and then prepared as a carbon paste electrode (CPE). Cyclic voltammetry was employed to probe the catalytic behavior of the biosensor. A linear calibration plot is obtained over a wide concentration range of glucose from 1×10^{-5} to 2×10^{-3} M. Under optimal conditions, the biosensor exhibits high sensitivity ($0.34\,\mu A \cdot mM^{-1}$), low detection limit ($1 \times 10^{-4}$ M), high affinity to glucose ($K_m = 0.02$ mM), and also good reproducibility (R.S.D. 2.8%, $n = 10$) and a stability of about ten days when stored dry at +4°C. Besides, the effects of pH value, scan rate, mediator effects on the glucose current, and electroactive interference of the biosensor were also discussed. As a result, the biosensor exhibited an excellent electrocatalytic response to glucose as well as unique stability and reproducibility.

1. Introduction

The determination of the glucose level is significant in chemical samples, biological and clinical, as well as in food processing and fermentation [1]. Diabetes mellitus especially is one of the principal causes of demise and disability in the world and is highly responsible for kidney failure, heart disease, and sightlessness. About 200 million people in the world are afflicted with diabetes mellitus [2]. So the determination of blood glucose levels rapidly, conveniently, precisely, and economically is vital for its diagnosis and effective management [3]. Therefore, the development and fabrication of a cost-effective, simple, accurate, portable, and rapid sensor for glucose are socially crucial for diabetes mellitus [4]. For this aim, electrochemical biosensors have been applied successfully for the determination of glucose.

Glucose oxidase (GOD) has been significantly used to check blood glucose levels in diabetes patients mainly due to it being inexpensive, stable, and of practical use [5] and its catalytic ability to glucose. It has a flavin adenine dinucleotide (FAD) redox point, which is entirely surrounded in the apoenzyme. It acts upon by catalysis transfer of the electron from glucose to gluconolactone. The space among its two FAD/$FADH_2$ midpoints and surface the electrode is so long that explicit transfer of the electron from the enzyme to the electrode is tricky to be realized [6]. Most glucose

Table 1: Comparison between current market CGMS/sensor results devices [26–28].

Brand	Guardian Real-Time	MiniMed 530G with Enlite	Dexcon G4 Platinum	FreeStyle
Company	Medtronic	Medtronic	Dexcom	Abbott
FDA approved date	2006	2007	2007	2008
Sensor life (d)	3	6	7	5
Sensor style	Insertion under skin	Insertion under skin	Insertion under skin	Insertion under skin
Startup initialization time (h)	2	2	2	10
Calibration (Y/N) with finger-stick test	Y, 2 h after insertion, the first 6 h, and then every 12 h	Y, 2 h after insertion, the first 6 h, and then every 12 h	Y, every 12 h	Y, approximately 1, 2, 10, 24, and 72 h after insertion

CGMS = continuous glucose monitor system; FDA = Food and Drug Administration.

measurements are based on the immobilization of glucose oxidase (GOD) for detecting H_2O_2 concentration which is obtained from the GOD enzymatic reaction [7, 8] as illustrated by the following steps:

$$\begin{aligned} {}^{GOX}FAD + \text{glucose} &\rightarrow {}^{GOX}FADH_2 + \text{gluconolactone (enzymatic)} \\ {}^{GOX}FADH_2 &\rightarrow {}^{GOX}FAD + 2e + 2H^+ \text{ (electrochemical)} \\ {}^{GOX}FADH_2 + O_2 &\rightarrow {}^{GOX}FAD + H_2O_2 \text{ (enzymatic)} \\ {}^{GOX}FAD + 2e + 2H^+ &\rightarrow {}^{GOX}FADH_2 \text{ (electrochemical)} \end{aligned} \tag{1}$$

There are many methods for the GOD immobilization such as covalent cross-linking [9–11], electrochemical polymerization [12], sol-gel encapsulating [13], and adsorption methods [14]. Among these immobilization methods, adsorption is the simplest. Adsorption can retain the bioactivities of the immobilized enzyme well because their action needs no chemical reagents. Biosensors based on the adsorption of protein, however, are limited by the amount of the immobilized enzyme on the electrode surface and are therefore unstable. The sensitivity of the biosensor is directly proportional to the enzyme surface density on the electrode; hence, an increase in enzyme loading on the biosensor surface may result in a significant improvement in its sensitivity [15]. But it is well known that the direct electron communication between the protein and the electrode transducer is difficult because the redox sites of the protein are deeply seated in the protein shell. In order to improve the direct electrochemistry of the redox protein, nanomaterials, such as carbon nanotubes [4, 5, 16, 17], metal nanoparticles (gold, iron, nickel, etc.) [2, 18–20], ZnO nanotubes [21], platinum or sputtered platinum [22, 23], glassy carbon [24], and CdS nanoparticles [25], have been used widely due to their catalytic ability and good biocompatibility properties. A comparison of the current market glucose sensor results with published articles is given in Table 1 [26–28]. Nanoparticles can offer many advantages, such as large surface-to-volume ratio, high surface reaction activity, and strong adsorption ability to immobilize the desired biomolecules. Many metal and semiconductor nanoparticles have been chosen to prepare modified electrodes [29]. Nanomaterials, which were used as a carrier in enzyme sensor, not only increase the stability and the amount of the immobilized enzyme but also improve the catalytic activity of the protein and the responsibility of the sensor [18].

Also newly, a sequence of inorganic porous materials such as mud [30], montmorillonite [31], and mesoporous silicate [6, 32–34] has been proven to be certifying as the immobilization patterns because of their high chemical, mechanical, and thermal solidity as well as good adsorption due to its large specific surface area and absorbency. So the combination of mesoporous molecular sieves into redox enzymes could deliver a bioactive compound [6].

In this paper, we describe a simple route to the production of the Se-MCM-41 mesoporous electrode. The preparation method is simple, and selenium nanoparticles may be readily formed on a carbon paste electrode surface constructing a simple, economical, and accurate amperometric sensor for glucose. The electrode current response of glucose shows excellent stability and reproducibility. The prepared sensors can be used for the determination of glucose in human blood.

2. Experimental

2.1. Materials. Glucose oxidase (GOD from *Aspergillus niger*, E. C. 1.1.3.4), glutaraldehyde (50 wt. %), and glucose were purchased from Sigma. Ferrocene (98%, Merck) was used as received. All other reagents were of analytical grade and used without further purification. Stock solutions of glucose were prepared in 0.05 M phosphate buffer (pH 7.0) and mutarotated for at least 24 h before use. All solutions were prepared and made up with double-distilled water (DDW). Carbon graphite powder (Ultra F, 200 meshes, Johnson Matthey) and paraffin oil (from Fluka) has been used for the preparation of carbon paste electrode (CPE).

2.2. *Preparation of Se-MCM-41/GOD Particles.* The composite material was added in the flask containing 25 mL glutaraldehyde solution (5% in 0.05 M phosphate buffer, pH 6.62). The container containing the mixture of reactants was stirred at room temperature for three hours. The solution was then filtered and washed several times with deionized water to remove excess glutaraldehyde. Tollens' reagent was used to test the presence of glutaraldehyde. For this purpose, Tollens' reagent (the silver mirror test) test was carried out until a colorless solution is produced, thus indicating the absence of excess glutaraldehyde.

The glutaraldehyde cross-linked Se-MCM-41 mesopore composite has mixed with 3 mL of 2 mg·mL^{-1} GOD solution in a phosphate buffer. The reaction mixture was incubated at 0–5°C with shaking (100 rpm) for six hours. The supernatant was removed, and the composite was washed three times with phosphate buffer solution (pH 6, 0.05 M). The GOD immobilized material was recovered from the solution and stored at +4°C for subsequent uses.

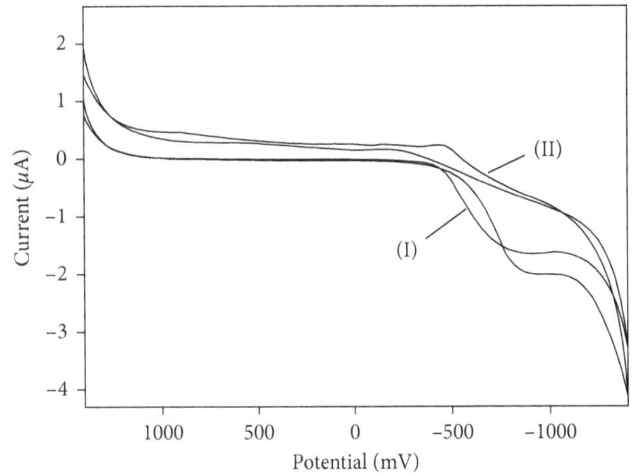

FIGURE 1: Cyclic voltammogram of (I) Se-MCM-41 and (II) immobilized GOD to the Se-MCM-41 in 0.05 M phosphate buffer (pH 7) at a scan rate of 20 mV·s^{-1}.

2.3. *GOD Activity Assay.* Activity assay of the enzyme was carried out using *o*-dianisidine dye. The colorless dye in the reduced form reacts in the presence of peroxidase (POD) and oxidizing agent to give a brown colored solution, measurable at 500 nm using UV-Vis [35]. The GOD catalyzes the oxidation of glucose to δ-gluconolactone, producing H_2O_2 which is then measured indirectly by the oxidation of the dye in the presence of POD, following the reaction as shown in (2) and (3). The specific activity of glucose oxidase was calculated according to (2), and it was found that the 0.116 U/mg solid produced is higher than the one reported previously [11]:

$$\beta\text{-D-glucose} + O_2 + H_2O \xrightarrow{\text{GOD}} \delta\text{-glucono-1,5-lactone} + H_2O_2 \tag{2}$$

$$H_2O_2 + o\text{-dianisidine (reduced)} \xrightarrow{\text{POD}} o\text{-dianisidine (oxidized)} \tag{3}$$

$$\beta\text{-D-glucose} + O_2 + H_2O \xrightarrow{\text{GOD}} \delta\text{-glucono-1,5-lactone} + H_2O_2 \tag{4}$$

The amount of enzyme is calculated using the following formula: enzyme (units/ml) = ($\Delta A_{500\,\text{nm}}$/min test − $\Delta A_{500\,\text{nm}}$/min blank) (3.1)·(df)/(7.5)·(0.1), where 3.1 = volume (in milliliters) of assay; df = dilution factor; 7.5 = millimolar extinction coefficient of oxidized *o*-dianisidine at 500 nm; 0.1 = volume (in milliliters) of enzyme used.

2.4. *Electrode Preparation and Electrochemical Instrumentation.* The electrochemical behavior of the Se-MCM-41-GOx mesopore composite sample was studied using a working electrode made of carbon paste (CPE) modified with the solid particles. This electrode was prepared by mixing 5 mg of the obtained composite with 50 mg of graphite (1 : 10). After adding a few drops of paraffin oil, the mixture was homogenized using a pestle in an agate mortar. Then, the combination was housed in a polyethylene tube (inner diameter 2.4 mm) and polished on a smooth paper layer before conducting each of the experiments. An electric contact was made by a copper wire through the back of the electrode.

Electrochemical experiments were carried out in a conventional three-electrode cell comprising the modified carbon paste as the working electrode, an Ag/AgCl/(KCl 3 M) reference electrode, and a platinum wire as the auxiliary electrode. Experiments of all electrochemical were done with 757 VA Computrace (Metrohm) using a potential scan of −1400 mV to +1400 mV to fit the window.

All experiments performed at room temperature in 0.05 M (pH 7.0) phosphate buffer as the supporting electrolyte. Electrolyte solutions were deoxygenated with nitrogen bubbling for at least 5 min and a nitrogen atmosphere kept over the solution during electrochemical measurements.

3. Results and Discussion

3.1. *Cyclic Voltammograms of GOx to the Se-MCM-41 Composite.* To verify the electrochemical properties of the Se-MCM-41/GOD/CPE, we carried out the direct electrochemical measurement of the immobilized GOD on the Se-MCM-41/CPE using typical CPE electrodes without GOD. Figure 1 displays the typical cyclic voltammograms (CV) for Se-MCM-41/CPE (I) and Se-MCM-41/GOD/CPE (II) in 0.05 M deoxygenated phosphate buffer solution (pH 7.0) over the potential range from 1.4 to −1.4 V at 20 mV·s^{-1} scan

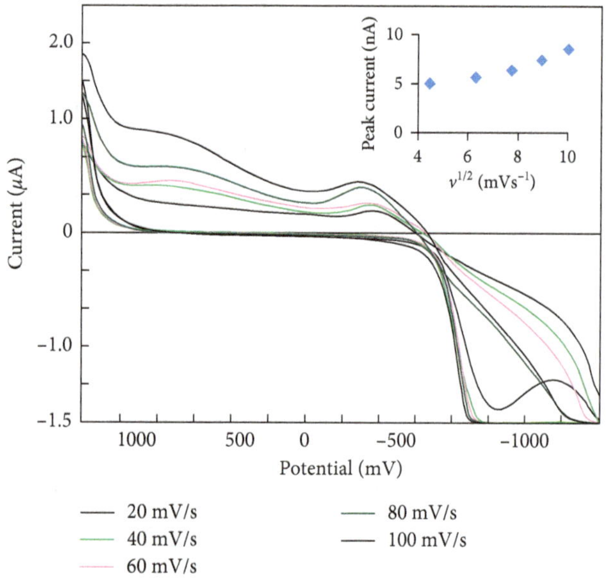

FIGURE 2: Cyclic voltammograms of Se-MCM-41/GOD/CPE in pH 7.0 PBS at 20, 40, 60, 80, and 100 mV·s^{-1} scan rates (from inner to outer). Inset: peak currents of plots versus $v^{1/2}$.

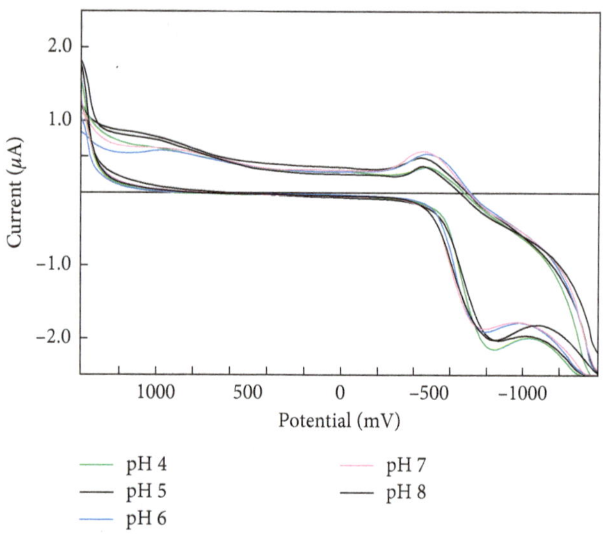

FIGURE 3: Effect of pH on the response of the biosensor (Se-MCM-41/GOD/CPE) in the presence of 2 mM glucose.

rate. As seen from the plot (I), no redox peaks were observed using the typical CPE electrode which indicated Se-MCM-41 as being electroinactive in the potential window. Plot II shows the CV of GOD immobilized on Se-MCM-41/CPE, and a pair of oxidation-reduction peak appears. The results indicated that the reaction of catalytic oxidation of glucose by GOD occurred and the immobilized GOD retained its electrocatalytic activity for the oxidation of glucose [36, 37].

Flavin adenine dinucleotide is a part of the GOD molecule and is known to undergo a redox reaction, where two electrons and two protons were exchanged, and the electrochemical response of GOD immobilized on the solid surface is due to the redox reaction of FAD [38]. The FAD redox potentials in various enzymes ranged from 190 to −490 mV. In plot (II), the FAD peak appeared about −470 mV, which was acceptable, and suggested that GOD bound to the Se-MCM-41 composite was active in the electron transfer [11].

The main cause for the transfer electron between the silica-based electrode and the enzyme might be due to the electrostatic relations, such as hydrogen bonding and hydrophilic attraction between MCM-41 and glucose oxides (GOD). The communication between MCM-41 and GOD is much stronger than that between CPE and GOD due to the presence of Si-OH groups on the external surface of MCM-41. In our study, the glucose oxides (GOD) should be immobilized on the external surface of Se-MCM-41 by physical adsorption because the pore diameter of GOD (about 4–6 nm) is bigger than the size of Se-MCM-41 (about 2.7 nm) [6]. Due to the presence of many acidic silicon hydroxyl (Si-OH) groups on the outer surface of the Se-MCM-41, the oxidation reaction of GOD will be not easy thermodynamically.

3.2. Effect of Scan Rate. The effect of scan rate on the electrochemistry of the immobilized GOD is shown in Figure 2. With increasing scan rate, the anodic peak currents of the GOD increased linearly, and the anodic peak potential of GOD is shifted to a more positive value. The dependence of the current (I_p) on the square root of the scan rate ($v^{1/2}$) (inset in Figure 2) is an essential diagnostic criterion for establishing the type of reaction mechanism by cyclic voltammetry. It suggests that the Se-MCM-41 composite is sufficiently thick and the electron transfer between selenium MCM-41 composite is slow, leading to the system to mimic semi-infinite linear diffusion [39].

3.3. Effect of pH on the Peak Current. The effect of pH on the anodic peak current was investigated over the range pH 4.0–8.0 (buffer solution of K_2HPO_4–KH_2PO_4) in the presence of 2 mM glucose at 25°C. The resulting I versus pH (4–8) data are illustrated in Figure 3. As can be seen, the peak potential of GOD is dependent on the solution pH with the maximum response observed at pH 7.0, which is consistent with that of most GOD-based glucose biosensors. Therefore, we fixed the solution pH at 7.0 for the investigation of the analytical performance of Se-MCM-41 carbon paste electrode covered with GOD. Either the strongly acidic solution or alkaline solution would decrease the bioactivity of the enzyme. Thus, the proposed biosensor was practical to detect glucose levels in real samples at neutral pH value [40]. These results are similar to previously reported values of optimum pH for the use of glucose oxidase with other artificial electron acceptors [5, 29, 41–43].

3.4. Effect of the Glucose Concentration. The cyclic voltammograms of Se-MCM-41/GOD/CPE with successive addition of glucose to air-saturated 0.05 M PBS (pH 7.0) are shown in Figure 4. The peak current of Se-MCM-41/GOD/CPE increased linearly with increasing concentration of glucose up to 2 mM. The calibration range of glucose concentration was prepared from 0.01 to 12.0 mM. The linear response range of the sensor

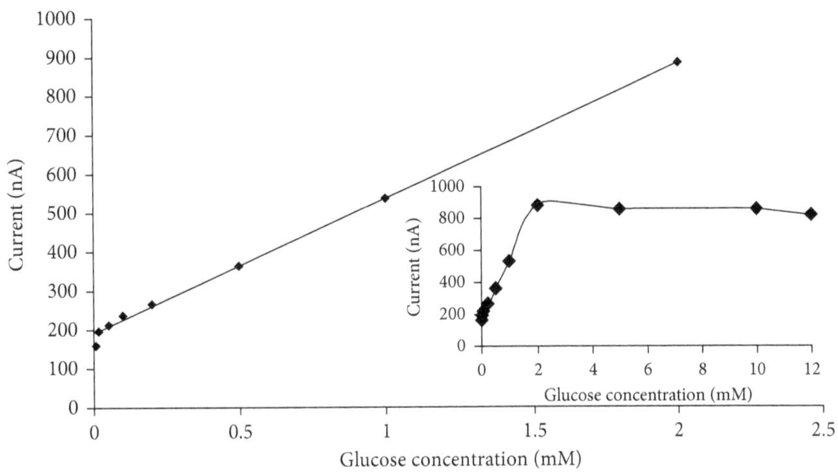

FIGURE 4: Correlation between the maximum current and the concentration of glucose.

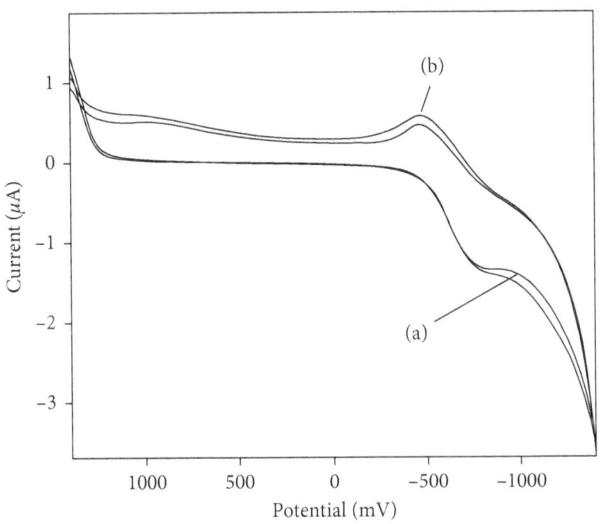

FIGURE 5: Cyclic voltammograms of Se-MCM-41/GOD/CPE in 0.05 M PBS (pH 7.0) containing 0.2 mM ferrocene in the absence (a) and presence (b) of 2.0 mM glucose.

to glucose concentration was from 0.01 to 2.0 mM with a correlation coefficient of 0.997. The sensitivity of Se-MCM-41/GOD/CPE to glucose was found to be 0.34 μA·mM^{-1}, which is higher than values found in the literature for GOx/PtNPs/CNTs/carbon paste (0.28 μA·mM^{-1}), Nafion/GOx film electrode (51.0 nA·mM^{-1}), and poly(*o*-phenylenediamine) covered screen-printed electrode (16.6 nA·mM^{-1}) [44–47].

According to the Lineweaver–Burk form of the Michaelis–Menten equation, the relation between the reciprocal of the response current and the reciprocal of glucose concentration can be obtained. The apparent Michaelis–Menten constant (K_m), an indicator of enzyme-substrate reaction kinetics, can be used to evaluate the biological activity of the immobilized enzyme, and this constant can be calculated from the Lineweaver–Burk equation, given below:

$$\frac{1}{I_{ss}} = \left(\frac{K_m}{I_{max}}\right)\left(\frac{1}{C_g}\right) + \frac{1}{I_{max}}, \quad (5)$$

where C_g is the substrate concentration, I_{ss} is the steady-state current, and I_{max} is the maximum current measured under substrate saturation [48].

Here, Michaelis-Menten constant K_m and the maximum response current of the biosensor based on the Se-MCM-41/GOD electrode are calculated and equal to 0.02 mM and 0.45 μA, respectively. The low K_m value of 0.02 mM indicates the affinity of the enzyme to the electrode, which was smaller than those of 5.20, 22.8, 14.4, 21, and 37.6 mM, for immobilized GOD [5, 45–48], indicating that the GOD immobilized on Se-MCM-41/GOD has high affinity to glucose. These results imply that the biosensor based on the Se-MCM-41/GOD electrode is valuable and sensitive.

3.5. Glucose Bioelectrocatalytic Oxidation. Although the direct transfer electron of GOD was achieved, some intermediaries are still used to accelerate the electron transfer rate between electrode and GOD [8, 9]. To increase the response current, ferrocene, a satisfactory electron transfer mediator has been used in the biosensor based on Se-MCM-41/GOD/CPE [44]. Figure 5 has shown cyclic voltammograms of the Se-MCM-41/GOD/CPE in N_2-saturated, pH 7.0 PBS containing 0.2 mM ferrocene as the facilitator. When 2.0 mM glucose was added to the solution, the anodic peak current increased, using ferrocene as the mediator in N_2-saturated solutions, consequently, demonstrating that Se-MCM-41/GOD/CPE is able to electrocatalyze the oxidation of glucose. These results can be explained from the following equations:

$$\begin{gathered}\text{Glucose} + \text{GOD (ox)} \rightarrow \text{gluconolactone} + \text{GOD (red)}\\ \text{GOD (red)} + 2\ \text{ferrocene} \rightarrow \text{GOD (ox)} + 2\ \text{ferrocene} + 2\text{H}^+\\ 2\ \text{ferrocene} \rightarrow 2\ \text{ferrocene} + 2e^-\end{gathered} \quad (6)$$

3.6. Analytical Performance. The amperometric response characteristics of the enzyme electrode are affected by the electroactive interferents. In this section, the effects of these factors on the behavior of the GOD electrode, reproducibility, and stability have been investigated in detail and discussed accordingly.

3.6.1. Reproducibility. The reproducibility of the glucose biosensor was investigated by successively detecting 0.10 mM glucose ten times; the relative standard deviation (RSD) calculated was 2.8%. The RSD for the detection of 1.0 mM glucose with three sensors prepared independently under the same conditions was 3.6%, demonstrating a good reproducibility of the measurements performed.

3.6.2. Stability. The stability of the enzyme electrode was also investigated by amperometric measurements. It retained 91% of its initial current response, when the enzyme electrode was stored at +4°C for glucose after intermittent use over a 10 days interval. So, it can be presumed that the presence of nanoparticle selenium incorporated MCM-41 is very efficient in retaining the enzyme activity of GOD.

3.6.3. Interferences. The effects of interference have been observed by testing the amperometric response of 0.1 mM glucose in the presence of their normal physiological concentration (0.1 mM for ascorbic acid and 0.1 mM for citric acid) [48]. When the glucose-to-interferent concentration ratio was 1 : 1, a decrease of 14% for ascorbic acid or an increase of 25% for citric acid in the current value of the 0.5 mM glucose was observed, respectively. In addition, when the ratio of glucose to interferent was 1 : 2 for the ascorbic acid or citric acid, the response current decreased significantly as compared to the response for the 1 : 1 ratio. These results suggested that the presence of interferences such as ascorbic or citric acid did not affect the current measurement of the GOD significantly using the Se-MCM-41/GOD. The results also suggested that the proposed glucose has high selectivity and compatibility on the Se-MCM-41/GOD as the support material for the sensor.

4. Conclusion

The preparation of a new type of glucose oxidase electrode using nanoselenium particles and the MCM-41 mesoporous composite was well demonstrated. The preparation of the electrode is easy, fast, and reproducible. Cyclic voltammetric results confirmed that the prepared electrode presents a highly electrocatalytic activity for the oxidation of glucose. The optimum pH was obtained at pH 7.0 which is the optimal environment herein. Since the pH value of human blood was around 7.4, the potentiometric glucose biosensor was suitable for measuring the concentration of glucose in human blood. Under the optimized experimental conditions, the catalytic currents are linear to the concentrations of glucose from 1×10^{-5} to 2×10^{-3} M. The detection limit was 1×10^{-4} M with a signal to noise ratio of 3. The sensor exhibits good reproducibility and stability during the measurement. At the same time, the biosensor demonstrates high sensitivity ($0.34\,\mu A \cdot mM^{-1}$). Besides, the biosensor possesses high sensitivity and excellent chemical and mechanical stability. All the results show that the prepared Se-MCM-41/GOD can provide a promising material for biosensor designs and other biological applications.

Conflicts of Interest

The authors declare that they have no conflicts of interest.

Authors' Contributions

Sabriye Yusan performed the experiments, analysed the data, and prepared the manuscript. (Late) Alias Mohd Yusof and Mokhlesur M. Rahman designed, supervised, analysed the data, and edited the manuscript. Tengku M. Ariff, Nasir Mohamad, Ahmad Zubaidi A. Latif, Mohd Aznan Md Aris, and Wan Sani B. Wan Nik provided financial support and performed critical analysis.

Acknowledgments

The authors are profoundly grateful for the financial support given by the Faculty of Science, Universiti Teknologi Malaysia (UTM), Research Management Center of International Islamic University Malaysia, and Universiti Sultan Zainal Abidin through the Project no. EDW B 14-217-1102 and NRGS/PR057-1, respectively.

References

[1] S. Liu and H. Ju, "Reagentless glucose biosensor based on direct electron transfer of glucose oxidase immobilized on colloidal gold modified carbon paste electrode," *Biosensors and Bioelectronics*, vol. 19, no. 3, pp. 177–183, 2003.

[2] M. Rahman, A. J. S. Ahammad, J.-H. Jin, S. J. Ahn, and J.-J. Lee, "A comprehensive review of glucose biosensors based on nanostructured metal-oxides," *Sensors*, vol. 10, no. 5, pp. 4855–4886, 2010.

[3] A. E. G. Cass, G. Davis, G. D. Francis et al., "Ferrocene-mediated enzyme electrode for amperometric determination of glucose," *Analytical Chemistry*, vol. 56, no. 4, pp. 667–671, 1984.

[4] M. M. Rahman, A. Umar, and K. Sawada, "Development of amperometric glucose biosensor based on glucose oxidase co-immobilized with multi-walled carbon nanotubes at low potential," *Sensors and Actuators B: Chemical*, vol. 137, no. 1, pp. 327–333, 2009.

[5] X. Kang, Z. Mai, X. Zou, P. Cai, and J. Mo, "A novel glucose biosensor based on immobilization of glucose oxidase in chitosan on a glassy carbon electrode modified with gold–platinum alloy nanoparticles/multiwall carbon nanotubes," *Analytical Biochemistry*, vol. 369, no. 1, pp. 71–79, 2007.

[6] Z. H. Dai, J. Ni, X. H. Huang, G. F. Lu, and J. C. Bao, "Direct electrochemistry of glucose oxidase immobilized on a hexagonal mesoporous silica-MCM-41 matrix," *Bioelectrochemistry*, vol. 70, no. 2, pp. 250–256, 2007.

[7] J. Li and X. Lin, "Glucose biosensor based on immobilization of glucose oxidase in poly(o-aminophenol) film on polypyrrole-Pt nanocomposite modified glassy carbon electrode," *Biosensors and Bioelectronics*, vol. 22, no. 12, pp. 2898–2905, 2007.

[8] A. Vaze, N. Hussain, C. Tang, D. Leech, and J. Rusling, "Biocatalytic anode for glucose oxidation utilizing carbon nanotubes for direct electron transfer with glucose oxidase,"

Electrochemistry Communications, vol. 11, no. 10, pp. 2004–2007, 2009.

[9] Z. Lin, J. Chen, and G. Chen, "An ECL biosensor for glucose based on carbon-nanotube/nafion film modified glass carbon electrode," *Electrochimica Acta*, vol. 53, no. 5, pp. 2396–2401, 2008.

[10] C.-W. Liao, J.-C. Chou, T.-P. Sun, S.-K. Hsiung, and J.-H. Hsieh, "Preliminary investigations on a glucose biosensor based on the potentiometric principle," *Sensors and Actuators B: Chemical*, vol. 123, no. 2, pp. 720–726, 2007.

[11] S. Akella and C. K. Mitra, "Electrochemical studies of glucose oxidase immobilized on glutathione coated gold nanoparticles," *Indian Journal of Biochemistry & Biophysics*, vol. 44, pp. 82–87, 2007.

[12] M.-Q. Liu, J.-H. Jiang, Y.-L. Feng, G.-L. Shen, and R.-Q. Yu, "Glucose biosensor based on immobilization of glucose oxidase in electrochemically polymerized polytyramine film and overoxidised polypyrrole film on platinized carbon paste electrode," *Chinese Journal of Analytical Chemistry*, vol. 35, no. 10, pp. 1435–1438, 2007.

[13] G. Changa, Y. Tatsu, T. Goto, H. Imaishi, and K. Morigaki, "Glucose concentration determination based on silica sol–gel encapsulated glucose oxidase optical biosensor arrays," *Talanta*, vol. 83, no. 1, pp. 61–65, 2010.

[14] H. Dai, X. Wu, H. Xu, Y. Wang, Y. Chi, and G. Chen, "A highly performing electrochemiluminescent biosensor for glucose based on a polyelectrolyte-chitosan modified electrode," *Electrochimica Acta*, vol. 54, no. 19, pp. 4582–4586, 2009.

[15] Q. Gao, Y. Guo, W. Zhang, H. Qi, and C. Zhang, "An amperometric glucose biosensor based on layer-by-layer GOx-SWCNT conjugate/redox polymer multilayer on a screen-printed carbon electrode," *Sensors and Actuators B: Chemical*, vol. 153, no. 1, pp. 219–225, 2010.

[16] C. Deng, J. Chen, X. Chen, C. Xiao, L. Nie, and S. Yao, "Direct electrochemistry of glucose oxidase and biosensing for glucose based on boron-doped carbon nanotubes modified electrode," *Biosensors and Bioelectronics*, vol. 23, no. 8, pp. 1272–1277, 2008.

[17] J. Zhang, M. Feng, and H. Tachikawa, "Layer-by-layer fabrication and direct electrochemistry of glucose oxidase on single wall carbon nanotubes," *Biosensors and Bioelectronics*, vol. 22, no. 12, pp. 3036–3041, 2007.

[18] X. Zhi-Gang, L. Jian-Ping, T. Li, and C. Zhi-Qiang, "A novel electrochemiluminescence biosensor based on glucose oxidase immobilized on magnetic nanoparticles," *Chinese Journal of Analytical Chemistry*, vol. 38, no. 6, pp. 800–804, 2010.

[19] X. Wang, Y. Zhanga, C. E. Banks, Q. Chen, and X. Ji, "Non-enzymatic amperometric glucose biosensor based on nickel hexacyanoferrate nanoparticle film modified electrodes," *Colloids and Surfaces B: Biointerfaces*, vol. 78, no. 2, pp. 363–366, 2010.

[20] F. N. Comba, M. D. Rubianes, P. Herrasti, and G. A. Rivas, "Glucose biosensing at carbon paste electrodes containing iron nanoparticles," *Sensors and Actuators B: Chemical*, vol. 149, no. 1, pp. 306–309, 2010.

[21] T. Kong, Y. Chen, Y. Ye, K. Zhang, Z. Wang, and X. Wang, "An amperometric glucose biosensor based on the immobilization of glucose oxidase on the ZnO nanotubes," *Sensors and Actuators B: Chemical*, vol. 138, no. 1, pp. 344–350, 2009.

[22] S. Y. Lu, C. E. Li, D. D. Zhang et al., "Electron transfer on an electrode of glucose oxidase immobilized in polyaniline," *Journal of Electroanalytical Chemistry*, vol. 364, no. 1-2, pp. 31–36, 1994.

[23] R. J. H. J. Os van, A. Bult, C. G. J. Koopal, and W. P. Bennekom van, "Glucose detection at bare and sputtered platinum electrodes coated with polypyrrole and glucose oxidase," *Analytica Chimica Acta*, vol. 335, no. 3, pp. 209–216, 1996.

[24] X. Xu, J. Chen, W. Li, Z. Nie, and S. Yao, "Surface nanocrystallization of glassy carbon electrode: application in direct electrochemistry of glucose oxidase," *Electrochemistry Communications*, vol. 10, no. 10, pp. 1459–1462, 2008.

[25] Y. Huang, W. Zhang, H. Xiao, and G. Li, "An electrochemical investigation of glucose oxidase at a CdS nanoparticles modified electrode," *Biosensors and Bioelectronics*, vol. 21, no. 5, pp. 817–821, 2005.

[26] G. McGarraugh, "The chemistry of commercial continuous glucose monitors," *Diabetes Technology & Therapeutics*, vol. 11, no. 1, pp. 17–24, 2009.

[27] T. D. Mall, "Comparison of current continuous glucose monitors (CGMs)," 2014, http://www.diabetesnet.com/diabetes-technology/meters-monitors/continuous-monitors/compare-current-monitors.

[28] S. Colberg, *50 Secrets of the Longest Living People with Diabetes*, Da Capo Press, Boston, MA, USA, 1st edition, 2008.

[29] B. Zheng, S. Xie, L. Qian, H. Yuan, D. Xia, and M. M. F. Choi, "Gold nanoparticles-coated eggshell membrane with immobilized glucose oxidase for fabrication of glucose biosensor," *Sensors and Actuators B: Chemical*, vol. 152, no. 1, pp. 49–55, 2010.

[30] C. Lei, F. Lisdat, U. Wollenberger, and F.W. Scheller, "Cytochrome c/clay modified electrode," *Electroanalysis*, vol. 11, no. 4, pp. 274–276, 1999.

[31] C. Fan, Y. Zhuang, G. Li, J. Zhu, and D. Zhu, "Direct electrochemistry and enhanced activity for hemoglobin in a sodium montmorillonite film," *Electroanalysis*, vol. 12, no. 14, pp. 1156-1157, 2000.

[32] Z. Dai, S. Liu, H. Ju, and H. Chen, "Direct electron transfer and enzymatic activity of hemoglobin in a hexagonal mesoporous silica matrix," *Biosensors and Bioelectronics*, vol. 19, no. 8, pp. 861–867, 2004.

[33] L. Zhang, Q. Zhang, and J. Li, "Direct electrochemistry and electrocatalysis of hemoglobin immobilized in bimodal mesoporous silica and chitosan inorganic-organic hybrid film," *Electrochemistry Communications*, vol. 9, no. 7, pp. 1530–1535, 2007.

[34] J. D. Webb, S. MacQuarrie, K. McEleney, and C. M Crudden, "Mesoporous silica-supported Pt catalysts: an investigation into structure, activity, leaching and heterogeneity," *Journal of Catalysis*, vol. 252, no. 1, pp. 97–109, 2007.

[35] H. U. Bergmeyer, K. Gawehnand, and M. Grassl, "*Methods of Enzymatic Analysis*," H. U. Bergmeyer, Ed., vol. 1, pp. 457-458, Academic Press Inc., New York, NY, USA, 2nd edition, 1974.

[36] L. Xu, Y. Zhu, Y. Li, X. Yang, and C. Li, "Bienzymatic glucose biosensor based on co-immobilization of glucose oxidase and horsedish peroxidase on gold nanoparticles-mesoporous silica matrix," in *Proceedings of 2nd International Nanoelectrics Conference IEEE*, pp. 390–393, Shanghai, China, 2008.

[37] J. Wang, *Analytical Electrochemistry*, John Wiley & Sons, Inc., Hoboken, NJ, USA, 3rd edition, 2006.

[38] X. Kang, J. Wang, H. Wu, I. A. Aksay, J. Liu, and Y. Lin, "Glucose oxidase–graphene–chitosan modified electrode for direct electrochemistry and glucose sensing," *Biosensors and Bioelectronics*, vol. 25, no. 4, pp. 901–905, 2009.

[39] A. Mugweru, B. L. Clark, and M. V. Pishko, "Electrochemical sensor array for glucose monitoring fabricated by rapid immobilization of active glucose oxidase within photochemically

polymerized hydrogels," *Journal of Diabetes Science and Technology*, vol. 1, no. 3, pp. 366–371, 2007.

[40] Y. Liu, S. Wu, H. Ju, and L. Xu, "Amperometric glucose biosensing of gold nanoparticles and carbon nanotube multilayer membranes," *Electroanalysis*, vol. 19, no. 9, pp. 986–992, 2007.

[41] Z. Wen, D. Lju, B. Ye, and X. Zhou, "Development of disposable electrochemical sensor with replaceable glucose oxidase tip," *Analytical Communications*, vol. 34, no. 1, pp. 27–30, 1997.

[42] R. Antiochia and L. Gorton, "Development of a carbon nanotube paste electrode osmium polymer-mediated biosensor for determination of glucose in alcoholic beverages," *Biosensors and Bioelectronics*, vol. 22, no. 11, pp. 2611–2617, 2007.

[43] W. Liang and Y. Zhuobin, "Direct electrochemistry of glucose oxidase at a gold electrode modified with single-wall carbon nanotubes," *Sensors*, vol. 3, no. 12, pp. 544–554, 2003.

[44] M. Yang, Y. Yang, Y. Liu, G. Shen, and R. Yu, "Platinum nanoparticles doped sol–gel/carbon nanotubes composite electrochemical sensors and biosensors," *Biosensors and Bioelectronics*, vol. 21, no. 7, pp. 1125–1131, 2006.

[45] K. Balasubramanian and M. Burhard, "Biosensors based on carbon nanotubes," *Analytical and Bioanalytical Chemistry*, vol. 385, no. 3, pp. 452–468, 2006.

[46] M. E. Ghica and C. M. A. Brett, "Development of a carbon film electrode ferrocene-mediated glucose biosensor," *Analytical Letters*, vol. 38, no. 6, pp. 907–920, 2005.

[47] M. Yuqing, C. Jianrong, and W. Xiaohua, "Construction of a Glucose biosensor by immobilizing glucose oxidase within a poly(ophenylenediamine) covered screen-printed electrode," *Online Journal of Biological Sciences*, vol. 6, no. 1, pp. 18–22, 2006.

[48] D. Pan, J. Chen, S. Yao, L. Nie, J. Xia, and W. Tao, "Amperometric glucose biosensor based on immobilization of glucose oxidase in electropolymerized o-aminophenol film at copper-modified gold electrode," *Sensors and Actuators B: Chemical*, vol. 104, no. 1, pp. 68–74, 2005.

4

"Turn-Off" Fluorescent Sensor for Pamidronate Disodium and Zoledronic Acid based on Newly Synthesized Carbon Dots from Black Tea

Peng Li[1] **and Yongcheng Hu**[2]

[1]*The Graduate School, Tianjin Medical University, 22 Qixiangtai Road, Heping District, Tianjin 300071, China*
[2]*Department of Orthopedic Oncology, Tianjin Hospital, 406 Jiefang South Road, Tianjin 300210, China*

Correspondence should be addressed to Yongcheng Hu; yongchenghu@126.com

Academic Editor: Chih-Ching Huang

As common bisphosphonates drugs, pamidronate disodium and zoledronic acid have been widely investigated for bone metastases. In this paper, a new "turn-off" model based on carbon dots (CDs) from black tea was established to analyze the two kinds of bisphosphonate drugs, pamidronate disodium and zoledronic acid. Through the new sensor, both of drugs can be quantitative, respectively, with the limit of detection of 5×10^{-9} mol·L^{-1} and 6×10^{-9} mol·L^{-1}. In addition, the fluorescence of newly prepared CDs can be quenched by two drugs with various degrees via photoinduced electron transfer, which can be perfectly used to distinguish them. Most importantly, this turn-off method has been employed to analyze the two drugs under the influence of foreign interference factors. This method provides a new view and guidance for the rapid analysis and recognition of drugs for bone metastases *in vitro* and *in vivo*.

1. Introduction

Due to their excellent photo stability, high luminous intensity, wide range of excitation spectrum, and narrow emission spectrum, carbon dots (CDs) have been applied as fluorescent sensor in many fields, such as biochemistry, food science, and medical analysis [1–5]. Recently, a kind of food-based carbon dots have attracted more and more attention because of its good biocompatibility, low toxicity, and easy preparation [6–8]. In the past, many reports of fluorescence detecting mode "turn-off" based on CDs have already been reported, such as the turn-off fluorescent sensor based on nitrogen-doped carbon dots applied to detect Hg^{2+} ions by Zhang and Chen [9] and the turn-off fluorescence sensor for the detection of ferric ion in water using green synthesized N-doped carbon dots [10].

As a significant kind of drugs, bisphosphonates play an important role in the treatment of metabolic bone diseases such as bone metasis and have been expanded widely to inhibit bone resorption and management of skeletal disorder including osteoporosis and malignant hypercalcemia [11, 12]. Meanwhile, as two representative drugs among the bisphosphonates, pamidronate disodium and zoledronic acid have been studied by more and more researchers [13, 14]. However, despite their frequent use, the rapid analytical method of them whether *in vitro* or *in vivo* has largely remained very few, and the traditional chromatographic methods were mostly time consuming and expensive and, what's more, have complicated pretreatments [15–17]. On the other hand, known as a fast, sensitive, and stable analytical technology, the fluorescent sensor based on quantum dots, especially carbon dots, have a good application prospect in the drug analysis [18, 19].

In this paper, a new turn-off sensor based on carbon dots from black tea was established to quantitatively characterize the two kinds of bisphosphonate drugs, pamidronate disodium and zoledronic acid. The morphology and chemical composition, stability, and fluorescence properties of the

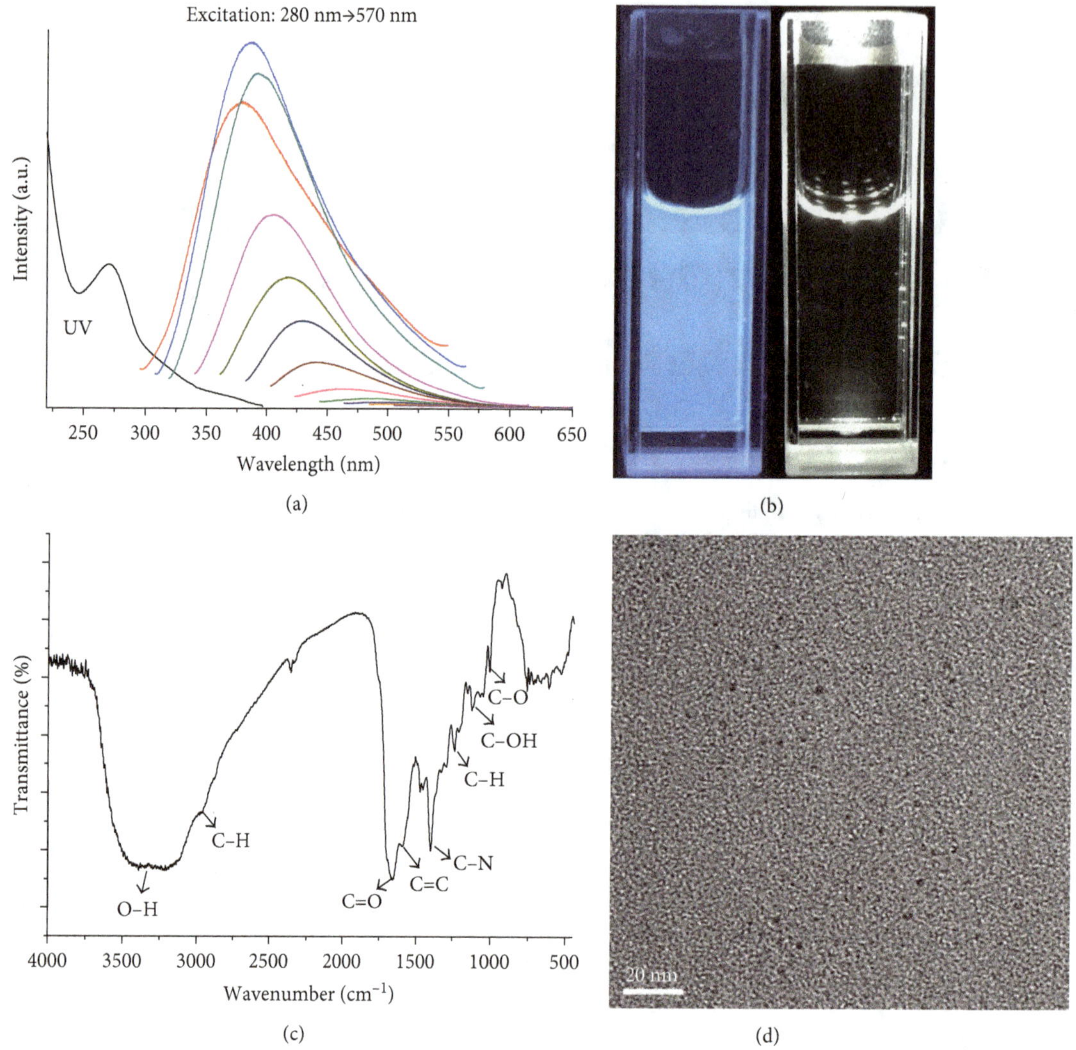

FIGURE 1: (a) UV-Vis spectrum of CDs (black line) and excitation-dependent of the CD fluorescence in water at room temperature. (b) The photograph of the CD solution under UV light of 365?nm (left) and under visible light (right). (c) FTIR spectra of CDs and (d) HRTEM images of CDs.

newly prepared CDs were analyzed in details. Through using the different binding ability between CDs and different drugs, this turn-off sensor can be successfully used to quantitatively analyze pamidronate disodium and zoledronic acid no matter under the influence of foreign interference factors or in the human plasma. Most importantly, the fluorescent sensor has been successfully used for the distinction of the two bisphosphonate drugs. This method illustrates its promising opportunities in the studies of other drugs *in vivo* or *in vitro* using the fluorescent sensor based on CDs.

2. Experimental

2.1. Experimental Materials and Reagents. The black tea was bought from Minghuang Natural Food Development Co., Ltd. Tris (hydroxymethyl) amino methane, HCl, KCl, NaCl, $MgCl_2$, $ZnCl_2$, $CuCl_2$, $CaCl_2$, and $FeCl_3$ were purchased from Sinopharm Chemical Reagent Co., Ltd. CDs from the black tea via the hydrothermal route were synthesized in our laboratory. Deionized distilled water prepared from a Mole water purification system was used.

2.2. Apparatus and Procedures. UV-Vis absorption spectra were examined by using a U-3900 UV-Vis spectrometer (Hitachi). The fluorescence spectra were acquired by an FL-7000 luminescence spectrometer (Hitachi). High-resolution transmission electron microscopy (HRTEM) was examined by using a Tecnai G2 F30 S-Twin microscope (Philips-FEI). Samples containing different concentrations of drugs and CDs were made up to 1 mL in 50 $mmol{\cdot}L^{-1}$ Tris-HCl (pH = 6.0). The concentration of CDs was set at $8.0 \times 10^{-5}\,g{\cdot}mL^{-1}$. The emission spectrum of the solution was then measured 5 min later. All optical measurements were performed at room temperature under ambient conditions, and the excitation wavelength (λ_{ex}) was 360 nm.

2.3. Preparation of CDs from the Black Tea. The CDs were synthesized by using a facile, green, and low-cost

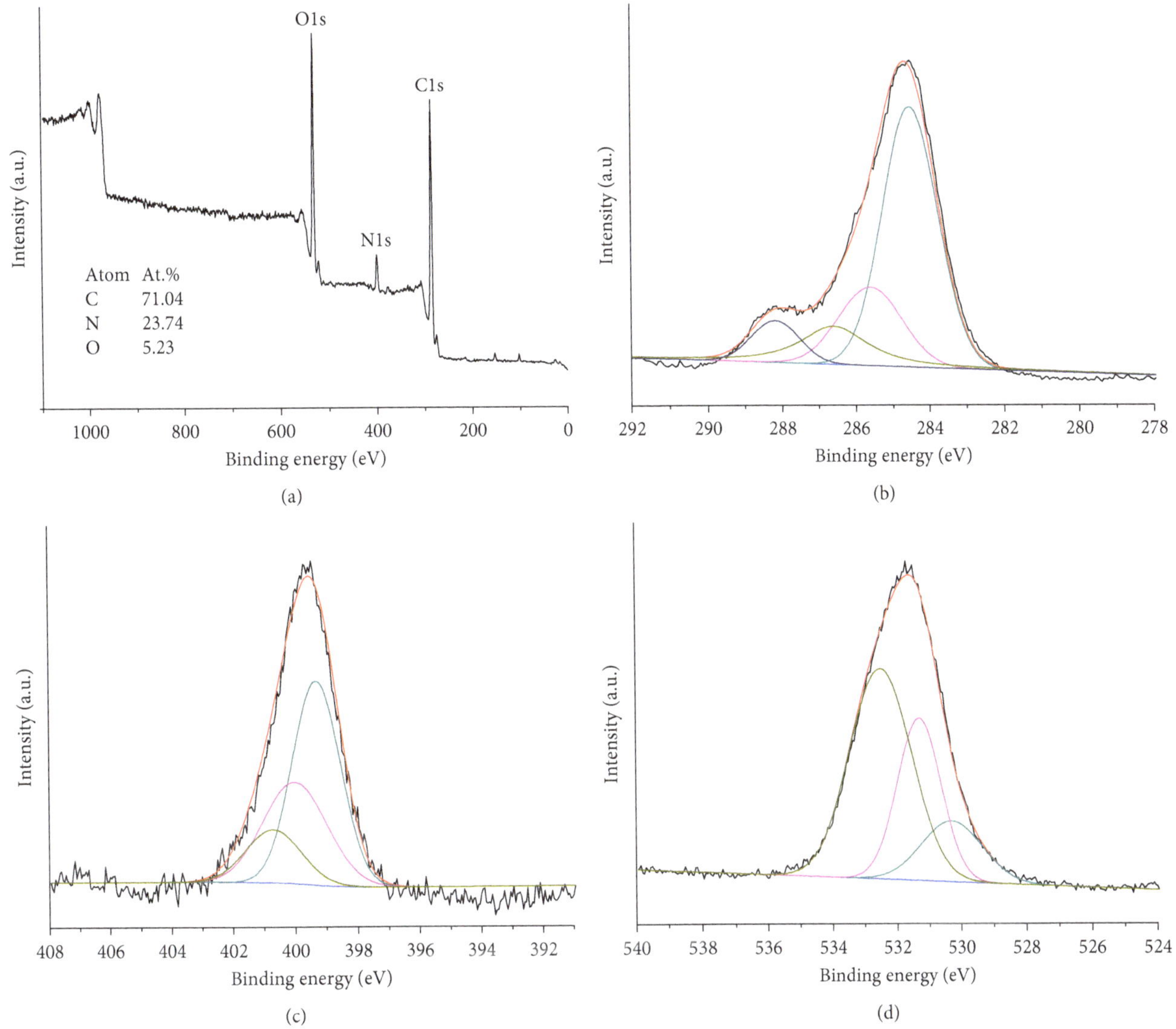

FIGURE 2: (a) XPS survey spectrum of CDs. (b) C1s, (c) O1s, and (d) N1s high-resolution XPS spectra of the CQDs.

hydrothermal treatment based on the black tea as the carbon source as follows: 0.5 g the black tea and 40 mL water were added into 50 mL reaction kettle and then heated at 180°C for 3 h. The obtained CDs were centrifuged for 15 min at a rotating speed of 5000 r·min^{-1} and then dialyzed in a dialysis bag (molecular weight cut off = 3500) for 12 h; finally, the product were freeze dried for further analysis.

2.4. Quantum Yield Measurement. The quantum yields (QYs) of the CDs were measured by comparing the integrated fluorescence intensity (320 nm excitation) and the absorbance value (320 nm) of CD and quinine sulfate was used as a reference. QY was calculated as follows:

$$Q_X = Q_{ST} \cdot \left(\frac{I_X}{I_{ST}}\right)\left(\frac{A_{ST}}{A_X}\right)\left(\frac{\eta_X^2}{\eta_{ST}^2}\right), \quad (1)$$

where I is the integrated fluorescence intensity, A represents the absorbance value, and η means the refractive index of the solvent (both 1.33). The subscripts "X" and "ST" were CDs and quinine sulfate, respectively. The QY_{ST} was 54% while quinidine sulfate was in 0.1 mol·L^{-1} H_2SO_4. In order to minimize the absorption effect, the absorbance value was set as 0.1 [9].

3. Results and Discussion

3.1. Fluorescence Properties and Characterization of CDs. As shown in Figure 1(a), it can be seen that the absorbance band of CDs was centered at around 270 nm in the UV spectrum which was from the n-π^* transition. The CDs emitted strong blue light under UV irradiation of 365 nm (Figure 1(b)), and while the λ_{ex} increased from 280 to 570 nm, the emission spectra of the CDs displayed a typical excitation wavelength-dependent characteristic, which had been suggested to be a result of varied fluorescence characteristics of CDs with various sizes or emissive sites existed on the surfaces of the CDs [2]. Chemical and structural information about the CDs using the FTIR spectroscopy was shown in Figure 1(c); the peak around

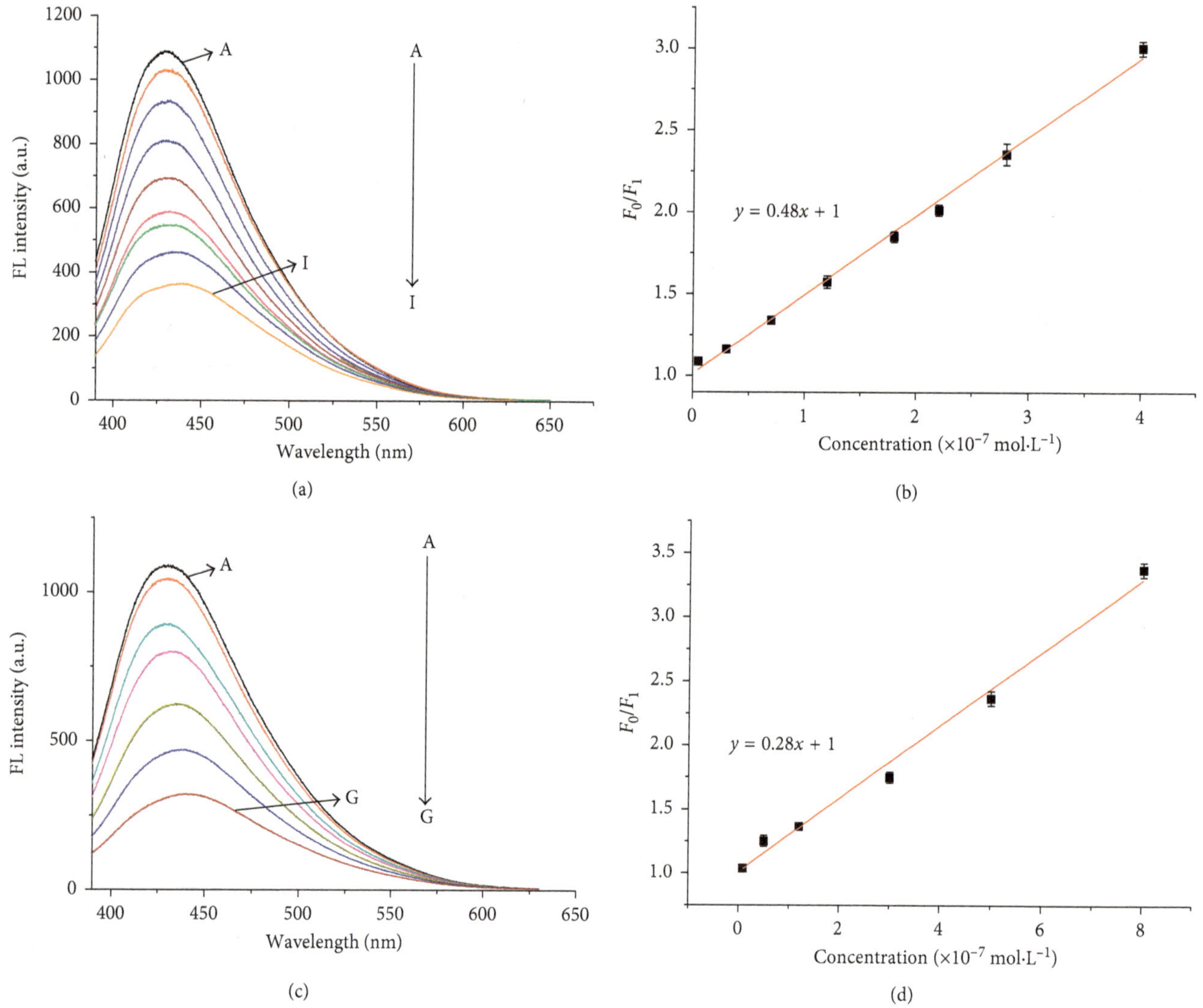

Figure 3: Fluorescence quenching behavior of CDs after the addition of various concentrations (5×10^{-9} mol·L^{-1} to 4×10^{-7} mol·L^{-1}) of pamidronate disodium (A-I (a)) and the addition of various concentrations (6×10^{-9} mol·L^{-1} to 8×10^{-7} mol·L^{-1}) of zoledronic acid (A-G (c)). The linear plot of relative fluorescence intensity (F_0/F_1) of CDs as a function of pamidronate disodium (5×10^{-9} mol·L^{-1} to 4×10^{-7} mol·L^{-1} (b) and zoledronic acid (6×10^{-9} mol·L^{-1} to 8×10^{-7} mol·L^{-1} (d)). Every data point was the mean of three measurements. The error bars were the standard deviation.

3214 cm^{-1} indicated the existence of –OH; the peaks around 2940 and 1240 cm^{-1} were assigned to the C–H; the peaks at 1660 cm^{-1} and 1620 cm^{-1} were attributed to the stretching vibration of C=O and C=C, respectively; the peaks at 1401 cm^{-1} were C–N stretching vibration modes; and the peaks at 1123 and 1007 cm^{-1} indicated C–OH and C–O, respectively. Figure 1(d) shows the high-resolution TEM (HRTEM) image of the CDs which reveals that the CDs had a size distribution around 2.2 nm. According to (1), the QY of the CDs based on the black tea was calculated to be 10.23% with quinine sulfate as a reference.

In order to further study the structure of CDs, the XPS analysis was conducted. Figure 2(a) shows the XPS spectrum of the CDs, and the three predominant peaks of C1s, N1s, and O1s at 284.61, 399.49, and 531.91 eV were corresponded to the binding energies. As can be seen from the result that the as-prepared CDs mainly contained C, O, and N, the contents of the three elements were calculated to be 73.8% (C), 21.7% (O), and 4.5% (N). The C1s spectrum (Figure 2(b)) can be deconvoluted into four peaks at around 284.6, 285.6, 286.6, and 287.9 eV, which were attributed to C–C, C–N, C–O, and C=O groups, respectively [9]. In addition, the N1s spectrum (Figure 2(c)) can be deconvoluted into three peaks at 399.3, 400.1, and 400.7 eV which can be assigned to N–$(C)_3$, N–H, and C–N–H groups, respectively. From the O1s spectrum shown in Figure 2(d), the three fitted peaks at 530.3, 531.3, and 532.7 eV were ascribed to C–OH/C–O–C, adsorbed oxygen, and –C=O groups, respectively. The result from XPS showed that the surface of the CDs from the black tea was functionalized by multiple oxygenated and nitrous groups which may lead the interactions with different drug molecules.

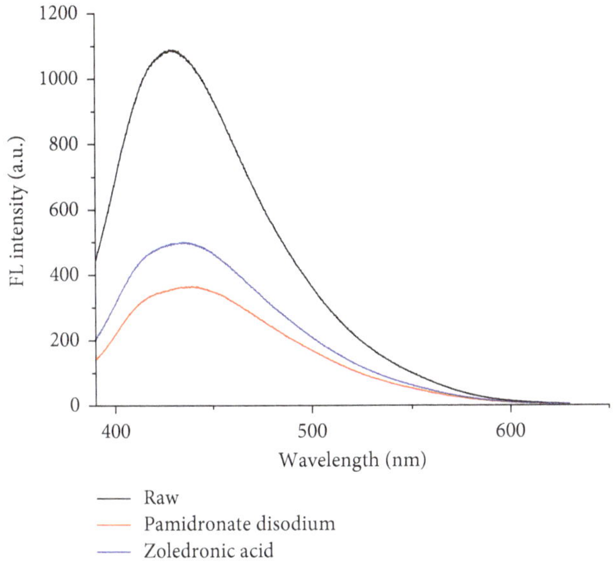

FIGURE 4: Fluorescence turn-off behavior of CDs in the presence of the same concentration of pamidronate disodium or zoledronic acid (4.0×10^{-7} mol·L^{-1}).

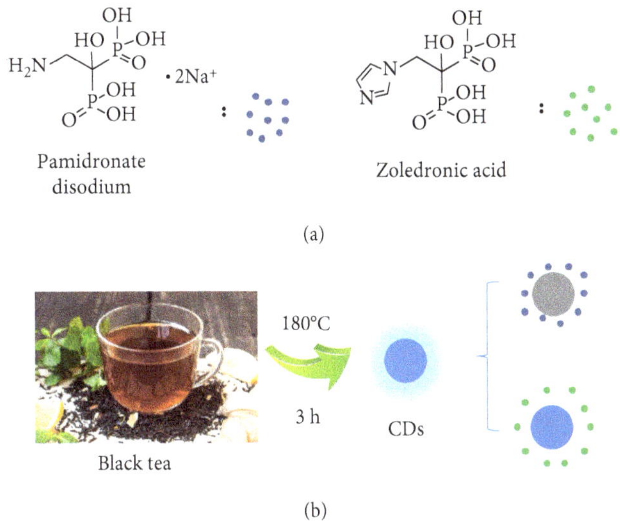

SCHEME 1: The chemical structure of pamidronate disodium or zoledronic acid and the illustration of turn-off fluorescent sensor for pamidronate disodium and zoledronic acid based on carbon dots from black tea.

3.2. Effective Quenching of Fluorescence of CDs by Pamidronate Disodium and Zoledronic Acid: Turn-Off Process. As shown in Figure 3(a), the increased addition of pamidronate disodium led to the gradual decreased fluorescence of CDs. And the relationship between the concentration of pamidronate disodium and the fluorescence intensities of CDs follows the "Stern-Volmer" equation:

$$\frac{F_0}{F_1} = 1 + K_{sv}[M], \quad (2)$$

where F_0 and F_1 were the fluorescence intensities of CDs in the absence and presence of the drugs, respectively; [M] was the concentration of pamidronate disodium, and the quenching constant K_{sv} defines the quenching efficiency of pamidronate disodium. The equation shows that F_0/F_1 was directly proportional to the concentration of pamidronate disodium. As shown in the Stern-Volmer plot in the inset of Figure 1, the K_{sv} of pamidronate disodium to the fluorescence of CDs was 4.8×10^6 L·mol^{-1}, with linear range from 5×10^{-9} mol·L^{-1} to 4×10^{-7} mol·L^{-1} and a correlation coefficient $R = 0.996$. On the other hand, as shown in Figure 3(c), when the concentration of zoledronic acid was increasing from 6×10^{-9} mol·L^{-1} to 8×10^{-5} mol·L^{-1}, the fluorescent intensity of the CDs at around 430 nm also decreases gradually, which indicating that zoledronic acid can effectively quench the fluorescence of CDs. Meanwhile, through the Stern-Volmer equation (Figure 3(d)), the K_{sv} of zoledronic acid to the fluorescence of CDs was 2.8×10^6 L·mol^{-1} with the linear range from 6×10^{-9} mol·L^{-1} to 8×10^{-7} mol·L^{-1} and a correlation coefficient $R = 0.991$. The detect limit of pamidronate disodium and zoledronic acid was, respectively, 5×10^{-9} mol·L^{-1} and 6×10^{-9} mol·L^{-1}. Obviously, pamidronate disodium exhibits better quenching ability to the fluorescence of CDs than zoledronic acid with the value of K_{sv} almost twice higher than that of zoledronic acid and seen in Figure 4, the addition of pamidronate disodium shows a significantly weaker fluorescence quenching phenomenon than the same concentration of zoledronic acid, which may be caused by the different intermolecular binding capacities. As shown in Scheme 1, compared with the imidazolyl group contained in zoledronic acid, the amino group contained in pamidronate disodium has the smaller space obstruction and a better affinity to the polycarboxyl group modified on the surface of the CDs, leading to much stronger quenching phenomenon to the fluorescence via photoinduced electron transfer and exhibiting excellent selectivity in detecting the two drugs with similar structures. To explain the sensing mechanism in depth, lifetimes of CDs before and after adding bisphosphonates were investigated and the

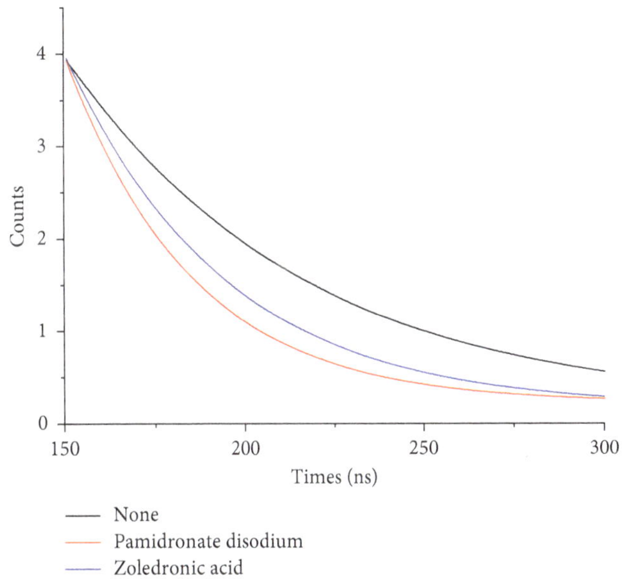

FIGURE 5: The lifetime of CDs after adding different bisphosphonates.

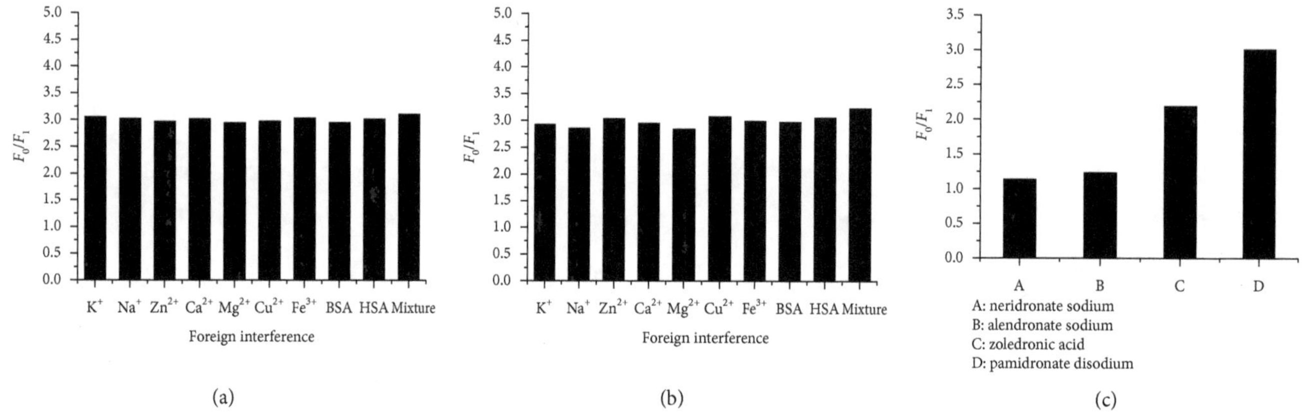

FIGURE 6: The influence of naturally abundant foreign metal ions and biomolecules (1.0×10^{-5} mol·L^{-1}) for the turn-off sensor ((a) pamidronate disodium: 4×10^{-7} mol·L^{-1} and (b) zoledronic acid: 8×10^{-7} mol·L^{-1}) and the comparison of different quenching abilities towards CDs of different bisphosphonates at the concentration of 4×10^{-7} mol·L^{-1} (c).

result was shown in Figure 5. The lifetime of CDs after adding different bisphosphonates was reduced by different degrees which can act as a proof for the photoinduced electron transfer mechanism. And what's more, the reduction fluorescence lifetime of CDs caused by pamidronate disodium was more than that caused by zoledronic acid. This also demonstrated the different quenching capabilities of bisphosphonates towards the carbon dots.

3.3. Selectivity of the Sensor for Pamidronate Disodium and Zoledronic Acid Detection. As to evaluate the selectivity of the CDs sensor, 10 different biologically relevant metal ions including Na^+, K^+, Mg^{2+}, Ca^{2+}, Fe^{3+}, Zn^{2+}, Cu^{2+}, Bovine serum albumin (BSA), human serum albumin (HSA), and the mixture of all the above were selected as foreign interference with a concentration at 1×10^{-5} mol·L^{-1}. The comparison is shown in Figure 6 which can indicate that the interference effect of all the 10 interferences were slight. In addition, even in the spiked samples of the human plasma and urine, a recovery of 101.3% and 99.4% for pamidronate disodium (4×10^{-7} mol·L^{-1}) and a recovery of 102.1% and 100.8% for zoledronic acid (8×10^{-7} mol·L^{-1}) were obtained. What's more, other kinds of bisphosphonates such as neridronate sodium and alendronate sodium also failed to have a significant impact on the fluorescence of CDs. It was obvious that a special effect appeared between CDs and pamidronate disodium or zoledronic acid, so the as-prepared CDs had a high selectivity for pamidronate disodium and zoledronic acid analysis.

4. Conclusion

In this study, a new turn-off model based on newly prepared carbon dots (CDs) from the black tea was established to analyze pamidronate disodium and zoledronic acid. By using the new sensor, both of drugs can be quantitative, respectively, with high sensitivity and stability. Meanwhile, the fluorescence of the CDs can be quenched by two drugs with various degrees via photoinduced electron transfer, which provides the sensor a high selectivity to the two structurally similar drugs. What's more, this turn-off method shows good stability and has been employed to analyze the two drugs under the influence of common foreign interference factors which indicating its good application prospects for drugs analysis *in vitro* and *in vivo*.

Conflicts of Interest

The authors declare that they have no conflicts of interest regarding the publication of this paper.

Acknowledgments

This work was financially supported by the Tianjin Natural Science Foundation (no. 12JCYBJC16400) and the Tianjin Municipal Health Bureau Fund Project (nos. 15KG124, 12KG121).

References

[1] S. Zhu, Q. Meng, L. Wang et al., "Highly photoluminescent carbon dots for multicolor patterning, sensors, and bioimaging," *Angewandte Chemie*, vol. 125, no. 14, pp. 4045–4049, 2013.

[2] S. Zhu, Y. Song, X. Zhao, J. Shao, J. Zhang, and B. Yang, "The photoluminescence mechanism in carbon dots (graphene quantum dots, carbon nanodots, and polymer dots): current state and future perspective," *Nano Research*, vol. 8, no. 2, pp. 355–381, 2015.

[3] Y. Dong, H. Pang, H. B. Yang et al., "Carbon-based dots co-doped with nitrogen and sulfur for high quantum yield and excitation-independent emission," *Angewandte Chemie International Edition*, vol. 52, no. 30, pp. 7800–7804, 2013.

[4] U. Baruah, N. Gogoi, A. Konwar et al., "Carbon dot based sensing of dopamine and ascorbic acid," *Journal of Nanoparticles*, vol. 2014, Article ID 178518, 8 pages, 2014.

[5] K. Yang, S. Wang, Y. Wang, H. Miao, and X. Yang, "Dual-channel sensor of carbon dots cooperating with gold nanoclusters employed for assaying multiple targets," *Biosensors and Bioelectronics*, vol. 91, no. 15, pp. 566–573, 2017.

[6] J. Shen, S. Shang, X. Chen, D. Wang, and Y. Cai, "Facile synthesis of fluorescence carbon dots from sweet potato for

Fe^{3+} sensing and cell imaging," *Materials Science and Engineering: C*, vol. 76, pp. 856–864, 2017.

[7] L. Wang and H. S. Zhou, "Green synthesis of luminescent nitrogen-doped carbon dots from milk and its imaging application," *Analytical Chemistry*, vol. 86, no. 18, pp. 8902–8905, 2014.

[8] S. Zhao, M. Lan, X. Zhu et al., "Green synthesis of bifunctional fluorescent carbon dots from garlic for cellular imaging and free radical scavenging," *ACS Applied Materials & Interfaces*, vol. 7, no. 31, pp. 17054–17060, 2015.

[9] R. Zhang and W. Chen, "Nitrogen-doped carbon quantum dots: facile synthesis and application as a "turn-off" fluorescent sensor for detection of Hg^{2+} ions," *Biosensors and Bioelectronics*, vol. 55, pp. 83–90, 2014.

[10] T. N. J. I. Edison, R. Atchudan, J. J. Shim, S. Kalimuthu, B.-C. Ahn, and Y. R. Lee, "Turn-off fluorescence sensor for the detection of ferric ion in water using green synthesized N-doped carbon dots and its bio-imaging," *Journal of Photochemistry and Photobiology B: Biology*, vol. 158, pp. 235–242, 2016.

[11] Y. Sakamoto, "Bilateral typical femoral fractures in a patient with metastatic breast cancer on long-term bisphosphonate therapy: a case report," *Journal of Osteoporosis and Physical Activity*, vol. 2, no. 1, pp. 1-4, 2014.

[12] V. Viereck, G. Emons, V. Lauck et al., "Bisphosphonates pamidronate and zoledronic acid stimulate osteoprotegerin production by primary human osteoblasts," *Biochemical and Biophysical Research Communications*, vol. 291, no. 3, pp. 680–686, 2002.

[13] L. Yang and S. Du, "Efficacy and safety of zoledronic acid and pamidronate disodium in the treatment of malignant skeletal metastasis: a meta-analysis," *Medicine*, vol. 94, no. 42, p. e1822, 2015.

[14] Y. Orita, I. Sugitani, S. Takao, K. Toda, J. Manabe, and S. Miyata, "Prospective evaluation of zoledronic acid in the treatment of bone metastases from differentiated thyroid carcinoma," *Annals of Surgical Oncology*, vol. 22, no. 12, pp. 4008–4013, 2015.

[15] R. Karthikeyan, O. S. A. I. Koushik, and P.S. Babu, "Stability indicating method development and validation of zoledronic acid in injectables by RP–HPLC," *Journal of International Research in Medical and Pharmaceutical Sciences*, vol. 7, no. 4, pp. 159–174, 2016.

[16] Ł. Matuszewski, A. Matuszewska, T. Mazurkiewicz, and S. Ohga, "Determination of bisphosphonates by ion-pair HPLC," *Journal of the Faculty of Agriculture Kyushu University*, vol. 56, no. 2, pp. 213–216, 2011.

[17] E. Ban, J. Y. Park, H. T. Kim, and C.-K. Kim, "Determination of alendronate in low volumes of plasma by column-switching high-performance liquid chromatography method and its application to pharmacokinetic studies in human plasma," *Archives of Pharmacal Research*, vol. 34, no. 12, pp. 2079–2086, 2011.

[18] T. Feng, X. Ai, G. An, P. Yang, and Y. Zhao, "Correction to charge-convertible carbon dots for imaging-guided drug delivery with enhanced in vivo cancer therapeutic efficiency," *ACS Nano*, vol. 10, no. 5, p. 5587, 2016.

[19] Z. Yang, M. Xu, Y. Liu et al., "Nitrogen-doped, carbon-rich, highly photoluminescent carbon dots from ammonium citrate," *Nanoscale*, vol. 6, no. 3, pp. 1890–1895, 2014.

5

Improvement in Analytical Methods for Determination of Sugars in Fermented Alcoholic Beverages

Ayalew Debebe,[1,2] Shibru Temesgen,[3] Mesfin Redi-Abshiro,[1] Bhagwan Singh Chandravanshi,[1] and Estifanos Ele[1]

[1]*Department of Chemistry, Addis Ababa University, P.O. Box 1176, Addis Ababa, Ethiopia*
[2]*Department of Chemistry, Haramaya University, P.O. Box 138, Dire Dawa, Ethiopia*
[3]*Department of Statistics, Addis Ababa University, P.O. Box 1176, Addis Ababa, Ethiopia*

Correspondence should be addressed to Mesfin Redi-Abshiro; mesfin.redi@aau.edu.et

Academic Editor: Hana Sklenarova

The main objective of this study was to improve the performance of analytical methods for the determination of sugars in fermented alcoholic beverages based on mid-infrared-partial least squares (MIR-PLS), high-performance liquid chromatography with the ultraviolet detector (HPLC-UV), high-performance liquid chromatography with the refractive index detector (HPLC-RI), and sulfuric acid methods. The MIR-PLS method was found to give good prediction of individual sugars: glucose, fructose, sucrose, and maltose in the alcoholic beverages with less than 4% error. The HPLC-UV method can be used for the determination of glucose in alcoholic beverages after derivatization with *p*-aminobenzoic acid ethyl ester. The HPLC-RI method was found to be applicable for the determination of individual sugars: glucose, fructose, sucrose, and maltose in the alcoholic beverages. The limit of detection (%, w/w) and recovery (%) of the individual sugars by the HPLC-RI method were fructose 0.001, 89.4–106; glucose 0.002, 92.4–109; and sucrose 0.002, 94.2–95.1. The sulfuric acid method was found to be useful for the determination of total sugar in the alcoholic beverages. The limit of detection (%, w/w) and recovery (%) of the total sugar by the sulfuric acid method were found to be 0.009, 98.2–109. The HPLC-RI method was applied to determine the level of individual sugars, while the sulfuric acid method was used to determine total sugar in Ethiopian traditional fermented alcoholic beverages: *Tella*, *Netch Tella*, *Filter Tella*, *Borde*, *Tej*, *Korefe*, *Keribo*, and *Birz*. The sugar contents in the real samples were found in the ranges (%): glucose 0.07–5.60, fructose 0.09–8.50, sucrose and maltose 0.08–3.00, and total sugar 12.0–64.5. The levels of sugars in Ethiopian traditional fermented alcoholic beverages were found to be comparable with literature data.

1. Introduction

Carbohydrates are one of the most abundant compounds in foods [1–6]. They are classified into five major classes, which are monosaccharides, disaccharides, oligosaccharides, polysaccharides, and nucleotides [7]. Mono- and disaccharides with a sweet flavor are commonly called sugars [8].

Alcoholic beverages contain sugars and other soluble solids [9]. In beverages, especially in beer, 80–85% of nonvolatile materials (30–40 g/L) are sugars [10]. Fructose, glucose, maltose, sucrose, and maltotriose (but not lactose) are fermentable sugars [10]. However, oligosaccharides containing more than three glucose units cannot be fermented [10]. Thus, when fermentation is completed, only small amounts of lower sugars are found in alcoholic beverages [10].

The determination of sugars in alcoholic beverages has been reported using a large number of analytical techniques such as chromatographic techniques, spectroscopic techniques, colorimetric methods, iodometric methods, and enzymatic techniques [11]. Iodometric methods [11] and traditional colorimetric methods [12] are unable to quantitate sugars individually. Chromatography and capillary electrophoresis methods require derivatization (indirect method) because carbohydrates are nonvolatile [1, 13] and lack both charge and strong chromophore [11]. Although derivatization provides high sensitivity, it is complicated and time-consuming [11]. Both gas and liquid chromatographic techniques have used different derivatizing reagents. For gas chromatography, trimethylsilyl (TMS) ethers are the most popular derivatizing agent [14]. However, TMS derivatization has limitations such as it is unstable, forms multiple peaks, and has difficulty in preparation [13, 14]. Likewise, among different UV active and fluorophore-derivatizing agents, *p*-aminobenzoic acid ethyl ester (ABEE) is the most widely known one [1]. The derivatization of reducing sugars with ABEE is easy and required no special equipment; moreover, the method showed higher sensitivity and elimination of the possible doublet that could be formed by mutarotation of the free reducing end of sugars [1]. In the direct method, high-performance liquid chromatography (HPLC) has been used with pulsed amperometric, evaporative light-scattering, and refractive index detectors [1, 15] and ultraviolet detectors at 195 nm [3]. A direct method with the sugar-borate complex using capillary zone electrophoresis (CZE) has been used. However, it takes about 20 min to analyze several carbohydrates because the sugar-borate complexes migrate against the direction of the EOF [11]. Chromatographic methods have high accuracy, but they are often time-consuming, labor intensive, and require tedious and complex processing. Recently, near-infrared (NIR) and middle-infrared (MIR) spectrometries have become popular. They have been applied in a low-cost, rapid, and non-destructive way [16]. However, different techniques have their own advantages and disadvantages.

The determination of sugars in alcoholic beverages has considerable importance [1–4, 17, 18]. It is useful to know the contribution of sugars to flavor and sensory characteristics to evaluate their nutritive (caloric) value [1], to evaluate how they are formed and assimilated [10], and to know their health impacts [19].

The literature survey revealed that only a few studies have been conducted on the determination of chemical composition of Ethiopian traditional fermented beverages. These include determination of phenolics [20], alcohol contents [21–23], and minerals [24]. However, no study has been reported on the sugar contents of traditional fermented beverages.

Therefore, the objectives of this study are (i) to develop new analytical methods for the determination of sugars in fermented alcoholic beverages based on mid-infrared-partial least squares (MIR-PLS), high-performance liquid chromatography with the refractive index detector (HPLC-RI), high-performance liquid chromatography with the ultraviolet detector (HPLC-UV), and sulfuric acid methods, (ii) to compare the analytical parameters of the newly developed methods, and (iii) to determine the level of sugars in Ethiopian traditional fermented alcoholic beverages: *Tella*, *Netch Tella*, *Filter Tella*, *Borde*, *Tej*, *Korefe*, *Keribo*, and *Birz*.

2. Materials and Methods

2.1. Instrumentation. UV-Vis spectrophotometer (Lambda 950; PerkinElmer, UK) with a 1 cm path length quartz cuvette was used to determine total carbohydrates. The Fourier-transform infrared spectrometer (Spectra 65; PerkinElmer, UK) with ZnSe window as a sample holder, and HPLC-UV and HPLC-RI (Agilent Technologies, Germany) were used for the determination of individual sugars.

2.2. Reagents and Chemicals. Ethanol (99.99%; Fisher Scientific, UK), glucose (laboratory reagent; Merck Extra Pure, England), fructose (laboratory reagent; Pharmacos Ltd, England), sucrose (analytical reagent; Guangdong Guanghua Chemical Factory Co. Ltd, China), and maltose (laboratory reagent; The British Drug Houses Ltd, Poole, England) were used to prepare synthetic calibration and validation sets. *p*-Amino benzoic acid ethyl ester (ABEE) (Riedel-de Haen AG, Seelze, Hanover, Germany), acetic acid (99.5%; BDH Chemicals Ltd, Poole, England), and sodium borohydride (BDH Chemicals Ltd, Poole, England) were used for derivatizing sugars. Sulfuric acid (Sigma-Aldrich, Germany) was used for the determination of total carbohydrate. Chloroform (Carlo Erba Reagents, France) was used for purifying the derivatized compound. Distilled deionized water was used for the preparation of standards and dilution of samples.

2.3. Preparation of Standard Solutions. Calibration and validation sets were prepared by mixing ethanol with sugars for sugar determination with ethanol standards by the MIR-PLS method. The compositions of the standards were ethanol (2–12%, w/w), fructose (0–5%, w/w), glucose (0–5%, w/w), sucrose (0–5%, w/w), and maltose (0–5%, w/w), whereas for sugar standards without ethanol, the compositions of the standards were glucose (0–14.2%, w/w), fructose (0–17.3%, w/w), sucrose (0–23.4%, w/w), and maltose (0–15.01%, w/w).

For the HPLC-RI method, average calibration curves were developed with standard solutions of glucose, fructose, and sucrose from 0.03% (w/w) to 0.2% (w/w).

For the HPLC-UV method, the average calibration curve was constructed with standard solutions of glucose from 0.0002 to 0.002 mg/L which were derivatized using ABEE.

For the sulfuric acid method, the average calibration curve was constructed with the series of glucose standard solutions in the range 0.01–0.1 g/L.

2.4. Sampling and Sample Preparation. Eight most popular Ethiopian traditional fermented beverages, *Tej* (honey wine), *Tella* (a malt beverage like beer), *Korefe*, *Keribo*, *Birz*, *Netch Tella*, *Filter Tella*, and *Borde*, were selected for this study. A total of 57 samples: 15 *Tej*, 15 *Tella*, 6 *Korefe*, 6 *Keribo*, 4 *Birz*, 4 *Netch Tella*, 4 *Filter Tella*, and 3 *Borde*, were collected randomly from vending houses at different subcities of Addis Ababa (Ethiopia) and nearby towns (Sebeta, Dukem, Sululta, Sendafa, and Burayu) of Oromia Regional State. The characteristics of alcoholic samples (the pH value and ethanol content) and brief information about raw materials and processes used for the production of Ethiopian traditional fermented beverages are given in Table 1. A 500 mL aliquot of each type of the beverages was collected from the

TABLE 1: Physicochemical properties, raw materials, and production process of some Ethiopian traditional alcoholic beverages.

S. no.	Samples (% alcohol)*	Physicochemical properties	Raw materials	Production process
1	*Tella* (2.9 ± 0.3), *Korefe* (4.6 ± 0.4)	Dark brown in color, with pH 3.56 ± 0.02 and 4.28 ± 0.02	*Kita* (a thin, 5–10 mm thick, pancake-like bread), *enkuro* (a dark brown toasted flour), *bikil* (germinated grain), and powdered *gesho* (*Rhamnus prinoides*) [25, 26]	A four-phase fermentation for 10–12 days [25, 26]
2	*Tej* (9.1 ± 0.3)	Yellow, sweet, effervescent, and cloudy, with pH 3.65 ± 0.01	Honey or mixture of sugar with honey and leaves of *gesho* (*Rhamnus prinoides*) [25, 27]	Mixing boiled must with *gesho* (*Rhamnus prinoides*) and unboiled must and then allowing to ferment for 5 days in warm weather or for 15–20 days in colder weather [25, 27]
3	*Birz* (6.5 ± 0.8)	Yellow, sweet, effervescent, and cloudy, with pH 3.40 ± 0.06	Honey or mixture of sugar with honey [25]	Has a short fermentation period, usually overnight [25]
4	*Borde* (1.8 ± 0.4)	Opaque, effervescent, and whitish-grey to brown colored with a thick consistency and a sweet-sour taste, with pH 5.77 ± 0.03	Unmalted maize (*Zea mays*), barley (*Hordeum vulgare*), wheat (*Triticum aestivum*), finger millet (*Eleusine coracana*), sorghum (*Sorghum bicolor*), and/or *tef* (*Eragrostis tef*) and their malt; additional ingredients garlic, fresh chili (*Capsicum minimum*), ginger, and salt [25, 28, 29]	A four-phase fermentation for less than 4 days [25, 28, 29]
5	*Keribo* (1.7 ± 0.3)	Dark brown colored, with pH 3.72 ± 0.03	Unmalted roasted barley (*Hordeum vulgare*), sugar, and yeast [30]	Has a short fermentation period, usually overnight [30]

*% alcohol was determined by the method in [22].

three sites of each of the subcities of Addis Ababa and nearby towns. A 1000 mL bulk sample was prepared for each sample type from one specific sampling site. This was done by taking 333.3 mL of the beverage from each of the three samples from one place and mixing well in a 1 L volumetric flask. All the samples were collected using glass amber bottles and kept at 4°C until the analysis time. Ethiopian traditional fermented alcoholic beverages are either liquid or semiliquid. The liquid samples such as *Tella, Tej, Birz, Netch Tella, Filter Tella,* and *Keribo* were filtered before analysis. The sugars were extracted from the semiliquid samples such as *Borde* and *Korefe* by an optimized procedure.

2.5. Derivatization of Sugar for the HPLC-UV Method. The derivatization procedures of sugar with ABEE reported by Gomis et al. [1] and Munegumi and Goto [31] were different. Thus, a modified procedure was used in this study. An ABEE methanol solution (4 mL, 0.5 g/mL) and glacial acetic acid (310 mL) were dissolved at 40–50°C in a polypropylene tube. A 0.60 g sodium borohydride was added to the tube, which was sealed with a screw cap and vortexed to give an ABEE stock solution. An aliquot (2 mL) of the ABEE stock solution and a standard saccharide solution or sample (500 μL) were mixed by vortexing, and the resulting solution was heated at 80°C for 6 h. After cooling and centrifuging for 1 min, the filtrate was treated with 3 mL water and centrifuged. Again, the filtrate was treated with chloroform (2 × 5 mL). Finally, the upper layer (the aqueous phase) was used for HPLC analysis after filtration using a micromembrane (0.45 mm pore size).

2.6. Procedure for Total Sugars. A 1 mL aliquot of carbohydrate solution was rapidly mixed with 3 mL of concentrated sulfuric acid in a test tube and vortexed for 30 s. The temperature of the mixture was raised rapidly within 10–15 s after addition of sulfuric acid. The solution was cooled in ice for 2 min to bring it to room temperature. Finally, UV light absorption at 315 nm was measured using a UV spectrophotometer. Reference (reagent blank) solutions were prepared following the same procedure as above, except that the carbohydrate aliquot was replaced with distilled deionized water [12].

2.7. HPLC Conditions for Indirect (Derivatized) Determination. The derivatized sample was analyzed to determine the sugars using HPLC with the UV detector at λ_{max} 230 nm. The chromatographic separation was achieved on a C_{18} column maintained at 45°C. A binary solvent system comprising 0.5% aqueous trifluoroacetic acid as solvent A and acetonitrile as solvent B was used under the gradient mode. The gradient condition was 0–5 min solvent A, 5–10 min 0–20% B, 10–35 min 20–25% B, 35–50 min 25–45% B, and 50–60 min 45–100% B. The mobile-phase flow rate was 0.5 mL·min^{-1}, and sample injection volume was 3 μL.

2.8. HPLC Conditions for Direct (Nonderivatized) Determination. The sugar determination was done using HPLC with the RI detector. The chromatographic separation was achieved in the Hi-Plex H column (7.7 × 300 mm) maintained at 35°C. The solvent used was distilled-deionized water. The mobile-phase flow rate and sample injection volumes were 0.5 mL·min^{-1} and 10 μL, respectively.

3. Results and Discussion

3.1. Extraction of Sugars. For the extraction of sugars from the semiliquid samples, the types and amounts of the extracting solvent and extraction time were optimized. 80% (v/v) methanol, 80% (v/v) ethanol, and water were considered as the extracting solvent. Among the solvents checked, 80% (v/v) methanol showed better efficiency than others (the data are not presented). The extraction time was varied in the range 30–120 min. The optimum time for extraction of sugars was found to be 90 min. Hence, 90 min was used to extract sugars from the beverages (Table 2).

Table 2: Optimized extraction time for sugar extraction from semiliquid samples.

Extraction time (min)	30	60	90	120
Total carbohydrate (Glu·(g/L))	14.9 ± 0.1	15.1 ± 0.1	25.8 ± 0.2	13.4 ± 0.2

3.2. Determination of Individual Sugars Using MIR-PLS. Carbohydrates have strong absorption bands which overlap in the spectral region, 850–1200 cm^{-1} (Figure 1). This seriously hinders the quantification of individual carbohydrates [32, 33]. Hence, MIR-PLS was proposed as a method for the determination of individual sugars. The C–O stretching band in the spectral region 850–1200 cm^{-1} was selected due to higher sensitivity of the spectra to develop a calibration model [33]. The MIR spectra of selected fermented alcoholic beverages are shown in Figure 2. It can be seen that the MIR spectrum of *Korefe* resembles very much the spectrum of ethanol given in Figure 1. This is because of relatively low sugar content in the *Korefe* compared to the sugar content in the *Birz* and *Tej*.

Two models using sugars (fructose, sucrose, glucose, and maltose) with and without ethanol were used by modifying Rambla et al. [32] and Leopold et al. [33] models. To determine sugars using the MIR-PLS method, a method without ethanol was chosen to avoid interference. To avoid overfitting, the number of principal components (PCs) was fixed to 6 for glucose, fructose, sucrose, and maltose in the model.

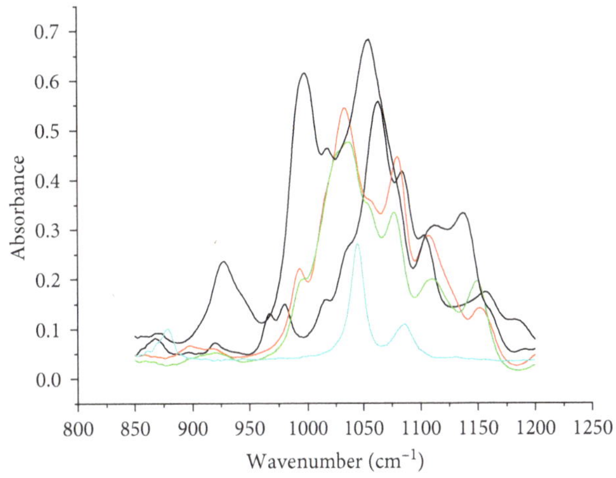

Figure 1: MIR spectra of 5% (w/w) ethanol and 10% (w/w) individual sugars.

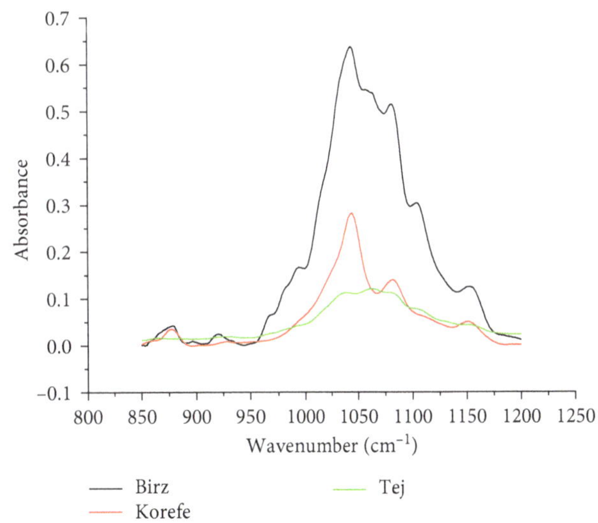

Figure 2: MIR spectra of selected fermented alcoholic beverages.

3.3. Pretreatment Methods. In order to find the model with best prediction capacity, PLS regression was applied to different spectra such as raw spectra, first derivative spectra, second derivative spectra, and others. Leopold et al. [33] also applied PLS regression to raw spectra, first derivative spectra, and second derivative spectra. In their study, lower RMSEP values were provided by the first derivative spectra. But in the present study, the second derivative spectra provided the lower RMSEP values (Table 3). This is because the second derivative preprocessing removes background and increases spectral resolution [34]. Hence, the second derivative spectra were used for the prediction of sugar concentration in the alcoholic beverages.

3.4. Method Validation. To avoid concentration data overfitting, the cross-validation method, leaving out one sample at a time, was used. Accordingly, the validation was done using 50 synthetic samples. The obtained validations showed that the PLS calibration model of each has a very good performance (Table 3). It was also found comparable with the report of Irudayaraj and Tewari [35] in terms of correlation (R^2), RMSEP, and the number of factors used. In addition, the predicted amounts were evaluated and compared with the measured values at the 99% confidence level. Overall, no significant variations were obtained between the measured and predicted amounts.

Table 3: Results of MIR-PLS calibration models for the determination of sugars in alcoholic beverages.

Data treatment	Analytes	Principal components	Calibration		Validation	
			R^2	RMSEE	R^2	RMSEP
Second derivative	Glucose	6	0.992	0.18	0.987	0.55
Second derivative	Fructose	6	0.997	0.06	0.996	0.28
Second derivative	Sucrose	6	0.989	0.22	0.996	0.38
Second derivative	Maltose	6	0.999	0.04	0.997	0.29

RMSEE: root mean square error of estimation; RMSEP: root mean square error of prediction.

The correlations between actual and predicted values of the analytes are shown in Figure 3. From the correlation graphs (Figure 3), almost all the curves have shown better correlations. The model was validated in terms of % error using the following equation:

$$\% \text{ error} = \frac{\text{PLS predicted values} - \text{actual values}}{\text{actual values}} \times 100. \quad (1)$$

The % error was found in the range 0.21–3.7% (3.7% glucose, 1.1% fructose, 0.21% sucrose, and 0.23% maltose). This showed that the proposed method has less than 4% error. Thus, the MIR-PLS method was found to give good prediction of sugars in the alcoholic beverages.

3.5. Comparisons of MIR-PLS with HPLC-RI. Comparison of sugar content obtained by MIR-PLS and HPLC-RI was made. The comparison is illustrated in Figure 4. Fructose showed the best comparison, while glucose showed the least comparison among all the sugars. The two techniques were found to be comparable and showed no significance difference at the 95% confidence level. Hence, MIR-PLS was found to be a promising method for sugar determination in the alcoholic beverages. However, to apply for the analysis of real samples, the samples should be purified from potential interferences such as ethanol, phenolic compounds, proteins, amino acids, and others [36, 37].

3.6. Determination of Individual Sugars with Derivatization Using HPLC-UV. Carbohydrates lack chromophore [11], and they cannot be detected by the UV detector in the HPLC-UV. Thus, they require derivatization. In this study, the derivatization was done using glucose, fructose, and sucrose with *p*-aminobenzoic acid ethyl ester (ABEE), acetic acid, and sodium borohydride. The imine formation was preceded for aldose, not for ketoses and sucrose. Thus, the obtained results are in agreement with different literature reports [1, 31] (Figure 5).

It can be clearly seen in the chromatogram (Figure 5) that only glucose and mixture have peaks at 21 min retention time due to the glucose labeling, while all the chromatograms (a–e) have peaks at 42 min, owing to excess ABEE. The structure of derivatized glucose (the formed compound) is shown in Scheme 1.

Glucose derivatization was further confirmed by ^{1}H NMR and ^{13}C NMR analysis. ^{1}H NMR spectrum of the compound Glu-ABEE showed the presence of four protons on 1,4-disubstituted aromatic ring appearing at δ 6.73 ppm (2H on C_6 and C_8; $J = 8.4$ Hz) and 7.69 ppm (2H on C_5 and C_9; $J = 8.4$ Hz). This suggested that the four protons are on different chemical environments (i.e., two in one and the remaining two in another environment). In addition, based on the J values, they are protons which are coupled. The proton signals between δ 3 ppm and 3.6 ppm are due to the protons of OH glucose. The signals at δ 4.2 ppm (2H on C_2; quartet) and δ 1.3 ppm (3H on C_1; triplet) are due to CH_2 and CH_3 of the ester, respectively. The signals at δ 2.5 ppm are due to DMSO.

The ^{13}C NMR analyzed with the DEPT-135 spectrum revealed the presence of three quaternary (from the benzene ring at δ 152 ppm (C_7) and δ 118 ppm (C_4) and from the carboxyl group of ester at δ 166 ppm (C_3)) and three methylene carbon atoms (at δ 59.94 ppm (C_{10}), δ 60.14 ppm (C_{15}), and δ 60.95 ppm (C_2)). The signal at δ 112 (C_6 and C_8) and δ 131 (C_5 and C_9) was from methine carbons of the ring. The ^{13}C NMR signals at δ 76 ppm (C_{11}), 74.69 ppm (C_{12}), 70.42 ppm (C_{13}), and 84.98 ppm (C_{14}) were from the Cs of the glucose part. The signal at δ 14.78 ppm (C_1) was from the methyl of the ester group. Moreover, the absence of signal around $\delta = 100$ ppm confirmed that the glucose ring was opened and further reaction occurred. Thus, the ^{1}H NMR and ^{13}C NMR spectra of the compound confirmed that the derivatization has occurred.

HPLC chromatogram and NMR data confirmed that the labeling is only for glucose. Therefore, only glucose can be determined. To construct a calibration curve, a series of glucose standards from 0.0002 mg/L to 0.002 mg/L were prepared in triplicate. Accordingly, the average calibration equation $y = 6.41 \times 10^5 x - 84$ (where y = peak area and x = glucose in g/mL) with $R^2 = 0.9997$ was obtained. Therefore, for the determination of glucose, this alternative approach was chosen for two reasons. First, the reducing agent $NaBH_4$ is less toxic and easily available than the usual reducing agent $NaBH_3CN$, and second, the method has a wider linear range and better correlation.

3.7. Determination of Sugars Using HPLC-RI and Sulfuric Acid Method. The chromatograms of sugars and ethanol are illustrated in Figure 6. For the determination of sugars in the real samples, peak identifications were made in the same chromatographic system by comparing each peak's retention time (t_R) with the reference (standards) and by spiking the sample with the standards. Standards solutions of 0.03–0.2% (w/w) for glucose, fructose, and sucrose and standards solution of 0.01–0.1 g/L for the total sugars were prepared in triplicate. The calibration equations $y = 1.87 \times 10^6 x + 1012$, $y = 2.02 \times 10^6 x - 1836$, and $y = 2.00x + 2241$ (where y = peak area and x = concentration of

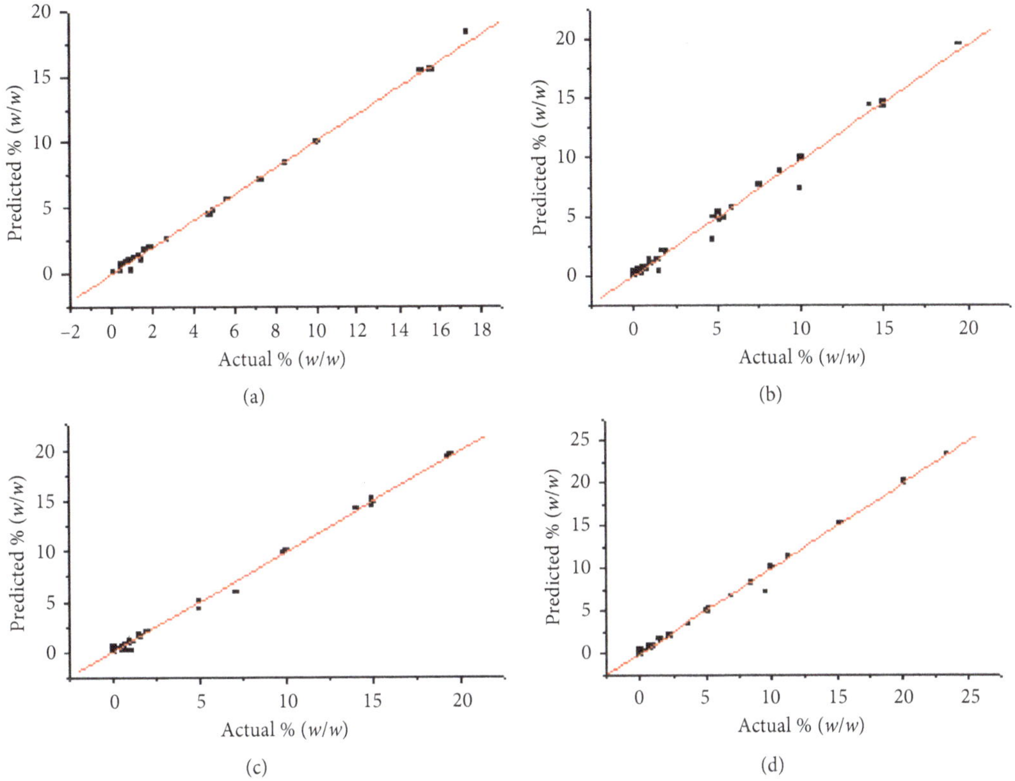

FIGURE 3: Correlation statistics between actual and predicted values for fructose (a), glucose (b), maltose (c), and sucrose (d).

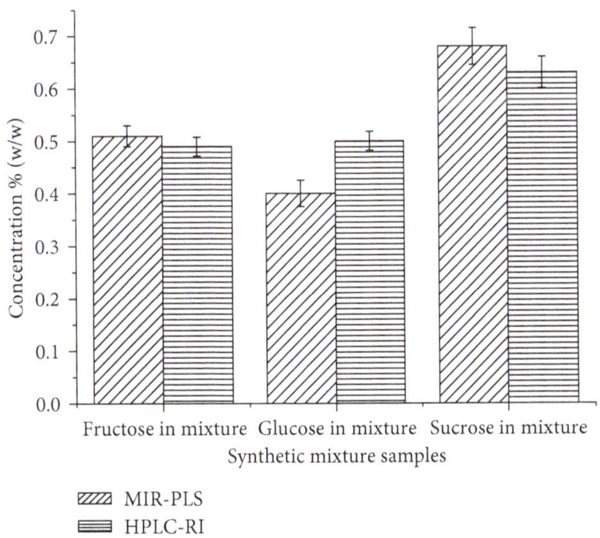

FIGURE 4: Comparison of MIR-PLS with HPLC-RI.

analytes in % (w/w)) were assigned for fructose, glucose, and sucrose, respectively. The calibration equation used for the total sugars was $y = 10.6x - 0.022$ (where $y =$ peak height and $x =$ glucose concentration in g/L).

3.8. Method Validation. The proposed method was validated using the recovery test. For recovery determination of fructose, glucose, and sucrose, 0.05 and 0.1% (w/w) from each type were spiked. But for the recovery determination of total sugars, 6.5 and 13.0 g/L glucose were spiked. The results are presented in Table 4. LOD = 3σ of the residues (y-intercepts)/slope [38] was also determined and given in Table 4. The recovery percentages obtained for fructose (89–106%), glucose (92–109%), sucrose (94–95%), and total carbohydrate (98–109%) are in the acceptable ranges. This implies that the matrix effects of the samples were not considerable. Therefore, the proposed techniques are appropriate to quantify individual and total sugar contents in the fermented alcoholic beverages.

3.9. Comparison of Advantages and Disadvantages of the Four Methods. In this study, four methods (MIR-PLS, HPLC-UV, HPLC-RI, and sulfuric acid methods) have been developed for the determination of sugars in the fermented alcoholic beverages. A comparison has been made on the advantages and disadvantages of the individual methods, and the results are summarized in Table 5. The MIR-PLS method is applicable to determine individual sugars: glucose, fructose, sucrose, and maltose, with good precision and accuracy. But the method suffers from interferences from ethanol, phenolic compounds, proteins, amino acids, and others. The HPLC-UV method is only applicable to determine glucose but not other individual sugars. It also requires derivatization which is time-consuming. The HPLC-RI method is applicable to determine individual sugars: glucose, fructose, and sucrose, without the matrix effect but not other individual sugars. The sulfuric acid method is applicable to determine total sugars with good

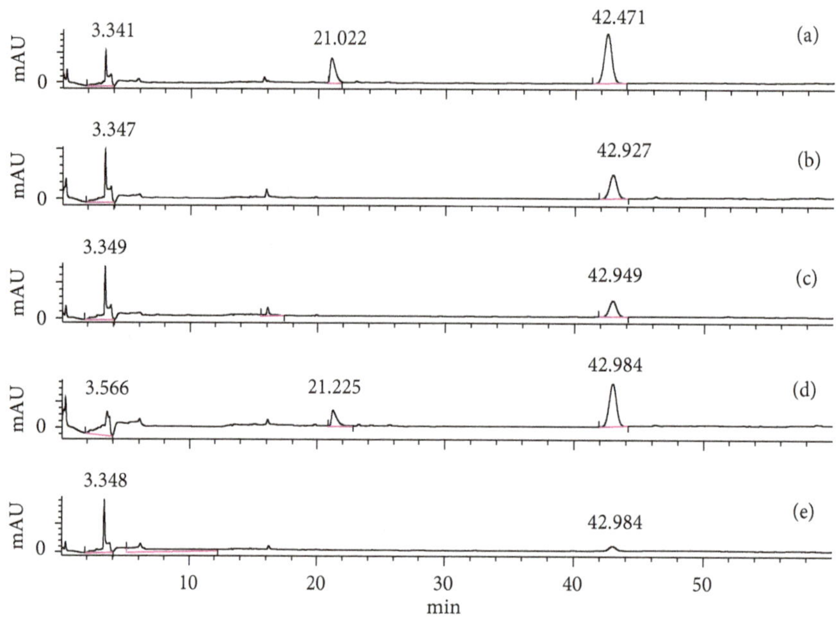

FIGURE 5: Chromatograms of sugar derivatives: (a) glucose, (b) fructose, (c) sucrose, (d) mixture, and (e) ABEE.

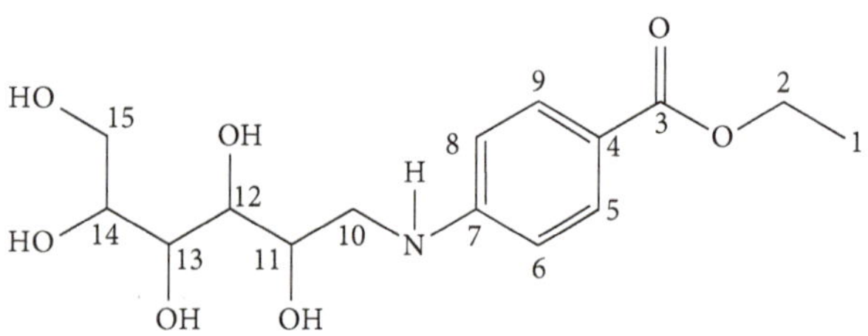

SCHEME 1: The structure of derivatized glucose.

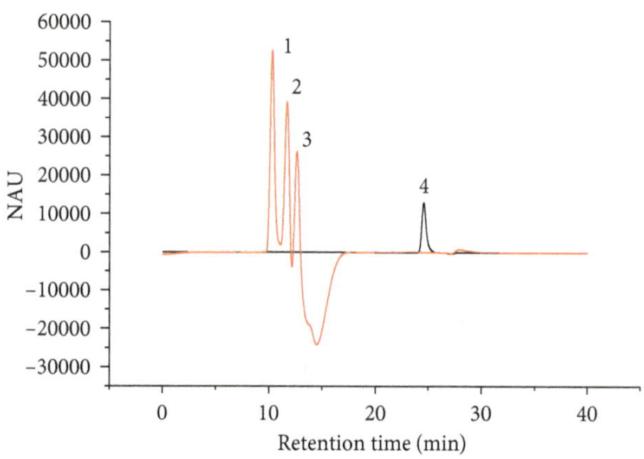

FIGURE 6: Chromatograms of the mixture of sugars and ethanol: 1 = sucrose, 2 = glucose, 3 = fructose, and 4 = ethanol. Column: Hi-Plex H 7.7 × 300 mm ID (35°C). Mobile phase: water. Flow rate: 0.5 mL/min. Injection volume: 10 μL.

precision and accuracy. However the method is not applicable to determine individual sugars.

3.10. Analysis of Real Samples. The amounts of sugars found in the real samples (fermented alcoholic beverages) are presented in Table 6. The amounts of individual sugars are expressed as % (w/w) and the total sugar as g/L in Ethiopian traditional fermented alcoholic beverages. This is because in the literature, the individual sugar contents are expressed as % (w/w) and the total sugar content as g/L to make the comparison of results of the present study with the results reported in the literature meaningful. The order of the beverages based on the total sugar contents (in g/L) was *Birz* > *Tej* > *Keribo* > *Korefe* > *Netch Tella* > *Filter Tella* > *Tella* > *Borde*. *Birz* and *Tej* were found to be the leading ones. Particularly for *Birz*, it is mainly by the unfermented sugars left after fermentation. In *Tej*, additional honey was added, while it is ready to serve. For the rest of the beverage types, the major reason of variation is the differences in the composition of the raw materials (Table 1) and fermentation time used [25, 30].

Although it is not reasonable to compare beverages of different types, the comparisons were done without considering the factors that can influence their levels of sugars. In this study, *Birz* was the first in the levels of glucose and fructose, while *Keribo* was the first in the level of sucrose. Both controlled *Tella* and *Tej* have comparable glucose, fructose, and sucrose with the average values of the corresponding samples. However, there were still variations among beverages. This showed that the variation in the composition of the raw materials, preparation process, and fermentation time used (Table 1) are the crucial factors for the difference in the level of sugars in the traditional fermented beverages.

The total sugar contents with average values of 37.0 g/L in nonalcoholic beer [39], 42.1 g/L in alcoholic beer [39], 90.2 g/L in wort from phases I–V [40], and 6.75 g/L [41] and 48.8 g/L [42] in wine were reported. The Ethiopian traditional fermented alcoholic beverages have been found to contain less sugar than in wort [40] and beer [39] except

TABLE 4: Limit of detection and recovery percentage for individual sugars by the HPLC-RI method and for total sugar by the sulfuric acid method.

S. no.	Sample	Fructose		Glucose		Sucrose		Total sugar	
		LOD (%, w/w)	Recovery (%)	LOD (%, w/w)	Recovery (%)	LOD (%, w/w)	Recovery (%)	LOD (g/L)	Recovery (%)
1	*Keribo*	0.001	106	0.002	93.2	0.002	94.3	0.009	98.2
2	*Tej*	0.001	89.4	0.002	109	0.002	94.2	0.009	100
3	*Tella*	0.001	90.3	0.002	92.4	0.002	95.1	0.009	109

TABLE 5: Comparison of advantages and disadvantages of the four newly developed methods for the determination of sugars in the fermented alcoholic beverages.

Method	Advantages	Disadvantages
MIR-PLS method	The method is applicable to determine individual sugars: glucose, fructose, sucrose, and maltose, with good precision and accuracy.	To apply for the analysis of real samples, the samples should be purified from potential interferences such as ethanol, phenolic compounds, proteins, amino acids, and others.
HPLC-UV method	The method is applicable to determine glucose, and it has a wider linear range and better correlation.	The method is not applicable to determine individual sugars except glucose, and it requires derivatization which is time-consuming.
HPLC-RI method	The method is applicable to determine individual sugars: glucose, fructose, and sucrose, without the matrix effect.	The method is applicable to determine individual sugars: glucose, fructose, and sucrose, but not other sugars.
Sulfuric acid method	The method is applicable to determine total sugar (carbohydrate) with good precision and accuracy.	The method is not applicable to determine individual sugars.

TABLE 6: The amount of sugars found in Ethiopian traditional fermented alcoholic beverages (individual sugars (%, w/w) by the HPLC-RI method and total sugar (g/L) by the sulfuric acid method).

S. no.	Beverage types	Number	Concentration of individual sugars (%, w/w)			Total carbohydrate (g/L)
			Fructose	Glucose	Sucrose and maltose	
1	*Tella*	15	0.09 ± 0.12	0.07 ± 0.09	0.08 ± 0.12	12.0 ± 4.9
2	*Tej*	15	2.98 ± 1.10	1.66 ± 0.60	0.19 ± 0.20	64.5 ± 24
3	*Korefe*	6	0.30 ± 0.10	0.43 ± 0.30	0.18 ± 0.15	34.4 ± 13.5
4	*Keribo*	6	0.10 ± 0.05	0.60 ± 0.03	3.00 ± 0.30	50.0 ± 9.1
5	*Birz*	4	8.50 ± 0.50	5.60 ± 1.40	1.00 ± 0.70	131 ± 18
6	*Borde*	3	0.09 ± 0.01	4.33 ± 0.01	0.40 ± 0.01	11.3 ± 1.25
7	*Netch Tella*	4	0.70 ± 1.33	0.79 ± 1.35	0.08 ± 0.03	23.3 ± 2.8
8	*Filter Tella*	4	1.40 ± 2.70	0.90 ± 1.60	0.36 ± 0.60	19.0 ± 4.1
9	Control *Tella*	3	0.04 ± 0.01	0.03 ± 0.01	0.12 ± 0.01	102 ± 1.8
10	Control *Tej*	3	1.77 ± 0.05	1.11 ± 0.04	0.14 ± 0.01	270 ± 5.7

Birz, *Tej*, and *Korefe*. But they have higher sugar than that reported by Matsuhiro et al. [41].

In some beverages, the concentration of glucose, fructose, and sucrose was reported. The concentration values of sucrose, glucose, and fructose were found in the ranges: 1–3.28 and 0–10.5, 0.13–2.17 and 0–6.9, and 0.3–3.72 and 0–6.7% (w/w), respectively, in flavored water and soft drinks, respectively [43, 44]. Since both the flavored water and soft drinks are not fermented beverages, their sugar contents are expected to be higher than those in fermented beverages. Even though the Ethiopian traditional fermented beverages have passed through some extent of fermentation, the obtained results confirmed that the Ethiopian fermented alcoholic beverages have comparable sugar content with that in the nonfermented beverages.

4. Conclusion

Improvement in the performance of analytical methods for the determination of sugars in alcoholic beverages using MIR-PLS, HPLC-UV with a derivatizing agent, and HPLC-RI has been presented. The MIR-PLS method is applicable to determine individual sugars: glucose, fructose, sucrose, and maltose, with good precision and accuracy. The HPLC-RI method is applicable to determine individual sugars: glucose, fructose, and sucrose, without the matrix effect. The HPLC-UV method is applicable to determine only glucose but not other sugars. The sulfuric acid method is applicable to determine total sugar with good precision and accuracy. The amount of individual sugars in Ethiopian traditional fermented alcoholic beverages was determined by using the

HPLC-RI method, and the total sugar was determined by using the sulfuric acid method. The sugar contents in the Ethiopian fermented alcoholic beverages are comparable with those in the commercial beverages. The differences in the level of sugars in the different types of traditional fermented beverages are due to the variation in the composition of the raw materials, preparation process, and fermentation time used in the traditional fermented beverages.

Conflicts of Interest

The authors declare that they have no conflicts of interest.

Acknowledgments

The authors express their gratitude to the Department of Chemistry, Addis Ababa University, Ethiopia, for providing the laboratory facilities. Ayalew Debebe is thankful to Haramaya University, Ethiopia, for sponsoring his PhD study.

References

[1] D. B. Gomis, D. M. Tamayo, and J. M. Alonso, "Determination of monosaccharides in cider by reversed-phase liquid chromatography," *Analytica Chimica Acta*, vol. 436, no. 1, pp. 173–180, 2001.

[2] J. H. Ko, H. Huang, G. W. Kang, and W. J. Cheong, "Simultaneous quantitative determination of monosaccharides including fructose in hydrolysates of yogurt and orange juice products by derivatization of monosaccharides with *p*-aminobenzoic acid ethyl ester followed by HPLC," *Bulletin of the Korean Chemical Society*, vol. 26, no. 10, pp. 1533–1538, 2005.

[3] F. Momenbeik and J. H. Khorasani, "Analysis of sugars by micellar liquid chromatography with UV detection," *Acta Chromatographica*, vol. 16, pp. 58–69, 2006.

[4] S. Chandraju, R. Mythily, and C. S. Chidan Kumar, "Extraction, isolation and identification of sugars from banana peels (*Musa sapientum*) by HPLC coupled LC/MS instrument and TLC analysis," *Journal of Chemical and Pharmaceutical Research*, vol. 3, no. 3, pp. 312–321, 2011.

[5] L. C. Anyika, S. I. Okonkwo, and E. N. Ejike, "Comparative analysis of monosaccharide and disaccharide using different instrument refractometer and polarimeter," *International Journal of Research in Chemistry and Environment*, vol. 2, no. 4, pp. 270–274, 2012.

[6] C. S. Chidan Kumar, R. Mythily, S. Chandraju, and R. Venkatachalapathi, "Analysis of sugar components extracted from pumpkin (*Cucurbita pepo*) peels, research & reviews," *Journal of Food Science and Technology*, vol. 1, no. 3, pp. 1–6, 2012.

[7] A. Skalska-Kaminska, G. Matysik, M. Wojciak-Kosior, H. Donica, and I. Sowas, "Thin-layer chromatography of sugars in plant material," *Annales*, vol. 22, pp. 17–24, 2009.

[8] F. J. Rambla, S. Garrigues, and M. Guardia, "PLS-NIR determination of total sugar, glucose, fructose and sucrose in aqueous solutions of fruit juices," *Analytica Chimica Acta*, vol. 344, no. 1–2, pp. 41–53, 1997.

[9] L. Gutierrez, A. Zapata, L. Co1l, and C. Diez, "Analytical study of the mineral and sugar fractions of peach liqueurs," *Food Chemistry*, vol. 54, no. 1, pp. 113–117, 1995.

[10] P. Lehtonen and R. Hurme, "Liquid chromatographic determination of sugars in beer by evaporative light scattering detection," *Journal of The Institute of Brewing*, vol. 100, no. 5, pp. 343–346, 1994.

[11] M. Karkacier, M. Erbas, M. K. Uslu, and M. Aksu, "Comparison of different extraction and detection methods for sugars using amino-bonded phase HPLC," *Journal of Chromatographic Science*, vol. 41, no. 6, pp. 331–333, 2003.

[12] A. A. Albalasmeh, A. A. Berhe, and T. A. Ghezzehei, "A new method for rapid determination of carbohydrate and total carbon concentrations using UV spectrophotometry," *Carbohydrate Polymers*, vol. 97, no. 2, pp. 253–261, 2013.

[13] W. Zhang, H. He, and X. Zhang, "Determination of neutral sugars in soil by capillary gas chromatography after derivatization to aldononitrile acetates," *Soil Biology and Biochemistry*, vol. 39, no. 10, pp. 2665–2669, 2007.

[14] A. I. Ruiz-Matute, O. Hernandez-Hernandez, S. Rodriguez-Sanchez, M. L. Sanz, and I. Martinez-Castro, "Derivatization of carbohydrates for GC and GC–MS analyses," *Journal of Chromatography B*, vol. 879, no. 17–18, pp. 1226–1240, 2011.

[15] D. B. Gomis, D. M. Tamayo, and J. J. Mangas Alonso, "Evolution of sugars in cider brandy aged in oak barrels: A contribution to its characterization," *Journal of Agriculture and Food Chemistry*, vol. 51, no. 4, pp. 923–926, 2003.

[16] F. Shen, Y. Ying, B. Li, Y. Zheng, and J. Hu, "Prediction of sugars and acids in chinese rice wine by mid-infrared spectroscopy," *Food Research International*, vol. 44, no. 5, pp. 1521–1527, 2011.

[17] S. Cortacero-Ramırez, A. Segura-Carretero, C. Cruces-Blanco, C. M. Hernainz-Bermudez, and A. Fernandez-Gutierrez, "Analysis of carbohydrates in beverages by capillary electrophoresis with precolumn derivatization and UV detection," *Food Chemistry*, vol. 87, no. 3, pp. 471–476, 2004.

[18] C. S. Chidan Kumar, S. Chandraju, R. Mythily, T. Ahmad, and N. M. Made Gowda, "Extraction of sugars from black gram peels by reversed-phase liquid chromatography systems and identification by TLC and mass analysis," *Advances in Analytical Chemistry*, vol. 2, no. 4, pp. 32–36, 2012.

[19] N. P. Evmiridis, N. K. Thanasoulias, and A. G. Vlessidis, "Determination of glucose and fructose in mixtures by a kinetic method with chemiluminescence detection," *Analytica Chimica Acta*, vol. 398, no. 2–3, pp. 191–203, 1999.

[20] A. Debebe, B. S. Chandravanshi, and M. Redi-Abshiro, "Total contents of phenolics, flavonoids, tannins and antioxidant capacity of selected traditional Ethiopian alcoholic beverages," *Bulletin of the Chemical Society of Ethiopia*, vol. 30, no. 1, pp. 27–37, 2016.

[21] A. Debebe, A. Anberbir, M. Redi-Abshiro et al., "Alcohol determination in distilled alcoholic beverages by liquid phase Fourier transform mid-infrared and near-infrared spectrophotometers," *Food Analytical Methods*, vol. 10, pp. 172–179, 2017.

[22] A. Debebe, M. Redi-Abshiro, and B. S. Chandravanshi, "Non-destructive determination of ethanol levels in fermented alcoholic beverages using Fourier transform mid-infrared spectrometry," *Chemistry Central Journal*, vol. 11, no. 1, 2017.

[23] A. Debebe, S. Temesgen, M. Redi-Abshiro, and B. S. Chandravanshi, "Partial least square-ultra violet near infrared spectrometric determination of ethanol in distilled alcoholic beverages and its comparison with reference method, gas chromatography," *Bulletin of the Chemical Society of Ethiopia*, vol. 31, no. 2, pp. 201–209, 2017.

[24] A. Debebe, B. S. Chandravanshi, and M. Redi-Abshiro,

"Assessment of essential and non-essential metals in Ethiopian traditional fermented alcoholic beverages," *Bulletin of the Chemical Society of Ethiopia*, vol. 31, no. 1, pp. 17–30, 2017.

[25] G. Debebe, "Determination of ethanol level in beverages," Master Thesis, Addis Ababa University, Addis Ababa, Ethiopia, 2006.

[26] T. Yohannes, F. Melak, and K. Siraj, "Preparation and physicochemical analysis of some Ethiopian traditional alcoholic beverages," *African Journal of Food Science*, vol. 7, no. 11, pp. 399–403, 2013.

[27] B. Bahiru, T. Mehari, and M. Ashenafi, "Chemical and nutritional properties of "Tej", an indigenous Ethiopian honey wine: variations within and between production units," *The Journal of Food Technology in Africa*, vol. 6, no. 3, pp. 104–108, 2001.

[28] K. Abegaz, F. Beyene, T. Langsrud, and J. A. Narvhus, "Indigenous processing methods and raw materials of borde, an Ethiopian traditional fermented beverage," *The Journal of Food Technology in Africa*, vol. 7, no. 2, pp. 59–64, 2002.

[29] M. Ashenafi, "A review on the microbiology of indigenous fermented foods and beverages of Ethiopia," *Ethiopian Journal of Biological Sciences*, vol. 5, no. 2, pp. 189–245, 2006.

[30] R. A. Abawari, "Indigenous processing methods and raw materials of Keribo: An Ethiopian traditional fermented beverage," *Journal of Food Resource Science*, vol. 2, no. 1, pp. 13–20, 2013.

[31] T. Munegumi and A. Goto, "Reversed-phase high-performance liquid chromatography of disaccharides labeled by 4-aminobenzoic acid ethyl ester," *Der Pharma Chemica*, vol. 6, no. 1, pp. 248–252, 2014.

[32] F. J. Rambla, S. Garriguesa, N. Ferrerb, and M. Guardiaa, "Simple partial least squares–attenuated total reflectance Fourier transform infrared spectrometric method for the determination of sugars in fruit juices and soft drinks using aqueous standards," *Analyst*, vol. 123, pp. 277–281, 1998.

[33] L. F. Leopold, N. Leopold, H.-A. Diehl, and C. Socaciu, "Quantification of carbohydrates in fruit juices using FTIR spectroscopy and multivariate analysis," *Spectroscopy*, vol. 26, no. 2, pp. 93–104, 2011.

[34] A. Moghimi, M. H. Aghkhani, A. Sazgarnia, and M. Sarmad, "Vis/NIR spectroscopy and chemometrics for the prediction of soluble solids content and acidity (pH) of Kiwi fruit," *Biosystems Engineering*, vol. 106, no. 3, pp. 295–302, 2010.

[35] J. Irudayaraj and J. Tewari, "Simultaneous monitoring of organic acids and sugars in fresh and processed apple juice by Fourier transform infrared–attenuated total reflection spectroscopy," *Applied Spectroscopy*, vol. 57, no. 12, pp. 1599–1604, 2003.

[36] A. Nose and M. Hojo, "Hydrogen bonding of water–ethanol in alcoholic beverages," *Journal of Bioscience and Bioengineering*, vol. 102, no. 4, pp. 269–280, 2006.

[37] R. R. Madrera and B. S. Valles, "Determination of volatile compounds in cider spirits by gas chromatography with direct injection," *Journal of Chromatographic Science*, vol. 45, no. 7, pp. 428–434, 2007.

[38] M. M. Sanagi, S. L. Ling, Z. Nasir, W. A. W. Ibrahim, and A. A. Naim, "Comparison of signal-to-noise, blank determination, and linera regression methods for the estimation of detection and quantification limits for volatile organic compounds by gas chromatography," *Journal of AOAC International*, vol. 92, no. 6, pp. 1833–1838, 2009.

[39] M. Jurková, Č. Pavel, K. Štěrba, and J. Olšovská, "Determination of total carbohydrate content in beer using its pre-column enzymatic cleavage and HPLC-RI," *Food Analytical Methods*, vol. 7, no. 8, pp. 1677–1686, 2014.

[40] Z. Hodžić, B. Banjanin, and J. Sadadinović, "Fermentable carbohydrates determination in different worts by HPLC-RI," *Journal of Engineering*, vol. 1, pp. 105–110, 2008.

[41] B. Matsuhiro, R. Torres, E. A. Zúñiga, M. J. Aguirre, L. Mendoza, and M. Isaacs, "Determination of low molecular weight carbohydrates in cabernet sauvignon red wines," *Journal of the Chilean Chemical Society*, vol. 54, no. 4, pp. 405–407, 2009.

[42] H. Kelebek, S. Selli, A. Canbas, and T. Cabaroglu, "HPLC determination of organic acids, sugars, phenolic compositions and antioxidant capacity of orange juice and orange wine made from a Turkish Cv. Kozan," *Microchemical Journal*, vol. 91, no. 2, pp. 187–192, 2009.

[43] M. A. Augustin and K. L. Khor, "Determination of sugars in soft drinks by high performance liquid chromatography," *Pertanika*, vol. 9, no. 1, pp. 119–123, 1986.

[44] A. Sapcanin, G. Jancan, R. Jadric et al., "Determination of sugar content in flavored waters," *Bulletin of Chemical Technology of Bosnia and Herzegovina*, vol. 41, pp. 29–32, 2013.

6

Rapid and Simple Determination of Sarafloxacin and Difloxacin in Beef by Capillary Electrophoresis Coupled with Solid-Phase Extraction

Qian Wang,[1,2] Rui Zhao,[3] Yan Cui,[3,4] Cuicui Liu,[3,4] Liya Liang,[3,4] and Shijie Yan[3,4]

[1]*Department of Animal Science and Veterinary Medicine, Tianjin Agricultural University, Tianjin 300384, China*
[2]*College of Food Science and Engineering, Northwest A&F University, Yangling, Shaanxi 712100, China*
[3]*Department of Food Science and Biological Engineering, Tianjin Agricultural University, Tianjin 300384, China*
[4]*Tianjin Engineering and Technology Research Center of Agricultural Products Processing, Tianjin 300384, China*

Correspondence should be addressed to Shijie Yan; yanshijie@126.com

Academic Editor: Pablo Richter

A simple and rapid capillary electrophoresis method with diode array detector was developed for determination of sarafloxacin and difloxacin in beef. In this study, the experimental parameters affecting the determination were systematically optimized, including wavelength, buffer system, pH and concentration, and separation temperature and voltage. Under the optimal conditions, sarafloxacin and difloxacin could be quantified within 4 min using $H_3BO_3/Na_2B_4O_7$ buffer (35 mmol/L, pH 8.8) as background electrolyte, 25 kV as separation voltage, and 22°C as the column temperature. The linear range of the method was 1–20 μg/mL with LOD 0.8 μg/mL for sarafloxacin and 0.5–20 μg/mL with LOD 0.3 μg/mL for difloxacin. The RSDs for the peak area of 8 μg/mL sarafloxacin were 4.8% (intraday) and 7.8% (interday), respectively. The proposed method has been applied to determine the residue of sarafloxacin and difloxacin in beef samples with the satisfactory recovery.

1. Introduction

Fluoroquinolone (FQ) is the most important group of broad-spectrum antibiotics and are widely used in clinical practice for human and veterinary to treat and prevent various bacterial infections [1]. Difloxacin (DIF) and its metabolite sarafloxacin (SAR) belong to FQ antibacterials and usually work by inhibition of bacterial DNA gyrase to inhibit cell reproduction [2, 3]. Now, they are extensively used for prophylaxis and the treatment of diseases and as feed additives for mass gain promotion. However, the widespread usage of these antibacterials may bring about other potential negative effects. For example, the misuse of these antibacterials may lead to their presence in foodstuffs of animal origin and consequently increase the resistant human pathogens constituting a public health hazard [4, 5]. In the European Union (EU), the presence of these drugs in foodstuffs has been regulated through the Commission Regulation (EU) number 37/2010, and maximum residue limits (MRLs) have been established for different food matrices of animal origin at community level under Council Regulation number 470/2009. The maximum residue limit (MRL) for DIF is 400 μg/kg in muscle of cattle, and the MRLs in other biological materials are between 100 and 1400 μg/kg. MRLs of 30 mg/kg are recommended for SAR in fish muscle, 100 mg/kg in chicken liver, and 10 mg/kg in skin and fat; however, there is no regulation, at the moment, for the SAR in beef. Therefore, it is important to develop accurate and effective method for detecting SAR and DIF.

Recently, the methods for FQ residue analysis in food matrices of animal origin were mainly based on high-performance liquid chromatography with fluorometry or mass spectrometry and immunoassay [6–15]. Most of them suffer from tedious procedures, high solvent, and sample consumption. As a modern separation technique, capillary electrophoresis (CE) is very promising. It can be applied to food analysis because of its advantages, such as higher separation efficiency, high analysis speed, and very small

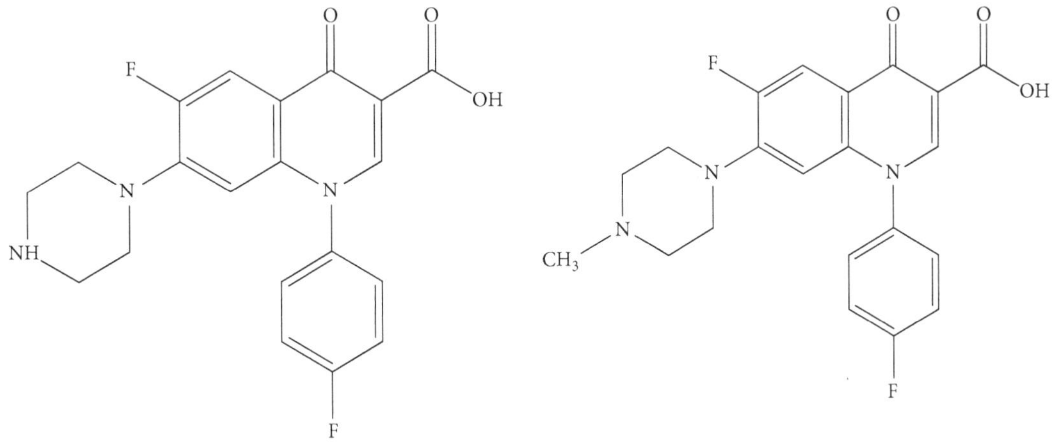

FIGURE 1: The chemical structure of (a) sarafloxacin and (b) difloxacin.

consumption of expensive reagents and toxic solvents [16, 17]. Solid-phase extraction (SPE), the most common and well-established sample pretreatment technique, is usually used for the cleanup of complex matrices and preconcentration of target analytes [18]. CE coupled with SPE technology has been validated with satisfying results via the determination of DIF and SAR in chicken muscle [2]. Thus, we have reasons to expect that CE-SPE will continue its contribution to detection of other foodstuffs.

In this paper, we established an effective and reliable method allowing the determination of DIF and its metabolite SAR in beef by CE-SPE. The effects of buffer system, pH and concentration, and separation temperature and voltage on the separation and determination of SAR and DIF were systematically studied. The present method was further applied for the determination of SAR and DIF in beef. The result indicated that the method possessed the potential for rapidly and simply detecting SAR and DIF residues in food samples.

2. Materials and Methods

2.1. Materials and Chemicals. Acetonitrile, methanol, dichloromethane, and hexane (HPLC grade) were provided by Fisher Company. SAR hydrochloride (99.6%) and DIF (99.9%) were obtained from China Institute of Veterinary Drugs Control. C_{18} Solid-phase extractions were purchased from Bonna-Agela Technologies. Sodium hydroxide (NaOH), boric acid, and borax (analytical grade) were purchased from Thermo Fisher Scientific. Doubly deionized water (DDW, 18 MΩ cm) was prepared by a Milli-Q system (Millipore Corporation, Billerica, MA, USA).

H_3BO_3-$Na_2B_4O_7$ buffer was prepared by mixing 35 mmol/L H_3BO_3 solution with 35 mmol/L $Na_2B_4O_7$ solution to the required pH. All the buffer solutions were made with DDW, filtered through 0.22 μm membrane filters, and degassed in an ultrasonic bath for 1 min before use.

2.2. Instrumentation. FLUKO FA25 high speed homogenizer (Darmstadt, Germany) was used to homogenize the beef sample. A high-speed refrigerated centrifuge (Thermo Fisher, Waltham, MA, USA) was used for centrifugation. The pH of the buffer was measured with a PB-10 potentiometer (Mettler Toledo, Zurich, Switzerland). CE analyses were run on a P/AGE system (Beckman Instruments, Palo Alto, CA, USA) with a diode array detector. Beckman 32.0 Kraft software was used for system control, data collection, and electropherogram integration. An untreated fused-silica capillary with an inner diameter of 50 μm and total length of 75 cm (65 cm to the detector) was purchased from the Yongnian Optic Fiber Plant (Hebei, China). A vacuum rotary evaporator (Rongsheng, Minhang District, Shanghai, China) was used for concentration of the extracts.

New capillaries were successively pretreated with methanol, H_2O, 0.1 mol/L NaOH, H_2O, 0.1 mol/L HCl, and H_2O for 5, 5, 30, 5, 30, and 5 min, respectively. Each day before the measurements, the capillary was conditioned with 0.1 mol/L NaOH and H_2O for 10 min, followed by rinsing with the running buffer for 10 min. In order to keep the capillary wall in good condition, 0.1 mol/L NaOH filled it when it was not being used.

2.3. Spiking Control Sample and Extraction Procedure. SAR and DIF are usually present in food matrices of animal origin (Chemical structures of the FQ are shown in Figure 1). In this study, beef was chosen and bought from local market. Before the spiking and recovery studies, the sample was verified to not contain targets.

10 g of homogenized beef spiked with 100 μL of a known concentration solution of mixed standard solution (10 μg/mL, 20 μg/mL, and 40 μg/mL) was added to a 50 mL polypropylene centrifuge tube, and then 20 mL of 4% ammonia-methanol was added. The mixture was shaken on an orbital shaker for 5 min and then centrifuged at 12000 rpm for 20 min. The residue was extracted once more. The supernatant was collected and concentrated to dryness in a rotary evaporator at 45°C. The dry residue was redissolved in 1 mL of running buffer (35 mmol/L, pH 8.8, H_3BO_3-$Na_2B_4O_7$). Then, the solution was filtered with a 0.22 μm membrane filters for analysis.

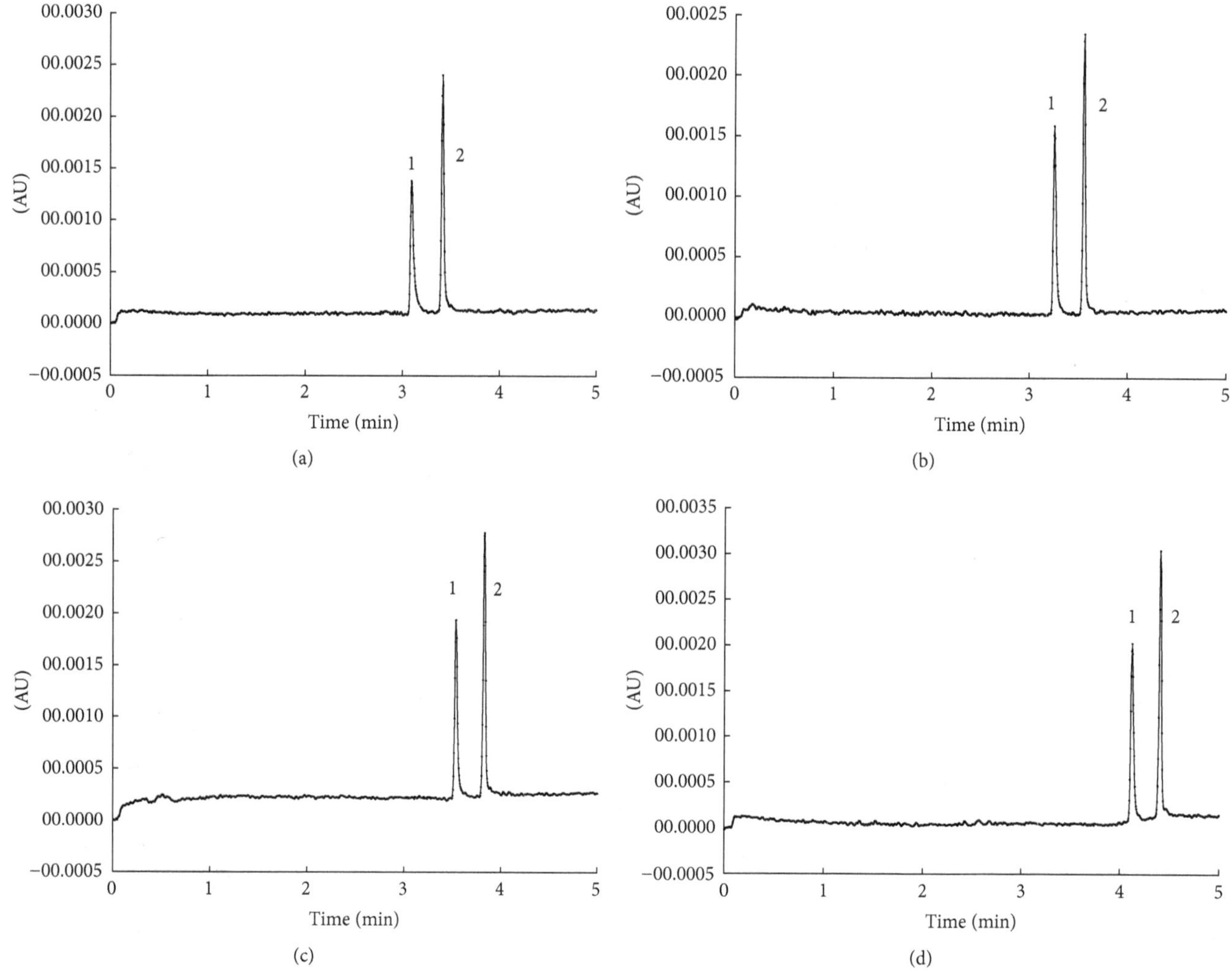

FIGURE 2: Effect of the different pH on the separation. Separation conditions: running buffer, 35 mmol/L H_3BO_3-$Na_2B_4O_7$; separate voltage, 25 kV; operating temperature, 22°C; the detection wavelength, 275 nm; injection, 0.5 psi, 4 s. Chromatogram identification: (a) 8.6; (b) 8.8; (c) 9.0; (d) 9.2. Peak identification: 1: sarafloxacin; 2: difloxacin.

2.4. SPE Preconcentration. The C_{18} (500 mg/6 mL) cartridge was activated with 12 mL methanol and 6 mL water to wash the impurities away. Then the fluoroquinolone was eluted with the eluent, collected into 50 mL distillation flask, and evaporated to dryness at 45°C under reduced pressure. The residue was redissolved with 1 mL of running buffer (35 mmol/L, pH 8.8, H_3BO_3-$Na_2B_4O_7$). Then, the solution was filtered with a 0.22 μm membrane filters for analysis.

3. Results and Discussion

3.1. Establishment of Fluoroquinolone CE Standard Method

3.1.1. Wavelength. In order to measure the detection wavelength for DIF and SAR, 10 μg/mL standard sample solution was scanned ranging from 190 nm to 300 nm. The maximum absorption value of the two analytes was at 275 nm, so 275 nm was selected as the detection wavelength.

3.1.2. Buffer pH. The pH is one of the important factors influencing the charge of targets, further affecting their migration time, so the effect of pH on the separation efficiency of SAR and DIF was investigated with pH ranging from 8.6 to 9.2. As shown in Figure 2, SAR and DIF could be baseline separated with pH increasing from 8.4 to 9.0. However, the retention time was prolonged, which might be as a result of the increase of electromigration, whose direction was opposite to electroosmotic flow. Considering the retention time, three levels of pH (pH = 8.6, 8.8, 9.0) were applied to finish the orthogonal test.

3.1.3. Buffer Concentration. The influence of buffer concentrations on the separation of the SAR and DIF was also investigated from 10 to 40 mmol/L. As shown in Figure 3, the concentration of the buffer directly affected the retention time and resolution of the two analytes. When the buffer concentration was 10 mmol/L, SAR and DIF could not be baseline separated. With the buffer concentration increasing,

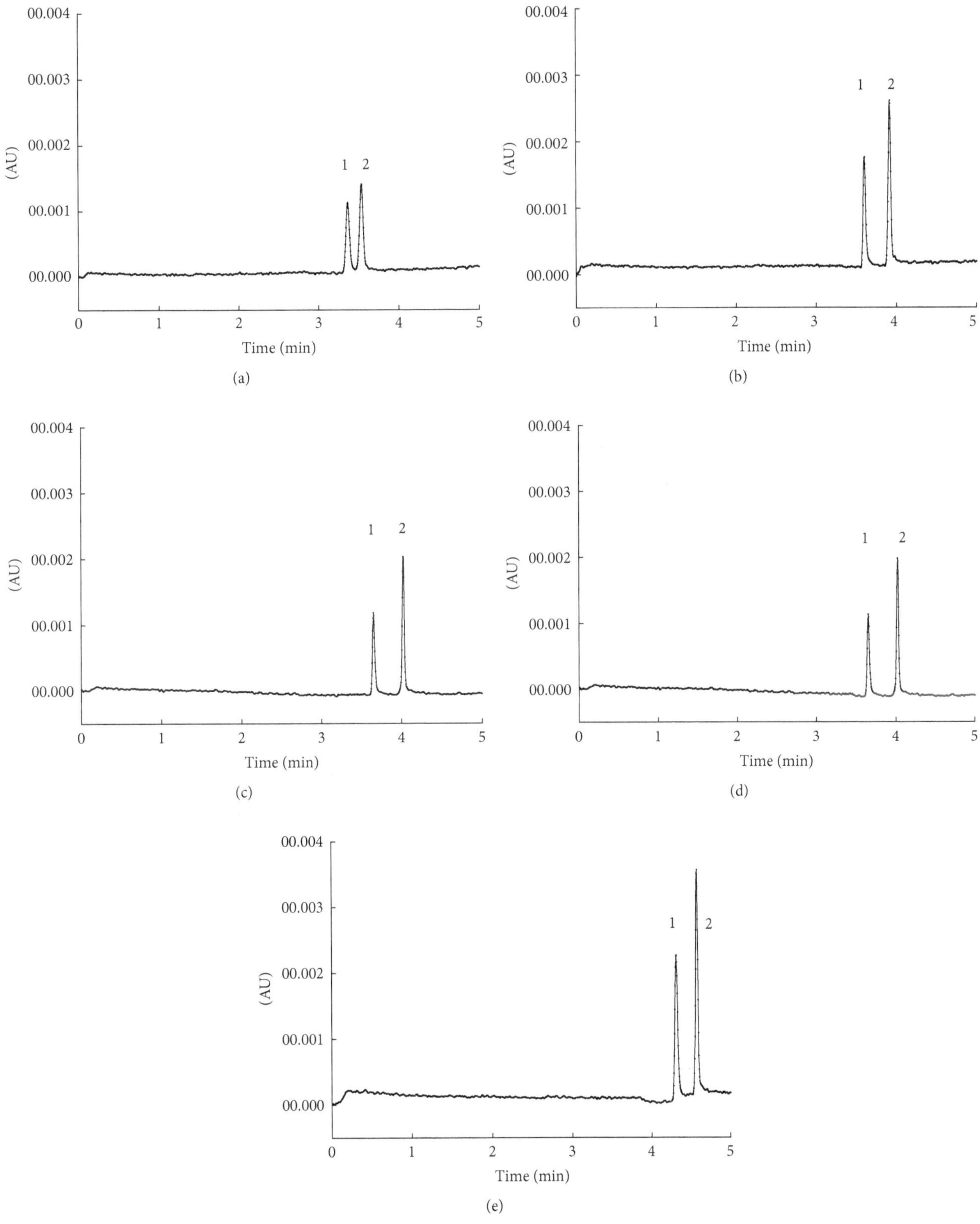

FIGURE 3: Effect of different buffer concentration on the separation. Chromatogram identification: (a) 10 mmol/L; (b) 20 mmol/L; (c) 30 mmol/L; (d) 35 mmol/L; (e) 40 mmol/L. Separation conditions and peak identification were as in Figure 2.

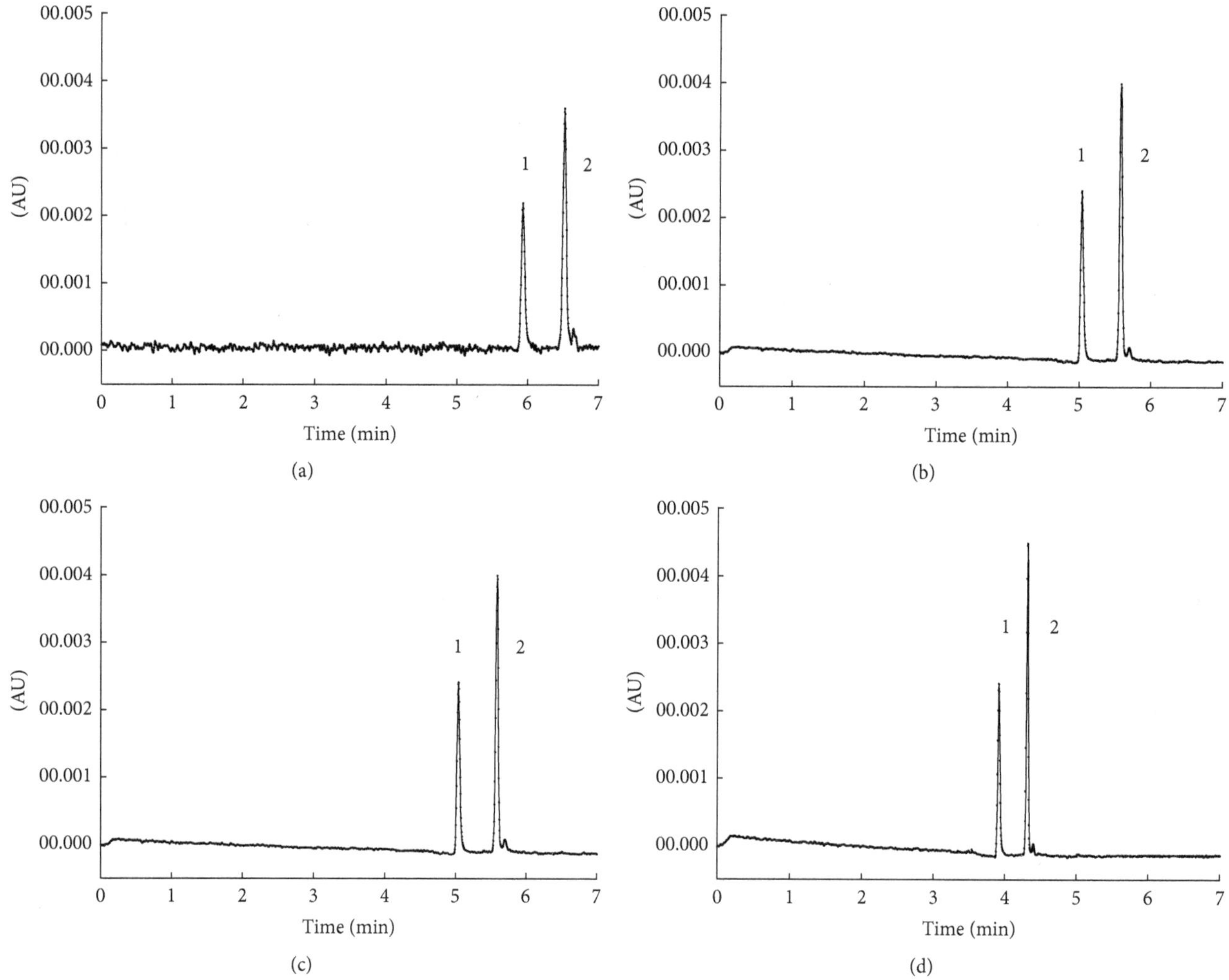

FIGURE 4: Effect of different separation voltage on the separation. Chromatogram identification: (a) 17 kV; (b) 20 kV; (c) 22 kV; (d) 25 kV. Separation conditions and peak identification were as in Figure 2.

the resolution of the two analytes improved greatly. What is more, the retention time was prolonged, which might be attributed to the fact that the higher buffer concentration resulted in the reduction of electroosmotic flow. Herein, the orthogonal test was carried out by three levels of H_3BO_3-$Na_2B_4O_7$ buffer (30, 35, and 40 mmol/L).

3.1.4. Separation Voltage. The comparison of the effects of different separation voltages on the separation performance can be seen in Figure 4; with the separation voltage increasing, the migration time of the two analytes was shortened, along with the peak shape improving. Thus, 20, 22, and 25 kV were chosen as separation voltage to perform the orthogonal test.

3.1.5. Separation Temperature. The separation temperature also plays an important role in the separation efficiency and separation repeatability. Figure 5 shows that the targets can be baseline separated with pH ranging from 20°C to 30°C, but the baseline at lower temperature is not stable. With the temperature increasing, the electroosmotic flow increased due to the decrease of viscosity, further shortening the retention time. Therefore, 22, 25, and 27°C were selected to carried out the orthogonal test.

3.1.6. Orthogonal Experiment. Resolution and analysis time are key parameters in high-performance capillary electrophoresis separation, which always affect the sensitivity, analysis efficiency, and potential application of the method. Herein, in order to obtain the optimal analysis performance, separation conditions, including buffer pH and concentration and separate voltage and temperature, were systemically optimized by orthogonal experiment. The results are shown in Table 1. In view of the analysis time and resolution, the optimal capillary electrophoresis conditions were established: running buffer 35 mmol/L H_3BO_3-$Na_2B_4O_7$ (pH 8.8); separate voltage: 25 kV; operating temperature: 22°C; the detection wavelength: 275 nm; injection pressure: 0.5 psi; injection time: 4 s (electrophoretogram of stand solution was shown in Figure 6).

3.2. Optimization of the Extraction Solution. Eluent type and concentration also played an important role in SPE

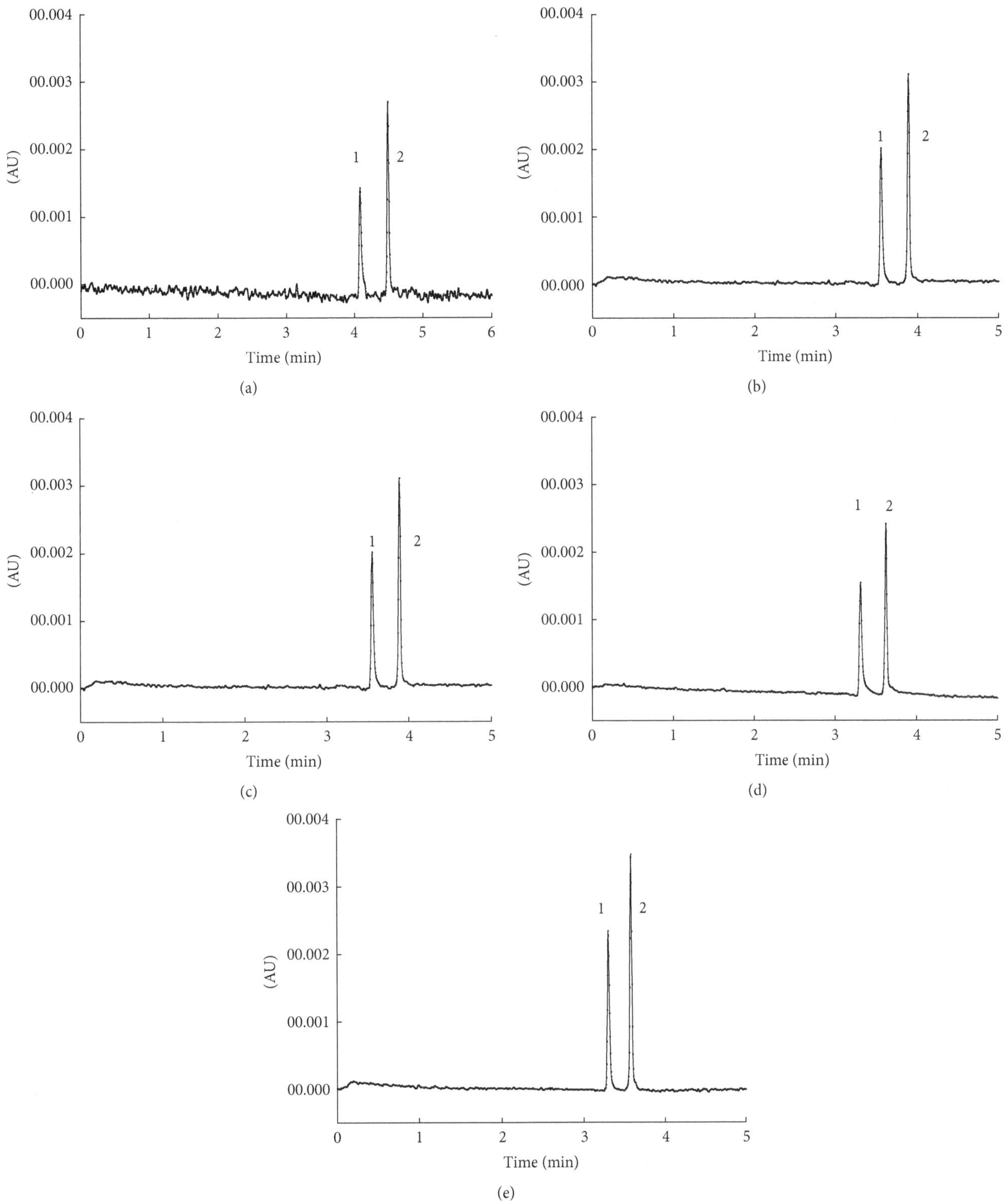

FIGURE 5: Effect of different temperature on the separation. Chromatogram identification: (a) 20°C; (b) 22°C; (c) 25°C; (d) 27°C; (e) 30°C. Separation conditions and peak identification were as in Figure 2.

Table 1: The results of orthogonal experiment.

	Ion concentration (mmol/L)	pH	Separation voltage (kV)	Temperature (°C)	Resolution (R)	Analysis time (min)
1	30	8.6	20	22	2.1	5.38
2	30	8.8	22	25	1.8	5.02
3	30	9.0	25	27	1.7	3.46
4	35	8.6	22	27	1.0	4.82
5	35	8.8	25	22	1.9	3.58
6	35	9.0	20	25	2.8	5.70
7	40	8.6	25	25	1.8	3.52
8	40	8.8	20	27	1.1	5.28
9	40	9.0	22	22	2.0	5.17

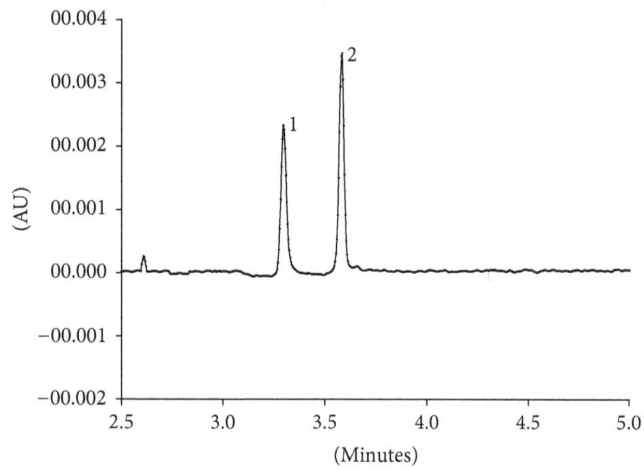

Figure 6: Electrophoretogram of stand solution (20 μg/mL). Separation conditions: running buffer, 35 mmol/L H_3BO_3-$Na_2B_4O_7$ (pH 8.8); separate voltage, 25 kV; operating temperature, 22°C; the detection wavelength, 275 nm; injection, 0.5 psi, 4 s. Peak identification was as in Figure 2.

preconcentration system. A favorable eluent not only should efficiently elute the target ions from the SPE column but also should not interfere with the determination of the targets. In this study, we systemically investigated the elution ability of methanol, acetonitrile, dichloromethane, n-hexane, and 4% ammonia-methanol (Table 2). The results showed that n-hexane and dichloromethane could not or partially elute the targets, which cannot meet the requirement of the experiment. When using acetonitrile and methanol as eluent, the elution efficiency was significantly improved. Particularly, the elution efficiency of 4% ammonia-methanol for the two drugs was achieved up to 98.5%–99.8%. As a result, 4% ammonia-methanol was selected as the eluent.

3.3. Method Evaluation. Under the optimal conditions, the analytical parameters of the current method were studied by standard solutions of the DIF and SAR, and the results were summarized in Table 3. Figure 6 showed an electrophoretogram of standard solution (20 μg/mL), obtained by adopting the previous extraction procedure and capillary electrophoresis conditions. The analytical calibration curves

Table 2: Recoveries of fluoroquinolone extracted with different solvent (%).

Eluent	SAR		DIF	
	Recovery (%)	SD	Recovery (%)	SD
Dichloromethane	54.0	0.95	58.0	0.70
Acetonitrile	85.4	0.98	82.0	1.41
Methanol	92.2	1.23	89.1	0.53
4% ammonia-methanol	98.5	1.18	99.8	1.45

of SAR and DIF were in the range of 1–20 μg/mL (R^2 = 0.9977) and 0.5–20 μg/mL (R^2 = 0.9936), respectively. Correlation coefficient r test was employed to test the linearity of the calibration curve, showing that the linearity of the proposed method was reliable at 99.9% confidence level. The limit of detection (LOD) was found to be 0.8 μg/mL for SAR and 0.3 μg/mL for DIF, defined as the concentration of target for which the signal-to-noise ratio was equal to three (S/N = 3). The limit of quantitation (LOQ) defined as the lower limit of linearity range for the obtained calibration curve was 1 μg/mL for SAR and 0.5 μg/mL for DIF. Herein, F test was used to test the interday differences of LOD and LOQ. The result showed that there are no significant differences at 95% confidence level. Besides, the reproducibility of the peak area for 8 μg/mL SAR and DIF was tested. The RSDs for interday and intraday assay were 7.8% and 4.8%, respectively, which indicates that the current method gives a repeatable quantification of DIF and SAR.

3.4. Sample Analysis. To evaluate the practicability of the method, blank samples were spiked with SAR and DIF standard solution at three different concentrations (100, 200, and 400 μg/kg) and then analyzed three times under the optimal conditions. Details were shown in Table 4. The recoveries were obtained in the range of 82.4–92.9% with SD (n = 3) lower than 6.0%. The LODs of SAR and DIF in beef were obtained at 100 μg/kg and 50 μg/kg, respectively. These results indicated that the current method was reliable and possessed potential application in the analysis of real samples.

TABLE 3: Analytical parameters of the current method.

Parameter	Sarafloxacin	Difloxacin
Calibration curves	$Y = 318.50X - 232.95$	$Y = 517.24X - 85.493$
Correlation coefficient (r^a, $n = 5$)	0.9988	0.9967
Linearity range (μg/mL)	1–20	0.5–20
LOD (μg/mL) (F^b, $n = 5$)	0.8 (1.60)	0.3 (3.40)
LOQ (μg/mL) (F^b, $n = 5$)	1 (2.12)	0.5 (4.00)

[a]Linearity test of the calibration curve. theoretical values at 99.9% confidence level: $r_{99.9\%,3} = 0.991$.
[b]Interday differences test, theoretical values at 95% confidence level: $F_{95\%,4} = 6.39$.

TABLE 4: Recovery study of sarafloxacin and difloxacin in spiked sample by CE (mean ± SD, $n = 3$).

Samples	Targets	Found (μg/kg)	Spiked level (μg/kg)	Detected (μg/kg)	Recovery (%)
Beef	Sarafloxacin	0	100	91.7 ± 1.5	91.7 ± 1.5
			200	185.8 ± 6.3	92.9 ± 3.2
			400	342.7 ± 5.1	85.7 ± 1.3
	Difloxacin	0	100	90.1 ± 5.3	90.1 ± 5.3
			200	168.0 ± 4.9	84.0 ± 2.5
			400	329.4 ± 9.4	82.4 ± 2.3

3.5. Methodology Evaluation. To ensure human food safety, the European Union (EU) has set maximum residue limits (MRLs) of veterinary medicinal products in foodstuffs of animal origin. However, no MRLs have been defined for SAR in beef. In this study, CE-SPE which has been validated with satisfying results [2] was introduced in determination of DIF and SAR in beef. In order to obtain a better analysis performance, separation conditions were systemically optimized by orthogonal experiment. In view of the analysis time and resolution, the optimal capillary electrophoresis conditions were established. Compared to previous report [2], the detection time was shortened to 3.58 min in this paper. The LODs of SAR and DIF in beef were obtained at 100 μg/kg and 50 μg/kg, respectively. It is demonstrated that the proposed method allows for the rapid and sensitive detection of SAR and DIF in beef at a concentration below the MRL of EU.

4. Conclusion

In this work, a rapid and simple method was established allowing the determination of DIF and its metabolite SAR in beef by CE coupled with SPE. The parameters affecting the performance of developed method were systematically studied. Under the optimal conditions, good recovery, precision, sensitivity, and analysis speed were obtained. These results indicated that the proposed method might serve as a reliable tool for the effective determination of SAR and DIF in beef and other food samples.

Conflicts of Interest

The authors declare that there are no conflicts of interest regarding the publication of this paper.

Acknowledgments

This work was financially supported by the Funds of Tianjin Agricultural Science and Technology Cooperation Project (0804130) and Tianjin Municipal College Students Innovation and Entrepreneurship Training Project (201410061070).

References

[1] B. N. Tochi, J. Peng, S. Song, L. Liu, H. Kuang, and C. Xu, "Determination of sarafloxacin and its analogues in milk using an enzyme-linked immunosorbent assay based on a monoclonal antibody," *Analytical Methods*, vol. 8, no. 7, pp. 1626–1636, 2016.

[2] D. Barrón, E. Jiménez-Lozano, S. Bailac, and J. Barbosa, "Determination of difloxacin and sarafloxacin in chicken muscle using solid-phase extraction and capillary electrophoresis," *Journal of Chromatography B*, vol. 767, no. 2, pp. 313–319, 2002.

[3] D. C. Hooper, "Mode of action of fluoroquinolones," *Drugs*, vol. 58, supplement 2, pp. 6–10, 1999.

[4] Z. Cao, R. Liu, and B. Yang, "Potential toxicity of sarafloxacin to catalase: spectroscopic, ITC and molecular docking descriptions," *Spectrochimica Acta. Part A: Molecular and Biomolecular Spectroscopy*, vol. 115, pp. 457–463, 2013.

[5] J. A. Murillo Pulgarín, A. Alañón Molina, F. Martínez Ferreras, and P. Fernández López, "Rapid and simple determination of sarafloxacin in egg by time-resolved chemiluminescence," *Food Analytical Methods*, vol. 6, no. 4, pp. 1153–1161, 2013.

[6] U. Hubicka, P. Zmudzki, B. Zuromska-Witek, P. Zajdel, M. Pawłowski, and J. Krzek, "Separation and characterization of ciprofloxacin, difloxacin, lomefloxacin, norfloxacin, and ofloxacin oxidation products under potassium permanganate treatment in acidic medium by UPLC-MS/MS," *Talanta*, vol. 109, pp. 91–100, 2013.

[7] L. Lucatello, P. Cagnardi, F. Capolongo, C. Ferraresi, F. Bernardi, and C. Montesissa, "Development and validation of an LC–MS/MS/MS method for the quantification of fluoroquinolones in several matrices from treated Turkeys," *Food Control*, vol. 48, pp. 2–11, 2015.

[8] M. Lombardo-AGüí, A. M. García-Campaña, C. Cruces-Blanco, and L. Gámiz-Gracia, "Determination of quinolones in fish by ultra-high performance liquid chromatography with fluorescence detection using QuEChERS as sample treatment," *Food Control*, vol. 50, pp. 864–868, 2015.

[9] G.-Y. Fan, R.-S. Yang, J.-Q. Jiang et al., "Development of a class-specific polyclonal antibody-based indirect competitive ELISA for detecting fluoroquinolone residues in milk," *Journal of Zhejiang University: Science B*, vol. 13, no. 7, pp. 545–554, 2012.

[10] X. Zhang and Z. Zhang, "Capillary electrophoresis-based immunoassay for the determination of brevetoxin-B in shellfish using electrochemical detection," *Journal of Chromatographic Science*, vol. 51, no. 2, pp. 107–111, 2013.

[11] C. C. Liu, X. Feng, H. L. Qian, G. Z. Fang, and S. Wang, "Determination of norfloxacin in food by capillary electrophoresis immunoassay with laser-induced fluorescence detector," *Food Analytical Methods*, vol. 8, no. 3, pp. 596–603, 2015.

[12] J. Cui, K. Zhang, Q. Huang, Y. Yu, and X. Peng, "An indirect competitive enzyme-linked immunosorbent assay for determination of norfloxacin in waters using a specific polyclonal antibody," *Analytica Chimica Acta*, vol. 688, no. 1, pp. 84–89, 2011.

[13] W. Sheng, Y. Li, X. Xu, M. Yuan, and S. Wang, "Enzyme-linked immunosorbent assay and colloidal gold-based immunochromatographic assay for several (fluoro)quinolones in milk," *Microchimica Acta*, vol. 173, no. 3-4, pp. 307–316, 2011.

[14] Y. Z. Liu, G. X. Zhao, P. Wang, J. Liu, H. C. Zhang, and J. P. Wang, "Production of the broad specific monoclonal antibody against sarafloxacin for rapid immunoscreening of 12 fluoroquinolones in meat," *Journal of Environmental Science and Health. Part B*, vol. 48, no. 2, pp. 139–146, 2013.

[15] X. W. Zhang and Z. X. Zhang, "Capillary electrophoresis-based immunoassay with electrochemical detection as rapid method for determination of saxitoxin and decarbamoylsaxitoxin in shellfish samples," *Journal of Food Composition and Analysis*, vol. 28, no. 1, pp. 61–68, 2012.

[16] Z. Zhang, X. Li, A. Ge, F. Zhang, X. Sun, and X. Li, "High selective and sensitive capillary electrophoresis-based electrochemical immunoassay enhanced by gold nanoparticles," *Biosensors and Bioelectronics*, vol. 41, no. 1, pp. 452–458, 2013.

[17] C. Liu, G. Fang, Q. Deng, Y. Zhang, J. Feng, and S. Wang, "Determination of metolcarb in food by capillary electrophoresis immunoassay with a laser-induced fluorescence detector," *Electrophoresis*, vol. 33, no. 9-10, pp. 1471–1476, 2012.

[18] G. Fang, J. Feng, Y. Yan, C. Liu, and S. Wang, "Highly selective determination of chrysoidine in foods through a surface molecularly imprinted sol-gel polymer solid-phase extraction coupled with HPLC," *Food Analytical Methods*, vol. 7, no. 2, pp. 345–351, 2014.

7

Discrimination and Geographical Origin Prediction of *Cynomorium songaricum* Rupr. from Different Growing Areas in China by an Electronic Tongue

Jiaji Ding [1,2] **Caimei Gu,** [2] **Linfang Huang** [2] **and Rui Tan** [1]

[1] *College of Medcine, Southwest Jiaotong University, Chengdu 610031, China*
[2] *Institute of Medicinal Plant Development, Chinese Academy of Medical Sciences & Peking Union Medical College, Beijing 100193, China*

Correspondence should be addressed to Linfang Huang; 15801545922@139.com and Rui Tan; tanrui@home.swjtu.edu.cn

Academic Editor: Jaroon Jakmunee

Cynomorium songaricum Rupr. is a well-known and widespread plant in China. It has very high medicinal values in many aspects. The study aimed at discriminating and predicting *C. songaricum* from major growing areas in China. An electronic tongue was used to analyze *C. songaricum* based on flavor. Discrimination was achieved by principal component analysis and linear discriminant analysis. Moreover, a prediction model was established, and *C. songaricum* was classified by geographical origins with 100% degree of accuracy. Therefore, the identification method presented will be helpful for further study of *C. songaricum*.

1. Introduction

Cynomorium songaricum Rupr. of the family *Cynomoriaceae* is a desert, holoparasitic perennial plant found in China, Mongolia, Iran, and Afghanistan [1, 2]. In China, *C. songaricum* grows in Xinjiang, Inner Mongolia, Ningxia, Qinghai, and Gansu [3] (Figure 1). *C. songaricum*, called *Suo Yang* in China, is a known food, nutrient, and a tonic herb for improving kidney and immunity function and treating constipation [4, 5]. This plant is one of the most popular herbs in the world and is documented in some famous medicinal works [1]. Various compounds, including flavonoids, organic acids, steroids, saccharides, terpenoids, phloroglucinol adducts, phenylpropanoids, and other types of compounds, have been isolated from *C. songaricum* to date [1, 6]. These chemical compounds exhibit numerous biological activities, including antiapoptosis, antifatigue, antioxidant, antiosteoporotic, antiaging, antidiabetic, anti-HIV protease, anti-HCV protease, and fertility promotion [1, 7–9].

In our previous work, we presented that the chemical constituents of *C. songaricum* from different producing areas vary, thus affecting the quality of the plant [2]. Genuine medicinal herb, which means Daodi yaocai in Chinese, is a unique definition in traditional Chinese medicine. Medicinal herbs growing in a specific place exhibit high quality [10]. Currently, chromatographic herbal fingerprints have become one of the most applied quality control tools for similarity analyses of herbal medicines [11]. However, it costs a relatively long time. Thus, a more convenient way for identification and quality control of herbs is needed.

Electronic tongues are analytical systems formed from an array of electrochemical sensors combined with data-processing tools intended to interpret electrochemical signals. Similar to human receptors, the sensors of an electronic tongue undergo a series of reactions. While the generated reactions differ from one another, the information acquired from each sensor is complementary. Then, the results combined by the sensors generate a unique fingerprint that can reflect the macroscopic characteristics of samples. In biological mechanisms, gustatory signals are transducted by brain nerves in the form of electric signals. Electronic tongue sensors approach flavors similarly, given that electric signals

FIGURE 1: The distribution of *Cynomorium songaricum* Rupr. in China.

are generated with potentiometric variations. The perception and recognition of taste quality are based on the recognition or building of activated sensory nerve patterns in the brain and the gustation fingerprint of a product. This step is accomplished by the statistical software of the electronic tongue that can translate sensor data into taste patterns [12–19]. In the recent years, electronic tongues have been commonly used to analyze food and beverages, given their advantages of short response time, strong objectiveness, human safety, and repeatability [20]. As for some herbs like *C. songaricum*, they have different tastes and flavors according to different places of origin and are ready to eat. Based on this, the simplicity and convenience of electronic tongues could be used in the analysis of the herbs.

In this work, we first developed a method to discriminate and predict the geographical origin of *C. songaricum* from different growing areas in China by using an electronic tongue. Pattern recognition techniques, including principal component analysis (PCA) and linear discriminant analysis (LDA), were used for data analysis in this research. In addition, this study provided a simple approach for identifying the geographical origins of *C. songaricum*, and the acquired information can be used for evaluating the quality of *C. songaricum* growing in China.

2. Materials and Methods

2.1. Samples. C. songaricum samples were collected from different areas in China (Kashgar in Xinjiang, Tarbagatay in Xinjiang, Jiuquan in Gansu, Guyuan in Ningxia, Hotan in Xinjiang, Haixi in Qinghai, Ejin Banner in Inner Mongolia, and Alxa Left Banner in Inner Mongolia) (Table 1). All of the samples were authenticated by Professor Linfang Huang in the Institute of Medicinal Plant Development, Chinese Academy of Medical Sciences, and Peking Union Medical College, Beijing, China.

2.2. Instrument. The electronic tongue system (taste sensing system Astree II, France) consists of a reference electrode and seven liquid sensors (ZZ, JE, BB, CA, GA, HA, and JB) with a cross-selection function, a fully automated sample injector, and a personal computer with a software for sample injection, data acquisition, and chemometric analysis.

2.3. Experimental Procedures. Pieces of each sample (10 g) were placed in a beaker, soaked with 200 mL of pure water for 30 min, and then decocted for 30 min. The solution was filtered immediately. The residue was processed according to the abovementioned method twice. Afterward, all filtrates were combined. The obtained solution was placed into the special beaker of the electronic tongue and detected at room temperature.

Each sensor collected data from each sample for 120 s and was cleaned for 10 s. Then, data were recorded by the data acquisition system. All assays were carried out in triplicate.

2.4. Pattern Recognition. In this paper, PCA and LDA were used to differentiate *C. songaricum* originating from different places.

Table 1: Sample list of *Cynomorium songaricum* Rupr.

Sample	Place of origin
Discrimination	
KX-1	Kashgar, Xinjiang
KX-2	Kashgar, Xinjiang
TX-1	Tarbagatay, Xinjiang
TX-2	Tarbagatay, Xinjiang
TX-3	Tarbagatay, Xinjiang
JG-1	Jiuquan, Gansu
JG-2	Jiuquan, Gansu
GN-1	Guyuan, Ningxia
GN-2	Guyuan, Ningxia
GN-3	Guyuan, Ningxia
HX-1	Hotan, Xinjiang
HX-2	Hotan, Xinjiang
HX-3	Hotan, Xinjiang
HQ-1	Haixi, Qinghai
HQ-2	Haixi, Qinghai
HQ-3	Haixi, Qinghai
EBIM-1	Ejin Banner, Inner Mongolia
EBIM-2	Ejin Banner, Inner Mongolia
EBIM-3	Ejin Banner, Inner Mongolia
ALBIM-1	Alxa Left Banner, Inner Mongolia
ALBIM-2	Alxa Left Banner, Inner Mongolia
Prediction	
KX-3	Kashgar, Xinjiang
KX-4	Kashgar, Xinjiang
TX-4	Tarbagatay, Xinjiang
TX-5	Tarbagatay, Xinjiang
JG-3	Jiuquan, Gansu
JG-4	Jiuquan, Gansu
GN-4	Guyuan, Ningxia
GN-5	Guyuan, Ningxia
HX-4	Hotan, Xinjiang
HX-5	Hotan, Xinjiang
HQ-4	Haixi, Qinghai
HQ-5	Haixi, Qinghai
EBIM-4	Ejin Banner, Inner Mongolia
EBIM-5	Ejin Banner, Inner Mongolia
ALBIM-3	Alxa Left Banner, Inner Mongolia
ALBIM-4	Alxa Left Banner, Inner Mongolia

PCA is a multivariate statistical method that reduces the dimensionality of data while retaining most of the variation in the data [21]. This approach was created before World War II but became widely used during the "Quantitative Revolution" in the 1960s [22]. New linear combinations of variables are created to accomplish the reduction. The combinations, called principal components, characterize the objects studied and satisfy certain statistical and mathematical conditions. Thus, samples can be displayed by few variables, and assessment of similarities and differences among samples is simplified. Thus, PCA is a suitable method to discriminate different samples and is extensively applied in food and drug analysis.

LDA is another commonly used technique for data discrimination and dimensionality reduction. LDA is strongly linked to regression analysis and analysis of variance (ANOVA), which also aims at expressing one dependent variable as a combination of other measurements or features [23]. The method maximizes the ratio of the between-class distance to the within-class distance to guarantee maximum discrimination. LDA has been used in numerous applications, such as image retrieval, microarray data classification, face recognition, and food and beverage discrimination [24–27].

3. Results and Discussion

3.1. Radar Map. Figure 2 shows the radar map of samples from different places in China. The sensor response of the electronic tongue varied with the change in geographical origins. Evidently, sensors BB, CA, and ZZ show strong signals to the samples. In particular, sensor BB exhibits the strongest response to the samples. Figure 2(a) shows the distinction among all samples from different places clearly. Signals from different samples show a considerable difference. Figure 2(b) illustrates the signals of every sample separately. Samples KX, TX, and HX are different from others, with the signals of sensor ZZ of these samples not exceeding 1000. Shapes of the radar maps of other samples are similar.

3.2. Principal Component Analysis. The first discrimination model was established using PCA to visualize the different *C. songaricum* groups where possible. The accumulated explained variance was 89.6%, which was distributed in 79.5% (PC1) and 10.1% (PC2). Figure 3 shows the results of PCA score plot, and several trends are observed. Eight types of *C. songaricum* samples can be classified in general. Moreover, *C. songaricum* samples from Xinjiang are discriminated clearly between samples from other provinces. Similar samples appear in the same location of the graph. Thus, *C. songaricum* samples from Gansu, Ningxia, Qinghai, and Inner Mongolia are similar. The chemical constituents of these samples may be similar as well.

3.3. Linear Discriminant Analysis. Figure 4 shows the dispersion of *C. songaricum* samples by the LDA model. Compared with the PCA model, the LDA model shows a clearer discrimination among the eight types of *C. songaricum*. The explained variances by each discriminant function (DF) were 96.0% (DF1) and 2.9% (DF2). Each group of *C. songaricum* samples can be distinctly classified with others. As a result, the LDA model is a superior method to discriminate *C. songaricum* from different growing areas in China.

Given the good discrimination feature of the model, we used the LDA model to predict the geographical origin of unknown *C. songaricum* samples. As shown in Table 2, the prediction model can classify *C. songaricum* by geographical origins with 100% degree of accuracy.

4. Conclusions

We applied an electronic tongue to classify and predict *C. songaricum* samples from different places of origin. PCA and LDA were used for discrimination. The LDA model shows a clearer discrimination than the PCA model. The eight

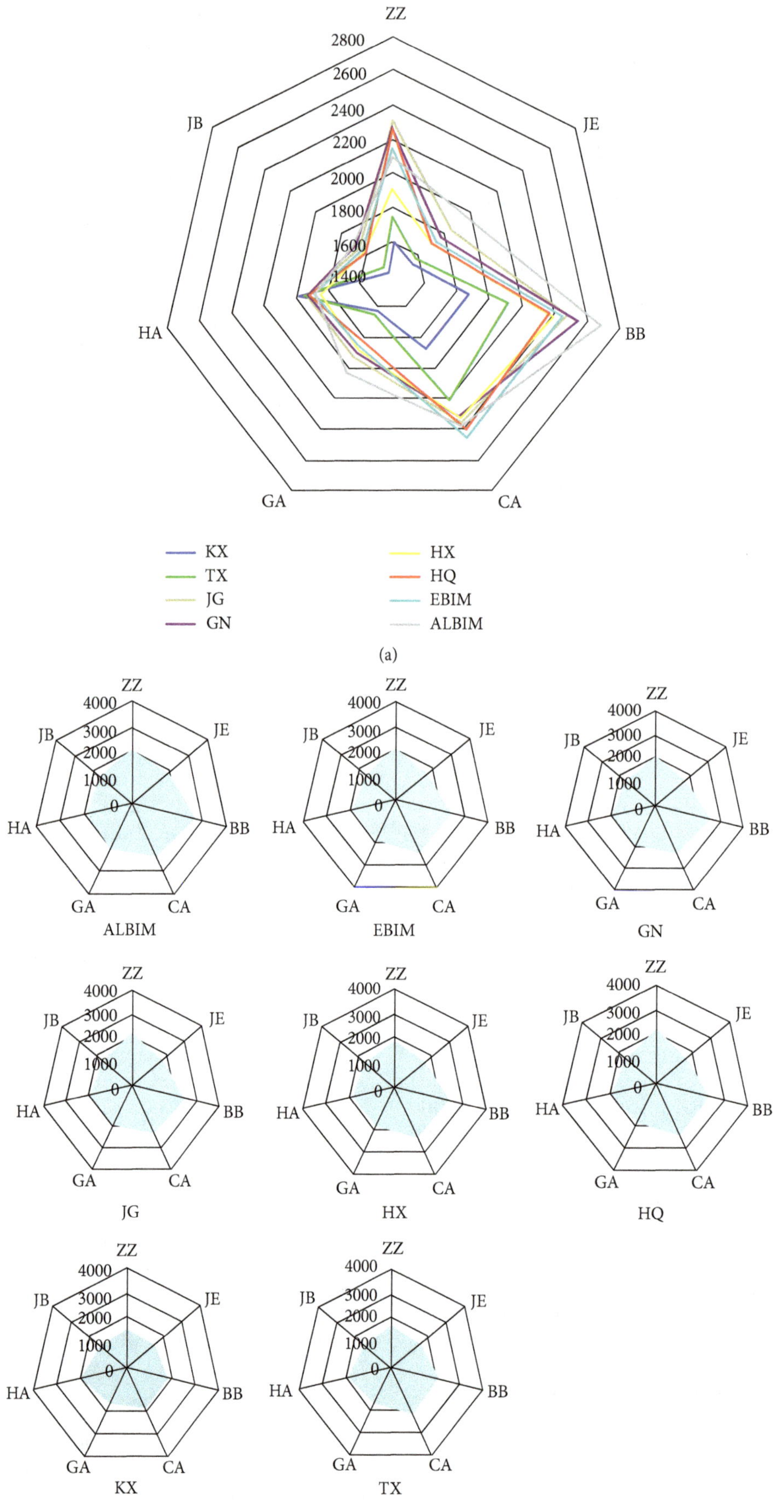

FIGURE 2: The radar map of *C. songaricum* from different producing areas in China.

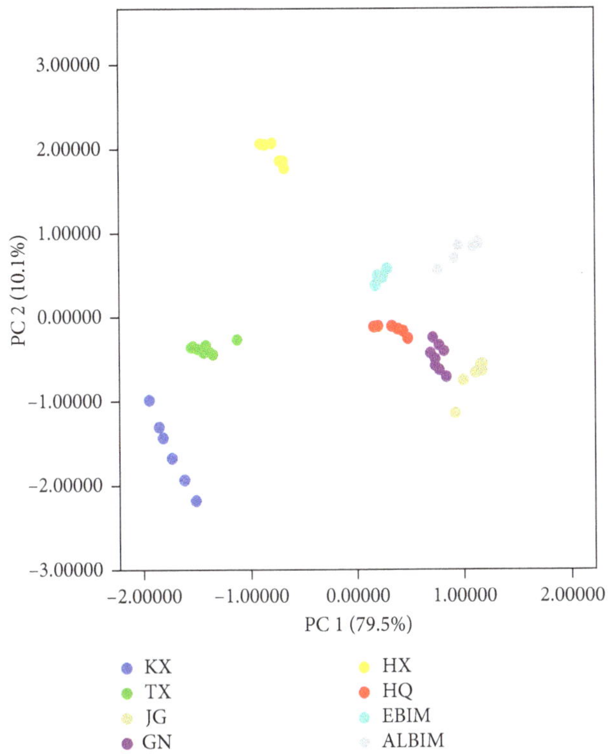

FIGURE 3: Discrimination of PCA plots for different *C. songaricum* samples.

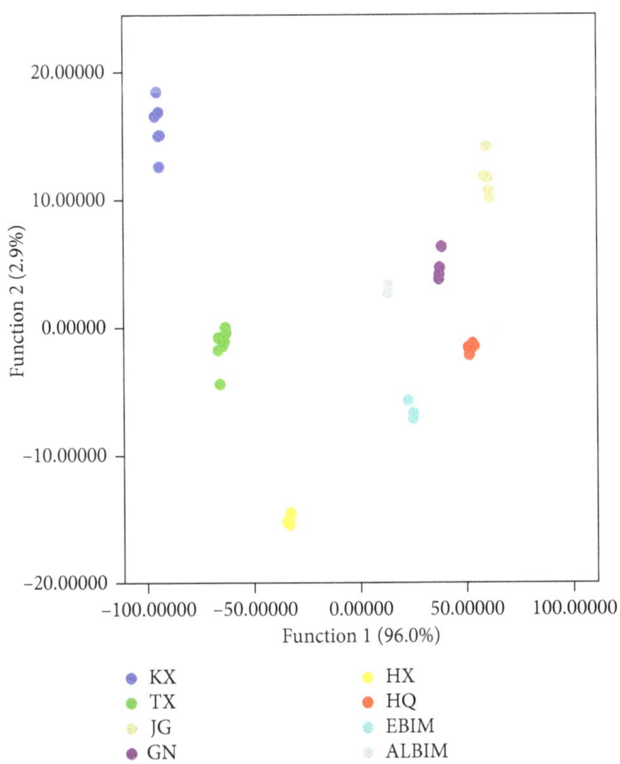

FIGURE 4: Discrimination of LDA plots for different *C. songaricum* samples.

groups of *C. songaricum* samples can be classified precisely using the LDA model. On this basis, we used the LDA model to predict the geographical origin of several unknown *C. songaricum* samples. A prediction model with 100% degree of accuracy was achieved. We present for the first time a method for the discrimination and geographical origin prediction of *C. songaricum* from different growing areas in China according to their flavor by an electronic tongue. The operations of data acquisition and processing are simpler and more convenient than the traditional chemical methods. The acquired information can be used for evaluating the quality of *C. songaricum* growing in China according to our previous work. Moreover, the identification and quality analysis method presented by us will be helpful for further study of *C. songaricum*. Further efforts should be focused on investigating the connection between the flavor and chemical constituents of *C. songaricum* samples and correlating electronic tongue signals with human perceptions of taste.

TABLE 2: Confusion matrix for LDA prediction method of *C. songaricum* samples from different producing areas.

Actual	Predicted							
	KX	TX	JG	GN	HX	HQ	EBIM	ALBIM
KX	6							
TX		6						
JG			6					
GN				6				
HX					6			
HQ						6		
EBIM							6	
ALBIM								6

Conflicts of Interest

The authors declare that they have no conflicts of interest regarding the publication of this article.

Acknowledgments

This study was supported by grants from the National Natural Science Foundation of China (81473315), Public Welfare Scientific Research Project of State Administration of Traditional Chinese Medicine (201507004-2-1), CAMS Innovation Fund for Medical Sciences (no. 2016-I2M-3-015), and State Administration of Foreign Experts Affairs P.R. China (no. T2017052).

References

[1] Z. Cui, Z. Guo, J. Miao et al., "The genus *Cynomorium* in China: an ethnopharmacological and phytochemical review," *Journal of Ethnopharmacology*, vol. 147, no. 1, pp. 1–15, 2013.

[2] X. Zhang, C. Gu, B. Ahmad, and L. Huang, "Optimization of extract method for *Cynomorium songaricum* Rupr. By response surface methodology," *Journal of Analytical Methods in Chemistry*, vol. 2017, Article ID 6153802, 7 pages, 2017.

[3] J. Chen, H. Wong, and K. Ko, "Ursolic acid-enriched herba cynomorii extract induces mitochondrial uncoupling and glutathione redox cycling through mitochondrial reactive

oxygen species generation: protection against menadione cytotoxicity in H9c2 cells," *Molecules*, vol. 19, no. 2, pp. 1576–1591, 2014.

[4] S. Jin, Eerdunbayaer, A. Doi et al., "Polyphenolic constituents of *Cynomorium songaricum* Rupr. And antibacterial effect of polymeric proanthocyanidin on *Methicillin-resistant Staphylococcus aureus*," *Journal of Agricultural and Food Chemistry*, vol. 60, no. 29, pp. 7297–7305, 2012.

[5] S. Tuvaanjav, H. Shuqin, M. Komata et al., "Isolation and antiviral activity of water-soluble *Cynomorium songaricum* Rupr. Polysaccharides," *Journal of Asian Natural Products Research*, vol. 18, no. 2, pp. 159–171, 2016.

[6] H. C. Meng, S. Wang, Y. Li, Y. Y. Kuang, and C. M. Ma, "Chemical constituents and pharmacologic actions of *Cynomorium* plants," *Chinese Journal of Natural Medicines*, vol. 11, no. 4, pp. 321-329, 2013.

[7] C. Ma, N. Nakamura, H. Miyashiro, M. Hattori, and K. Shimotohno, "Inhibitory effects of constituents from *Cynomorium songaricum* and related triterpene derivatives on HIV-1 protease," *Chemical & Pharmaceutical Bulletin*, vol. 47, no. 2, pp. 141–145, 1999.

[8] C. M. Ma, Y. Wei, Z. G. Wang, and M. Hattori, "Triterpenes from *Cynomorium songaricium*—analysis of HCV protease inhibitory activity, quantification, and content change under the influence of heating," *Journal of Natural Medicines*, vol. 63, no. 1, pp. 9–14, 2009.

[9] F. R. Yu, Y. Liu, Y. Z. Cui et al., "Effects of a flavonoid extract from *Cynomorium songaricum* on the swimming endurance of rats," *American Journal of Chinese Medicine*, vol. 38, no. 1, pp. 65–73, 2010.

[10] S. Zheng, W. Ren, and L. Huang, "Geoherbalism evaluation of radix *Angelica sinensis* based on electronic nose," *Journal of Pharmaceutical and Biomedical Analysis*, vol. 105, pp. 101–106, 2015.

[11] M. Goodarzi, P. J. Russell, and Y. V. Heyden, "Similarity analyses of chromatographic herbal fingerprints: a review," *Analytica Chimica Acta*, vol. 804, pp. 16–28, 2013.

[12] X. Cetó, C. Apetrei, M. del Valle, and M. L. Rodríguez-Méndez, "Evaluation of red wines antioxidant capacity by means of a voltammetric e-tongue with an optimized sensor array," *Electrochimica Acta*, vol. 120, pp. 180–186, 2014.

[13] X. Cetó, J. Capdevil, A. Puig-Pujor, and M. del Valle, "Cava wine authentication employing a voltammetric electronic tongue," *Electroanalysis*, vol. 26, no. 7, pp. 1504–1512, 2014.

[14] X. Cetó, J. M. Gutiérrez, M. Gutiérrez et al., "Determination of total polyphenol index in wines employing a voltammetric electronic tongue," *Analytica Chimica Acta*, vol. 732, pp. 172–179, 2012.

[15] M. del Valle, "Electronic tongues employing electrochemical sensors," *Electroanalysis*, vol. 22, pp. 1539–1555, 2010.

[16] L. Escuder-Gilabert and M. Peris, "Review: highlights in recent applications of electronic tongues in food analysis," *Analytica Chimica Acta*, vol. 665, no. 1, pp. 15–25, 2010.

[17] J. M. Gutiérrez, Z. Haddi, A. Amari et al., "Hybrid electronic tongue based on multisensor data fusion for discrimination of beers," *Sensors and Actuators B: Chemical*, vol. 177, pp. 989–996, 2013.

[18] F. Winquist, "Voltammetric electronic tongues—basic principles and applications," *Microchimica Acta*, vol. 163, no. 1-2, pp. 3–10, 2008.

[19] Y. Vlasov, A. Legin, A. Rednitskaya, C. Di Natale, and A. D'Amico, "Nonspecific sensor arrays ("electronic tongue") for chemical analysis of liquids (IUPAC Technical Report)," *Pure and Applied Chemistry*, vol. 77, no. 11, pp. 1965–1983, 2009.

[20] A. K. Deisingh, D. C. Stone, and M. Thompson, "Applications of electronic noses and tongues in food analysis," *International Journal of Food Science and Technology*, vol. 39, no. 6, pp. 587–604, 2004.

[21] M. Ringner, "What is principal component analysis," *Nature Biotechnology*, vol. 26, no. 3, pp. 303-304, 2008.

[22] A. Maćkiewicz and W. Ratajczak, "Principal components analysis (PCA)," *Computers & Geosciences*, vol. 19, no. 3, pp. 303–342, 1993.

[23] R. A. Fisher, "The use of multiple measurements in taxonomic problems," *Annals of Eugenics*, vol. 7, no. 2, pp. 179–188, 1936.

[24] P. N. Belhumeur, J. P. Hespanha, and D. J. Kriegman, "Eigenfaces vs. Fisherfaces: recognition using class specific linear projection," *IEEE Transactions on Pattern Analysis and Machine Intelligence*, vol. 19, no. 7, pp. 711–720, 1997.

[25] S. Dudoit, J. Fridltand, and T. P. Speed, "Comparison of discrimination methods for the classification of tumors using gene expression data," *Journal of the American Statistical Association*, vol. 97, no. 457, pp. 77–87, 2002.

[26] D. L. Swets and J. J. Weng, "Using discriminant eigenfeatures for image retrieval," *IEEE Transactions on Pattern Analysis and Machine Intelligence*, vol. 18, no. 8, pp. 831–836, 1996.

[27] C. A. Blanco, R. de la Fuenta, I. Caballero, and M. L. Roderíguez-Méndez, "Beer discrimination using a portable electronic tongue based on screen-printed electrodes," *Journal of Food Engineering*, vol. 157, pp. 57–62, 2015.

8

Detection of Polycyclic Aromatic Hydrocarbons in Water Samples by Annular Platform-Supported Ionic Liquid-Based Headspace Liquid-Phase Microextraction

Xiaojie Sun,[1] Jie Tan,[1,2] Haiyan Ding,[1] Xiaojie Tan,[3] Jun Xing,[4] Lihong Xing,[1] Yuxiu Zhai,[1] and Zhaoxin Li[1]

[1]*Key Laboratory of Testing and Evaluation for Aquatic Product Safety and Quality, Ministry of Agriculture and Rural Affairs, Yellow Sea Fisheries Research Institute, Chinese Academy of Fishery Sciences, Qingdao 266071, China*
[2]*Qingdao National Laboratory for Marine Science and Technology, Qingdao 266000, China*
[3]*The Affiliated Hospital of Qingdao University, Qingdao 266000, China*
[4]*Key Laboratory of Analytical Chemistry for Biology and Medicine, Ministry of Education, College of Chemistry and Molecular Sciences, Wuhan University, Wuhan 430072, China*

Correspondence should be addressed to Jie Tan; tanjie@ysfri.ac.cn and Zhaoxin Li; lizx@ysfri.ac.cn

Academic Editor: Luca Campone

In this paper, a new method of annular platform-supported headspace liquid-phase microextraction (LPME) was designed using ionic liquid as an extraction solvent, wherein extraction stability and efficiency were improved by adding an annular platform inside the extraction bottle. The ionic liquid 1-silicyl-3-benzylimidazolehexafluorophosphate was first synthesized and proved to be an excellent extraction solvent. Coupled with liquid chromatography, the proposed method was employed to analysis of polycyclic aromatic hydrocarbons (PAHs) in water and optimized in aspects of extraction temperature, extraction solvent volume, extraction time, pH, stirring rate, and salt effect of solution. The results indicated that this method showed good linearity ($R^2 > 0.995$) within 0.5 μg·L^{-1} to 1000 μg·L^{-1} for PAHs. The method was more suitable for extraction of volatile PAHs, with recoveries from 65.0% to 102% and quantification limits from 0.01 to 0.05 μg·L^{-1}. It has been successfully applied for detection of PAHs in seawater samples.

1. Introduction

PAHs have been included in the environmental control legislation by many agencies, such as the US Environmental Protection Agency (EPA) [1] and the European Union. The maximum admissible concentration of naphthalene (1200 ng·L^{-1}), anthracene (100 ng·L^{-1}), fluoranthene (100 ng·L^{-1}), and benzo[a]pyrene (100 ng·L^{-1}) for superficial water was restricted by the European Union [2, 3]. To better evaluate the impact of PAHs on ecosystems, an efficient and sensitive analysis method must be developed. However, PAHs are found at very low levels (ng·L^{-1} or pg·L^{-1}) in environment samples. It is necessary to perform an enrichment step in sample preparation due to the lack of sensitivity obtained with conventional analytical instrumentation.

Sample pretreatment has been the bottleneck of analytical chemistry. In recent decades, LPME has been developed rapidly as a novel pretreatment technology [4–7], reducing analysis time and large amounts of toxic solvent. Due to the low miscibility with water [8–10], ionic liquids (ILs) have become a new extraction solvent for LPME [11,12] and been extensively applied in sample pretreatment prior to chromatography [13–24], overcoming the toxicity and reducing the consumption of ordinary organic solvents. In addition, characteristics of ILs can be tuned by a combination of different anions and cations for task-specific extraction of analytes in various solvent mediums.

According to extraction patterns, application of ILs in the LPME technology can be classified into the following: dispersive liquid-liquid microextraction (DLLME) [25], hollow-fiber LPME [26], and suspended single-drop microextraction (SDME). Dispersive LPME requires large extraction amount with limited extraction efficiency; hollow-fiber LPME features low recovery with poor reproducibility; SDME has drawn much concern owing to its high enrichment efficiency. However, research studies showed that the sample solution should be stirred continuously during SDME, so extraction droplets easily fall off, affecting extraction stability and reproducibility. Furthermore, the droplets' volume of SDME was restricted, although adding a silicon rubber tube [12], a polytetrafluoroethylene sleeve [15], or a plastic tube [27] on a syringe needle seriously limits the improvement of extraction efficiency. Therefore, the SDME device and pattern should be optimized to promote the development of SDME technology. Furthermore, ILs have played crucial roles in the aspect. 1-Octyl-3-methylimidazolehexafluorophosphate ([C_8MIM]PF_6) was first used as an SDME solvent coupled with liquid chromatography to extract PAHs and aniline compounds in water samples [12]. Since then, some other types of ILs were used as an extraction solvent for SDME [28].

In this study, an IL-based headspace LPME device was designed using the homemade IL 1-silicyl-3-benzylimidazolehexafluorophosphate ([SBnIM]PF_6) as an extraction solvent. The device was optimized by adding an annular platform inside the extraction bottle; then, IL was slowly pressed dropwise down to the annular platform by a syringe. The developed method was named "annular platform-supported IL-based headspace LPME" (APS-IL-LPME). The method altered the suspension state of extracted droplets, overcame the influence of droplet gravity, and significantly improved extraction stability and efficiency. Meanwhile, the large extraction platform can flexibly regulate addition of extraction solvent to control extraction capacity and enrichment efficiency. Considering that fluoranthene and naphthalene showed larger partition coefficients in the system of [C_nMIM][PF_6]/water [9,29], liquid chromatography was combined with APS-IL-LPME to determine PAHs in water samples. Simultaneously, the IL [SBnIM]PF_6 containing the benzyl group in the structure of cation was used as an extraction solvent, which will be more suitable for extraction of analytes including benzene rings. The influence of different parameters on extraction was investigated and optimized in aspects of extraction temperature, extraction solvent volume, extraction time, pH, stirring rate, and salt effect of solution. Moreover, the proposed method was compared with other preconcentration methods for determination of PAHs in water samples.

2. Materials and Methods

2.1. Apparatus and Reagents. A 2695 high-performance liquid chromatograph was equipped with an automatic sampler and a 2475 fluorescence detector (Waters Corp., US). An XW-80A vortex mixer (Shanghai Medical University Instrument Factory), a Milli-Q ultrapure water meter (Millipore Corp., US), and a Sorvall Biofuge Primo centrifuge (Thermo Fisher Scientific Corp., US) were used.

1-Benzylimidazole (99%) was purchased from Fluka Corp. (US). γ-Chloropropyl-trimethylsilane, naphthalene (99%, Nap), acenaphthene (98%, Acp), fluorene (98%, Flu), anthracene (98.5%, Ant), fluoranthene (98%, FL), and pyrene (98%, Pyr) were purchased from AccuStandard Corp. (US). Acetonitrile was purchased from Shanghai ANPEL Corp. (China). Other solvents and reagents were of analytical grade (SCRC, China). The water used in this experiment was redistilled. With a brown glass bottle, water samples were collected from seawater in Shazikou, Qingdao, Shandong Province, and preserved under 4°C. Concentrations of six kinds of standard PAH stock solutions were separately 1.0 $mg{\cdot}mL^{-1}$, which were dissolved in acetonitrile, and the mixed standard stock solution was diluted to 10.0 $\mu g{\cdot}mL^{-1}$ by acetonitrile. Subsequently, the solutions were stored at −20°C and diluted into standard working solutions of different concentrations by acetonitrile as needed.

2.2. The Device of APS-IL-LPME. The APS-IL-LPME device was self-manufactured in the lab. As shown in Figure 1, the whole device includes a thermostatic water bath ((2) in Figure 1), a heatable magnetic stirrer ((3) in Figure 1), and an extraction bottle with a sealing bottle cap ((1) in Figure 1). The thermostatic water bath is laid on the heatable magnetic stirrer, and the extraction bottle (made of glass) with a sealing bottle cap (made of polytetrafluoroethylene) is placed in the thermostatic water bath and fixed on an iron stand. The extraction bottle mainly consisted of the upper bottle body (ID 4.0 cm, height 4.5 cm; (5) in Figure 1), annular extraction platform (width 0.75 cm; (7) in Figure 1), and lower bottle body (ID 2.5 cm, height 4.0 cm; (6) in Figure 1), all of which are integrated as a whole from top to bottom. Inner diameter of the upper bottle body is larger than that of the lower bottle body, and the annular extraction platform is a platform with high inner and outer sides and a sunken middle part, which is used to bear the weight of extracted droplets.

2.3. Preparation of ILs. The IL [SBnIM]PF_6 was prepared as shown in Figure 2. Firstly, 1-benzylimidazolium was synthesized [30]. Then, [SBnIM]Cl (A) was prepared through a reaction between γ-chloropropyl-trimethylsilane and benzylimidazolium, which was according to the references [31,32]. Lastly, the chloride anion was exchanged for the hexafluorophosphate anion (PF_6^-) [33].

The general procedures for the synthesis of 1-benzylimidazole were as follows: formaldehyde water solution (36%, 8.35 g) and glyoxal water solution (32%, 18.1 g) were added to a 250 mL, three-necked flask provided with a stirrer and a reflux condenser. While the mixture was heated to 50°C, a mixture of benzylamine solution (10.7 g in 50 mL methanol) and ammonia solution (28%, 6.05 g) was added dropwise. After the mixture refluxed for an additional 4 h at 50°C, the methanol and water were removed under reduced

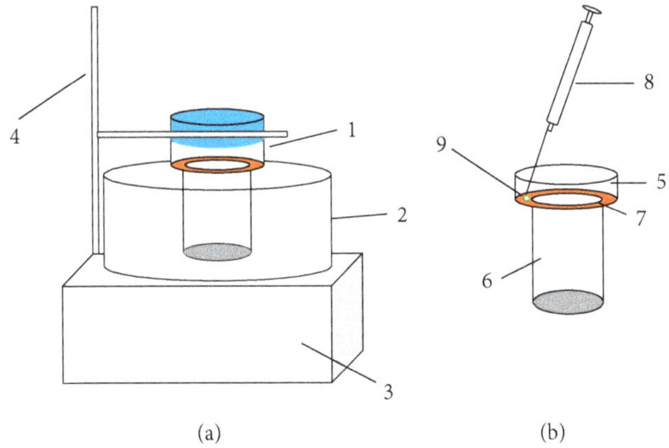

Figure 1: The device for APS-IL-LPME. (a) (1) Extraction bottle; (2) thermostatic water bath; (3) heating magnetic stirrer; (4) iron stand. (b) (5) The upper bottle; (6) the lower bottle; (7) annular extraction platform; (8) microsyringe; (9) ionic liquid droplets.

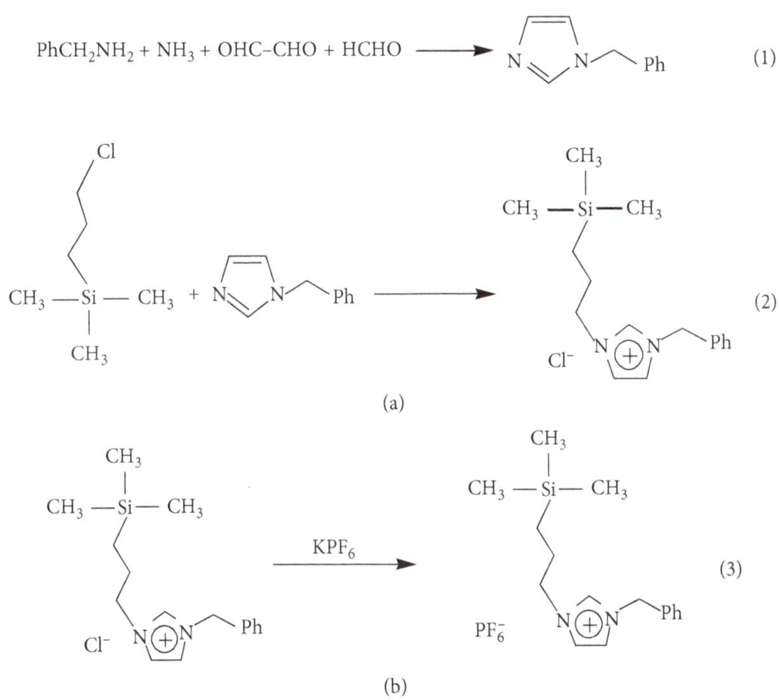

Figure 2: Synthetic routes of the ionic liquid $[SBnIM]PF_6$.

pressure. The residue was then concentrated and dried in a vacuum desiccator at 60°C. The product of 1-benzylimidazole was obtained with the yield about 75%.

Secondly, 1-benzylimidazole (0.05 mol, 4.60 g) was added dropwise to a solution of γ-chloropropyltrimethylsilane (0.05 mol, 17.3 g in 25 mL toluene). The solution was then refluxed for 72 h under N_2 atmosphere. The resultant was washed with deionized water to remove the residue of reactants. After concentrating and drying, the light-yellow transparent and viscous compound [SBnIM]Cl (A) was obtained and the yield was about 90%.

Then, the IL $[SBnIM]PF_6$ (B) was prepared through anion-exchange reactions. The process was as follows: the compound [SBnIM]Cl (A) was dissolved in water, and then 1 M equivalent of KPF_6 in water was added dropwise. After stirring for 24 h at room temperature, the resultant was dissolved in dichloromethane and washed with deionized water. The silver nitrate test was used to confirm absence of chloride. Finally, the product was concentrated and dried, forming a light-yellow viscous compound $[SBnIM]PF_6$ (B), and the yield reached 90%.

The final product $[SBnIM]PF_6$ was confirmed by proton nuclear magnetic resonance (^{1}HNMR) spectroscopy and Fourier transform infrared (FT-IR) spectroscopy. (i) ^{1}HNMR (D_2O; 400 MHz; δ (ppm)): 8.76 (s, 1H), 7.42–7.43 (d, 2H), 7.33–7.41 (m, 5H), 5.28 (s, 2H), 4.09 (t, 2H), 3.44 (m, 9H), 1.83 (m, 2H), and 0.56 (m, 2H). (ii) FT-IR (KBr, υ (cm^{-1})): 3127 and 3059 [υ(Ar–H)]; 2958, 2932, and 2868 [υ(C–H)]; 1558, 1494, and 1452 [υ(C=C), υ(C=N)]; 1156 [υ(Si–C)]; 1083 and 1023 [υ(Si–O)]; and 836 [υ(P–F)].

2.4. Extraction Procedures. First, 10 mL of the PAH-spiked water sample (5.0 μg·L^{-1}) was transferred to the extraction bottle, and sodium chloride was added to saturated concentration. The solution was inside the lower bottle body and remained below the annular extraction platform, which was headspace LPME. The sealing bottle cap was opened; the front end of the microsyringe (100 μL) was placed in the middle position of the annular extraction platform (as shown in Figure 1(b)). Then, the syringe was slowly pressed down to release ILs (50 μL) on the annular platform. The sealing bottle cap was fastened down; extraction was then implemented through stirring with the required water bath temperature. After extraction, the sealing bottle cap was loosened, the microsyringe was directly used to absorb IL on the annular platform, the solution was diluted to 100 μL using acetonitrile, and 10 μL of the solution was absorbed through the automatic sampler for high-performance liquid chromatography (HPLC) analysis.

2.5. Analysis Conditions. The analytes were separated on a Waters PAHC$_{18}$ chromatographic column (250 mm × 4.6 mm, ID 5 μm, Waters Corp.) at a column temperature of 40°C. The sample injection volume was 10 μL. The mobile phase consisted of acetonitrile (A) and water (B). The total run time was 20.0 min with a flow rate of 1.0 mL min^{-1}. The gradient elution steps were as follows: 0–10.0 min, A increased from 60% to 100%, then remained at 100% until 16.0 min, and thereafter returned to 60% in 1.0 min, followed by re-equilibration time for 3.0 min. The parameters of emission wavelength (λ_{em}) and excitation wavelength (λ_{ex}) for PAHs on a fluorescence detector are shown in Table 1.

3. Results and Discussion

3.1. Optimization of Extraction Conditions. Extraction efficiency of ILs for the analytes was mainly influenced by several parameters. In this study, optimized parameters included volume of the extraction solvent, extraction temperature, extraction time, pH, stirring rate, and salt effect of solution. During optimization of parameters, the PAH-spiked water sample (5.0 μg·L^{-1}) was used as an extraction substrate. Extraction efficiencies of methods were evaluated through enrichment factors (EFs) and recoveries. Computational formulas were expressed as follows:

$$\mathrm{EFs} = \frac{C_t}{C_0} = \frac{(n_t/v_t)}{C_0}, \quad (1)$$

$$\mathrm{recovery} = \left(\mathrm{EFs} \times \frac{V_t}{V_0}\right) \times 100\%, \quad (2)$$

where C_t and n_t refer to the concentration and molar quantity of PAHs in the extraction phase, respectively; C_0 is the initial concentration of PAHs in the water sample before extraction; V_t corresponds to the final volume of the extraction phase; and V_0 stands for the volume of the water sample before extraction.

3.2. Volume of ILs. Volume of the extraction solvent is an important factor influencing extraction efficiency. This study assessed influences of different volumes (10, 20, 50, 80, and 100 μL) of the IL [SBnIM]PF_6 on extraction efficiency of six kinds of PAHs (Figure 3). When volume of IL was within 10 μL to 50 μL, extraction efficiency gradually improved as volume enlarged; enrichment efficiency remained within 50 μL to 80 μL but decreased when IL volume increased to 100 μL. Thus, as volume of the extraction solvent enlarged, extraction amount (n_t in Formula (1)) of analytes correspondingly increased. However, when volume (V_t) of the extraction solvent escalated too rapidly, analyte concentration (C_t) and EFs decreased. Therefore, to obtain maximum enrichment efficiency and good reproducibility, volume of the IL extraction solvent was set at 50 μL.

Table 1: The fluorescence detection wavelengths for PAHs.

Time (min)	λ_{ex} (nm)	λ_{em} (nm)
0	275	325
6.85	252	370
7.80	250	390
9.00	280	460
10.0	320	380
11.5	270	390

3.3. Extraction Temperature. To obtain satisfactory enrichment efficiency, the added water sample should not be higher than the level of the annular extraction platform. The extraction process involved headspace LPME. Extraction temperature directly influenced velocity and concentration of analytes released from the water sample and then influenced extraction efficiency indirectly. The increase in extraction temperature will shorten equilibrium time, accelerate analysis, and increase concentration of analytes in the gas phase. However, extraction for analytes was an absorption process. Thus, high temperature will reduce the distribution coefficient of analytes in the extraction solvent and also diminish extraction amount. Therefore, this study also assessed influences of different temperatures on extraction efficiency (Figure 4). When the temperature increased from room temperature to 50°C, EFs of six kinds of PAHs increased rapidly but slowed down under 50°C to 60°C and decreased under 70°C. Thus, in this study, 50°C was selected as an optimal extraction temperature.

3.4. Extraction Time. Theoretically, before reaching the equilibrium during extraction, prolonged duration will increase contents of analytes absorbed by the extraction solvent. This study also evaluated the influence of extraction time on extraction efficiency under 50°C (Figure 5). EFs of different kinds of PAHs improved rapidly before 30 min and improved slowly afterward. In this study, the extraction time was set to 30 min to guarantee EFs of analytes. However, according to Figure 5, if larger EFs are desired, an extraction time up to 50 min would be adopted.

3.5. pH of Solution. To study the influence of pH on extraction efficiency, 1 mol·L^{-1} HCl or NaOH was used to regulate pH of the extracted water sample within 2 to 13. The

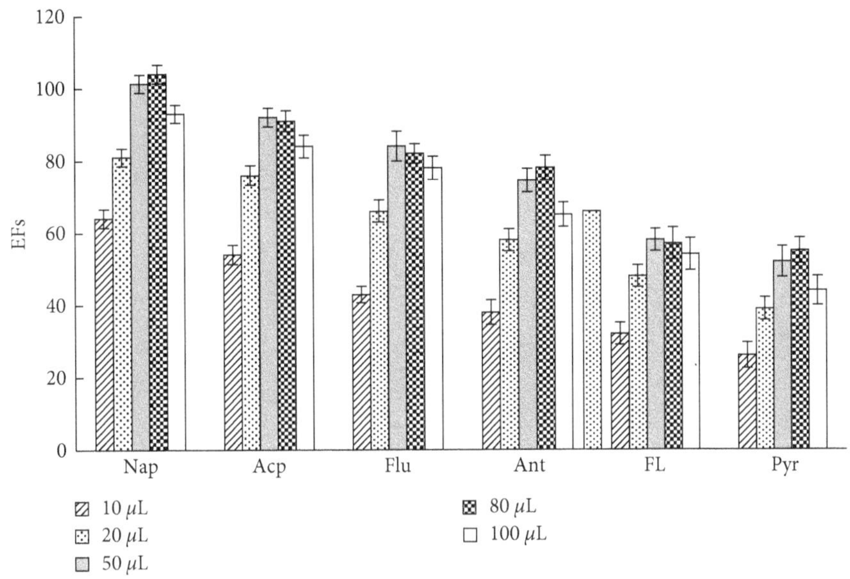

FIGURE 3: The effect of IL volume on EFs. Other extraction conditions: 10 mL of aqueous standards (containing PAHs at a concentration of $5\,\mu g{\cdot}L^{-1}$); extraction temperature, 50°C; extraction time, 30 min; the concentration of NaCl, 25%. Values are mean of $n = 3$ replicates.

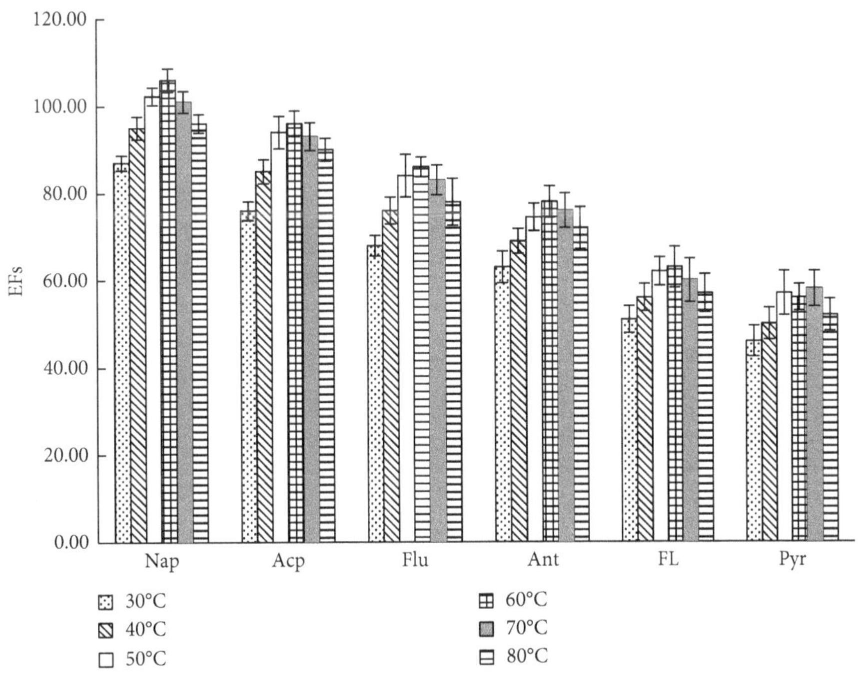

FIGURE 4: The effect of temperature on EFs. Other extraction conditions: 10 mL of aqueous standards (containing PAHs at a concentration of $5\,\mu g{\cdot}L^{-1}$); extraction temperature, 50°C; extraction solvent volume, 50 μL; extraction time, 30 min; the concentration of NaCl, 25%. Values are mean of $n = 3$ replicates.

results indicated that pH exerted minor influence on extraction efficiency of PAHs. Maybe, it was due to the fact that PAHs had no ionizable groups and they were stable at the molecular state, a condition not influenced by pH of the solution. Therefore, for convenience and reproducibility, water samples can be directly extracted and enriched without any assistance only if their pH were within 2 to 13.

3.6. Stirring Rate. In order to evaluate the influence of stirring rate, 10 mL of $5\,\mu g{\cdot}L^{-1}$ PAH mixed aqueous solution was extracted at 50°C for 30 min using 50 μL $[SBnIM]PF_6$ with stirring rates from 200 to 1500 rpm. Theoretically, before reaching the equilibrium, accelerating the stirring rate will increase contents of analytes absorbed by the extraction solvent. The results showed that extraction efficiency of all

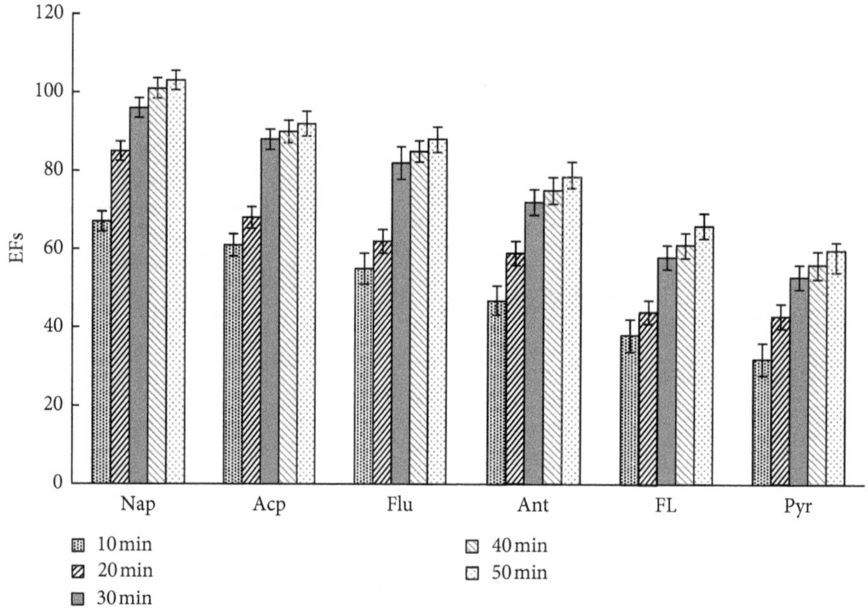

FIGURE 5: The effect of extraction time on EFs. Other extraction conditions: 10 mL of aqueous standards (containing PAHs at a concentration of 5 $\mu g{\cdot}L^{-1}$); extraction temperature, 50°C; extraction solvent volume, 50 μL; extraction time, 30 min; the concentration of NaCl, 25%. Values are mean of $n = 3$ replicates.

analytes slightly increased as the increase of stirring rate was up to 1000 rpm, while the reproducibility was poor after the stirring rate greater than 1000 rpm. A stirring rate of 1000 rpm was selected in this work.

3.7. Salt Effect. In headspace LPME, adding NaCl in the sample solution can reduce solubility of analytes in the water phase, improve the gas-liquid partition coefficient, and enlarge concentration of analytes in the gas phase to increase extraction amount indirectly. However, extremely high NaCl concentration will hinder transport of analytes in the sample solution and influence their diffusion in the gas phase. This study assessed the effects of NaCl addition (0%–36%, w/v) on extraction efficiency (Figure 6). The results showed that enrichment efficiency was the highest when the NaCl solution was saturated, indicating that reducing solubility of PAHs in the water phase exerted a significant influence. Therefore, the optimum concentration for salt addition was saturated NaCl solution.

3.8. Method Evaluation. To evaluate practical applicability of the APS-IL-LPME method, reproducibility, limit of detection (LOD), and limit of quantification (LOQ) were investigated with PAH-spiked ultrapure water under the optimum conditions, the results of which are summarized in Table 2. It showed that the method presented good linearity ($R^2 > 0.995$) within 0.5 $\mu g{\cdot}L^{-1}$ to 1000 $\mu g{\cdot}L^{-1}$ for the six kinds of PAHs. Replicate extractions of spiked ultrapure water solution (5.0 $\mu g{\cdot}L^{-1}$) were carried out on intraday and interday to show precision of the method. The relative standard deviation (RSD) of PAHs ranged from 4.7% to 11.6% in all cases, and LOQ was within 0.01 $\mu g{\cdot}L^{-1}$ to 0.05 $\mu g{\cdot}L^{-1}$. The recoveries of Nap, Acp, Flu, Ant, and FL were within 70.2% to 102%, whereas only that of Pyr was lower than 70%. Meanwhile, EFs ranged from 65.0 to 102, which were comparable with those reported by Liu et al. [12].

The comparison of different methods for determination of PAHs in water is shown in Table 3. LLE [34] used less extraction time and exhibited low LOQ, but it was relatively solvent-consuming and had lower extraction efficiency. The SPE [35] method gave higher extraction efficiency. However, it exhibited high LOQ and needed more sample volume, while the operation of SPME [36] was complex and time-consuming. The LPME combined with gas chromatography [25, 37, 38] showed higher LOQs. The IL-SDME method [12] showed high EFs, while had lower extraction efficiency (<50%), calculated according to Formula (2) in Section 3.1. The previous reported IL-LPME methods [12, 22, 39, 40] all used $[C_8MIM]PF_6$ as an extraction solvent, and the LODs of most of them were higher than those of the proposed method. In conclusion, the proposed APS-IL-LPME method is simple, stable, and an advantageous alternative, the extraction efficiency and LOQ of which are better than or comparable with those of other methods.

3.9. Application to Real Water Samples. Seawater samples collected from Shazikou in Qingdao and corresponding PAH-spiked seawater samples were extracted and detected according to the optimization method. As shown in Figure 7(c), no PAHs were found in the seawater sample. To evaluate this method, the seawater sample was spiked with PAHs at two levels (5.0 and 10 $\mu g{\cdot}L^{-1}$). As can be seen in Table 4, the recoveries for six PAHs were from 67.2% to 106% with the RSD between 3.5% and 10.4%. Compared with the recoveries of PAHs in spiked ultrapure water solution, the effects of matrix constituents were minor.

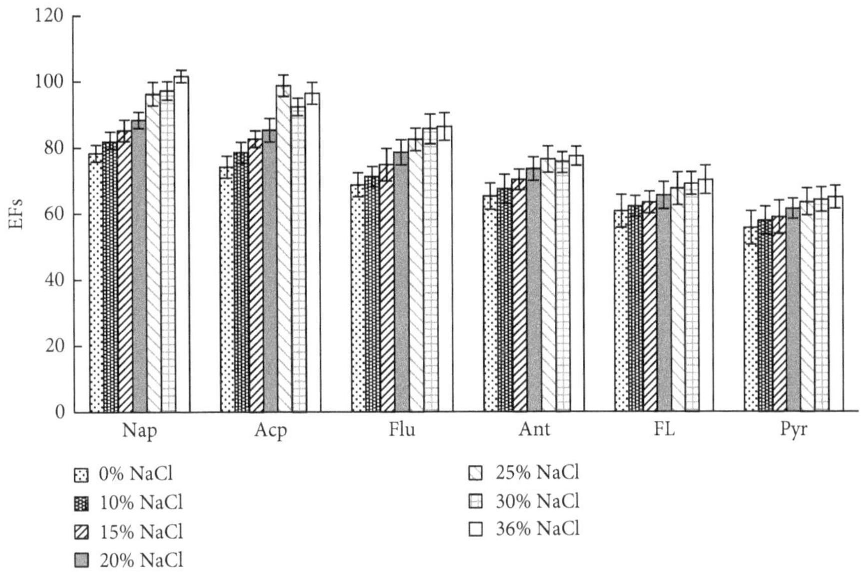

FIGURE 6: The effect of NaCl% on EFs. Other extraction conditions: 10 mL of aqueous standards (containing PAHs at a concentration of 5 $\mu g \cdot L^{-1}$); extraction temperature, 50°C; extraction solvent volume, 50 μL; extraction time, 30 min. Values are mean of $n = 3$ replicates.

TABLE 2: Main analytical parameters for the enrichment of PAHs with the proposed method.

Analytes	Linear range ($\mu g \cdot L^{-1}$)	R^2	LODs ($\mu g \cdot L^{-1}$)	LOQs ($\mu g \cdot L^{-1}$)	Precision (%RSD)	
					Intraday ($n = 6$)	Interday ($n = 3$)
Nap	0.5–1000	0.9981	0.003	0.01	4.7	7.9
Acp	0.5–1000	0.9976	0.003	0.01	4.8	7.4
Flu	0.5–1000	0.9954	0.006	0.02	5.1	8.1
Ant	0.5–1000	0.9988	0.010	0.03	5.5	10.7
FL	0.5–1000	0.9963	0.015	0.05	6.8	11.6
Pyr	0.5–1000	0.9972	0.003	0.01	5.3	9.6

TABLE 3: Comparison of the proposed method with other preconcentration methods for determination of PAHs in water.

Preconcentration technique	Analytical technique	Extraction solvent	Sample volume (mL)	Solvent volumea (mL)	Extraction time (min)	Extraction efficiency (%)	LOQ ($\mu g \cdot L^{-1}$)	Reference
LLE	HPLC-Flu	CH_2Cl_2	250	90	6	51–104	0.1–4.4	[34]
SPE	HPLC-Flu	C_6H_{14}-CH_2Cl_2	50	5	50	67–99	0.7–56.2	[35]
SPME	GC-MS/MS	—	18	—	70	60–102	0.1–1.0	[36]
DLLME	GC-FID	C_2Cl_4	5	0.008	Seconds	60–111	23.3–100.0	[25]
LLME-FA	GC-FID	Toluene	22	0.05	5	100–108	14–41 (LOD)	[37]
IL-DLLME	GC-MS	$[P_{6,6,6,14}{}^+][Ni(II)(hfacac)_3{}^-]$	25	0.025	10	84–115	0.5–8.7 (LOD)	[38]
IL-DLLME	HPLC-UV	$[C_8MIM]PF_6$	10	0.055	30	83.5–118	0.0005–0.88 (LOD)	[39]
IL-DLLME	HPLC-Flu	$[C_8MIM]PF_6$	10	0.05	2	90–102	0.1–7.0	[22]
IL-SDME	HPLC-Flu	$[C_8MIM]PF_6$	15	0.003	30	—	—	[12]
IL-HS-LPME	HPLC-Flu	$[C_8MIM]PF_6$	5	0.003	60	88–110	0.03–0.1 (LOD)	[40]
APS-IL-LPME	HPLC-Flu	$[SBnIM]PF_6$	10	0.05	30	65.0–102	0.01–0.05 (LOQ)/0.003–0.015 (LOD)	This work

—, not required or not available. [a]The volume of extraction solvent needed in the extraction process.

Figure 7(b) shows the typical chromatogram of the sample spiking with 10 $\mu g \cdot L^{-1}$ of the PAHs. After APS-IL-LPME, PAHs in seawater were enriched effectively, and the baseline was well separated. Combined with the results from Figures 3–6 and Table 4, EFs and recoveries decreased gradually with increasing boiling point and molecular weight due to decreased molecule diffusion and gas-liquid partition coefficient. The results were consistent with the

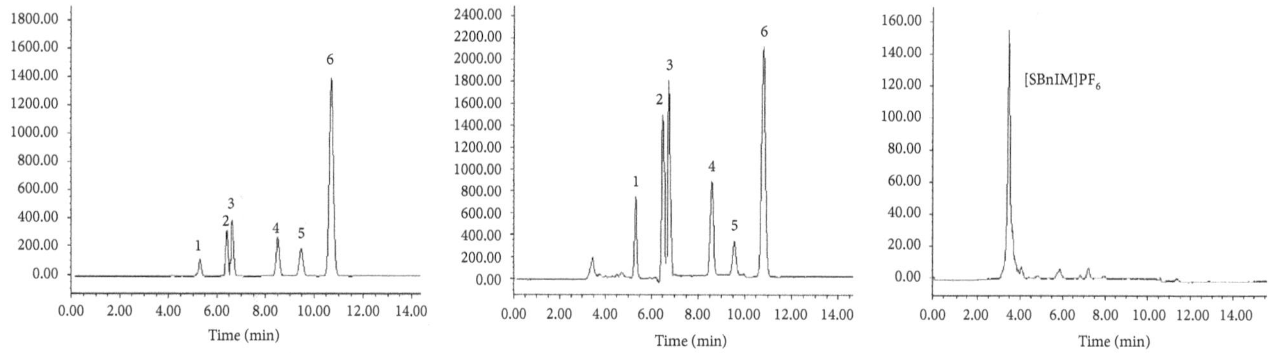

FIGURE 7: Chromatograms of (a) standard solution (200 $\mu g{\cdot}L^{-1}$), (b) ionic liquid extracted spiked seawater (10 $\mu g{\cdot}L^{-1}$), and (c) ionic liquid-extracted blank seawater. Peaks: (1) naphthalene; (2) acenaphthene; (3) fluorene; (4) anthracene; (5) fluoranthene; (6) pyrene.

TABLE 4: Spiked recoveries of the six PAHs in real seawater samples.

Analytes	Spiked amounts ($\mu g{\cdot}L^{-1}$)	Average recovery (%)	RSD (%) ($n = 5$)
Nap	5	106	4.2
	10	105	3.5
Acp	5	97.8	5.5
	10	95.3	7.5
Flu	5	86.2	9.7
	10	81.4	6.1
Ant	5	77.3	6.3
	10	79.8	6.0
FL	5	72.5	6.5
	10	70.3	4.9
Pyr	5	68.3	10.4
	10	67.2	8.2

findings of headspace LPME reported by other studies [12], indicating that this method is more suitable for enrichment of volatile PAHs in water samples due to the nonvolatile property of IL.

4. Conclusions

This work proposed a new method of APS-IL-LPME using the IL [SBnIM]PF_6 as an extraction solvent, which overcame the influence of IL gravity and significantly improved extraction stability and efficiency. The IL [SBnIM]PF_6 was homemade and proved to be an excellent extraction solvent for LPME. Combined with liquid chromatography, the method was employed to detect PAHs in the water sample. The optimized conditions for extraction were under 50°C for 30 min with 50 μL extraction solvent and saturated NaCl concentration in the sample solution. The results indicated that the extraction efficiency and LOQ of the method were better than or comparable with the previously reported ones and suitable for detection of PAHs, such as Nap, Acp, Flu, Ant, FL, and Pyr, in water solutions, with EFs ranging from 65.0 to 102 and LOQ from 0.01 to 0.05 $\mu g{\cdot}L^{-1}$. The APS-IL-LPME method was first used for enrichment of PAHs in real seawater samples, indicating that it will have broad potential in enrichment and analysis of organic pollutants in the environment and food samples. This method is especially advantageous for the enrichment of volatile compounds and shows higher extraction stability than suspended SDME by adding an annular platform. However, as the headspace LPME mode, the amount of water samples is limited and not higher than the level of annular platform. Moreover, in this work, nonvolatile compounds are not transferred to the headspace and not extracted by the ionic liquids. In the future work, we will continue to develop a new extraction mode to apply this method in immersed LPME for extraction of nonvolatile compounds.

Conflicts of Interest

The authors declare that they have no financial and commercial conflicts of interest in connection with the work submitted.

Acknowledgments

This work was kindly supported by the Special Scientific Research Funds for Central Non-Profit Institutes, Yellow Sea Fisheries Research Institute, Chinese Academy of Fishery Sciences (No. 20603022016007), National Key R&D

Program of China (No. 2017YFC1600705), National Natural Science Foundation of China (No. 21207162), and Key Laboratory of Separation Sciences for Analytical Chemistry, Dalian Institute of Chemical Physics, Chinese Academy of Sciences (No. KL-1604).

References

[1] United States Environmental Protection Agency, *National Primary Drinking Water Standards*, United States Environmental Protection Agency, Washington, DC, USA, 2003.

[2] European Commission, *Directive 2006/0129 of the European Parliament and of the Council on Environmental Quality Standards in the Field of Water Policy and Amending Directive 2000/60/EC*, European Commission, Brussels, Belgium, 2006.

[3] European Commission, *Directive 2008/105/EC, 2008, Official Journal of the European Communities L-348/84 (24th December, 2008), Council Directive (16th December, 2008) on Environmental Quality Standards in the Field of Water Policy, Amending and Subsequently Repealing Council Directives 82/176/EEC, 83/513/EEC, 84/156/EEC, 84/491/EEC, 86/280/EEC and Amending Directive 2000/60/EC of the European Parliament and of the Council*, European Commission, Brussels, Belgium, 2008.

[4] D. Nagaraju and S. D. Huang, "Determination of triazine herbicides in aqueous samples by dispersive liquid–liquid microextraction with gas chromatography–ion trap mass spectrometry," *Journal of Chromatography A*, vol. 1116, no. 1-2, pp. 89–97, 2007.

[5] N. Fattahi, Y. Assadi, M. R. M. Hosseini, and E. Z. Jahromi, "Determination of chlorophenols in water samples using simultaneous dispersive liquid–liquid microextraction and derivatization followed by gas chromatography-electron-capture detection," *Journal of Chromatography A*, vol. 1157, no. 1-2, pp. 23–29, 2007.

[6] A. Papadopoulou, I. P. Román, A. Canals, K. Tyrovola, and E. Psillakis, "Fast screening of perfluorooctanesulfonate in water using vortex-assisted liquid–liquid microextraction coupled to liquid chromatography–mass spectrometry," *AnalyticaChimicaActa*, vol. 691, no. 1-2, pp. 56–61, 2011.

[7] F. Pena-Pereira, N. Cabaleiro, I. Calle et al., "Directly suspended droplet microextraction in combination with microvolume UV–vis spectrophotometry for determination of phosphate," *Talanta*, vol. 85, no. 2, pp. 1100–1104, 2011.

[8] J. G. Huddleston, H. D. Willauer, R. P. Swatloski, A. E. Visser, and R. D. Rogers, "Room temperature ionic liquids as novel media for 'clean' liquid–liquid extraction," *Chemical Communications*, vol. 16, pp. 1765-1766, 1998.

[9] D. W. Armstrong, L. He, and Y. S. Liu, "Examination of ionic liquids and their Interaction with molecules, when used as stationary phases in gas chromatography," *Analytical Chemistry*, vol. 71, no. 17, pp. 3873–3876, 1999.

[10] J. G. Huddleston, A. E. Visser, W. M. Reichert, H. D. Willauer, G. A. Broker, and R. D. Rogers, "Characterization and comparison of hydrophilic and hydrophobic room temperature ionic liquids incorporating the imidazoliumcation," *Green Chemistry*, vol. 3, no. 4, pp. 156–164, 2001.

[11] C. F. Poole, "Chromatographic and spectroscopic methods for the determination of solvent properties of room temperature ionic liquids," *Journal of Chromatography A*, vol. 1037, no. 1-2, pp. 49–82, 2004.

[12] J. Liu, G. Jiang, Y. Chi, Y. Cai, Q. Zhou, and J. Hu, "use of ionic liquids for liquid-phase microextraction of polycyclic aromatic hydrocarbons," *Analytical Chemistry*, vol. 75, no. 21, pp. 5870–5876, 2003.

[13] E. Aguilera-Herrador, R. Lucena, S. Cárdenas, and M. Valcárcel, "Direct coupling of ionic liquid based single-drop microextraction and GC/MS," *Analytical Chemistry*, vol. 80, no. 3, pp. 793–800, 2008.

[14] Q. X. Zhou and C. L. Ye, "Ionic liquid for improved single-drop microextraction of aromatic amines in water samples," *MicrochimicaActa*, vol. 162, no. 1-2, pp. 153–159, 2008.

[15] J. F. Liu, Y. G. Chi, G. B. Jiang, C. Tai, J. F. Peng, and J. T. Hu, "Ionic liquid-based liquid-phase microextraction, a new sample enrichment procedure for liquid chromatography," *Journal of Chromatography A*, vol. 1026, no. 1-2, pp. 143–147, 2004.

[16] Y. C. Fan, Z. L. Hu, M. L. Chen, C. S. Tu, and Y. Zhu, "Ionic liquid based dispersive liquid–liquid microextraction of aromatic amines in water samples," *Chinese Chemical Letters*, vol. 19, no. 8, pp. 985–987, 2008.

[17] F. Q. Zhao, J. Li, and B. Z. Zeng, "Coupling of ionic liquid-based headspace single drop microextraction with GC for sensitive detection of phenols," *Journal of Separation Science*, vol. 31, no. 16-17, pp. 3045–3049, 2008.

[18] M. J. Trujillo-Rodríguez, P. Rocío-Bautista, V. Pino, and A. M. Afonso, "Ionic liquids in dispersive liquid-liquid microextraction," *Trends in Analytical Chemistry*, vol. 51, pp. 87–106, 2013.

[19] Ł. Marcinkowski, F. Pena-Pereira, A. Kloskowski, and J. Namieśnik, "Liquid-phase microextraction techniques based on ionic liquids for preconcentration and determination of metals," *Trends in Analytical Chemistry*, vol. 61, pp. 54–66, 2014.

[20] D. Ge and H. K. Lee, "Ionic liquid based hollow fiber supported liquid phase microextraction of ultraviolet filters," *Journal of Chromatography A*, vol. 1229, pp. 1–5, 2012.

[21] J. Zhang and H. K. Lee, "Headspace ionic liquid-based microdrop liquid-phase microextraction followed by microdrop thermal desorption-gas chromatographic analysis," *Talanta*, vol. 81, no. 1-2, pp. 537–542, 2010.

[22] M. T. Pena, M. C. Casais, M. C. Mejuto, and R. Cela, "Development of an ionic liquid based dispersive liquid–liquid microextraction method for the analysis of polycyclic aromatic hydrocarbons in water samples," *Journal of Chromatography A*, vol. 1216, no. 36, pp. 6356–6364, 2009.

[23] A. Sarafraz-Yazdi and F. Mofazzeli, "Ionic liquid-based submerged single drop microextraction: a new method for the determination of aromatic amines in environmental water samples," *Chromatographia*, vol. 72, no. 9-10, pp. 867–873, 2010.

[24] C. Yao, P. Twu, and J. L. Anderson, "Headspace single drop microextraction using micellar ionic liquid extraction solvents," *Chromatographia*, vol. 72, no. 5-6, pp. 393–402, 2010.

[25] M. Rezaee, Y. Assadi, M. R. M. Hosseini, E. Aghaee, F. Ahmadi, and S. Berijani, "Determination of organic compounds in water using dispersive liquid–liquid microextraction," *Journal of Chromatography A*, vol. 1116, no. 1-2, pp. 1–9, 2006.

[26] G. Shem and H. K. Lee, "Hollow fiber-protected liquid-phase microextraction of triazine herbicides," *Analytical Chemistry*, vol. 74, no. 3, pp. 648–654, 2002.

[27] J. L. Manzoori, M. Amjadi, and J. Abulhassani, "Ultra-trace determination of lead in water and food samples by using ionic liquid-based single drop microextraction-electrothermal atomic absorption *spectrometry*," *AnalyticaChimicaActa*, vol. 644, no. 1-2, pp. 48–52, 2009.

[28] Ł. Marcinkowski, F. Pena-Pereira, A. Kloskowski, and J. Namiesnik, "Opportunities and shortcomings of ionic liquids in single-drop microextraction," *Trends in Analytical Chemistry*, vol. 72, pp. 153–168, 2015.

[29] M. H. Abraham, A. M. Zissimos, J. G. Huddleston, H. D. Willauer, R. D. Rogers, and W. E. Acree, "Ionic liquids: solvation ability and polarity," *Industrial & Engineering Chemistry Research*, vol. 42, no. 3, pp. 413–418, 2003.

[30] X. Cao and Z. Le, "Synthesis of chiral ionic liquids," *Chinese Journal of Organic Chemistry*, vol. 2010, no. 30, pp. 816–832, 2010.

[31] X. J. Sun, C. Y. Wu, and J. Xing, "Ionic liquid-bonded polysiloxane as stationary phase for capillary gas chromatography," *Journal of Separation Science*, vol. 33, no. 20, pp. 3159–3167, 2010.

[32] X. J. Sun, Y. L. Zhu, P. Wang, J. Li, C. Y. Wu, and J. Xing, "High temperature and highly selective stationary phases of ionic liquid bonded polysiloxanes for gas chromatography," *Journal of Chromatography A*, vol. 1218, no. 6, pp. 833–841, 2011.

[33] S. Chun, S. V. Dzybua, and R. A. Bartsch, "Influence of structural variation in room-temperature ionic liquids on the selectivity and efficiency of competitive alkali metal salt extraction by a crown ether," *Analytical Chemistry*, vol. 73, no. 15, pp. 3737–3741, 2001.

[34] L. Pensado, E. Blanco, M. C. Casais et al., "Strategic sample composition in the screening of polycyclic aromatic hydrocarbons in drinking water samples using liquid chromatography with fluorimetric detection," *Journal of Chromatography A*, vol. 1056, no. 1-2, pp. 121–130, 2004.

[35] O. Delhomme, R. Rieb, and M. Millet, "Solid-phase extraction and LC with fluorescence detection for analysis of PAHsin rainwater," *Chromatographia*, vol. 65, no. 3-4, pp. 163–171, 2007.

[36] V. Fernandez, E. Concha, S. Muniategui, P. Lopez, and D. Prada, "Solid-phase microextraction–gas chromatographic–tandem mass spectrometric analysis of polycyclic aromatic hydrocarbons: towards the European Union water directive 2006/0129 EC," *Journal of Chromatography A*, vol. 1176, no. 1-2, pp. 48–56, 2007.

[37] M. H. Hosseini, M. Rezaee, S. Akbarian, F. Mizani, M. R. Pourjavid, and M. Arabieh, "Homogeneous liquid–liquid microextraction via flotation assistance for rapid and efficient determination of polycyclic aromatic hydrocarbons in water samples," *AnalyticaChimicaActa*, vol. 762, pp. 54–60, 2013.

[38] J. L. Benedé, J. L. Anderson, and A. Chisvert, "Trace determination of volatile polycyclic aromatic hydrocarbons in natural waters by magnetic ionic liquid-based stir bar dispersive liquid microextraction," *Talanta*, vol. 176, pp. 253–261, 2018.

[39] Q. Zhou and Y. Gao, "Determination of polycyclic aromatic hydrocarbons in water samples by temperature controlled ionic liquid dispersive liquid–liquid microextraction combined with high performance liquid chromatography," *Analytical Methods*, vol. 6, no. 8, pp. 2553–2559, 2014.

[40] J. Xiao, J. Cheng, F. Guo et al., "Ultrasound-assisted headspace ionic-liquid microextraction of polycyclic aromatic hydrocarbons at elevated temperatures," *MicrochimicaActa*, vol. 177, no. 3-4, pp. 465–471, 2012.

9

A Molecular Imprinted Polymer as a Flow-Through Optical Sensor for Oxazepam

Roberta G. Machicote,[1] Marcela A. Castillo,[1] Maria E. Pacheco,[2] and Liliana Bruzzone[2]

[1]*División Química Analítica, Departamento de Química, Facultad de Ciencias Exactas, Universidad Nacional de La Plata, Calle 47, Esq. 115, 1900 La Plata, Argentina*
[2]*Laboratorio de Investigación y Desarrollo de Métodos Analíticos (LIDMA), Facultad de Ciencias Exactas, Universidad Nacional de La Plata, Calle 47, Esq. 115, 1900 La Plata, Argentina*

Correspondence should be addressed to Roberta G. Machicote; rmachicote@quimica.unlp.edu.ar

Academic Editor: Krishna K. Verma

A flow-through optosensing system for oxazepam recognition with fluorescence detection was performed by means of a molecular imprinted polymer based on its acid hydrolysis product, 2-amino-5-chlorobenzophenone. The synthesis was conducted via a noncovalent imprinting methodology, using methacrylic acid as a functional monomer and ethylene glycol dimethacrylate as a cross-linking agent. Hydrolysis (types and concentration of acids), polymer retention capacity, binding properties, and elution (selectivity and reversibility) conditions were optimized. The selected molecular imprinted polymer had a molar ratio composition of 1 : 6 : 45 (template : functional monomer : cross-linker). The proposed method was applied to the determination of oxazepam in a pharmaceutical formulation. External standard calibration, standard additions calibration, and Youden's calibration were carried out in order to evaluate constant and proportional errors due to the matrix. The developed metabolite-based recognition system for benzodiazepines is an innovative procedure that could be followed in routine and quality control assays.

1. Introduction

Oxazepam (OXA), 7-chloro-l,3-dihydro-3-hydroxy-5-phenyl-2H-1,4-benzodiazepin-2-one, is a therapeutically short-to-intermediate-acting benzodiazepine used to produce sedation, induce sleep, relieve anxiety, and muscle spasms [1]. Due to its sedative properties, it has a high potential for abuse specially when it is combined with other depressants such as alcohol or opiates [2]. Urine has been the most widely used biological matrix for the determination of benzodiazepines and metabolites in drug abuse testing and toxicology [3].

Several analytical methods have been proposed for OXA determination: most of them use chromatographic techniques and others use fluorometry and capillary electrophoresis; also, some electrochemical sensors have been developed [4–13]. Since benzodiazepine metabolites are converted to benzophenone, a sensitive determination method could be performed using the hydrolyzed product [5, 14].

Molecularly imprinted polymers (MIPs) are synthetic materials that act as artificial biomimetic receivers capable to recognize and specifically interact with the analyte used as a template in the polymerization process. The specificity and selectivity gained with MIPs make them suitable as recognition elements for chemical and biochemical sensors [15–28].

In this work, a rapid, simple, and low-cost analytical method is proposed for the indirect determination of OXA in pharmaceutical formulations based on the development of an optical sensor for its acid hydrolysis product, 2-amino-5-chlorobenzophenone (BZF). A MIP has been synthesized via the noncovalent imprinting methodology as a recognition element in a flow-through optosensing system with fluorescence detection.

2. Materials and Methods

2.1. Materials. Methacrylic acid (MAA) was purchased from Fluka (Buchs, Switzerland), and ethylene glycol dimethacrylate

(EGDMA) and α,α'-azoisobutyronitrile (AIBN) were acquired from Aldrich (Milwaukee, WI, USA). They were purified as described in a previous work [29].

All other chemicals were of analytical grade and used as received. Oxazepam (OXA) was kindly supplied by a pharmaceutical store. 2-amino-5-chlorobenzophenone (BZF) (purity 98%) was acquired from Aldrich (Milwaukee, WI, USA). A pharmaceutical formulation (tablets) containing OXA was purchased from a local store. Double-distilled water was used.

Stock solutions of OXA and BZF were prepared in adequate solvents as needed for the planned assays.

2.2. Apparatus. All recordings of fluorescence spectra and fluorimetric measurements were carried out on a PerkinElmer LS-50B luminescence spectrometer (Beaconsfield, England) equipped with a pulsed xenon lamp (half peak height <10 μs, 60 Hz), an R928 photomultiplier tube, and a computer working with FL Winlab software.

The optosensing manifold used for the luminescence measurements was similar to the one employed in a previous work [29]. A 25 μL, 1.5 mm path length, conventional flow-through quartz cell (Hellma, model 176.052-QS, Mullheim, Germany) was loaded with the synthesized MIP [29]. The as-packed active phase was used with satisfactory readings for several months. (All the experiments were carried out without changing the active phase from the flow cell.)

A four channel Gilson's Minipuls 3 peristaltic pump (Villiers-le-Bel, France) fitted with organic solvent resistant Tygon tubing was used to generate the flowing streams. A Rheodyne 5020 rotatory valve (Rohnert Park, CA, USA) provided with a 500 μL loop was employed for sample introduction. PTFE tubes (0.8 mm internal diameter) were used throughout the flow system.

Gas chromatographic analysis (GC) were performed with a Shimadzu GC-2014 (Japan) provided with an AOC-20i automatic injector and FID detector working with a Restek RTx-65TG capillary column.

Absorption spectra were recorded on a Shimadzu UV-240 (Japan) recording spectrophotometer, equipped with a 1.0 cm path length quartz cell. Derivative absorption spectra were performed applying simple mathematical calculations.

2.3. Oxazepam Hydrolysis. As temperature alone does not produce the OXA hydrolysis, aliquots of OXA working solutions were initially mixed with 0.1 M acid solutions (sulphuric, acetic, and hydrochloric acids) in screw-capped tubes and hydrolyzed at 100°C for 60 minutes to form BZF. After cooling, absorption spectra were registered and their first-order derivative spectra were obtained.

2.4. MIP/NIP Synthesis. A molecular imprinted polymer (MIP) with BZF was synthesized according to a standard method previously studied in our research group [29]. As BZF has an amino group, methacrylic acid (MAA) has been selected as a functional monomer. In a screw-capped tube, a total of 21 mg of BZF (0.09 mmol), template, were dissolved in 2 mL of chloroform, porogen, by sonication for 15 minutes. Afterwards, 531 μL (2.8 mmol) of EGDMA, a cross-linking agent, and 32 μL (0.37 mmol) of MAA were added and shaken in a Vortex during another 15 minutes. With the tube immersed in an ice bath, 3 mg (0.019 mmol) of AIBN, a radical initiator, was added and sonicated for 5 minutes. The mixture was degassed by bubbling N_2 for 10 minutes. The tube was heated at 60°C for 16 hours. The polymer was removed from the tube, ground in an agatha mortar, and sieved to an average particle size of 70–100 mesh. The particles were washed in the Soxhlet extractor with acetic acid : methanol (10% v/v) until BZF could no longer be detected by UV absorption in the washing solution. Finally, the particles were washed with methanol and dried at 40°C during 24 hours. A nonimprinted polymer (NIP) was synthesized using the same procedure without the addition of BZF.

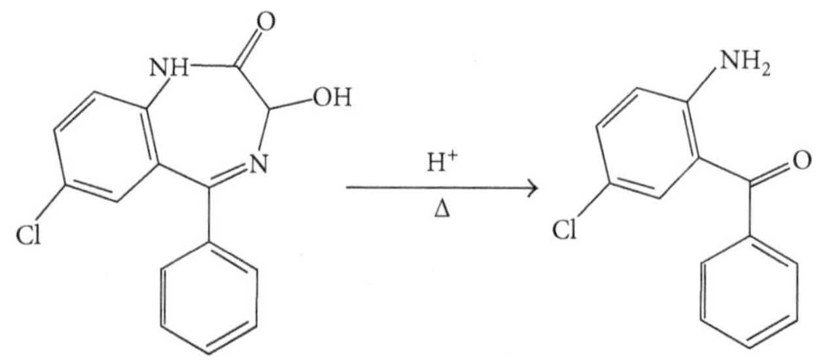

FIGURE 1: Oxazepam acid hydrolysis reaction.

2.5. Binding Assays. Binding assays were performed contacting each MIP/NIP during 48 hours with BZF chloroformic standard solutions. After this time, the mixture was centrifuged and the supernatant was analyzed by gas chromatography (GC). A calibration graph with standard BZF solutions was prepared in order to quantify BZF concentration. The BZF concentration bonded to the polymer was calculated by subtracting the BZF concentration in the supernatant to the initial one.

2.6. Oxazepam Determination. The proposed method was applied to the determination of OXA in a pharmaceutical formulation (Pausafren T, 15 mg OXA, tablets). Five tablets were accurately weighed, powdered in an agatha mortar, and homogenized. An average mass of 1.150 g per tablet was calculated.

3. Results and Discussion

3.1. Oxazepam Hydrolysis. The acid hydrolysis of benzodiazepines to benzophenones has been previously reported [30, 31]. The hydrolysis product of oxazepam (OXA), 2-amino-5-chlorobenzophenone (BZF), is shown in Figure 1.

As absorption spectra of OXA and its respective benzophenone are overlapped, it is difficult to determine the BZF concentration in the presence of OXA when a partial hydrolysis takes place. The OXA and BZF first-order derivative spectra have two wavelengths at which dA/dλ are zero: 236.9 nm and 229.1 nm (Figure 2). These wavelengths are suitable for the quantification of both species after different hydrolysis treatments: BZF at 229.1 nm and OXA at 236.9 nm.

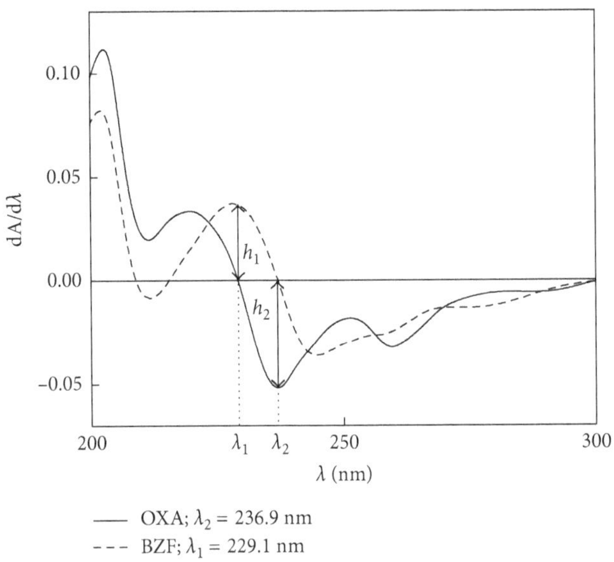

Figure 2: OXA and BZF first-order absorption derivative spectra.

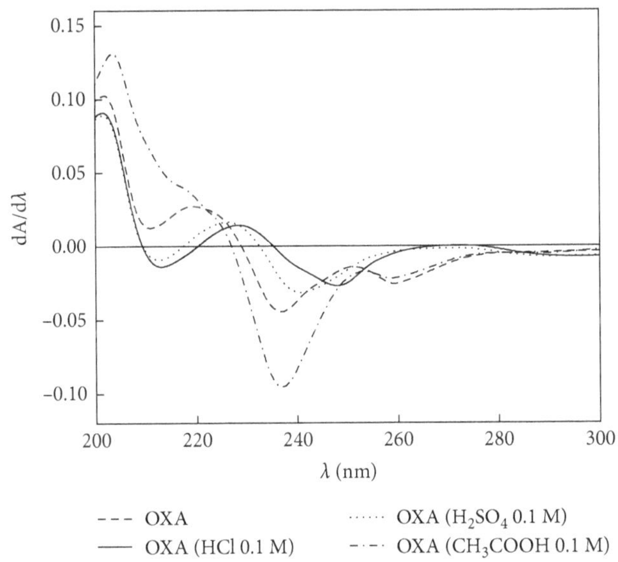

Figure 3: OXA first-order absorption derivative spectra after partial hydrolysis. [OXA] = 3.50×10^{-5} M.

Absorbances were measured at the mentioned wavelengths (Figure 3). From these values, it was deduced that the hydrolysis takes place with major efficiency in the presence of hydrochloric acid (HCl). The tested concentration was inadequate to produce complete hydrolysis. Afterwards, the hydrolysis reaction was performed with a series of increasing HCl molar concentration solutions. A 3 M HCl concentration was selected as the optimum one for the complete hydrolysis (Figure 4).

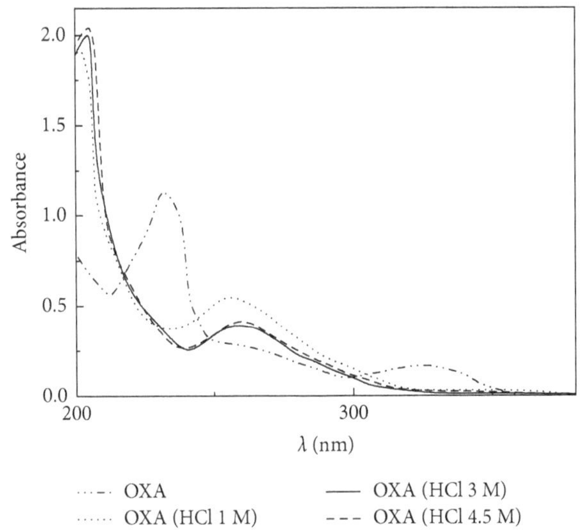

Figure 4: OXA first-order absorption derivative spectra after complete hydrolysis. [OXA] = 3.02×10^{-5} M.

Table 1: Polymers synthesized maintaining a 7.5 EGDMA/MAA molar ratio.

MIP	[MAA]/[BZF]	(mmol) BZF	MAA	EGDMA
1	2	0.043	0.086	0.647
2	4	0.043	0.172	1.290
3	6	0.043	0.258	1.935
4	8	0.043	0.344	2.580

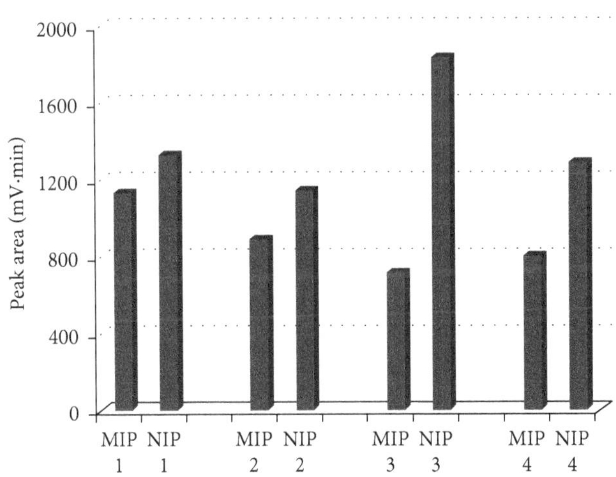

Figure 5: BZF peak area for each MIP/NIP after incubation time.

3.2. Binding Assays. In order to select the best retention capacity of the MIP, a series of polymers were synthesized maintaining a 7.5 EGDMA/MAA molar ratio (Table 1). The polymers were synthesized, ground, sieved, washed, and dried as previously stated. 2 mL of a 2.52×10^{-4} M BZF chloroformic solution was added to 13 mg of each MIP in a capped test tube. The tubes were shaken in a Vortex and let to stand for 48 hours at room temperature. After this time, the tubes were centrifuged at 4000 rpm for 5 minutes and the solutions were GC analyzed. Results are shown in Figure 5. All MIPs retain a greater BZF quantity than the NIPs. MIP #3 shows the best retention with an optimal MAA/BZF molar ratio equal to 6.

Table 2: Polymers synthesized maintaining a 6 MAA/BZF molar ratio.

MIP	[EGDMA]/[MAA]	(mmol)		
		BZF	MAA	EGDMA
1	2.5	0.043	0.258	0.645
2	5	0.043	0.258	1.290
3	7.5	0.043	0.258	1.935
4	10	0.043	0.258	2.580

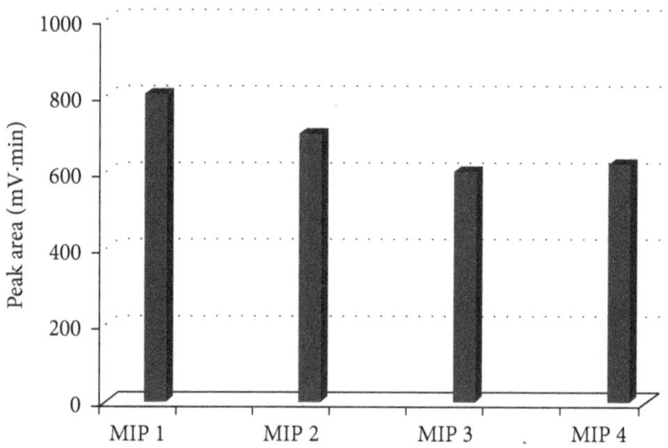

Figure 6: BZF peak area for each MIP after incubation time.

In order to evaluate the optimum EGDMA/MAA molar ratio, a series of polymers were synthesized maintaining a 6 MAA/BZF molar ratio (Table 2). The polymers were synthesized, ground, sieved, washed, and dried. Binding assays were performed as abovementioned. Results are shown in Figure 6. MIP #3 presents the major retention capacity with a 7.5 EGDMA/MAA molar ratio.

3.3. MIP/NIP Optimized Synthesis: Optosensing System Evaluation. The working polymers (MIP and NIP) were synthesized under the previously stated conditions (Section 3.2).

The MIP was introduced in the flow cell of the optosensing system and was evaluated in a simple way with a methanol carrier solution containing 10% v/v acetic acid. The fluorescent signal was monitored setting the excitation and emission wavelengths at 220 nm and 336 nm, respectively, with 10 nm of excitation and emission bandwidths. A carrier flow rate of 4 mL/min was used, and a stable signal was obtained. When a 1.28×10^{-5} M BZF methanolic solution was injected, a negative peak was obtained returning to the baseline after a definite period of time. Certainly, the MIP recognizes and retains the BZF. The inverted peaks suggest that the interaction produces a quenching of the MIP fluorescence (Figure 7). No appreciable changes were detected when NIP was assayed (Figure 8).

Under the conditions previously established, solutions of different BZF concentrations were injected in the system. Figure 9 shows the diagram when different concentrations of BZF methanolic solutions were injected into the flow system using the synthesized MIP as the sensing material. The regeneration with the carrier solution is also shown.

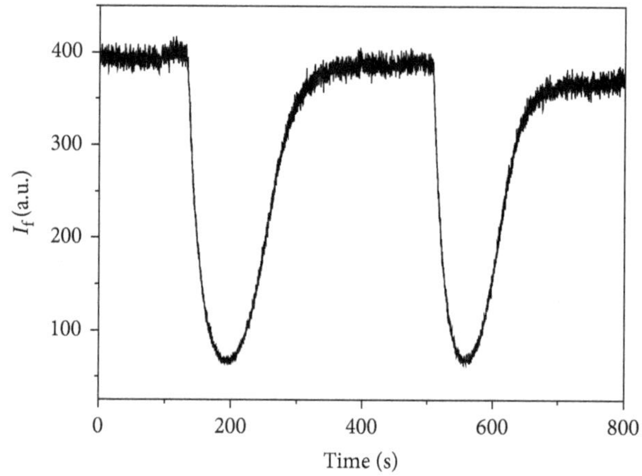

Figure 7: MIP response profile for a 1.28×10^{-5} M BZF methanolic solution.

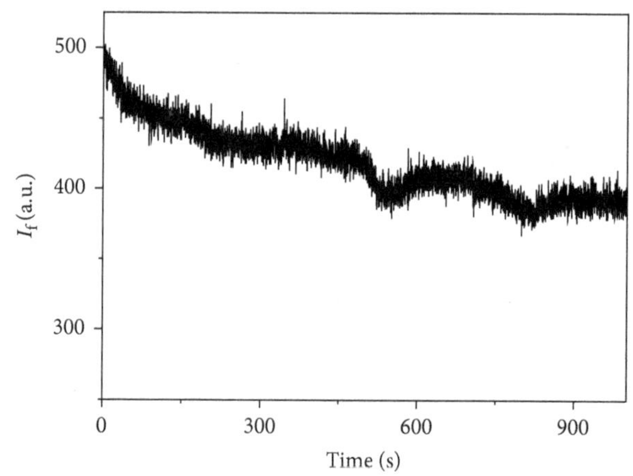

Figure 8: NIP response profile for a 1.28×10^{-5} M BZF methanolic solution.

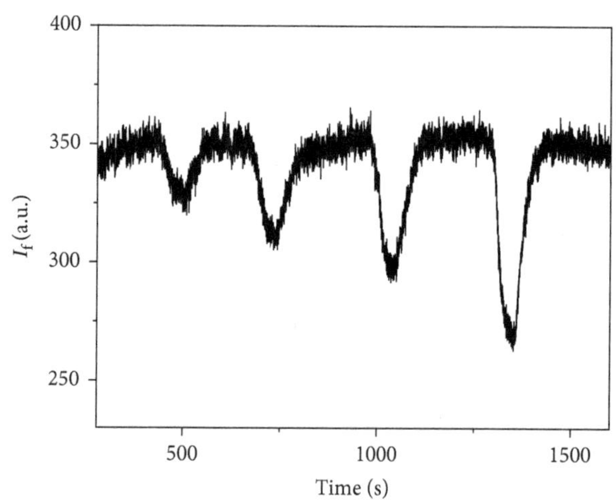

Figure 9: MIP response profile for different BZF methanolic solution concentrations: 1.78×10^{-5} M, 3.40×10^{-5} M, 4.98×10^{-5} M, and 7.12×10^{-5} M.

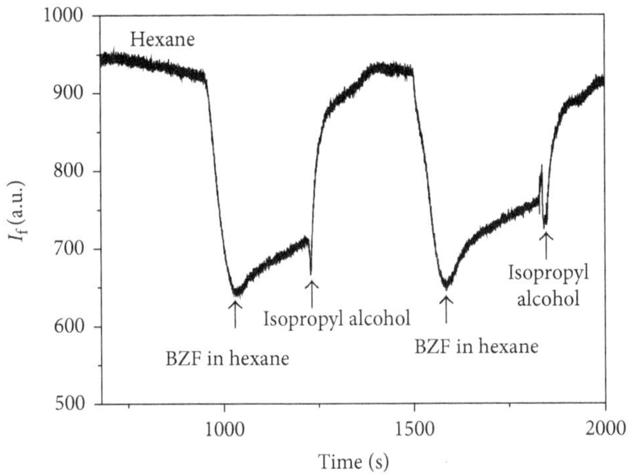

FIGURE 10: Fluorescence response profile for BZF removal/rebinding in a MIP.

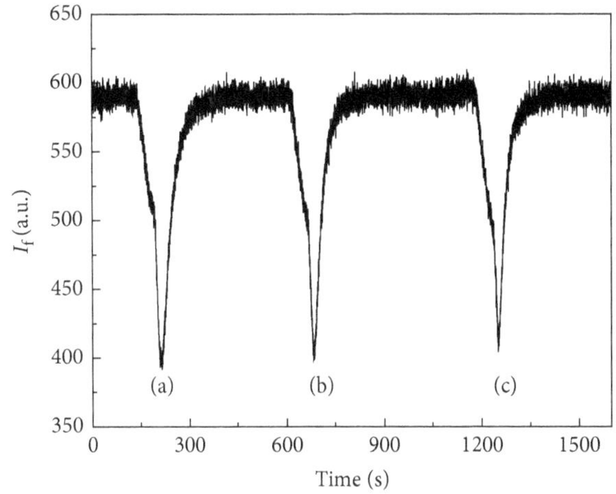

FIGURE 11: MIP response profile for BZF with different hexane/isopropyl alcohol mixtures as a carrier: (a) 90 : 10 (v/v), (b) 70 : 30 (v/v), and (c) 30 : 70 (v/v).

The polarity of the carrier solvent affects the BZF binding properties. A solvent with high polarity competes with the sample molecules for binding at the recognition sites. The result is a weakness in the specific polar interaction leading BZF to go to the carrier solution. Otherwise, in nonpolar solvent, the molecule stays into the polymer. Thus, a nonpolar solvent should be used for detecting the quenching of the fluorescence caused by the BZF retention. A polar solvent should be then used for eluting the molecule from the polymer leaving the MIP in the unbound situation for receiving a new injection.

A first attempt was made with hexane as a nonpolar carrier solvent injecting a 3.40×10^{-5} M BZF solution in hexane followed by an isopropyl alcohol injection. Experimental working parameters were optimized; excitation and emission wavelengths were set at 410 and 515 nm, respectively, with 10 nm and 5 nm corresponding bandwidths. This situation is shown in Figure 10. As can be seen, the change of solvents requires a too long time to restore the baseline leading to a nonsymmetrical peak and making the study unpractical. The next step was searching for a solvent that could let the BZF interact with the polymer (selectivity) and simultaneously be capable of reaching the initial conditions without changing it (reversibility). As hexane and isopropyl alcohol are miscible, different proportions of both were assayed. A 3.40×10^{-5} M BZF solution in hexane was injected in 90 : 10, 70 : 30, and 30 : 70 (v/v) hexane/isopropyl alcohol mixtures. The obtained results showed that the peaks widths were diminished when the content of isopropyl alcohol was increased (Figure 11). On the other hand, the baseline was quickly restored, but the intensity was diminished. A 90 : 10 (v/v) ratio of hexane/isopropyl alcohol was selected for further studies. The carrier flow rate showed a similar behavior: an increase of its velocity produced a narrow peak width with low intensity on account of the lesser retention. A carrier flow rate of 5 mL/min was chosen.

3.4. Oxazepam Determination. Two kinds of calibration graphs were performed: (i) with BZF standards and (ii) with a solution of previously hydrolyzed OXA.

(i) Solutions of BZF were prepared in 3 M HCl. 3 mL of each one were extracted with the same volume of hexane.

(ii) Solutions prepared with the obtained product after OXA acid hydrolysis were treated in the same way as in (i).

The organic phases were separated and injected in the flow system, and the peak heights were measured. Calibration graphs were obtained for the two cases and were analyzed using a t-test in order to compare both slopes [32]. It could be concluded that there is not any significant difference between the two calibration graphs. Therefore, calibration with BZF standards may be utilized for OXA quantification in a real sample (previous acid hydrolysis).

External standard and standard additions calibration graphs were carried out.

(i) External standard calibration graph: adequate aliquots of 1.24×10^{-3} M BZF stock solution were transferred to a series of 10.00 mL calibrated flasks attaining a final concentration range of 1.25×10^{-5} M–7.50×10^{-5} M after dilution with 3 M HCl.

(ii) Standard additions calibration graph: portions of 0.300 g of the sample were spiked with different amounts of OXA, diluted to 10.00 mL with 3 M HCl, and hydrolyzed; one milliliter of the so-obtained solution was diluted to 10.00 mL with 3 M HCl resulting in a mass range of 0–0.0125 g.

In both cases, extraction with hexane was performed and solutions were injected in the flow system. The slopes of both calibration graphs were compared by means of a t-test [32]. The results indicated the presence of an analyte/matrix interaction effect.

As matrix effects were found, the quantitative evaluation of the analyte was carried out by the standard additions and Youden calibration [33].

TABLE 3: Results of the least-squares regression analysis.

	Standard additions	Youden's calibration
n	10	8
Intercept	23.6	7.2
Slope	$4165\,g^{-1}$	$49\,g^{-1}$
s_a	0.9	0.7
s_b	$116\,g^{-1}$	$1\,g^{-1}$
$s_{y/x}$	20.52	3.22
r	0.9969	0.9984

n: number of calibration data points; s_a: standard deviation of the intercept; s_b: standard deviation of the slope; $s_{y/x}$: standard deviation of the regression; r: correlation coefficient.

TABLE 4: Figures of merit.

SEN†	$0.42\,mL{\cdot}\mu g^{-1}$
γ	$0.26\,mL{\cdot}\mu g^{-1}$
LOD	$11.2\,\mu g{\cdot}mL^{-1}$
LOQ	$34.2\,\mu g{\cdot}mL^{-1}$
Dynamic range	$11.2–123.3\,\mu g{\cdot}mL^{-1}$
Linear range	$34.2–123.3\,\mu g{\cdot}mL^{-1}$

†SEN: calibration sensitivity; γ: analytical sensitivity; LOD: limit of detection; LOQ: limit of quantification; LOD = $3.28 \times s_0$; LOQ = $10 \times s_0$; $s_0 = (s_{y/x}/b)\sqrt{(1/\sum_i n_i) + (1/m) + (\overline{x}^2/\sum_i (x_i - \overline{x})^2)}$.

(i) Standard additions calibration graph was obtained by weighing five equal portions of the ground tablets, spiked with different amounts of OXA and individually used to prepare the solutions required for the flow injection system.

(ii) Youden's calibration graph was obtained by weighing four different amounts of the ground tablets in the range of 0.25–0.85 g. Each of them was individually used to prepare the solutions required for the flow injection system.

The results of both calibration graphs are shown in Table 3.

Considering the mass of each tablet and the amount used to obtain the standard additions graph, the OXA content was calculated from the ratio of the difference between the ordinate intercepts of the standard additions and the Youden plots to the slope of the standard additions calibration graph. A value of 15.1 ± 0.9 mg OXA per tablet was obtained, in agreement with the content provided by the manufacturer (15 mg OXA per tablet). Matrix effects may be possible due to the presence of β-alanine in the sample. β-alanine has an amino group that can interact with MAA and thus interfere with the BZF recognition. Taking into account that the product contains 500 mg of β-alanine per tablet, the obtained result is certainly acceptable. The extraction with hexane may be another cause for the matrix effect. Table 4 summarizes the analytical figures of merit of the proposed methodology.

4. Conclusions

A novel optosensor for oxazepam has been developed based on its acid hydrolysis product, 2-amino-5-chlorobenzophenone. The novel use of the metabolite is beneficial in terms of the unnecessary handling of benzodiazepines. The proposed analytical methodology was applied for oxazepam determination in a pharmaceutical formulation, avoiding matrix effects by means of the Youden calibration method. The developed fluorescence flow-through optosensing system has wider impacts on other domains in society other than academics as it is suitable for routine assays and quality control in pharmaceutical industries and drug abuse divisions as an alternative for chromatographic methodologies. It opens up possibilities for knowledge to be used in unexpected, creative, and innovative ways beyond normal professional research.

Conflicts of Interest

The authors declare that there are no conflicts of interest regarding the publication of this paper.

Acknowledgments

The authors thank Facultad de Ciencias Exactas, Universidad Nacional de La Plata, Argentina, for partial financial support. Maria E. Pacheco is a fellow of Consejo Nacional de Investigaciones Científicas y Técnicas (CONICET).

References

[1] C. Page, C. Michael, M. Sutter, M. Walker, and B. B. Hoffman, *Integrated Pharmacology*, C.V. Mosby, Amsterdam, Netherlands, 2nd edition, 2002.

[2] B. Brands, B. Sproule, and J. Marshman, *Drugs & Drug Abuse*, Addiction Research Foundation, Toronto, Canada, 3rd edition, 1998.

[3] R. L. Hawks and C. Nora Chiang, *Urine Testing for Drugs of Abuse*, National Institute on Drug Abuse Research Monograph 73, Rockville, MD, USA, 1986.

[4] A. Zapardiel, J. A. Perez Lopez, E. Bermejo, L. Hernandez, and M. Chicharro, "Voltammetric studies on the interactions between camazepam metabolic series and human serum albumin. Determination of oxazepam using adsorptive stripping voltammetry," *Analytica Chimica Acta*, vol. 244, pp. 49–57, 1991.

[5] J. DolejSová, P. Solich, C. K. Polydorou, M. A. Koupparis, and C. E. Efstathiou, "Flow-injection fluorimetric determination of 1,4-benzodiazepines in pharmaceutical formulations after acid hydrolysis," *Journal of Pharmaceutical and Biomedical Analysis*, vol. 20, no. 1-2, pp. 357–362, 1999.

[6] S. McClean, E. O'Kane, J. Hillis, and W. F. Smyth, "Determination of 1,4-benzodiazepines and their metabolites by capillary electrophoresis and high-performance liquid chromatography using ultraviolet and electrospray ionisation mass spectrometry," *Journal of Chromatography A*, vol. 838, no. 1-2, pp. 273–291, 1999.

[7] C. Pham-Huy, G. Villain-Pautet, H. Hua et al., "Separation of oxazepam, lorazepam, and temazepam enantiomers by HPLC on a derivatized cyclodextrin-bonded phase: application to the determination of oxazepam in plasma," *Journal of Biochemical and Biophysical Methods*, vol. 54, no. 1–3, pp. 287–299, 2002.

[8] S. Pirnay, F. Hervé, S. Bouchonnet, B. Perrin, F. J. Baud, and I. Ricordel, "Liquid chromatographic–electrospray ionization mass spectrometric quantitative analysis of buprenorphine, norbuprenorphine, nordiazepam and oxazepam in rat plasma," *Journal of Pharmaceutical and Biomedical Analysis*, vol. 41, no. 4, pp. 1135–1145, 2006.

[9] M. E. Lozano-Chaves, J. M. Palacios-Santander, L. M. Cubillana-Aguilera, I. Naranjo-Rodríguez, and J. L. Hidalgo-Hidalgo-de-Cisneros, "Modified carbon-paste electrodes as sensors for the determination of 1,4-benzodiazepines: application to the determination of diazepam and oxazepam in biological fluids," *Sensors and Actuators B: Chemical*, vol. 115, no. 2, pp. 575–583, 2006.
[10] A. M. Gil Tejedor, P. Fernández Hernando, and J. S. Durand Alegría, "A rapid fluorimetric screening method for the 1,4-benzodiazepines: determination of their metabolite oxazepam in urine," *Analytica Chimica Acta*, vol. 591, no. 1, pp. 112–115, 2007.
[11] K. Bastos Borges, E. Figueiredo Freire, I. Martins, and M. E. Pereira Bastos de Siqueira, "Simultaneous determination of multibenzodiazepines by HPLC/UV: investigation of liquid-liquid and solid-phase extractions in human plasma," *Talanta*, vol. 78, no. 1, pp. 233–241, 2009.
[12] L. Magrini, A. Cappiello, G. Famiglini, and P. Palma, "Microextraction by packed sorbent (MEPS)-UHPLC-UV: a simple and efficient method for the determination of five benzodiazepines in an alcoholic beverage," *Journal of Pharmaceutical and Biomedical Analysis*, vol. 125, pp. 48–53, 2016.
[13] F. Jiang, Y. Rao, R. Wang et al., "Sensitive, automatic method for the determination of diazepam and its five metabolites in human oral fluid by online solid-phase extraction and liquid chromatography with tandem mass spectrometry," *Journal of Separation Science*, vol. 39, no. 10, pp. 1873–1883, 2016.
[14] R. Roškar and M. Sollner Dolenc, "Determination of benzodiazepines in urine via benzophenone derivatives using liquid chromatography-tandem mass spectrometry," *Archives of Industrial Hygiene and Toxicology*, vol. 61, no. 4, pp. 381–388, 2010.
[15] G. Wulff, "Molecular imprinting in cross-linked materials with the aid of molecular templates-a way towards artificial antibodies," *Angewandte Chemie International Edition*, vol. 34, no. 17, pp. 1812–1832, 1995.
[16] A. G. Mayes and K. Mosbach, "Molecularly imprinted polymers: useful materials for analytical chemistry?," *Trends in Analytical Chemistry*, vol. 16, no. 6, pp. 321–332, 1997.
[17] B. Sellergren, "Noncovalent molecular imprinting: antibody-like molecular recognition in polymeric network materials," *Trends in Analytical Chemistry*, vol. 16, no. 6, pp. 310–320, 1997.
[18] S. Al-Kindy, R. Badía, J. L. Suárez-Rodríguez, and M. E. Díaz-García, "Molecularly imprinted polymers and optical sensing applications," *Critical Reviews in Analytical Chemistry*, vol. 30, no. 4, pp. 291–309, 2000.
[19] B. Sellergren, *Molecularly Imprinted Polymers. Man-Made Mimics of Antibodies and their Applications in Analytical Chemistry*, Elsevier, Amsterdam, Netherlands, 2001.
[20] V. B. Kandimalla and H. Ju, "Molecular imprinting: a dynamic technique for diverse applications in analytical chemistry," *Analytical and Bioanalytical Chemistry*, vol. 380, no. 4, pp. 587–605, 2004.
[21] K. Karim, F. Breton, R. Rouillon, E. V. Piletska, A. Guerreiro, I. Chianella et al., "How to find effective functional monomers for effective molecularly imprinted polymers?," *Advanced Drug Delivery Reviews*, vol. 57, no. 12, pp. 1795–1808, 2005.
[22] A. Kloskowski, M. Pilarczyk, A. Przyjazny, and J. Namiesnik, "Progress in development of molecularly imprinted polymers as sorbents for sample preparation," *Critical Reviews in Analytical Chemistry*, vol. 39, no. 1, pp. 43–58, 2009.
[23] J. Li, Y. Li, Y. Zhang, and G. Wei, "Highly sensitive molecularly imprinted electrochemical sensor based on the double amplification by an inorganic prussian blue catalytic polymer and the enzymatic effect of glucose oxidase," *Analytical Chemistry*, vol. 84, no. 4, pp. 1888–1893, 2012.
[24] A. L. Jenkins, M. W. Ellzy, and L. C. Buettner, "Molecularly imprinted polymer sensors for detection in the gas, liquid, and vapor phase," *Journal of Molecular Recognition*, vol. 25, no. 6, pp. 330–335, 2012.
[25] H. F. Hawari, N. M. Samsudin, A. Y. Shakaff et al., "Highly selective molecular imprinted polymer (MIP) based sensor array using interdigitated electrode (IDE) platform for detection of mango ripeness," *Sensors and Actuators B: Chemical*, vol. 187, pp. 434–444, 2013.
[26] X. Wang, J. Luo, C. Yi, and X. Liu, "Paracetamol sensor based on molecular imprinting by photosensitive polymers," *Electroanalysis*, vol. 25, no. 8, pp. 1907–1916, 2013.
[27] L. Li, Y. Liang, and Y. Liu, "Designing of molecularly imprinted polymer-based potentiometric sensor for the determination of heparin," *Analytical Biochemistry*, vol. 434, no. 2, pp. 242–246, 2013.
[28] O. Gurtova, L. Ye, and F. Chmilenko, "Potentiometric propranolol-selective sensor based on molecularly imprinted polymer," *Analytical and Bioanalytical Chemistry*, vol. 405, no. 1, pp. 287–295, 2013.
[29] H. A. Andreetta and L. Bruzzone, "Fluorescence detection of atenolol using a molecular imprinted polymer," *Analytical Letters*, vol. 41, no. 1, pp. 36–45, 2008.
[30] A. Morales-Rubio, J. V. Julián-Ortiz, A. Salvador, and M. de la Guardia, "Hydrolysis of benzodiazepines in a microwave oven and ultraviolet derivative analysis of their benzophenones," *Microchemical Journal*, vol. 49, no. 1, pp. 12–19, 1994.
[31] D. Gambart, S. Cárdenas, M. Gallego, and M. Valcárcel, "An automated screening system for benzodiazepines in human urine," *Analytica Chimica Acta*, vol. 366, no. 1–3, pp. 93–102, 1998.
[32] D. L. Massart, B. G. M. Vandeginste, L. M. C. Buydens, S. De Jong, P. J. Lewi, and J. Smeyers-Verbeke, *Handbook of Chemometrics and Qualimetrics: Part A*, Elsevier, Amsterdam, Netherlands, 1997.
[33] R. C. Castells and M. A. Castillo, "Systematic errors: detection and correction by means of standard calibration, Youden calibration and standard additions method in conjunction with a method response model," *Analytica Chimica Acta*, vol. 423, no. 2, pp. 179–185, 2000.

10

The Potential of Graphene as an Adsorbent for Five Pesticides from Different Classes in Rape Oil Samples using Dispersive Solid-Phase Extraction

Katarzyna Madej,[1] Katarzyna Janiga,[1] and Wojciech Piekoszewski[1,2]

[1]*Department of Analytical Chemistry, Faculty of Chemistry, Jagiellonian University, Krakow 30-387, Poland*
[2]*Department of Food Science and Technology, School of Biomedicine, Far Eastern Federal University, Vladivostok, Russia*

Correspondence should be addressed to Wojciech Piekoszewski; wpiekosz@tlen.pl

Academic Editor: Mohamed Abdel-Rehim

Isolation conditions for five pesticides (metazachlor, tebuconazole, λ-cyhalothrin, chlorpyrifos, and deltamethrin) from rape oil samples were examined using the dispersive solid-phase graphene extraction technique. To determine the optimal extraction conditions, a number of experimental factors (amount of graphene, amount of salt, type and volume of the desorbing solvent, desorption time with and without sonication energy, and temperature during desorption) were studied. The compounds of interest were separated and detected by an HPLC-UV employing a Kinetex XB-C18 column and a mobile phase consisting of acetonitrile and water flowing in a gradient mode. The optimized extraction conditions were: the amount of graphene 15 mg, desorbing solvent (acetonitrile) 5 mL, time desorption 10 min at 40°C, and amount of NaCl 1 g. The detection limit for metazachlor, tebuconazole, λ-cyhalothrin, and chlorpyrifos was 62.5 $ng \cdot g^{-1}$, and for deltamethrin, it was 500 $ng \cdot g^{-1}$. The obtained results lead to the conclusion that graphene may be successfully used for the isolation of the five pesticides from rape oil. However, their determination at low concentration levels, as they occur in real oil samples, requires the employment of appropriately highly sensitive analytical methods, as well as a more suitable graphene form (e.g., magnetically modified graphene).

1. Introduction

Due to the potential toxicity, bioaccumulating properties, and wide use of pesticides for supporting and intensification of crops almost all over the world, these substances belong to the most often controlled ones in the environment and in food. Every year, many publications devoted to the development of new methods for pesticide determination in various food matrices appear. However, pesticide residue analysis in high-fat samples is still a challenging analytical task. Edible oil samples belong to such analytically difficult matrices. Due to complex matrices, as well as low concentrations of pesticide residues, frequently originated from various classes, these samples are analyzed using such advanced coupled techniques as LC-MS/MS [1, 2], LC-TOF-MS [3], UHPLC-MS/MS [4, 5], GC-MS/MS [2, 6], and GC-MS [7]. The employment of GC-ECD [8, 9], HPLC/UV [10], or HPLC/DAD [11] in this analytical area has also been reported. Generally, in all cases of oil matrix analysis, an appropriate sample preparation procedure, usually consisting of two or more stages, was required. In most cases, QuEChERS or modified QuEChERS-based procedures as the cleanup step were involved [1–6, 8]. In some cases, these were preceded by the freezing step for oil precipitation [2, 3, 8, 10]. The separate review article concerning recent developments and trends in the QuEChERS sample preparation approach, including the analysis of pesticides in various fatty food samples, was also presented [12].

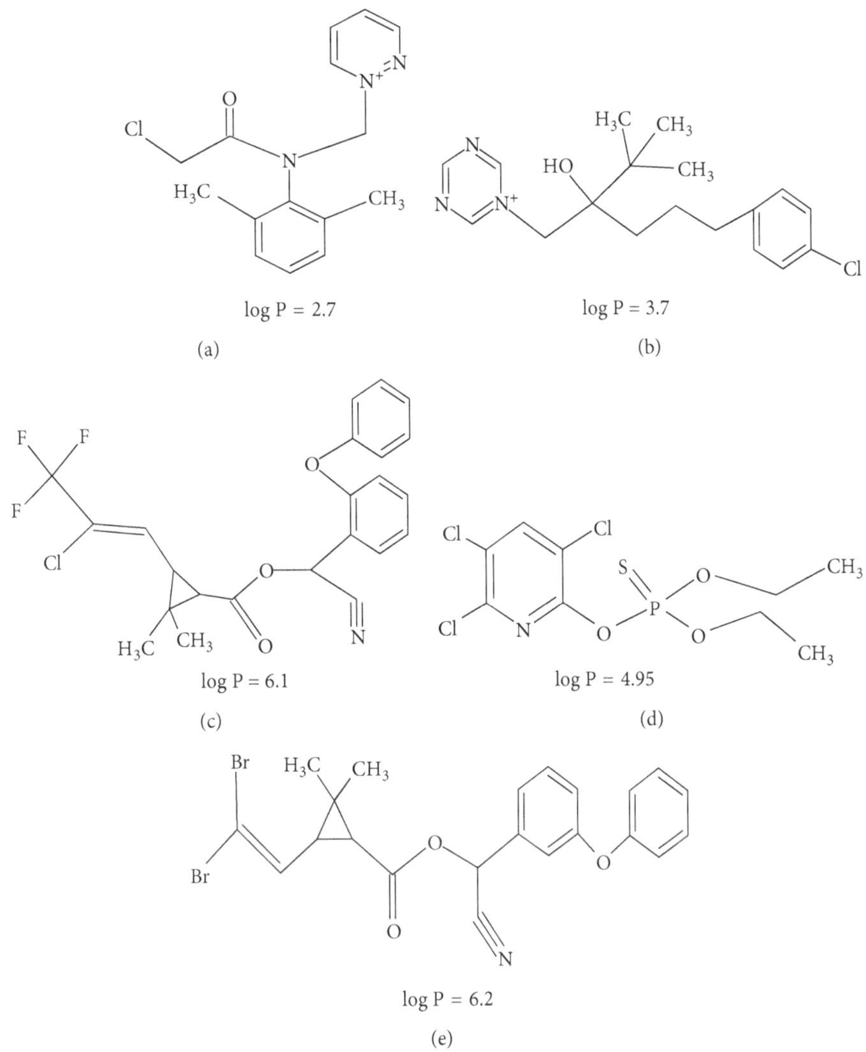

FIGURE 1: Chemical structures of the studied five pesticides with their partition coefficients (log *P*). Metazachlor (log *P* = 2.7) (a); tebuconazole (log *P* = 3.7) (b); λ-cyhalothrin (log *P* = 6.1) (c); chlorpyrifos (log *P* = 4.95) (d); deltamethrin (log *P* = 6.2) (e).

Graphene is a carbon nanomaterial which has been widely used (especially in modified forms) in determination of organic compounds including pesticides, usually in relatively simple matrices. These include environmental waters [13–16], fruit juices [17], fruits [18], vegetables [18–20], and teas [21, 22]. The potential unique adsorption properties of graphene resulted mainly from its large surface area and strong π-π interactions between its large delocalized electron system and the aromatic rings of the target molecules [23].

In this work, the main focus was to evaluate the adsorption potential of nonmodified graphene for the isolation of selected pesticides from various groups (chloroacetamide herbicides, triazole fungicides, pyrethroids, and organophosphorus pesticides) present in rape oil samples, which are commonly used for supporting of rape crops. To our best knowledge, this is the first time that graphene has been tried as the adsorbent for isolation of pesticides from such a difficult matrix as edible oil. The chemical structures of the studied components are presented in Figure 1.

Dispersive solid-phase extraction (d-SPE) combined with graphene was selected as the sample treatment method, and the HPLC-UV technique was used for the separation and detection of the compounds of interest in the resulting oil extracts.

2. Materials and Methods

2.1. Preparation of Standard Samples and Graphene Suspension. The stock of methanolic solutions of metazachlor, tebuconazole, λ-cyhalothrin, chlorpyrifos, and deltamethrin (1 mg·mL^{-1}) and the internal standard chlorfenvinphos (1 mg·mL^{-1}) were stored in a refrigerator at 4°C. The standard solutions were prepared by appropriate dilution of the stock solutions with methanol. The graphene suspension (3 mg·mL^{-1}) was prepared by weighing an appropriate amount of solid graphene and dispersing it in a suitable amount of deionized water using sonication for 6 h. The oil samples were prepared by weighing 1 gram of a commercial, ratified rape oil and then spiking it with appropriate amounts of standard solutions of the five or six pesticides and then left for equilibration at room temperature for one hour.

2.2. Apparatus. For oil sample preparation, the following laboratory equipment was used: technical balance (Radwag WPX 250, Poland), vortex TK 35 (Techno-Kartel, Germany), centrifuge MPW-260R (DHN, Poland), thermoblock TB-941U (Poland), and sonic bath Vibra Cell (Sonics & Materials Inc., USA). Chromatographic analyses were performed using the chromatographic system (Merck-Hitachi LaChrom) consisting of an L-7100 pump and an L-7455 UV spectrophotometric detector (Darmastadt, Germany) equipped with a Kinetex XB-C18 column (150 × 4.6 mm, 5 μm) which was supplied by Phenomenex (USA). The column temperature was 25°C. A mixture of water (A) and acetonitrile (B) was used as the mobile phase. The following optimal gradient conditions of the mobile phase flow were applied: 0 min (70% A and 30% B), 20 min (24% A and 76% B), 25 min (0% A and 100% B), and 35 min (0% A and 100% B). The flow rate was 1 mL·min^{-1}. The pesticides were detected at $\lambda = 220$ nm. The injection volume of standard solutions as well as oil sample extracts was 40 μL.

2.3. General d-SPE (Graphene) Procedure. For initial isolation of the five pesticides, 4 mL of acetonitrile was added with 1 g pesticide-spiked oil, vortexed (1 min), and left to freeze for minimum 2 h at −32°C in a horizontal position. An acetonitrile supernatant was decanted into a clean centrifuge test tube (7 mL) from above the frozen fat layer. The supernatant was evaporated to almost dryness, and then, an appropriate amount of aqueous graphene suspension and solid sodium chloride (for salting out pesticides) was added. The whole content of the test tube was vortexed (1 min) and then centrifuged (10 min, 14,000 rpm). Using a Pasteur pipette, the upper layer was discarded, and 4 mL of desorbing solvent was added to the bottom layer of graphene with the adsorbed pesticides and vortexed for 1 min without previous sample heating or after increasing the sample temperature in a thermoblock. Subsequently, the sample was subjected to sonication for an appropriate time at an appropriate temperature. The content was centrifuged (10 min, 14,000 rpm), and the acetonitrile supernatant with desorbed compounds was successfully transferred into an Eppendorf tube (2 mL) and evaporated to dryness under nitrogen in a thermoblock at 40°C. The dried residue was reconstituted into 200 μL of methanol and was ready for injection onto the chromatographic column.

3. Results

In order to obtain the best extraction conditions for isolation of the five studied pesticides from oil samples, several experimental factors such as the volume of graphene suspension, amount of the salt (NaCl), type and volume of the desorbing solvent, desorption time with and without application of sonication energy, and desorption temperature were studied. The optimization process was conducted using the method of one independent variable. The extraction recovery (ER) for each pesticide was calculated from the following equation:

$$\mathrm{ER}(\%) = \frac{P_{\text{pesticide in sample extract}}}{P_{\text{pesticide standard}}} \cdot f_1 \cdot f_2 \cdot 100, \quad (1)$$

where $P_{\text{pesticide in sample extract}}$ is the area of the peak corresponding to the pesticide in a sample extract, $P_{\text{pesticide standard}}$ is the area of the peak corresponding to the pesticide in a standard solution, f_1 is the ratio of theoretical volume of an acetonitrile supernatant above the frozen fat layer to volume of an acetonitrile supernatant taken from above the frozen fat layer, f_2 is the ratio of theoretical volume of an acetonitrile supernatant after desorption to volume of an acetonitrile supernatant after desorption, taken for evaporation.

For standardizing the extraction procedure, the acetonitrile supernatant volumes were taken to be the same, that is, 3.5 mL of the supernatant from above the frozen fat layer and 4 mL of the supernatant after the desorption process. Thus, values of the coefficients f_1 and f_2 are equaled to 1.14 and 1.25, respectively. As the optimization criterion (OC), the maximum average recovery rate for the five pesticides was adopted:

$$\mathrm{OC} = \frac{1}{5} \cdot \sum_{i=1}^{n=5} \mathrm{ER}_i = \text{maximum}, \quad (2)$$

where i is the pesticide number and $n = 5$ is the number of the studied pesticides.

3.1. Effect of Graphene Suspension Volume. Graphene was used as an adsorbent for the five pesticides. This nanostructured material seems to be an excellent adsorbent due to its huge surface area, as well as the hexagonal arrays of carbon atoms that are suitable for strong interactions with organic molecules, especially those possessing aromatic rings including the pesticides. The very large delocalized π-electron system enables the formation of strong π-π stacking interactions with aromatic rings [23].

To reduce errors when adding graphene to samples and to make the operation easy, graphene was prepared in the form of aqueous suspension. The amount of graphene was optimized at the five volume levels of the graphene suspension: 3, 4, 5, 6, and 7 mL. The remaining conditions such as the mass of NaCl (2 g), type and volume of the desorbing solvent (4 mL), sonication, desorption time (10 min), and temperature (ambient) during the desorption process were left constant. From this optimization step, the extraction results are shown in Table 1.

The maximal average value of extraction recovery for the five compounds was obtained for 5 mL of aqueous graphene suspension, and this volume was used for further study.

3.2. Effect of Desorbing Solvent Type. Four different solvents such as acetonitrile, acetone, ethyl acetate, and a mixture of n-hexane and dichloromethane (2 : 1, v/v) were tested as the desorbing agents for the studied pesticides from graphene. The choice of the tested solvents was based on their

TABLE 1: Effect of the graphene amount on extraction efficacy of the five pesticides from rape oil samples.

Pesticide	Extraction recovery (%), $n = 3$				
	9 mg of graphene	12 mg of graphene	15 mg of graphene	18 mg of graphene	21 mg of graphene
Metazachlor	11 ± 3	64 ± 7	**68 ± 5**	26 ± 5	14 ± 5
Tebuconazole	38 ± 8	72 ± 5	**77 ± 5**	44 ± 7	29 ± 7
λ-Cyhalothrin	56 ± 7	52 ± 6	**60 ± 5**	38 ± 11	17 ± 8
Chlorpyrifos	12 ± 7	63 ± 9	**66 ± 7**	49 ± 7	23 ± 6
Deltamethrin	13 ± 5	63 ± 11	**68 ± 9**	18 ± 4	14 ± 5
OC[1]	26.0 ± 20.3	63.0 ± 7.1	**67.0 ± 6.2**	35.1 ± 12.6[2]	19.4 ± 6.5[2]

[1]Value of the optimization criterion; [2]statistically significant from the highest result.

TABLE 2: Effect of the desorbing solvent type on extraction efficacy of the five pesticides from rape oil samples.

Pesticide	Extraction recovery (%), $n = 3$			
	Acetonitrile	Acetone	Ethyl acetate	Dichloromethane/hexane (1 : 2,v/v)
Metazachlor	**68 ± 6**	33 ± 5	61 ± 8	43 ± 6
Tebuconazole	**74 ± 7**	22 ± 5	25 ± 5	34 ± 8
λ-Cyhalothrin	**60 ± 6**	20 ± 4	50 ± 6	42 ± 6
Chlorpyrifos	**66 ± 6**	42 ± 3	8 ± 7	63 ± 6
Deltamethrin	**66 ± 5**	21 ± 3	39 ± 5	34 ± 5
OC[1]	**66.7 ± 5.0**	27.7 ± 9.5[2]	36.9 ± 20.8[2]	43.5 ± 11.8[2]

[1]Value of the optimization criterion; [2]statistically significant from the highest result.

TABLE 3: Effect of the desorbing solvent volume on extraction efficacy of the five pesticides from rape oil samples.

Pesticide	Extraction recovery (%), $n = 3$				
	3 mL of acetonitrile	4 mL of acetonitrile	5 mL of acetonitrile	6 mL of acetonitrile	7 mL of acetonitrile
Metazachlor	18 ± 4	54 ± 6	**68 ± 6**	29 ± 8	15 ± 3
Tebuconazole	38 ± 5	70 ± 5	**76 ± 5**	48 ± 6	29 ± 5
λ-Cyhalothrin	34 ± 6	54 ± 7	**65 ± 5**	32 ± 9	16 ± 5
Chlorpyrifos	22 ± 6	61 ± 10	**69 ± 8**	51 ± 5	25 ± 7
Deltamethrin	16 ± 4	60 ± 5	**67 ± 6**	21 ± 8	12 ± 5
OC[1]	25.9 ± 9.8[2]	59.8 ± 6.7	**69.2 ± 4.2**	36.2 ± 12.9[2]	19.7 ± 7.1

[1]Value of the optimization criterion; [2]statistically significant from the highest result.

polar/hydrophobic properties. The hydrophobicity of the solvents along with their elution strength in relation to the studied pesticides, adsorbed on graphene, increases in the following sequence: acetone, acetonitrile, ethyl acetate, and a mixture of *n*-hexane and dichloromethane (2 1, v/v). The remaining conditions of the volume of the graphene suspension (5 mL), amount of NaCl (2 g), volume of the desorbing solvent (4 mL), sonication, desorption time (10 min), and temperature (ambient) during desorption were left constant. Considering the ease of experimental operations (very good separation of the organic supernatant and the graphene phase), the best solvent appeared to be ethyl acetate. The maximal extraction recovery for all of the pesticides was achieved with acetonitrile. Therefore, acetonitrile was selected as the desorbing solvent for further optimization process. The effect of the type of the used solvent is shown in Table 2.

3.3. Effect of Desorbing Solvent Volume. The effect of the five different volumes of acetonitrile (3, 4, 5, 6, and 7 mL) as the desorbing solvent on the extraction results of the five pesticides was examined. The remaining conditions of the volume of the graphene suspension (5 mL), amount of NaCl (2 g), type of the desorbing solvent (acetonitrile), sonication, desorption time (10 min), and temperature (ambient) during desorption were left constant. The highest extraction recovery for all the analytes was obtained using 5 mL of acetonitrile, and this volume was used in further experiments. The effect of this parameter is shown in Table 3.

3.4. Effect of Sonication Energy and Desorption Time. For desorbing the five pesticides from graphene, 5 mL of acetonitrile was added to the graphene phase (with the adsorbed analytes) at the bottom of a tube and vortexed for 1 min and then subjected to sonication energy for 5 and 10 minutes. Additionally, in a separate experiment, the graphene phase was vortexed (1 min) without applying sonication energy. The rest of the experimental conditions (volume of graphene suspension (5 mL) and ambient temperature during

TABLE 4: Effect of the sonication energy and desorption time on extraction efficacy of the five pesticides from rape oil samples.

Pesticide	Extraction recovery (%), $n = 3$		
	10 minutes subjected to sonication	5 minutes subjected to sonication	Without sonication energy
Metazachlor	**68 ± 8**	24 ± 4	37 ± 9
Tebuconazole	**74 ± 7**	9 ± 4	43 ± 6
λ-Cyhalothrin	**60 ± 7**	24 ± 7	32 ± 6
Chlorpyrifos	**66 ± 6**	42 ± 8	63 ± 7
Deltamethrin	**66 ± 7**	9 ± 5	10 ± 5
OC[1]	**67.0 ± 4.8**	21.8 ± 3.7[2]	37.0 ± 19.1[2]

[1]Value of the optimization criterion; [2]statistically significant from the highest result.

TABLE 5: Effect of the NaCl amount on extraction efficacy of the five pesticides from rape oil samples.

Pesticide	Extraction recovery (%), $n = 3$		
	2 g NaCl	1 g NaCl	Without addition of NaCl
Metazachlor	68 ± 5	**69 ± 6**	22 ± 6
Tebuconazole	74 ± 8	**73 ± 8**	19 ± 4
λ-Cyhalothrin	60 ± 7	**63 ± 4**	9 ± 5
Chlorpyrifos	66 ± 7	**70 ± 9**	24 ± 5
Deltamethrin	66 ± 7	**68 ± 7**	20 ± 5
OC[1]	66.8 ± 5.5	**68.5 ± 4.2**	19.0 ± 6.5[2]

[1]Value of the optimization criterion; [2]statistically significant from the highest result.

TABLE 6: Effect of the desorption temperature on extraction efficacy of the five pesticides from rape oil samples.

Pesticide	Extraction recovery (%), $n = 3$	
	At ambient temperature	At increased temperature (40°C)
Metazachlor	69 ± 10	**81 ± 5**
Tebuconazole	74 ± 8	**84 ± 7**
λ-Cyhalothrin	63 ± 5	**59 ± 5**
Chlorpyrifos	70 ± 6	**83 ± 5**
Deltamethrin	68 ± 7	**98 ± 8**
OC[1]	56.0 ± 3.8[2]	**80.9 ± 13.0**

[1]Value of the optimization criterion; [2]statistically significant from the highest result.

desorption) were kept constant. A desorption process of 10 min gave the greatest extraction recovery. The effects of these experiments on extraction results are given in Table 4.

3.5. Effect of the Salt Added. In all the above experiments, before introduction of the graphene suspension to an oil sample, 2 g of NaCl was added for salting out of the pesticides. In this optimization step, the extraction efficacy of the analytes was studied by decreasing the amount of salt added to 1 g or performing the sample preparation procedure without addition of NaCl. The rest of the experimental parameters (volume of graphene suspension (5 mL), acetonitrile volume (5 mL), sonication (10 min), and ambient temperature during desorption) were kept constant. The results are given in Table 5.

3.6. Effect of Desorption Temperature. The final experimental factor studied was the temperature of the pesticide desorption process from graphene. In the first step, the graphene phase (with the adsorbed pesticides) was heated with 5 mL of acetonitrile in a thermoblock to a temperature of 40°C; in the second step, the content of the tube was subjected to sonication for 10 min, maintaining the temperature at 40°C. The rest of the experimental parameters (volume of graphene suspension (5 mL), acetonitrile volume (5 mL), and sonication (10 min)) were kept constant. The effect of increasing the desorption temperature and how it considerably improves extraction efficiency are shown in Table 6.

3.7. Evaluation of Linearity Range and Limit of Detection of the d-SPE (Graphene)-HPLC/UV Method. At the optimized extraction conditions, the linearity range and detection limit for each pesticide were determined.

Representative chromatograms are presented in Figure 2, Calibration curves were designed on the basis of analyses of oil samples containing mixtures of the five pesticides at the

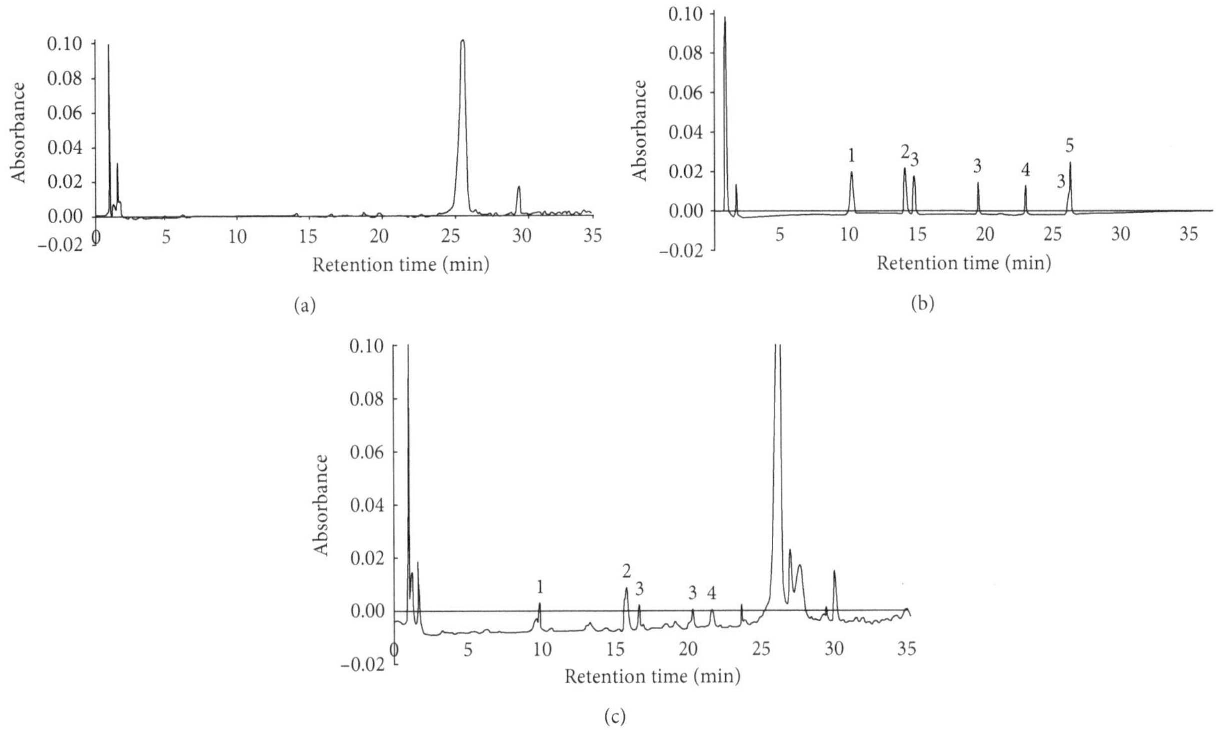

FIGURE 2: Chromatograms of the extract of the blank rape oil sample (a), five standard pesticides (1 μg/1 ml of methanol) (b), and the extract of the rape oil sample, spiked with the five standard pesticides (c) (concentration (62.5 ng/g) equal to LOQ). 1: metazachlor; 2: tebuconazole; 3: λ-cyhalothrin; 4: chlorpyrifos; 5: deltamethrin.

six concentrations (62.5, 125, 250, 500, 1000, and 2000 ng·g^{-1}) and the internal standard (chlorfenvinphos, 1 μg·mL^{-1}). A linearity range 62.5–2000 ng·g^{-1} was achieved for metazachlor, tebuconazole, λ-cyhalothrin, and chlorpyrifos with determination coefficients (r^2) 0.998, 0.991, 0.999, and 0.993, respectively. For deltamethrin, the linearity range was 500–2000 ng·g^{-1} with $r^2 = 0.999$. The detection limit was determined by estimation of the minimum concentration as equivalent to three times the background noise signal. The detection limit for metazachlor, tebuconazole, λ-cyhalothrin, and chlorpyrifos was 62.5 ng·g^{-1}, and for deltamethrin, it was 500 ng·g^{-1}.

4. Discussion

The potential adsorption properties of simple nonmodified graphene for the five pesticides (metazachlor, tebuconazole, λ-cyhalothrin, chlorpyrifos, and deltamethrin) analyzed from a rape oil matrix were explored. The fat content from the matrix was removed at two stages: (1) precipitation with acetonitrile and freezing at −32°C and (2) adsorption on the graphene surface (a smaller amount of oil left in the acetonitrile phase). The pesticides were adsorbed on graphene because of its high specific surface area, as well as through noncovalent interactions, especially the π-π stacking interaction with the aromatic rings of the studied compounds, and the hydrophobic effect.

The optimized d-SPE (graphene) extraction conditions for the five compounds present in rape oil samples were as follows: amount of graphene (15 mg), type of the desorbing solvent (acetonitrile), volume of the desorbing solvent (5 mL), desorption time using sonication energy (10 min), desorption temperature (40°C), and amount of salt (1 g NaCl). Under the optimized conditions, the following extraction recoveries for the examined pesticides were achieved: metazachlor (81%), tebuconazole (84%), λ-cyhalothrin (59%), chlorpyrifos (83%), and deltamethrin (98%). This is the first application of graphene-based materials in the analysis of pesticides in oil samples. There are relatively a small number of reports of carbon nanomaterials being used to extract pesticides from other less-complicated matrices including metazachlor from surface waters [24]; tebuconazole from environmental water [25], vegetable [20], and tea [22] samples; λ-cyhalothrin and deltamethrin from fruit [26] and vegetable [26, 27] samples; and chlorpyrifos from environmental water [14] and tea samples [22]. Depending on the kind of the pesticide or/and pesticide concentration and type of a matrix, the extraction recoveries of the pesticides of interest ranged from 79.0 to ca. 100%. The pesticides were separated and detected using the RP chromatographic system, which was adopted from literature [11] with slight modifications due to the tested pesticides. The detection limits of the five pesticides determined by the d-SPE (graphene)-HPLC/UV method were not quite satisfactory because some of them were above maximum residue levels (MRL; measured in rapeseeds [28]) for deltamethrin (MRL = 0.07 mg·kg^{-1}), metazachlor (MRL = 0.06 mg·kg^{-1}), and chlorpyrifos (MRL = 0.05 mg·kg^{-1}). For λ-cyhalothrin (MRL = 0.2 mg·kg^{-1}) and tebuconazole (MRL = 0.5 mg·kg^{-1}), the developed method may be used as the so-called "cutoff" method allowing to separate "good" samples from "bad" oil samples. However, in our investigations, HPLC-UV, which is a relatively not highly sensitive technique, was primarily used as the detection tool for estimating the adsorption potential of graphene for the studied pesticides present in the oil matrix. In the abovementioned works, the detection limits achieved were considerably lower (single nanogram or less), but in the vast majority cases (with the exceptions of [25] and [27] where 300 mL of the aqueous sample was analyzed by HPLC-UV and a 10 g portion of the homogenized plant sample was assayed by GC-ECD, resp.), highly sensitive detection methods like mass spectroscopy (MS) or tandem mass (MS-MS) spectroscopy were used.

5. Conclusion

In most cases of pesticide analysis involving graphene as an adsorbent, magnetic solid-phase extraction (MSPE) was applied where magnetically modified graphene composites as adsorbents were used. Only few works reported the employment of nonmodified graphene alone or in combination with other sorbents (in the modified QuEChERS method) using dispersive solid-phase extraction. This is the first time when graphene in its nonmodified form, using d-SPE, was studied as the adsorbent for the pesticides from four different groups present in such analytically difficult matrices as edible oil. Graphene may be successfully exploited in the preparation of edible oil samples for determination of pesticides from various groups. However, some improvements should be introduced in such analyses. For example, a more sensitive analytical method (e.g., GC-MS or LC-MS) may be used. Additionally, a more suitable modified graphene form, for example, magnetically modified graphene, which will prevent the nanomaterial aggregation, increase its dispersion in solvents, and improve its adsorption properties, would better enable separation of graphene from the isolated analytes in the supernatant using an external magnet.

Conflicts of Interest

The authors declare that they have no conflicts of interest.

References

[1] L. Polgár, B. Kmellár, J. F. García-Reyes, and P. Fodor, "Comprehensive evaluation of the clean-up step in QuEChERS procedure for the multi-residue determination of pesticides in different vegetable oils using LC-MS/MS," *Analytical Methods*, vol. 4, no. 4, p. 1142, 2012.

[2] C. Anagnostopoulos and G. E. Miliadis, "Development and validation of an easy multiresidue method for the determination of multiclass pesticide residues using GC-MS/MS and LC-MS/MS in olive oil and olives," *Talanta*, vol. 112, pp. 1–10, 2013.

[3] E. Sobhanzadeh, N. K. A. Bakar, A. M. R. Bin, and K. Nemati, "A simple and efficient multi-residue method based on QuEChERS for pesticides determination in palm oil by liquid chromatography time-of-flight mass spectrometry," *Environmental Monitoring and Assessment*, vol. 184, no. 9, pp. 5821–5828, 2012.

[4] D. Moreno-González, J. F. Huertas-Pérez, A. M. García-Campaña, and L. Gámiz-Gracia, "Determination of carbamates in edible vegetable oils by ultra-high performance liquid chromatography-tandem mass spectrometry using a new clean-up based on zirconia for QuEChERS methodology," *Talanta*, vol. 128, pp. 299–304, 2014.

[5] O. Lacina, M. Zachariasova, J. Urbanova, M. Vaclavikova, T. Cajka, and J. Hajslova, "Critical assessment of extraction methods for the simultaneous determination of pesticide residues and mycotoxins in fruits, cereals, spices and oil seeds employing ultra-high performance liquid chromatography-tandem mass spectrometry," *Journal of Chromatography A*, vol. 1262, pp. 8–18, 2012.

[6] P. Deme, T. Azmeera, B. L. A. P. Devi, P. R. Jonnalagadda, R. B. N. Prasad, and U. V. R. Vijaya Sarathi, "An improved dispersive solid-phase extraction clean-up method for the gas chromatography – negative chemical ionisation tandem mass spectrometric determination of multiclass pesticide residues in edible oils," *Food Chemistry*, vol. 142, pp. 144–151, 2014.

[7] T. Liu, Y. Ren, J. Xie, G. Song, and Y. Hu, "Determination of organophosphorus pesticides in edible oils by dispersive microextraction based on acetonitrile/water-coated Fe_3O_4," *Journal of the American Oil Chemists' Society*, vol. 90, no. 12, pp. 1937–1943, 2013.

[8] H. Muhamad, B. Hisyam, N. Kartini, and A. Bakar, "Comparative study of different clean-up techniques for the determination of λ-cyhalothrin and cypermethrin in palm oil matrices by gas chromatography with electron capture detection," *Food Chemistry*, vol. 134, no. 4, pp. 2489–2496, 2012.

[9] J. Li, D. Liu, T. Wu, W. Zhao, Z. Zhou, and P. Wang, "A simplified procedure for the determination of organochlorine pesticides and polychlorobiphenyls in edible vegetable oils," *Food Chemistry*, vol. 151, pp. 47–52, 2014.

[10] M. I. Ramli, Z. Zakaria, I. Sahid, T. Y. Ai, and H. Muhamad, "Determination of Herbicide diuron levels in palm oil matrices using HPLC-UV," *Sains Malaysiana*, vol. 41, pp. 1451–1459, 2012.

[11] T. Tuzimski and T. Rejczak, "Application of HPLC-DAD after SPE/QuEChERS with ZrO_2-based sorbent in d-SPE clean-up step for pesticide analysis in edible oils," *Food Chemistry*, vol. 190, pp. 71–79, 2016.

[12] T. Rejczak and T. Tuzimski, "A review of recent developments and trends in the QuEChERS sample preparation approach," *Open Chemistry*, vol. 13, no. 1, pp. 980–1010, 2015.

[13] Q. Zhou and Z. Fang, "Graphene-modified TiO_2 nanotube arrays as an adsorbent in micro-solid phase extraction for determination of carbamate pesticides in water samples," *Analytica Chimica Acta*, vol. 869, pp. 43–49, 2015.

[14] H. Razmi and M. Jabbari, "Development of graphene-carbon nanotube- coated magnetic nanocomposite as an efficient sorbent for HPLC determination of organophosphorus pesticides in environmental water samples," *International Journal of Environmental Analytical Chemistry*, vol. 95, no. 14, pp. 1353–1369, 2015.

[15] I. Montesinos, A. Sfakianaki, M. Gallego, and C. D. Stalikas, "Graphene-coated cotton fibers as a sorbent for the extraction of multiclass pesticide residues from water and their determination by gas chromatography with mass spectrometry," *Journal of Separation Science*, vol. 38, no. 5, pp. 836–843, 2015.

[16] A. Mehdinia, H. Khani, and S. Mozaffari, "Fibers coated with a graphene-polyaniline nanocomposite for the headspace solid-phase microextraction of organochlorine pesticides from seawater samples," *Microchimica Acta*, vol. 181, no. 1-2, pp. 89–95, 2014.

[17] Z. Shi, Q. Li, D. Xu, Q. Huai, and H. Zhang, "Graphene-based pipette tip solid-phase extraction with ultra high performance

liquid chromatography and tandem mass spectrometry for the analysis of carbamate pesticide residues in fruit juice," *Journal of Separation Science*, vol. 39, no. 22, pp. 4391–4397, 2016.

[18] X. Wu, R. Zhang, X. Liu et al., "Evaluation of graphene for effective cleanup of fruit and vegetable extracts in pesticide residue analysis," *Food Analytical Methods*, vol. 8, no. 1, pp. 243–253, 2015.

[19] N. Li, J. Chen, and Y. P. Shi, "Magnetic graphene solid-phase extraction for the determination of carbamate pesticides in tomatoes coupled with high performance liquid chromatography," *Talanta*, vol. 141, pp. 212–219, 2015.

[20] W. Guan, C. Li, X. Liu, S. Zhou, and Y. Ma, "Graphene as dispersive solidphase extraction materials for pesticides LC-MS/MS multi-residue analysis in leek, onion and garlic," *Food Additives & Contaminants: Part A*, vol. 31, no. 2, pp. 250–261, 2014.

[21] M. Zhang, H. Chen, L. Zhu, C. Wang, G. Ma, and X. Liu, "Solid-phase purification and extraction for the determination of trace neonicotinoid pesticides in tea infusion," *Journal of Separation Science*, vol. 39, no. 5, pp. 910–917, 2016.

[22] X. Liu, W. Guan, X. Hao, X. Wu, Y. Ma, and C. Pan, "Pesticide multi-residue analysis in tea using d-SPE sample cleanup with graphene mixed with primary secondary amine and graphitized carbon black prior to LC-MS/MS," *Chromatographia*, vol. 77, no. 1-2, pp. 31–37, 2014.

[23] R. Sitko, B. Zawisza, and E. Malicka, "Graphene as a new sorbent in analytical chemistry," *TrAC Trends in Analytical Chemistry*, vol. 51, pp. 33–43, 2013.

[24] M. C. Hurtado-Sánchez, R. Romero-González, and M. I. Rodríguez-Cáceres, "Rapid and sensitive on-line solid phase extraction-ultra high performance liquid chromatography–electrospray-tandem mass spectrometry analysis of pesticides in surface waters," *Journal of Chromatography A*, vol. 1305, pp. 193–202, 2013.

[25] W. Wang, X. Ma, Q. Wu, C. Wang, X. Zang, and Z. Wang, "The use of graphene-based magnetic nanoparticles as adsorbent for the extraction of triazole fungicides from environmental water," *Journal of Separation Science*, vol. 35, no. 17, pp. 2266–2272, 2012.

[26] M. Hou, X. Zang, C. Wang, and Z. Wang, "The use of silica-coated magnetic graphene microspheres as the adsorbent for the extraction of pyrethroid pesticides from orange and lettuce samples followed by GC-MS analysis," *Journal of Separation Science*, vol. 36, pp. 3242–3248, 2013.

[27] X. Wu, H. Zhang, L. Meng, X. Liu, and Y. Ma, "Graphene for cleanup in trace analysis of pyrethroid insecticides in cucumber and spinach," *Chromatographia*, vol. 75, no. 19-20, pp. 1177–1183, 2012.

[28] EU-Pesticides Database, 2017, http://ec.europa.eu/food/plant/pesticides/eu-pesticides.

11

A Pyridazine-Based Fluorescent Probe Targeting Aβ Plaques in Alzheimer's Disease

Yong Dae Park[1], **Jeum-Jong Kim**[2], **Sungbeom Lee**[1], **Chul-Hong Park**[1], **Hyoung-Woo Bai**[1], and **Seung Sik Lee**[1]

[1]*Research Division for Biotechnology, Advanced Radiation Technology Institute, Korea Atomic Energy Research Institute, Jeongeup 580-185, Republic of Korea*
[2]*Technology Innovation Support Team, Korea Research Institute of Chemical Technology (KRICT), Deajeon 305-600, Republic of Korea*

Correspondence should be addressed to Yong Dae Park; ydpark@kaeri.re.kr

Academic Editor: Subhankar Singha

Accumulation of β-amyloid (Aβ) plaques comprising Aβ40 and Aβ42 in the brain is the most significant factor in the pathogenesis of Alzheimer's disease (AD). Thus, the detection of Aβ plaques has increasingly attracted interest in the context of AD diagnosis. In the present study, a fluorescent pyridazine-based dye that can detect and image Aβ plaques was designed and synthesized, and its optical properties in the presence of Aβ aggregates were evaluated. An approximately 34-fold increase in emission intensity was exhibited by the fluorescent probe after binding with Aβ aggregates, for which it showed high affinity ($K_D = 0.35\,\mu M$). Moreover, the reasonable hydrophobic properties of the probe ($\log P = 2.94$) allow it to penetrate the blood brain barrier (BBB). In addition, the pyridazine-based probe was used in the histological costaining of transgenic mouse (APP/PS1) brain sections to validate the selective binding of the probe to Aβ plaques. The results suggest that the pyridazine-based compound has the potential to serve as a fluorescent probe for the diagnosis of AD.

1. Introduction

The misfolding and aggregation of proteins cause numerous neurodegenerative diseases, such as Alzheimer's disease (AD), prion disease (PrD), and Parkinson's disease (PD) [1]. AD, one of the most common protein misfolding diseases (PMDs), is characterized by the accumulation of misfolded β-amyloid (Aβ) peptides and neurofibrillary tangles (NFTs) containing tau protein in the brain. A recent report revealed that the buildup of Aβ plaques in the brain plays a significant role in the pathogenesis of AD [2, 3]. Therefore, approaches to visualize Aβ deposition might prove useful for diagnosing AD and evaluating the efficacy of AD therapeutics [4–6].

Several groups have reported novel positron emission tomography (PET) imaging agents targeting Aβ plaques to diagnose AD, including BAY94-9172, FDDNP, PIB, SB-13, AV-45, and IMPY [7–13]. However, these agents are hindered by factors such as long data acquisition processes, costly equipment, exposure to radioactivity, need for proficient personnel, and comparatively poor spatial resolution [14]. Interest in monitoring the progression of AD by imaging Aβ plaques using fluorescence spectroscopy has also increased [15, 16]. Compared to nuclear imaging methods, fluorescence imaging has many advantages, including providing real-time, nonradioactive, inexpensive, and high-resolution imaging, both in vivo and ex vivo. Consequently, various fluorescent probes for imaging Aβ plaques have been developed [17–22]. An excellent fluorescent probe for Aβ plaques must meet the following requirements [18, 21, 23]: (1) selective targeting of Aβ plaques, (2) acceptable lipophilicity ($\log P$ value between 1 and 3), (3) high-affinity binding, (4) straightforward synthesis, and (5) a significant change in fluorescent properties upon binding to Aβ deposits.

Based on these requirements, we developed and reported fluorescent pyridazine probes targeting Aβ plaques [24].

These pyridazine probes can be used for imaging through selective binding but lack the required binding affinity for $A\beta$ plaques. Here, we describe the optimization of pyridazine derivatives based on the conjugation of an electron acceptor with an electron donor.

To optimize these fluorescent probes, the electron-donating p-dimethylamino group and electron-accepting cyano group were introduced to construct a compound with a donor-π-acceptor structure (Figure 1). In this paper, we describe the synthesis and optical and biological properties of a cyano-based probe based on pyridazine. The ex vivo staining of $A\beta$ plaques in APP/PS1 mice brain sections by this fluorescent probe is also presented.

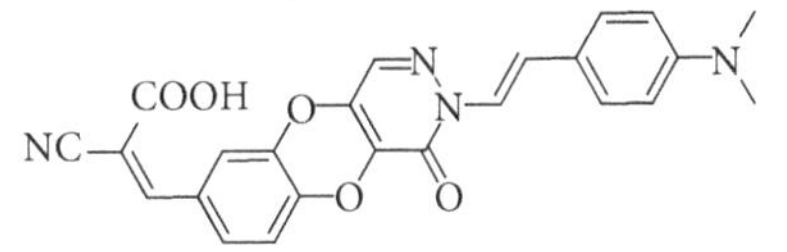

Figure 1: Chemical structure of fluorescent probe **3**.

2. Materials and Methods

2.1. General Experimental Methods. ^{1}H NMR spectra were recorded in $CDCl_3$ unless otherwise noted (values in ppm) using TMS as the standard with a JNM-ECA 500 spectrometer. Low resolution mass spectra were recorded using a Varian MAT 212 mass spectrometer. IR spectra (KBr) were measured with a Bruker-Vector 22 instrument (Bruker, Bremen). Flash column chromatography was performed using silica gel (70–230 mesh). All reagent-grade chemicals were purchased from Sigma-Aldrich (St. Louis, MO, USA), and synthetic $A\beta_{42}$ peptide was purchased from rPeptide (Bogart, GA, USA).

2.2. Synthesis and Characterization of Catechol Aldehyde (2). A mixture of **1** (300 mg, 0.97 mmol), 3,4-dihydroxybenzaldehyde (147 mg, 1.06 mmol), and K_2CO_3 (293 mg, 2.12 mmol) was dissolved in DMF (20 ml) and refluxed for 24 h. After evaporating the solvent under reduced pressure, H_2O (100 ml) and methylene chloride (50 ml) were added. The organic layer was separated and dried over $MgSO_4$. The pure product (**2**) was obtained by column chromatography on silica gel using CH_2Cl_2 as the eluent. Yield: 89%. IR (KBr) = 3091, 2920, 2852, 1691, 1671, 1647, 1605, 1590, 1526, 1499, 1364, 1280, 1188. ^{1}H NMR ($CDCl_3$) = 9.83 (s, 1H), 8.01 (d, 1H, J = 13.75 Hz), 7.67 (s, 1H), 7.54–7.38 (m, 4H), 7.18–6.99 (m, 2H), 6.69 (d, 2H, J = 8.70 Hz), 2.98 (S, 6H). MS (EI) m/z 375 $[M]^+$, 188, 159, 145, 117.

2.3. Synthesis and Characterization of Probe **3**. A mixture of **2** (100 mg, 0.27 mmol) and cyanoacetic acid (30 mg, 0.36 mmol) was vacuum dried, and $CHCl_3$ (50 ml) and piperidine were added. The solution was refluxed for 15 h. Then, H_2O (50 ml) was added. The organic layer was separated and dried over $MgSO_4$. The pure product (**3**) was obtained by column chromatography on silica gel (CH_2Cl_2: MeOH = 6 : 1). Yield: 58%. IR (KBr) = 3398, 3091, 2922, 2853, 2211, 1651, 1632, 1603, 1524, 1503, 1363, 1335, 1277, 1187, 1163, 1125. ^{1}H-NMR ($CDCl_3$) = 8.04 (d, 1H, J = 6.86 Hz), 7.81 (d, 1H, J = 14.25 Hz), 7.80 (s, 1H), 7.55–7.51 (m, 2H), 7.37–7.33 (dd, 2H, J = 8.56, 8.52 Hz), 7.18–7.09 (m, 1H), 6.99 (d, 1H, J = 14.29 Hz), 6.66 (d, 2H, J = 7.42 Hz), 2.89 (S, 6H). MS (EI) m/z 398 $[M\text{-}CO_2]^+$, 382, 256, 145, 129, 111, 97, 83, 78, 63.

2.4. UV/VIS and Fluorescence Analysis. UV/VIS and fluorescence spectra were recorded and analyzed. For the UV/VIS spectra, an Infinite M200 Pro Microplate reader (Tecan, Switzerland) equipped with cells with a 1.0 cm path length was used. The scan rate was 120 nm/min. The excitation and emission λ_{max} values of probe **3** (10 μM) were recorded with a detector (slit of 1 mm) and a data interval of 5 nm in DMF.

2.5. Preparation of Aβ42 Aggregates and Fluorescence Spectrum Measurement. Aggregated $A\beta$ peptide was prepared by diluting $A\beta42$ to a final concentration of 100 μM in PBS (pH 7.4). This solution was incubated at 200 rpm and 37°C for 3 days. The formation of $A\beta$ fibrils was confirmed by ThT assay. The excitation and emission λ_{max} values of probe **3** were measured using an Infinite M200 Pro Microplate reader (Tecan, Switzerland) equipped with a detector (slit 1 mm) with a data interval of 5 nm. The scan rate was 120 nm/min. Probe **3** (10 μM) was reacted with and without 20 μM $A\beta$ aggregates for 20 min in PBS at 37°C. The emission spectra and fluorescence intensity of the samples were measured. The fold increase was calculated by comparing the fluorescence intensity with and without 20 μM $A\beta$ aggregates.

2.6. Binding Constant (K_D) Measurement. A 10 μM solution of aggregated $A\beta42$ was combined with probe **3** (0.1, 0.5, 1, 2, 5, and 10 μM) in PBS (pH 7.4). The solutions were incubated for 10 min at 37°C, and then their fluorescence intensity was determined at 408 nm (excitation wavelength). K_D was determined as described previously [25].

2.7. Lipophilicity (log P). Probe **3** was added to a premixed suspension containing 500 μL of octanol and 500 μL of PBS solution, and the resulting suspension was vortexed vigorously for 10 min and centrifuged at 3000 rpm for 5 min. Two layers separated out, and 100 μL aliquots from octanol and the PBS solution layers were removed and analyzed for their fluorescence intensity. The log P value was calculated as the logarithm of the ratio of the fluorescence intensity in octanol versus that in PBS solution.

2.8. Maestro Images Analysis. An optical data study was performed using a Maestro 2.0 in vivo imaging system. The images were acquired as described previously [25]. Solutions of probe **3** (1 μM) were prepared with and without 20 μM $A\beta$ aggregates in PBS. Fluorescence emission was obtained by analyzing the resulting images with commercial software (Maestro™ 2.4).

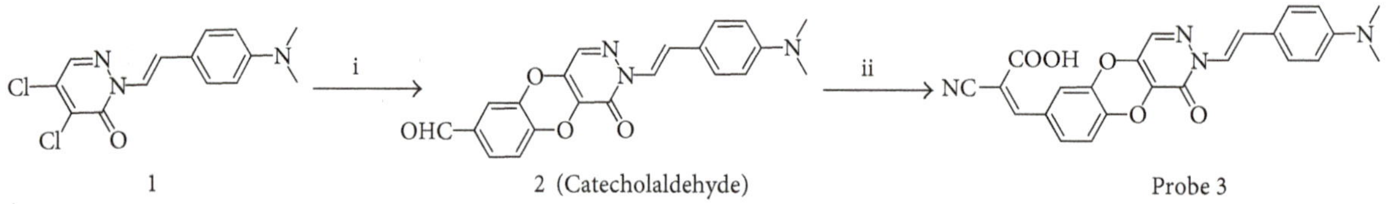

SCHEME 1: Reaction scheme for the synthesis of the pyridazine-based probe (**3**). (i) DMF, 3,4-dihydroxybenzaldehyde, K_2CO_3, refluxed for 24 h; (ii) $CHCl_3$, cyanoacetic acid, piperidine, refluxed for 15 h.

2.9. Histological Costaining with Aβ Antibody and Probe **3**. The brain from 12-month-old transgenic APP/PS1 mice was removed and cut into 5 μm sections. The mouse brain sections were stained with probe **3** and anti-Aβ using the following method: first, the brain sections were equilibrated in PBS solution for 10 min, washed with PBS containing 0.1% Tween 20 (PBS-T) and 5% BSA for 30 min, and washed again with PBS-T supplemented with 1% BSA for 5 min 3 times. Second, the washed sections were incubated with primary antibody (rabbit anti-Aβ, 1 : 100 dilution in PBS-T supplemented with 1% BSA) overnight at 4°C, washed with PBS-T supplemented with 1% BSA 3 times, and stained with secondary antibody (Alexa 555 goat antirabbit IgG, 1 : 100 dilution in PBS-T supplemented with 1% BSA). After washing with PBS, the prestained sections were stained with 10 μM probe **3** for 30 min. The stained section was washed with PBS and analyzed under an FV1000D (Olympus, Tokyo, Japan) confocal laser scanning microscope.

3. Results and Discussion

The synthesis of probe **3** is outlined in Scheme 1. First, commercially available 3,4-dihydroxybenzaldehyde was converted to the corresponding catechol aldehyde (**2**) by reacting it with compound **1**. The Knövenagel condensation of compound **2** with cyanoacetic acid afforded the final fluorescent probe (**3**).

The optical properties of the synthesized fluorescent probe (**3**) with aggregated Aβ42 peptides in PBS (pH 7.4) were analyzed, and the results are shown at Table 1. Probe **3** exhibited an excitation maximum at 408 nm and an emission maximum at 670 nm (Table 1 and Figure 2).

TABLE 1: Fluorescence profile and K_D and log P values of probe **3** with Aβ aggregates.

Optical properties	Probe **3**
λ_{ex} (nm)	408
λ_{em} (nm)	670
$\lambda_{ex}/\lambda_{em}$ with Aβ (nm)	408/604
Fold increase with Aβ	34.92
K_D (mean ± SD) (μM)	0.35 ± 0.03
log P (lipophilicity)	2.94

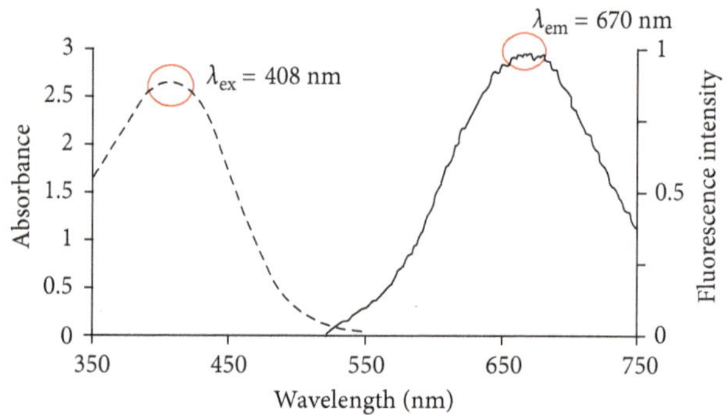

FIGURE 2: Absorbance and emission spectra of probe **3** in DMF. The maximum wavelengths in the absorbance and emission spectra are 408 nm and 670 nm, respectively.

To operate as a fluorescent probe targeting Aβ plaques, a compound must show a significant rise in fluorescence intensity upon binding with Aβ aggregates compared to the fluorescence intensity of free Aβ aggregates in solution [15]. Therefore, we compared the fluorescence intensity of probe **3** to the fluorescence intensity of the probe in the presence of Aβ aggregates (Figure 3(a)). As shown in Table 1, we observed a remarkable increase (35-fold) in the fluorescence intensity of probe **3** in the presence of Aβ aggregates. Additionally, the gain in fluorescence intensity was visually confirmed using a Maestro fluorescence imaging system (Figure 3(c)). This effect is due to conformational changes: When the probe in solution with Aβ aggregates is in the unbound state, free rotation through a single bond is permitted, whereas upon binding to Aβ aggregates, the probe exhibits a significant increase in fluorescence intensity due to restricted movement [26]. The binding of probe **3** to Aβ aggregates was also accompanied by a blueshift in the emission spectrum [15]. The emission wavelength of probe **3** exhibited significant blueshifts (66 nm, Table 1), indicating that probe **3** likely intercalated into the hydrophobic pocket of the Aβ aggregates. This result suggested that probe **3** could be "turned on" via an increase in fluorescence intensity and a blueshift in its emission wavelength upon interacting with Aβ aggregates.

Next, we measured the apparent binding constant (K_D) of fluorescent probe **3** to Aβ aggregates. The fluorescence intensity of solutions of probe **3** at various concentrations in the presence of Aβ aggregates was measured, revealing that the K_D value of probe **3** was 0.35 ± 0.03 μM (Table 1 and Figure 3(b)). This binding constant was significantly higher than that of our previously reported fluorescence probe, probe **1** (1.83 ± 0.31 μM) [24]. The lipophilicity (log P) of probe **3** was also evaluated to determine whether it could permeate through the blood brain barrier (BBB). The log P value of probe **3** was found to be 2.94 (Table 1), suggesting that probe **3** has desirable properties regarding BBB permeability [21].

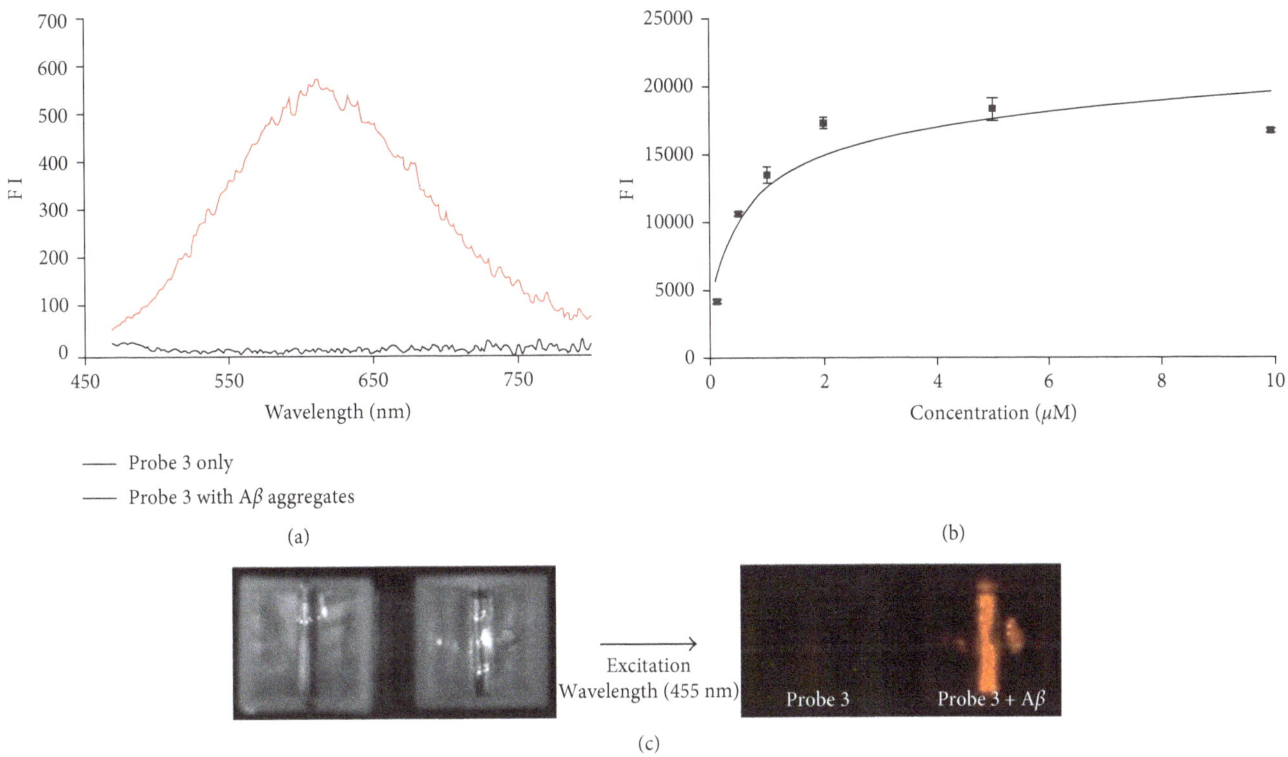

FIGURE 3: Emission spectra of probe **3** in the presence and absence of Aβ aggregates (a) and a plot of the fluorescence intensity (at $\lambda_{em} = 604$) as a function of the concentration of probe **3** in the presence of Aβ aggregates (10 μM) in PBS (b). The apparent dissociation constant (K_D) was 0.35 ± 0.03 μM. (c) Imaging of the fluorescence intensity of probe **3** and Aβ42 aggregates using a Maestro imaging system.

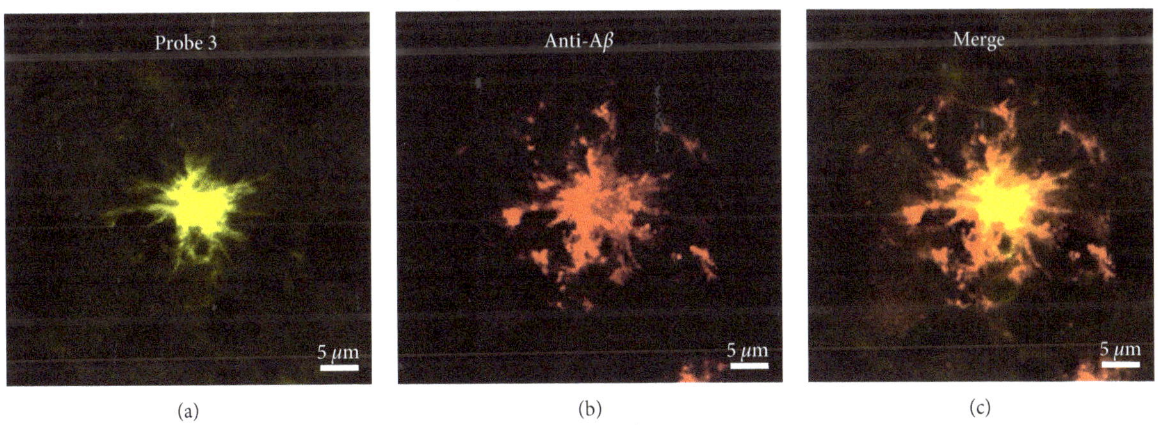

FIGURE 4: Histological double staining of 5 μm double sections from the cortex of APP/PS1 mouse brains with probe **3** and anti-Aβ. All of the images were acquired at a certain excitation wavelength (anti-Aβ: 555 nm and probe **3**: 408 nm) by a confocal laser scanning microscope.

The probe developed in this paper, probe **3**, meets the requirements for a fluorescence imaging probe for AD: high fluorescence receptivity, strong binding affinity, and hydrophobicity. To assess whether fluorescent probe **3** could stain Aβ plaques in mouse brain tissue, we further evaluated the histological costaining of Aβ plaques in APP/PS1 mouse brain sections with probe **3** and anti-Aβ. Aβ plaques in the mouse brain section were identified by staining with anti-Aβ as a control. As shown in Figure 4, the brain section exposed to probe **3** exhibited significant fluorescence. Notably, the merged images showed colocalization of the areas stained with probe **3** and anti-Aβ, which demonstrates the selective targeting of Aβ plaques by probe **3**.

4. Conclusions

In summary, we successfully synthesized probe **3** as a novel Aβ plaque-targeting fluorescent probe by applying the concept of a donor-π-acceptor structure to the scaffold of a previously reported pyridazine dye, probe **1**. Probe **3** exhibited a strong fluorescence response ($F_{A\beta}/F_0 > 34$-fold), high affinity for Aβ42 aggregates ($K_D = 0.35 \pm 0.03$ μM), and sufficient hydrophobicity to penetrate the BBB (log $P = 2.94$). Furthermore, probe **3** specifically stained the Aβ plaques in APP/PS1 mouse brain sections. These results indicate probe **3** as a novel fluorescence imaging agent for the study of AD.

Conflicts of Interest

The authors declare that they have no conflicts of interest.

Acknowledgments

This project was supported by the Ministry of Science and ICT (MIST), Republic of Korea.

References

[1] G. Edwards III, I. Moreno-Gonzalez, and C. Soto, "Amyloid-beta and tau pathology following repetitive mild traumatic brain injury," *Biochemical and Biophysical Research Communications*, vol. 483, no. 4, pp. 1137–1142, 2017.

[2] E. D. Roberson and L. Mucke, "100 years and counting: prospects for defeating Alzheimer's disease," *Science*, vol. 314, no. 5800, pp. 781–784, 2006.

[3] J. H. Kim, J. Lee, S. Lee, and E. J. Cho, "Quercetin and quercetin-3-β-d-glucoside improve cognitive and memory function in Alzheimer's disease mouse," *Applied Biological Chemistry*, vol. 59, no. 5, pp. 721–728, 2016.

[4] C. A. Mathis, Y. Wang, and W. E. Klunk, "Imaging beta-amyloid plaques and neurofibrillary tangles in the aging human brain," *Current Pharmaceutical Design*, vol. 10, no. 13, pp. 1469–1492, 2004.

[5] A. Nordberg, "PET imaging of amyloid in Alzheimer's disease," *The Lancet Neurology*, vol. 3, no. 9, pp. 519–527, 2004.

[6] D. J. Selkoe, "Imaging Alzheimer's amyloid," *Nature Biotechnology*, vol. 18, no. 8, pp. 823-824, 2000.

[7] E. D. Agdeppa, V. Kepe, J. Liu et al., "Binding characteristics of radiofluorinated 6-dialkylamino-2-naphthylethylidene derivatives as positron emission tomography imaging probes for beta-amyloid plaques in Alzheimer's disease," *Journal of Neuroscience*, vol. 21, no. 24, p. RC189, 2001.

[8] H. F. Kung, S. R. Choi, W. Qu, W. Zhang, and D. Skovronsky, "18F stilbenes and styrylpyridines for PET imaging of Aβ plaques in Alzheimer's disease: a miniperspective," *Journal of Medicinal Chemistry*, vol. 53, no. 3, pp. 933–941, 2010.

[9] C. A. Mathis, Y. Wang, D. P. Holt, G. F. Huang, M. L. Debnath, and W. E. Klunk, "Synthesis and evaluation of 11C-labeled 6-substituted 2-arylbenzothiazoles as amyloid imaging agents," *Journal of Medicinal Chemistry*, vol. 46, no. 13, pp. 2740–2754, 2003.

[10] N. Okamura, Y. Shiga, S. Furumoto et al., "In vivo detection of prion amyloid plaques using [(11)C]BF-227 PET," *European Journal of Nuclear Medicine and Molecular Imaging*, vol. 37, no. 5, pp. 934–941, 2010.

[11] M. Ono, A. Wilson, J. Nobrega et al., "11C-labeled stilbene derivatives as Aβ-aggregate-specific PET imaging agents for Alzheimer's disease," *Nuclear Medicine and Biology*, vol. 30, no. 6, pp. 565–571, 2003.

[12] C. C. Rowe, U. Ackerman, W. Browne et al., "Imaging of amyloid beta in Alzheimer's disease with 18F-BAY94-9172, a novel PET tracer: proof of mechanism," *The Lancet Neurology*, vol. 7, no. 2, pp. 129–135, 2008.

[13] Z. P. Zhuang, M. P. Kung, A. Wilson et al., "Structure-activity relationship of imidazo[1,2-a]pyridines as ligands for detecting beta-amyloid plaques in the brain," *Journal of Medicinal Chemistry*, vol. 46, no. 2, pp. 237–243, 2003.

[14] R. Weissleder and U. Mahmood, "Molecular imaging," *Radiology*, vol. 219, no. 2, pp. 316–333, 2001.

[15] W. M. Chang, M. Dakanali, C. C. Capule, C. J. Sigurdson, J. Yang, and E. A. Theodorakis, "ANCA: a family of fluorescent probes that bind and stain amyloid plaques in human tissue," *ACS Chemical Neuroscience*, vol. 2, no. 5, pp. 249–255, 2011.

[16] G. Yu, H. Jung, and H. Mok, "Indocyanine green-incorporated exosomes for improved in vivo imaging of sentinel lymph node," *Applied Biological Chemistry*, vol. 59, no. 1, pp. 71–76, 2016.

[17] M. Hintersteiner, A. Enz, P. Frey et al., "In vivo detection of amyloid-beta deposits by near-infrared imaging using an oxazine-derivative probe," *Nature Biotechnology*, vol. 23, no. 5, pp. 577–583, 2005.

[18] E. E. Nesterov, J. Skoch, B. T. Hyman, W. E. Klunk, B. J. Bacskai, and T. M. Swager, "In vivo optical imaging of amyloid aggregates in brain: design of fluorescent markers," *Angewandte Chemie International Edition*, vol. 44, no. 34, pp. 5452–5456, 2005.

[19] S. B. Raymond, J. Skoch, I. D. Hills, E. E. Nesterov, T. M. Swager, and B. J. Bacskai, "Smart optical probes for near-infrared fluorescence imaging of Alzheimer's disease pathology," *European Journal of Nuclear Medicine and Molecular Imaging*, vol. 35, no. S1, pp. S93–S98, 2008.

[20] Q. Li, J. S. Lee, C. Ha et al., "Solid-phase synthesis of styryl dyes and their application as amyloid sensors," *Angewandte Chemie International Edition*, vol. 43, no. 46, pp. 6331–6335, 2004.

[21] C. Ran, X. Xu, S. B. Raymond et al., "Design, synthesis, and testing of difluoroboron-derivatized curcumins as near-infrared probes for in vivo detection of amyloid-beta deposits," *Journal of the American Chemical Society*, vol. 131, no. 42, pp. 15257–15261, 2009.

[22] M. Ono, M. Ishikawa, H. Kimura et al., "Development of dual functional SPECT/fluorescent probes for imaging cerebral beta-amyloid plaques," *Bioorganic and Medicinal Chemistry Letters*, vol. 20, no. 13, pp. 3885–3888, 2010.

[23] W. E. Klunk, H. Engler, A. Nordberg et al., "Imaging brain amyloid in Alzheimer's disease with Pittsburgh Compound-B," *Annals of Neurology*, vol. 55, no. 3, pp. 306–319, 2004.

[24] Y. D. Park, J. H. Park, M. G. Hur et al., "Fluorescent 2-styrylpyridazin-3(2H)-one derivatives as probes targeting amyloid-beta plaques in Alzheimer's disease," *Bioorganic and Medicinal Chemistry Letters*, vol. 22, no. 12, pp. 4106–4110, 2012.

[25] S. J. Jung, S. H. Park, E. J. Lee et al., "Development of fluorescent probes that bind and stain amyloid plaques in Alzheimer's disease," *Archives of Pharmacal Research*, vol. 38, no. 11, pp. 1992–1998, 2015.

[26] M. Staderini, M. A. Martin, M. L. Bolognesi, and J. C. Menendez, "Imaging of beta-amyloid plaques by near infrared fluorescent tracers: a new frontier for chemical neuroscience," *Chemical Society Reviews*, vol. 44, no. 7, pp. 1807–1819, 2015.

12

Quantitative Analysis and Band Gap Determination for CIGS Absorber Layers using Surface Techniques

Yun Jung Jang,[1,2] Jihye Lee,[1] Kang-Bong Lee,[3] Donghwan Kim,[2] and Yeonhee Lee[1]

[1]*Advanced Analysis Center, Korea Institute of Science and Technology, Seoul 02792, Republic of Korea*
[2]*Department of Materials Science and Engineering, Korea University, Seoul 02841, Republic of Korea*
[3]*Green City Technology Institute, Korea Institute of Science and Technology, Seoul 02792, Republic of Korea*

Correspondence should be addressed to Yeonhee Lee; yhlee@kist.re.kr

Academic Editor: Chih-Ching Huang

Recently, $Cu(In_XGa_{(1-X)})Se_2$ (CIGS) absorber layers have been extensively studied by many research groups for thin-film solar cell technology. CIGS material is particularly promising due to its exceptionally high absorption coefficient and large band gap range, which is adjustable as a function of alloy stoichiometry. To enhance the conversion performance of CIGS solar cells, understanding the CIGS structure and composition is a crucial challenge. We conducted a quantitative study to determine the bulk composition of the major elements such as Cu, In, Ga, and Se of four different CIGS photovoltaic cells. The compositional information was obtained by X-ray fluorescence (XRF), inductively coupled plasma atomic emission spectroscopy (ICP-AES), and femtosecond laser ablation inductively coupled plasma mass spectrometry (*fs*-LA-ICP-MS). Then, the XRF concentration ratio was compared with the intensity ratio of *fs*-LA-ICP-MS to investigate the potential of accurate and rapid analysis using the *fs*-LA-ICP-MS technique. In contrast to the bulk information, the surface techniques can supply detailed information about the chemical composition across the depth profile. Here, elemental depth distributions of CIGS thin films were investigated using magnetic sector secondary ion mass spectrometry (SIMS) and Auger electron spectroscopy (AES). The atomic distributions of four different CIGS absorber layers exhibited a good agreement although they were obtained using two different surface instruments, AES and SIMS. Comparative analysis results of different CIGS absorber layers using SIMS, AES, and *fs*-LA-ICP-MS provide us with the appropriate technique for the information of accurate composition in a rapid analysis time. Thanks to a simple approach using the Ga/(In + Ga) ratio, the optical band gap energy of the $Cu(In_XGa_{(1-X)})Se_2$ quaternary layer was monitored in the entire CIGS layer. The elemental distribution and the band gap determination were then used to elucidate their relationship to the corresponding CIGS cell efficiency result.

1. Introduction

$Cu(In,Ga)Se_2$- (CIGS-)based photovoltaics is one of the most efficient thin-film solar cells due to their unique properties, such as high-absorption coefficient, tunable band gap, and long-term stability. CIGS solar cells have shown notable efficiencies between 21.7 and 22.6% in the laboratory when heavy alkali elements, rubidium and cesium, were incorporated into the substitution of sodium and potassium [1, 2]. Investigating the elemental concentration of CIGS absorber layers is an important challenge to understand the mechanisms that allow CIGS solar cells to exhibit high conversion efficiencies. Indeed, concentration plays a major role because it affects several properties of the CIGS layers, such as morphology, conductivity, carrier concentration, and, more importantly, band gap. For example, an excess of copper concentration in CIGS film induces electrical features such as high conductivity and high doping density from the $Cu_{2-X}Se$ secondary phase, which is not a single-phase solar cell, but plays a key role in grain growth [3]. The morphology varies with the copper concentration in the CIGS thin film, while the phase transformation depends on the substrate temperature [4, 5]. In addition, the Ga/(In + Ga) ratio is known to affect the band gap of absorber layers, which can be tuned by the Ga content [6–8]. CIGS thin films deposited by a three-step process increase the Ga composition toward the Mo layer, which causes an increase in the band energy. An increased band gap close to the Mo back contact reduces the recombination at the CIGS/Mo interface by accelerating the generated electrons to the PN-junction [9, 10]. Therefore, analysis of the composition of the absorber layers and characterization of the compositional distribution as a function of depth are essential.

Many research groups have studied the composition of CIGS absorber layers through inductively coupled plasma atomic emission spectroscopy (ICP-AES), X-ray fluorescence (XRF), and wavelength-dispersed electron probe micro-analyzer (EPMA) [11–14]. Recently, femtosecond laser ablation inductively coupled plasma mass spectrometry (*fs*-LA-ICP-MS) has been applied for the analysis of CIGS solar-cell film due to the sensitivity, rapid analysis, and minimal consumption of the region for an elemental analysis [15–18]. However, the above techniques have limitations in measuring the elemental concentration as a function of depth, and thus surface analytical techniques such as secondary ion mass spectrometry (SIMS) and Auger electron spectroscopy (AES) have been widely used [19–23]. The SIMS technique, with its high detection limit and ability to obtain the profiles of elements in parallel, is a powerful tool for investigating trace and constitutional elements in CIGS absorber thin films.

Here, four different CIGS absorber layers were fabricated by a three-step coevaporation process. The average compositions of the major elements in the CIGS absorber layers were measured by ICP-AES and XRF measurements. *fs*-LA-ICP-MS was also used for elemental analysis of the CIGS thin films, with the advantages of rapid analysis, minimal sample consumption, and no sample preparation. Surface techniques were compared with other analytical methods for the quantitative determination of the composition of the CIGS thin films. The bulk compositional techniques such as XRF, ICP-AES, and *fs*-LA-ICP-MS lead to various advantages for the analysis of the CIGS thin films. They are sensitive, accurate, and rapid methods without vacuum system so that they can be used complementarily to the surface techniques for the CIGS analysis. Depth profiles from SIMS and AES were used to compare the compositional distributions and to determine the band gap in the CIGS solar cells.

2. Materials and Methods

2.1. CIGS Solar-Cell Preparation. The $Cu(In,Ga)Se_2$ solar cells were produced at the solar-cell center of the Korea Institute of Science and Technology (KIST). Four different CIGS solar cells were fabricated with a similar stack structure, i.e., Al-doped ZnO (250 nm)/i-ZnO (50 nm)/CdS (50 nm)/CIGS (2 μm)/Mo (500 nm) on a soda-lime glass (SLG) substrate. The bilayer Mo film, consisting of a 100 nm thick porous bottom and a 400 nm thick dense top, was deposited by direct current sputtering. The CIGS absorbers were prepared with four different conditions to clarify the influence of the experimental conditions on the cell efficiency. Three absorber layers, B01, B02, and B03, were fabricated on a bilayer Mo film on an SLG substrate in a sequential three-stage evaporation process using Ga-In-Se, Cu-Se, and Ga-in-Se [24]. For the first sample (B01), the Se flux was maintained at 22 Å/s during CIGS growth, but this was changed to a 5 Å/s flux for the second sample (B02). The Se flux during three-stage coevaporation tended to influence the preferred orientation and morphology of the CIGS films. The B03 sample was prepared under experimental conditions similar overall to those of the first sample, but the 3rd stage and finishing step were performed at slightly lower temperature (540°C) instead of 570°C. It is considered to influence the fabrication of the CIGS absorber layer. Different processing temperatures are often attributed to the variation of device cell efficiency. For the B04 sample, the SiO_x film was deposited by plasma-enhanced chemical vapor deposition (PECVD) at 200°C on an SLG substrate to interrupt the diffusion of Na elements from the substrate. The process condition of B04 was correlated to the low content of alkali metals in the CIGS absorber. Instead of a porous bottom Mo layer (100 nm), a Mo : Na film (5 wt.%) was used as the first bottom Mo film grown on the SiO_x film. The CdS buffer layer was deposited by chemical bath deposition for 10 min at 75°C. The intrinsic ZnO layer and an Al-doped ZnO layer, with thicknesses of 50 nm and 250 nm, respectively, were made by radio frequency sputtering. The configurations and conditions of deposition of the CIGS solar cells are summarized in Table 1.

2.2. Instrumental Analysis. ICP-AES and XRF were used to determine the bulk elemental composition of the CIGS absorber layers. ICP-AES measurements were obtained using an Agilent 720-OES emission spectrometer (Agilent Technologies, Santa Clara, CA, USA). Calibration of the ICP-AES was performed using an aqueous solution of 5% nitric acid and 1 ppm, 5 ppm, and 10 ppm solutions of Cu, In, Ga, and Se, respectively. For ICP-AES measurement, a $0.5 \times 0.5\ cm^2$ CIGS film was digested with HNO_3 until the total dissolution of the CIGS films and the remaining Mo/glass substrate was removed from the solution. XRF measurements were completed using a ZSX Primus II (Rigaku, Tokyo, Japan), which was initially calibrated by a fundamental parameter (FP) based on the ratio of the four elements of the CIGS layer. For XRF measurement, CIGS standard was used and the entire 2 μm thick CIGS films were sampled. A femtosecond laser ablation instrument (J200, Applied Spectra, USA) was coupled to an ICP-MS (iCAP Q, Thermo, USA) using the following conditions: a fluence of pulsed laser at 5.1 J/cm^2 with a repetition time of 200 Hz and a 5 μm diameter spot size. Each sample was scanned by the laser beam in a zigzag pattern over an area of $200 \times 200\ \mu m^2$. At this condition, the complete ablation of a rectangular crater took about 3.85 s. The crater depth was more than 6 μm so that the Mo layer and part of the SLG substrate and CIGS film were sampled. The same *fs*-LA-ICP-MS measurement was repeated over a 3×3 crater array for each CIGS film, and the average from the nine rectangular craters were then used to represent the ICP-MS isotope signal of each CIGS sample.

XPS measurement was performed with a PHI-5000 VersaProbe instrument (ULVAC-PHI, Japan) using Al Kα

Table 1: Configurations and conditions of the deposition processes for the CIGS samples.

Sample number		B01	B02	B03	B04
Configuration	n-TCO	i-ZnO/ZnO : Al	i-ZnO/ZnO : Al	i-ZnO/ZnO : Al	i-ZnO/ZnO : Al
	n-buffer	CdS	CdS	CdS	CdS
	p-absorber (μm)	CIGS (1.70)	CIGS (1.98)	CIGS (2.24)	CIGS (2.21)
	Back contact	Mo	Mo	Mo	Mo/Mo : Na/SiO_x
Condition of CIGS deposition (Å/s) (°C)		Normal Se flux (~22) (570)	Low Se flux, (~5) (570)	Normal Se flux (~22) (540)	Normal Se flux (~22) (570)
Cell efficiency (%)		14.6	14.3	11.7	10.3
SEM cross-section image		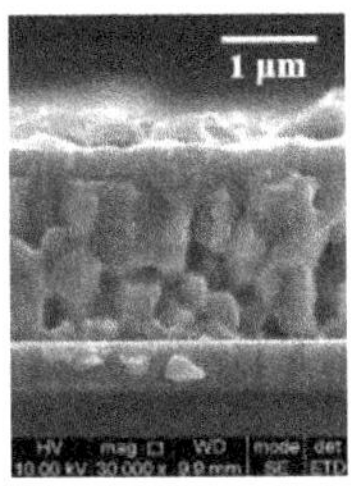	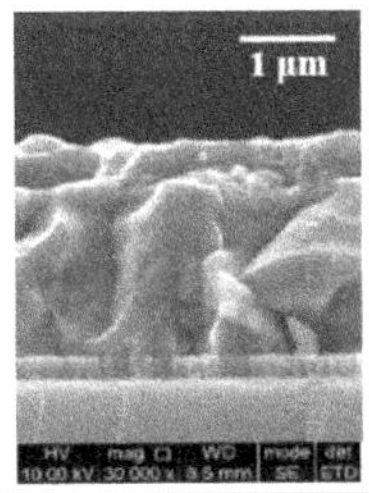	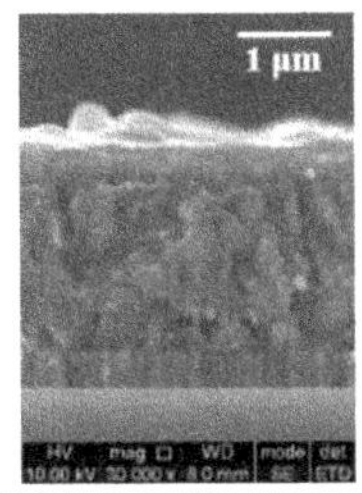	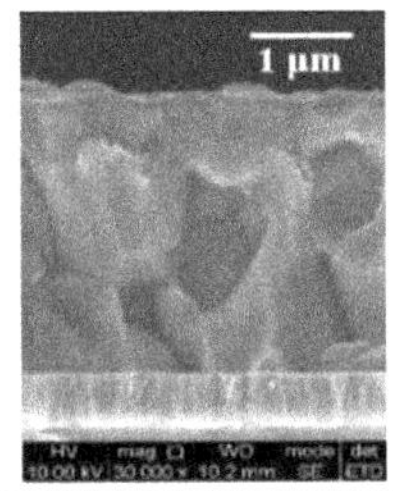

(1486.6 eV) radiation. The Na 1s spectra were detected at a low pass energy of 58.70 eV. The depth profiles of Cu, In, Ga, and Se in the CIGS absorber layers were obtained using two different surface analytical instruments. The magnetic sector SIMS was completed using an IMS-4fE7 instrument (CAMECA, France). The CIGS samples were measured under an analyzed current of 20 nA with a raster size of 200 × 200 μm^2 and a mass resolution of 1800, applying 10 kV to a Cs^+ primary ion gun and 4.5 kV to the sample to obtain an impact energy of 5.5 keV. AES data were obtained on a ULVAC-PHI 700 Scanning Auger microscope (ULVAC-PHI, Japan) under 5 kV at a current of 10 nA. AES intensities were acquired with the differentiated LMM transition at 920 eV, MNN transition at 405 eV, LMM transition at 1070 eV, and LMM transition at 1311 eV, for Cu, In, Ga, and Se, respectively. The AES depth profile was performed using a 3 kV Ar^+ ion beam under a sputter rate of 52.5 nm/min for the SiO_2 layer.

3. Results and Discussion

3.1. Composition of the CIGS Absorber Layers. The scanning electron microscopy (SEM) images of the four CIGS solar cells in cross section are shown in Table 1. The microstructures of all of the CIGS were faceted grains typical of absorbers grown in a three-stage process. The average thickness indicated in the table was calculated from several measurements collected on the same sample. Quantification of the bulk elemental concentrations of Cu, In, Ga, and Se and the depth distribution of these elements are essential to evaluate the accurate performances of the solar cells, which are related to solar-cell efficiency. The average concentrations of the four CIGS absorber layers were obtained by ICP-AES and XRF (Table 2). Differences between the atomic concentrations measured by ICP-AES and XRF were relatively low (within 1.4 at.%). The small difference between the results confirms the reliability of these analysis techniques to determine the bulk element compositions of CIGS absorber layers. Moreover, the measurement deviation can be mainly attributed to the analysis position and calibration method of each technique.

To find a fast, accurate, and reliable technique that can be readily applied for the quantification of CIGS absorber layers, *fs*-LA-ICP-MS was used and compared with other techniques. Calibration curves were generated for the *fs*-LA-ICP-MS signal ratios of the major elements with reference materials that were quantified by XRF [18]. Then, *fs*-LA-ICP-MS was performed to obtain the elemental concentration ratios of the samples. An *fs* laser was used to ablate the sample surface and to attain sequential access to the chemical composition. The *fs*-LA-ICP-MS measurements were performed for four different CIGS absorber layers, and the average values from the nine craters were used to represent the ICP-MS isotope signal of each absorber sample. The typical ICP-MS signals of ^{115}In, ^{71}Ga, ^{65}Cu, ^{77}Se, and ^{95}Mo isotopes for B04 are shown in Figure 1. These signals indicated the four major elements of the CIGS absorber layers from the nine different craters. The repetitive responses of the ICP-MS results were consistent over the nine craters, indicating that the CIGS absorber layers had a fairly uniform compositional distribution. The relative standard deviation (RSD) of the intensities of each isotope signal was measured in the range of 3.7–15.8%. The *fs*-LA-ICP-MS technique provided reproducible quantitative results for the CIGS layers. The signal of ^{95}Mo was also observed because the penetration depth of laser was deep enough to ablate both the CIGS layer and Mo layer under these experimental conditions. The intensities of each isotope signal are generated by the ablated amount of an element from the sample, which is related to the elemental concentration and the thickness of the CIGS layer. To improve the quantification of *fs*-LA-ICP-MS, the isotopic intensity ratios were calibrated with the relative concentration of the major elements in the CIGS layer. In Figure 2, the calibration curves for Ga/Cu, In/Cu, Ga/Se, In/Se, and Ga/In from the intensity ratios of the *fs*-LA-ICP-MS analysis are

TABLE 2: Average elemental composition of CIGS absorber layers by ICP-AES, XRF, SIMS, and AES.

	Determination technique	Atomic concentration (at.%)							
		Cu		In		Ga		Se	
B01	ICP-AES	26.4	—	16.4	—	8.9	—	48.3	—
	XRF	26.5	(0.4)[a]	16.1	(−1.8)	9.3	(4.5)	48.1	(−0.4)
	SIMS	27.4[b]	(3.8)	16.6	(1.2)	9.6	(7.9)	46.5	(−3.7)
	AES	25.9	(−1.9)	16.2	(−1.2)	8.2	(−7.9)	49.7	(2.9)
B02	ICP-AES	25.2	—	15.6	—	11.7	—	47.5	—
	XRF	24.6	(−2.4)	15.8	(1.3)	11.5	(−1.7)	48.1	(1.3)
	SIMS	25.3	(0.4)	16.0	(2.6)	12.1	(3.4)	46.6	(−1.9)
	AES	26.2	(4.0)	15.1	(−3.2)	9.2	(−21.4)	49.5	(4.2)
B03	ICP-AES	26.5	—	19.2	—	7.4	—	46.9	—
	XRF	26.2	(−1.1)	18.1	(−5.7)	7.4	(0.0)	48.3	(3.0)
	SIMS	25.7	(−3.0)	20.0	(4.2)	7.0	(−5.4)	47.3	(0.9)
	AES	26.3	(−0.8)	17.9	(−6.8)	7.2	(−2.7)	48.6	(3.6)
B04	ICP-AES	26.2	—	18.9	—	7.1	—	47.7	—
	XRF	26.4	(0.8)	18.3	(−3.2)	7.3	(2.8)	48.0	(0.6)
	SIMS	26.2	—	18.9	—	7.1	—	47.7	—
	AES	26.2	—	18.9	—	7.1	—	47.7	—

[a]The numbers in parentheses represent the percentage deviation from the ICP-AES value. [b]Calculated from the element peak intensities in the depth profile.

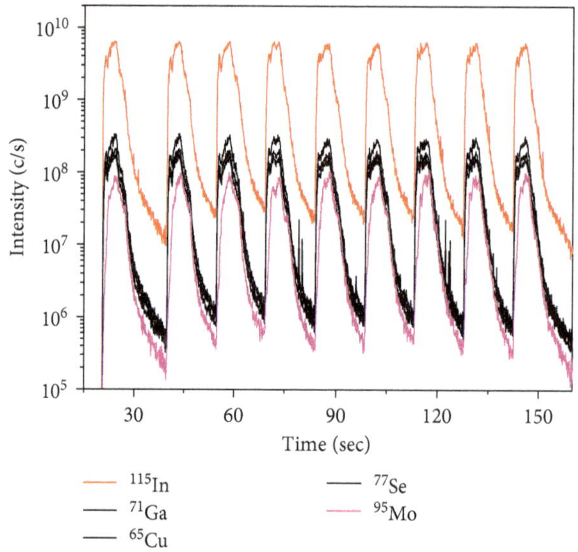

FIGURE 1: *fs*-LA-ICP-MS analysis of B04. Measurements were repeated nine times per sample, and the average value of the integral intensities was used (laser-repetition rate: 200 Hz, energy per pulse: 5.1 J/cm^2, and crater diameter: 5 μm).

obtained and compared with the XRF elemental concentration ratios. The figure exhibits a good linear correlation between two of these methods. The R^2 values of the linear fitting on the calibration curve for Ga/Cu and Ga/Se were close to 0.97, whereas the In/Se ratio had a lower value (0.85). Lower R^2 values in the calibration curve, including for In, were also observed in the previous study by Lee et al. [18]. This result might be due to the mass difference and physical properties of In compared to other major elements (Cu, Ga, and Se). Table 3 lists the average and RSD values of nine repeated measurements for elemental intensity ratios in *fs*-LA-ICP-MS. By using the intensity ratios instead of the absolute intensities, the RSD values of the nine measurements decreased to below 2% in average (maximum 2.87% for Ga/Se of B01 sample). These results imply that highly reproducible measurement can be performed with the elemental intensity ratios of *fs*-LA-ICP-MS at the optimal sampling condition. Therefore, the *fs*-LA-ICP-MS technique can be used to obtain accurate and rapid quantitative results for CIGS layer analysis.

3.2. Depth Profiling of the CIGS Absorber Layers. ICP-AES, XRF, and *fs*-LA-ICP-MS collect accurate and sensitive measurements, but these techniques cannot examine the depth distribution of elements. Thus, techniques such as SIMS and AES were used to investigate the elemental

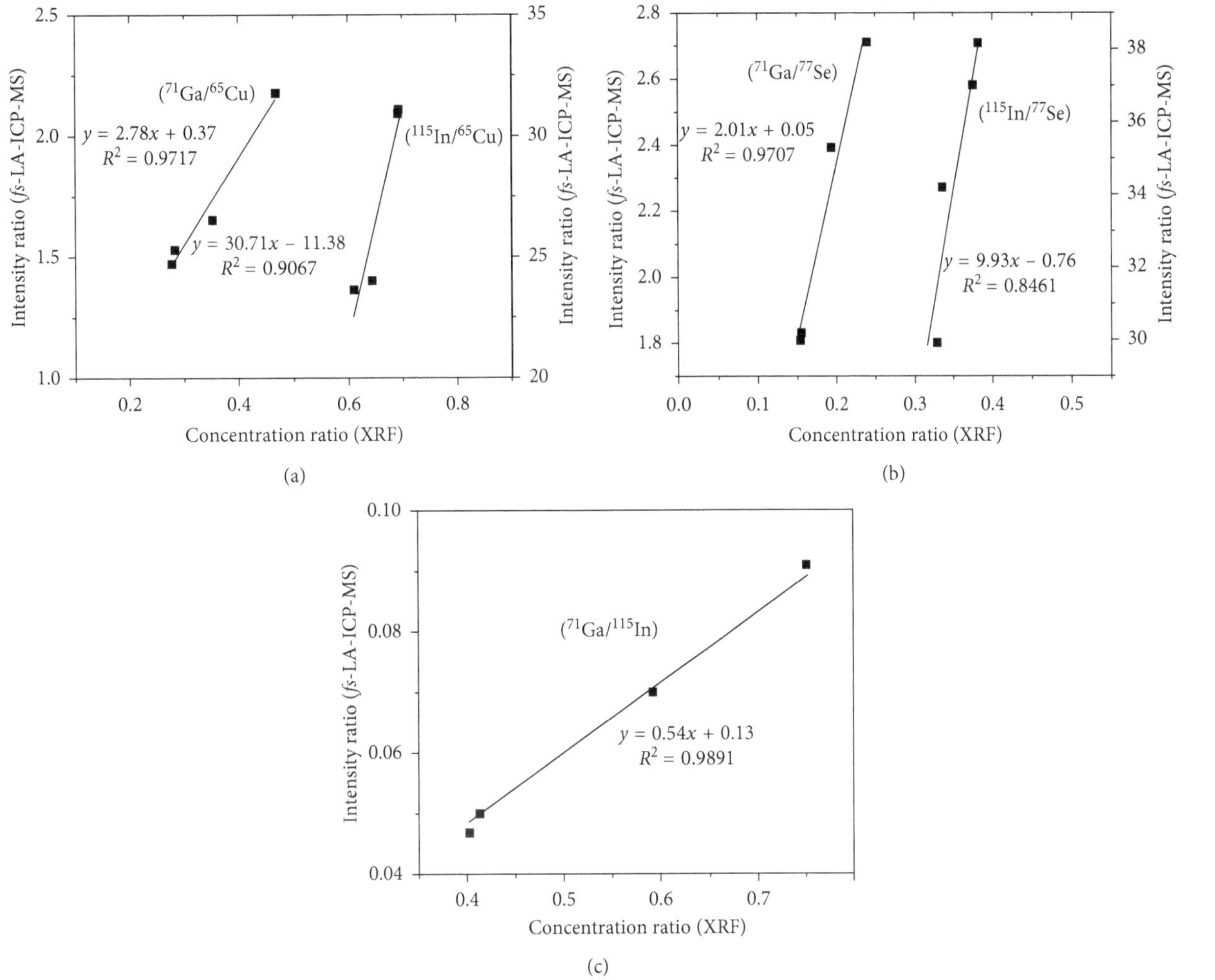

FIGURE 2: Calibration curves between the intensity ratios for *fs*-LA-ICP-MS and the concentration ratios for XRF measurement (^{71}Ga/^{77}Se, ^{115}In/^{77}Se, ^{71}Ga/^{65}Cu, ^{115}In/^{65}Cu, and ^{71}Ga/^{115}In ratios, respectively).

TABLE 3: Average and relative standard deviation (RSD) values of elemental intensity ratios using nine *fs*-LA-ICP-MS measurements.

Sample number	RSD (%)				
	Ga/Cu	In/Cu	Ga/Se	In/Se	Ga/In
B01	1.66[a] (±1.70%)[b]	23.71 (±2.15%)	2.41 (±2.87%)	34.48 (±2.30%)	0.07 (±2.59%)
B02	2.18 (±2.70%)	24.08 (±1.13%)	2.71 (±2.18%)	29.94 (±1.34%)	0.09 (±2.69%)
B03	1.43 (±2.10%)	29.25 (±2.81%)	1.80 (±1.55%)	36.91 (±2.22%)	0.05 (±1.25%)
B04	1.48 (±1.16%)	31.12 (±1.18%)	1.81 (±0.93%)	38.22 (±1.30%)	0.05 (±1.13%)
Average of RSD	(±1.92%)	(±1.82%)	(±1.88%)	(±1.79%)	(±1.92%)

[a]The numbers represent the average of intensity ratios. [b]The numbers in parentheses represent the RSD of intensity ratios.

composition in a depth profile. In particular, SIMS provides high depth resolution and high mass resolution. This technique also has the advantage of high sensitivity in identifying trace elements such as sodium (Na) and potassium (K) that are closely related to cell efficiency.

Figure 3 shows SIMS depth profiles of characteristic ions in the CIGS solar cells. Dashed lines identify the layer position in the CdS/CIGS/Mo structure. The Cu and Se intensities were fairly uniform throughout the sputter time. However, severe Ga grading was observed, especially for the B01 and B02 CIGS layers. The distribution of Na in the SIMS depth profiling was obtained at high intensity and is clearly distinguishable due to the low detection limit of SIMS for trace elements. In contrast with B01, B02, and B03, the Na secondary ion intensity in the SIMS depth profile appeared lower in the CIGS layer for B04. To confirm the Na concentration on the CIGS surface, the as-prepared sample of CIGS/Mo on the SLG substrate was placed in the XPS system. As shown in Figures 3(e)–3(h), the CIGS thin films had Na peaks, with the exception of B04, which had a SiO_2 barrier layer. No Na peak

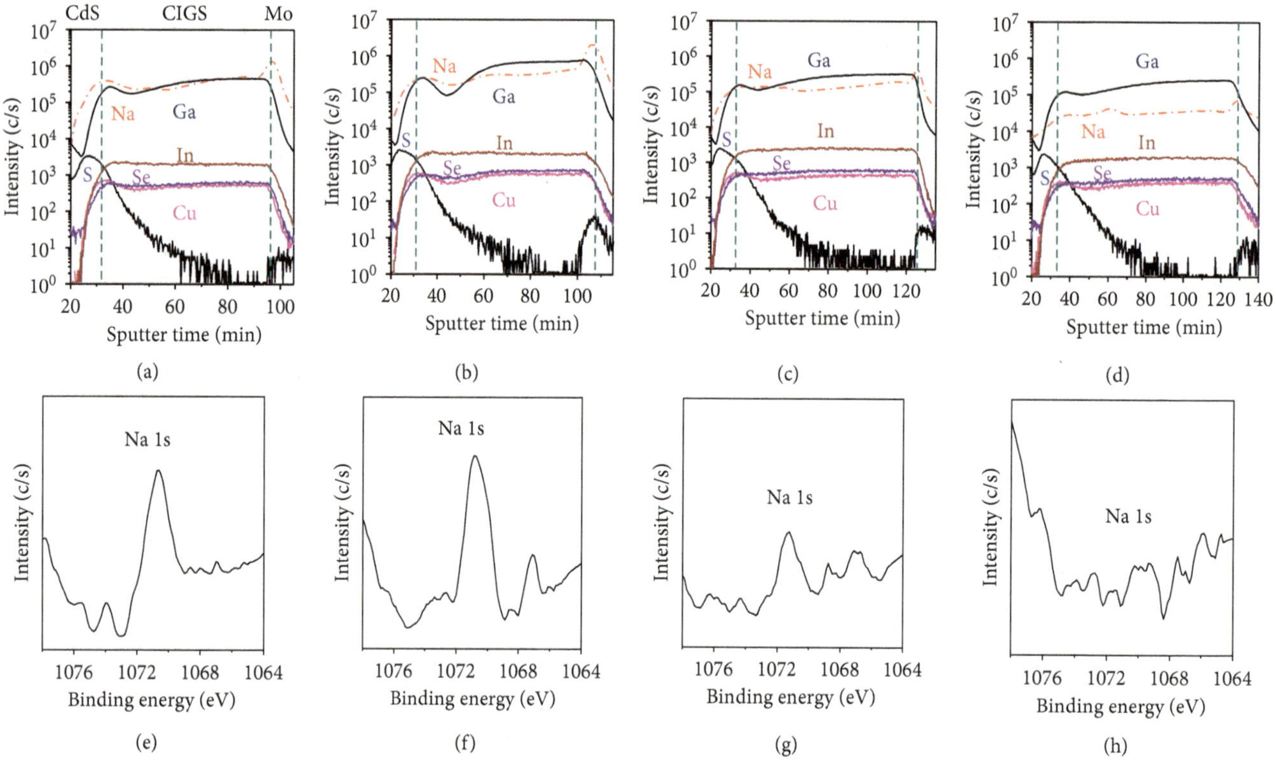

FIGURE 3: (a–d) SIMS depth profiles and (e–h) Na 1s peaks in the XPS spectra of the CIGS solar cells: (a, e) B01, (b, f) B02, (c, g) B03, and (d, h) B04.

was observed in the XPS spectrum of B04 even though a Mo : Na layer was present on the SiO_2 film. The decreasing Na intensity in B04 was mainly due to the presence of a SiO_x barrier layer on the SLG substrate, which limited Na diffusion during thermal treatment. The photovoltaic conversion efficiencies of B01 and B02, 14.6% and 14.3%, respectively, were greater than that of B04, measured at 10.3%. The Na distribution correlated with the change in conversion efficiency, which highlights the influence of alkali elements increasing the open-circuit voltage (V_{OC}) [25, 26].

All CIGS layers were quantitatively calculated using the ICP-AES concentration of B04 as a reference sample. The AES depth profiles were also quantitatively obtained in a way similar to the SIMS analysis of the CIGS layers. In Figure 4, the SIMS and AES results are overlaid to compare the compositional element distribution as a function of depth. The element distributions of SIMS and AES showed a good correspondence, except for the surface layer, which is the beginning of the profile in Figure 4. Large deviations in the interface layer appeared due to the surface effect, which is a classical artifact of the SIMS technique. The deviations can be reduced by detecting MCs^+ cluster ions as shown in a previous study [24].

All of the calculated elemental compositions using SIMS and AES are reported in Table 2. The values in parentheses indicate the deviations of the average concentrations from AES and SIMS with the measured values from ICP-AES. The deviations from AES are relatively low except for the Ga concentration, especially for B02. The quantification errors might be caused by the analysis position of the sample, which is not uniform at the macrolevel, differences in the chemical compositions of the reference and analytical samples, or differences in the calibration methods of the specific analysis techniques.

3.3. Determination of Band Gap of the CIGS Absorber Layers. The depth band gap variation in the CIGS thin film caused by changes in the Ga/(In + Ga) ratio is commonly referred to as Ga grading [27–29]. The AES and SIMS composition measurements were used to calculate the gallium ratios, Ga/(In + Ga), and band gap profiles, as shown in Figures 5(a) and 5(c) and Figures 5(b) and 5(d), respectively. The band gap gradients in the samples reflected the gradings in the gallium ratios. B01 and B02 showed a notch structure in their band gap profiles, and their solar cells showed noticeable double-Ga grading, as expected in high-efficiency solar cells. The B01 and B02 solar cells showed similar device performance although B02 had a steeper Ga distribution profile (Figures 5(a) and 5(c)). Generally, the depth profiles of Ga grading tend to influence the cell efficiency of a CIGS solar cell: Ga-grading profiles are directly related to the distribution of the conduction band minimum. Basically, a double-graded Ga distribution (rising from the center toward both the front and the back surface) is known to enhance the cell efficiency because the bidirectional grading enhances the

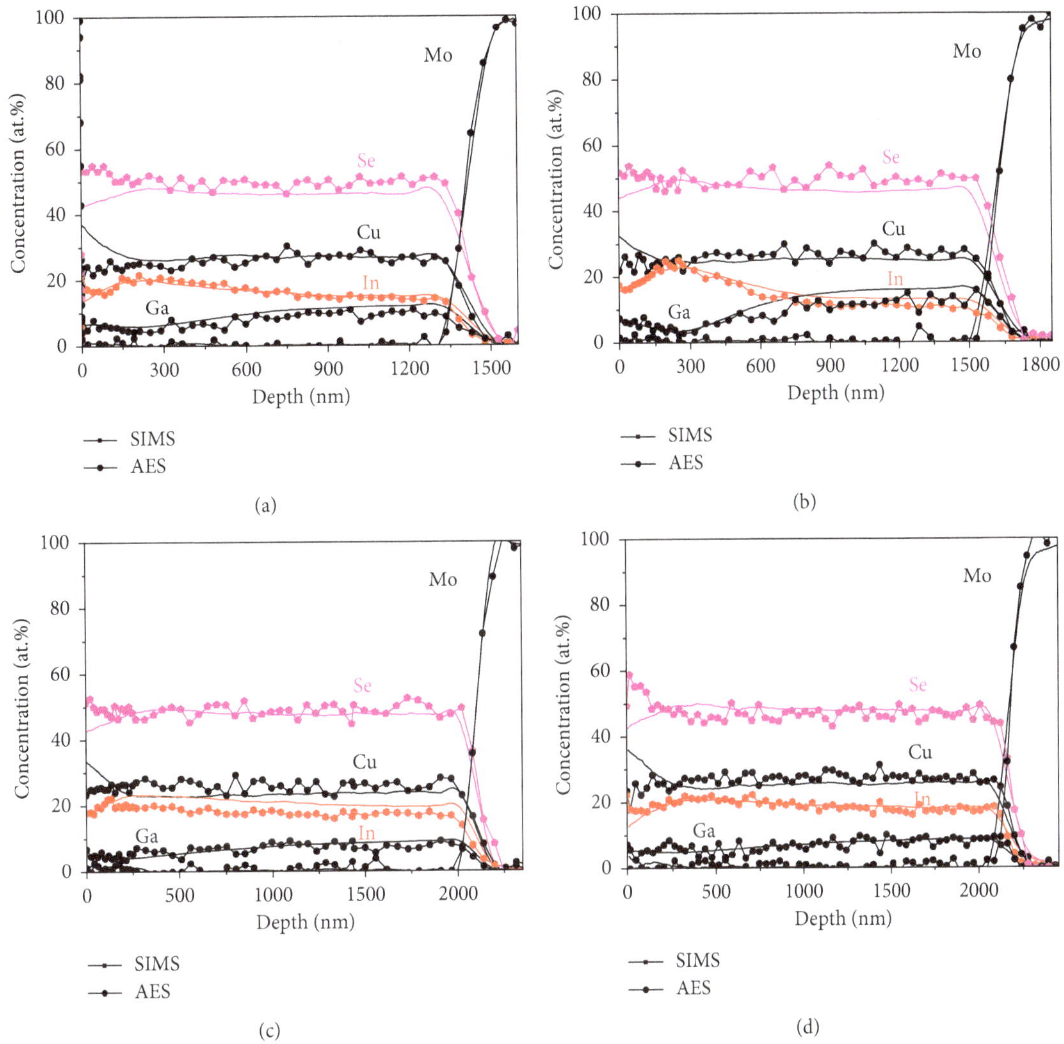

FIGURE 4: Overlaid compositional depth profiles of the CIGS absorber layers from SIMS and AES: (a) B01, (b) B02, (c) B03, and (d) B04.

open-circuit voltage and the notch in the middle enhances the short-circuit current by lowering the optical band gap. However, if the Ga grading becomes severe (i.e., a very low Ga notch is formed), carriers are easily entrapped, which in turn increases carrier recombination and prevents carrier transport. Although reducing Se flux made the Ga grading steeper in B02, its device performance did not substantially changed from that of B01. These results implied that these device performances were determined by factors other than Ga grading itself (i.e., CIGS interface qualities). The gallium ratios, Ga/(In + Ga), in B03 and B04 were relatively flat, and these gallium distributions influenced their low cell efficiencies.

4. Conclusions

To understand the high-efficiency of $Cu(In,Ga)Se_2$ solar cells, we accurately measured the compositions of Cu, In, Ga, and Se using ICP-AES and XRF, which have relatively low deviations within 1.4 at.%. The calibration curves were obtained by *fs*-LA-ICP-MS, which is a useful method for rapid elemental analysis and minimal sample consumption.

The depth profiles of the CdS/CIGS/Mo structure obtained through SIMS and XPS permitted us to highlight the minimal Na diffusion when a SiO_x layer was deposited between the SLG substrate and the Mo back contact layer. In contrast with B04, B01 and B02 were higher in photovoltaic efficiency, which is consistent with previous studies in which the alkali elements help to increase the open-circuit voltage.

The comparison of the SIMS depth profile with the AES depth profile confirmed the reliability of the depth analysis. Thanks to the accuracy of the depth distribution of the CIGS layers, the band gaps for four different CIGS absorber layers were determined using the Ga and In composition ratio along the film thickness. The B01 and B02 solar cells showed noticeable double-graded Ga distribution, which is known to enhance the cell efficiency by reducing the carrier recombination at the interface.

Comparative analyses of different CIGS absorber layers using various analytical techniques provide us with

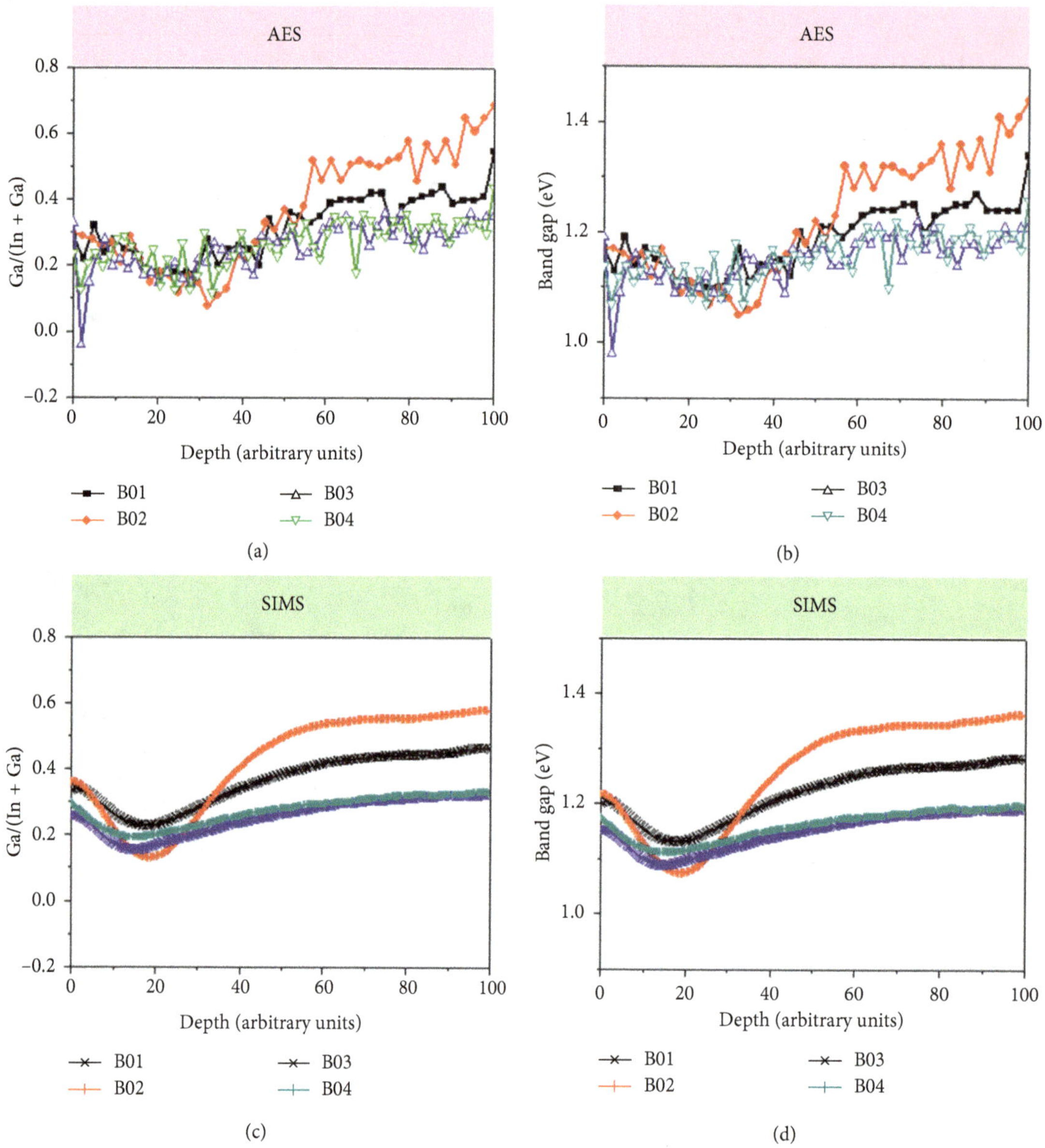

FIGURE 5: Determined band gap of the CIGS absorber layers from Ga/(In + Ga): (a) atomic ratios of Ga/(In + Ga) from AES; (b) band gap profiles calculated from AES; (c) atomic ratios of Ga/(In + Ga) from SIMS; (d) band gap profiles calculated from SIMS. Sputter time for each profile has been normalized to 100; actual sputter times varied from 65 to 95 min.

understanding of the accurate compositions, elemental distribution, and cell efficiency of CIGS solar cells. These results also give us the opportunity to select the proper analytical methods for the CIGS solar cells depending on the situations.

Conflicts of Interest

The authors declared that they have not conflicts of interest regarding the publication of this paper.

Acknowledgments

This work was supported by the Korean Ministry of Science and ICT Research Program (grant number 2014M3A9E1070337).

References

[1] P. Jackson, D. Hariskos, R. Wuerz, E. Lotter, W. Witte, and M. Powalla, "Effects of heavy alkali elements in Cu(In,Ga)Se2 solar cells with efficiencies up to 22.6%," *Physical Status Solidi Rapid Research Letters*, vol. 10, no. 8, pp. 583–586, 2016.

[2] P. Jackson, R. Wuerz, D. Hariskos et al., "Properties of Cu(In, Ga)Se_2 solar cells with new record efficiencies up to 21.7%," *Physical Status Solidi Rapid Research Letters*, vol. 9, no. 1, pp. 28–31, 2015.

[3] R. Klenk, T. Walter, D. Schmid, and H. W. Schock, "Growth mechanisms and diffusion in multinary and multilayer chalcopyrite thin films," *Japanese Journal of Applied Physics*, vol. 32, no. S3, pp. 57–61, 1993.

[4] H. Wang, Y. Zhang, X. L. Kou et al., "Effect of substrate temperature on the structural and electrical properties of CIGS films based on the one-stage co-evaporation process,"

Semiconductor Science and Technology, vol. 25, no. 5, article 055007, 2010.

[5] A. Han, Y. Huang, X. Liu, W. Xian, F. Meng, and Z. Liu, "Morthology, structure, and properties of Cu-poor and Cu-rich Cu(In,Ga)Se_2 films partially selenized using H_2Se gas," *Japanese Journal of Applied Physics*, vol. 55, no. 11, article 115502, 2016.

[6] B. Peace, J. Claypoole, N. Sun et al., "Characterization of Cu (In,Ga)Se_2 (CIGS) films with varying gallium ratios," *Journal of Alloys and Compounds*, vol. 657, pp. 873–877, 2016.

[7] M. Asaduzzaman, M. Hasan, and A. N. Bahar, "An investigation into the effects of band gap and doping concentration on Cu(In,Ga)Se_2 solar cell efficiency," *SpringerPlus*, vol. 5, no. 1, p. 578, 2016.

[8] M. Bär, W. Bohne, J. Röhrich et al., "Determination of the band gap depth profile of the penternary $Cu(In_{(1-X)}Ga_X)(S_YSe_{(1-Y)})_2$ chalcopyrite from its composition gradient," *Journal of Applied Physics*, vol. 96, no. 7, pp. 3857–3860, 2004.

[9] T. Dullweber, O. Lundberg, J. Malmström et al., "Back surface band gap gradings in Cu(In,Ga)Se_2 solar cells," *Thin Solid Films*, vol. 387, no. 1-2, pp. 11–13, 2001.

[10] U. Rau and H. W. Schock, "Electronic properties of Cu(In,Ga) Se_2 heterojunction solar cells–recent achievements, current understanding, and future challenges," *Applied Physics A: Materials Science & Processing*, vol. 69, no. 2, pp. 131–147, 1999.

[11] S. H. Kim, Y. J. Jang, J. H. Yoon, J. H. Jeong, and Y. Lee, "Quantitative analysis for CIGS thin films by surface analytical techniques," *Surface and Interface Analysis*, vol. 46, no. S1, pp. 276–280, 2014.

[12] C. L. Perkins, B. Egaas, I. Repins, and B. To, "Quantitative analysis of graded $Cu(In_{1-X},Ga_X)Se_2$ thin films by AES, ICP-OES, and EPMA," *Applied Surface Science*, vol. 257, no. 3, pp. 878–886, 2010.

[13] B. M. West, M. Stuckelberger, A. Jeffries et al., "X-ray fluorescence at nanoscale resolution for multicomponent layered structures: a solar cell case study," *Journal of Synchrotron Radiation*, vol. 24, no. 1, pp. 288–295, 2017.

[14] H. A. Al-Thani, S. A. Al-Shaibani, A. M. Al-Jaeedi et al., "Recent progress of CIGS thin film R&D at NEWRC," in *Proceedings of IEEE 43rd Photovoltaic Specialists Conference*, pp. 1408–1411, Portland, OR, USA, June 2016.

[15] S. H. Choi, J. S. Kim, J. Y. Lee et al., "Analysis of arsenic in rice grains using ICP-MS and fs LA-ICP-MS," *Journal of Analytical Atomic Spectrometry*, vol. 29, no. 7, pp. 1233–1237, 2014.

[16] G. Bauer and A. Limbeck, "Quantitative analysis of trace elements in environmental powders with laser ablation inductively coupled mass spectrometry using non-sample-corresponding reference materials for signal evaluation," *Spectrochimica Acta Part B: Atomic Spectroscopy*, vol. 113, pp. 63–69, 2015.

[17] R. Dargel, F. Heinemeyer, M. Köntges, J. Vogt, and C. Vogt, "Detection of trace impurities in Cu(In, Ga)Se_2 thin film solar cells by laser ablation ICP-MS," *Microchimica Acta*, vol. 165, no. 3-4, pp. 265–270, 2009.

[18] S. Lee, J. J. Gonzalez, J. H. Yoo, J. R. Chirinos, R. E. Russo, and S. Jeong, "Application of femtosecond laser ablation inductively coupled plasma mass spectrometry for quantitative analysis of thin Cu(In,Ga)Se_2 solar cell films," *Thin Solid Films*, vol. 577, pp. 82–87, 2015.

[19] M. J. Kim, J. Lee, S. H. Kim, H. Kim, K. B. Lee, and Y. Lee, "Detection of matrix elements and trace impurities in Cu(In, Ga)Se_2 photovoltaic absorbers using surface analytical techniques," *Journal of Nanoscience and Nanotechnology*, vol. 15, no. 10, pp. 7722–7726, 2015.

[20] A. Chirilă, P. Reinhard, F. Pianezzi et al., "Potassium-induced surface modification of Cu(In,Ga)Se_2 thin films for high-efficiency solar cells," *Nature Materials*, vol. 12, no. 12, pp. 1107–1111, 2013.

[21] J.-H. Yoon, T.-Y. Seong, and J.-H. Jeong, "Effect of a Mo back contact on Na diffusion in CIGS thin film solar cells," *Progress in Photovoltaics: Research and Applications*, vol. 21, no. 1, pp. 58–63, 2013.

[22] J. S. Jang, H. H. Hwang, H. J. Kang, H. C. Chae, Y. D. Chung, and K. J. Kim, "Accurate quantification of Cu(In,Ga)Se_2 films by AES depth profiling analysis," *Applied Surface Science*, vol. 282, pp. 777–781, 2013.

[23] A. Loubat, M. Bouttemy, S. Gaiaschi et al., "Chemical engineering of Cu(In,Ga)Se_2 surfaces: an absolute deoxidation studied by X-ray photoelectron spectroscopy and Auger electron spectroscopy signatures," *Thin Solid Films*, vol. 633, pp. 87–91, 2017.

[24] J. Lee, S. H. Kim, K.-B. Lee, B. K. Min, and Y. Lee, "Improved quantitative analysis of Cu(In,Ga)Se_2 thin films using MCs^+-SIMS depth profiling," *Applied Physics A: Materials Science & Processing*, vol. 115, no. 4, pp. 1355–1364, 2014.

[25] Y. M. Shin, D. H. Shin, J. H. Kim, and B. T. Ahn, "Effect of Na doping using Na_2S on the structure and photovoltaic properties of CIGS solar cells," *Current Applied Physics*, vol. 11, no. 1, pp. S59–S64, 2011.

[26] M. Theelen, V. Hans, N. Barreau, H. Steijvers, Z. Vroon, and M. Zeman, "The impact of alkali elements on the degradation of CIGS solar cells," *Progress in Photovoltaics: Research and Applications*, vol. 23, no. 5, pp. 537–545, 2015.

[27] A. M. Gabor, J. R. Tuttle, M. H. Bode et al., "Band-gap engineering in Cu(In,Ga)Se_2 thin films grown from $(In,Ga)_2Se_3$ precursors," *Solar Energy Materials and Solar Cells*, vol. 41-42, pp. 247–260, 1996.

[28] O. Lundberg, M. Edoff, and L. Stolt, "The effect of Ga-grading in CIGS thin film solar cells," *Thin Solid Films*, vol. 480-481, pp. 520–525, 2005.

[29] L. M. Mansfield, R. L. Garris, K. D. Counts et al., "Comparison of CIGS solar cells made with different structures and fabrication techniques," *IEEE Journal of Photovoltaics*, vol. 7, no. 1, pp. 286–293, 2017.

13

Development of Cellulosic Paper-Based Test Strips for Mercury(II) Determination in Aqueous Solution

Shoujuan Wang,[1] Zhen Xu,[2] Yongyi Fang,[2] Zhongming Liu,[1] Xin Zhao ,[1] Guihua Yang ,[1] and Fangong Kong [1]

[1] *State Key Lab of Biobased Material and Green Papermaking, Key Laboratory of Pulp and Paper Science and Technology Ministry of Education, Qilu University of Technology (Shandong Academy of Sciences), Jinan, Shandong 250353, China*
[2] *Xuancheng Product Quality Supervision and Inspection Institute, Xuan Cheng, Anhui 242000, China*

Correspondence should be addressed to Xin Zhao; zhaoxin_zixi@126.com, Guihua Yang; ygh2626@126.com, and Fangong Kong; kfgwsj1566@163.com

Academic Editor: Chih-Ching Huang

Titration method (dropping-on method) was introduced as an efficient approach for determining the mercury ion (Hg^{2+}) concentration in aqueous solution by using fabricated cellulosic paper-based test strips. In this study, dithizone used as a recognition reagent was physically loaded on cellulosic paper-based test strips for Hg^{2+} selective recognition. The sensing mechanism was established on the spectral absorption rate of the coordination compound that was formed by dithizone and Hg^{2+} under strong acidic conditions. The calibration curve was obtained by the absorbency of Hg^{2+}-dithizone complexes from different Hg^{2+} concentration solutions, and the correlation coefficient (R^2) reached 0.9971. The detection range of the test trip for Hg^{2+} was obtained at 0.1 μg/mL to 30 μg/mL. Moreover, these superior cellulosic paper-based test strips have a rapid color-forming time (1.5 min) and low volume demand (3.7 μL samples at 0.0127 g/L dithizone recognition concentration). This portable paper-based test strip can give potential applications for field screening or on-site semiquantitative analysis.

1. Introduction

Mercury is one of the most toxic elements to human health, causing several symptoms related to polyneuropathy mercurialis, neurasthenia, etc. [1]. Mercury ions can persistently exist in water environments and accumulate easily in organisms. Moreover, it has high toxicity, even in low concentrations [2]. Therefore, mercury pollution is an important issue that demands an ongoing development of analytical procedures to ensure its reliable determination [3, 4]. To fulfill the demands of the environmentally sustainable oriented goal for determining mercury in various matrices and at various levels of concentration, an essential tool is needed to assure accuracy and establish the traceability of the measured results. A lot of approaches have been used to develop optical or visual disposable sensors for mercury ion determination. The most common approaches are methods with spectrometric detectors [5–7]. Meanwhile, several compounds have been shown as selective for metal analysis and used as receptors for the design of sensing systems, ranging from alloys and amalgam [8] and synthetic metal ionophores [9, 10] to biological receptors [9, 11].

A fluorescence-based paper test strip is developed by using *N*-alkylaminopyrazole ligands for mercury detection in water, and a linear range from 10 μg/mL to 100 μg/mL mercury ion concentration ([Hg^{2+}]) was achieved [2]. Meanwhile, pyrazole-derived ligands have been determined to be suitable for the coordination of metal ions [12–14]. A polyester sheet with a sensing area containing plasticized polyvinyl chloride (PVC) was made by incorporating tetraarylborate salt as a selective recognition reagent and a porphyrin proton-selective fluoroionophore as the optical transducer for Hg^{2+} [15]. Later, Yallouz et al. incorporated cuprous iodide (Cu_2I_2) carboxymethyl cellulose membranes on paper for Hg^{2+} in fish samples [16]. Shi and Jiang used a pH indicator color for Hg^{2+} in waters with a detection limit of 0.2 ng/mL under the inhibition of the different types of enzymes [17]. Capitán-Vallvey et al. developed an irreversible optical test strip for mercury determination based on an ion-exchange mechanism [18]. A circular sensing zone containing the essential reagents was used to produce a selective response for Hg^{2+} and consisted of plasticized PVC incorporating the cation-selective neutral ionophore. Chandio et al. also developed a PVC membrane ion selective ported to determine the Hg^{2+} in waste water [19]. Gur'eva et al. developed a test method for the mercury(I, II) determination by putting an intensely colored insoluble compound on a polycaproamide membrane surface in weak acid solutions depending on the reaction between mercury (I, II) and 5-chloro-2-hydroxy-3-[(tetrahydro-2,4-dithioxo-1,3-thiazin-5-yl) azo] benzenesulphonic acid [20]. However, most of these methods require expensive chemicals and instruments, and the treating procedure is complex and not easy to handle. In addition, most of these test strips are based on the plastic membrane technology. The used test strips are not reusable and biodegradable after disposal. Also, as the diffusion of Hg^{2+}-containing solution in the plastic membrane is very slow even there are plasticizers in the membrane, the color-forming time usually needs a few minutes, even a few hours, which is not suitable for on-site screening analysis. Therefore, great emphasis still needs to be placed on the development of a low cost and convenient method for mercury determination.

Dithizone, as a kind of color-forming agent, has low cost and efficiency advantages, which can produce colored complexes between the color-forming agent in the paper strip and Hg^{2+} in the sample [21]. Based on the reaction between the dithizone and Hg^{2+} in the trichloromethane under strongly acidic conditions, the paper-based test strip for determining the concentration of Hg^{2+} in aqueous solutions could be developed. The test strip reacted rapidly with Hg^{2+} could lead to color formation of the test strip. Subsequently, the test color liquid diffused in indefinite directions due to liquid diffusivity. For the dipping-in method, the absorbency value continued to increase with the reaction time, and it was difficult to determine the result of color forming due to the reaction of the test strip and Hg^{2+} or liquid diffusivity.

The aim of this work is to introduce an efficient and simple approach to fabricate cellulosic paper-based test strips by physically impregnating paper strips in dithizone-containing solution to load dithizone into strips for determining the Hg^{2+} in aqueous solution. This preparation of the dithizone-loaded test strip is very simple and easily operated using biodegradable materials, paper strips, at low cost compared with other plastic-based test strips. In this study, dithizone is used as a recognition reagent to achieve selective recognition by Hg^{2+} with dithizone under strong acidic conditions. The calibration curve and pertinence are obtained by the absorbency from different concentrations of Hg^{2+}. Furthermore, all the experiments are performed towards the detection linear range of the test strip system and superior test conditions of cellulose paper-based test strips.

2. Experimental

2.1. Materials. The paper sample used for the test strip was quantitative filter paper of medium porosity, which was obtained from Fisher Scientific (Jinan, China). Dithizone was obtained from Sigma-Aldrich (Jinan, China). Mercury solutions with various concentrations were freshly prepared by dissolving a certain amount of mercury nitrate in nitric acid solution with pH 2.0. Other solutions, including mercury ions (Hg^{2+}), potassium ions (K^{+}), calcium ions (Ca^{2+}), sodium ions (Na^{+}), manganese ions (Mn^{2+}), magnesium ions (Mg^{2+}), copper ions (Cu^{2+}), barium ions (Ba^{2+}), aluminum ions (Al^{3+}), iron ions (Fe^{3+}), zinc ions (Zn^{2+}), phosphate ions (PO_4^{3-}), chloride ions (Cl^{-}), and sulfate ions (SO_4^{2-}), were prepared by using analytical reagent-grade chemicals and nitric acid solution with pH 2.0. The pH of solutions was adjusted by adding either diluted sodium hydroxide (NaOH) or nitric acid (HNO_3) solution. The spiked water sample of Hg(II) was prepared by adding Hg(II) into tap water with its pH adjusted to 2.0.

2.2. Apparatus. A UV-Vis spectrophotometer (UV-2550, Shimadzu, Nagoya, Japan) was used to determine the absorption of the colored complex solutions. A reflect spectrophotometer (i1Basic Pro2, X-Rite, Agilent, Foster City, California, USA) was used for the measurement of color intensity formed on the test strip.

2.3. Preparation of Cellulosic Paper-Based Test Strip. The dithizone was dissolved in trichloromethane (Fisher Scientific, Jinan, China) with a concentration of 0.0127 g/L. The filter paper was then immersed into this solution for 2 min at room temperature. The filter paper with dithizone was dried at 50°C in a nitrogen protected environment. The dried filter paper was then cut into 5 cm × 1 cm strips, sealed in a plastic bag, and kept until further use. The paper strip immersed in 0.0127 g/L dithizone solution had a 0.33 mg/cm^2 dithizone loading amount.

3. Methods

3.1. Spectral Reflectivity Measurement. The Hg^{2+} solution was dropped on the paper-based test strip. After the color-forming reaction finished, the color intensity developed on the test strip was measured using the reflect spectrophotometer at a fixed wavelength of 490 nm.

The color forming depended on the dithizone reaction with mercury(II) and the coordination compound (dithizone-mercury complex) was formed, and the linear relationship between the absorbency value and different Hg^{2+} concentrations was tested by spectrophotometer at a fixed wavelength of 490 nm [22].

3.2. Hg(II) Determination Using UV-Vis Spectroscopy Method. A 250 mL mercury solution with 10 μg/mL concentration was put into a separatory funnel (500 mL) by adding 1 mL sodium sulfite (20%). After mixing, 10 mL dithizone-trichloromethane solution was added, and the solution layered after 1 min. The mercury solution after contacting with the dithizone was dissolved in the trichloromethane and was transferred to a cuvette of the spectrophotometric measurements. A UV-Vis spectrophotometer (UV-2550, Shimadzu, Nagoya, Japan) was used to measure the absorbance value of the dithizone-mercury complex in the solution [22]. Then, the concentration of Hg(II) was calculated based on the calibration curve obtained in this study:

$$Y = 0.0558X + 0.0391,\quad R^2 = 0.9932, \tag{1}$$

where Y is the absorbance value and X is the mercury ion concentration (μg/mL).

4. Results and Discussion

4.1. Development of Paper-Based Test Strip. Due to the high specific surface area and plenty of capillary pores, paper sheets, such as filter paper, have high absorbability of inorganic or organic liquids [23]. By loading the effective chemicals onto a paper sheet, a paper-based test strip specific for the determination of Hg(II) concentration is expected to be portable, user-friendly, and handy. From this viewpoint, the selection of a suitable reagent to extract the Hg(II) content puts emphasis on the following considerations. This paper-based test strip would have specific features of (i) giving a color response so that the test strip, to be used with a color chart or miniature device, can quantitatively determine the Hg(II) concentration in the field screening or on-site semiquantitative analysis without excessive laboratory instruments and (ii) having a selectivity for Hg(II) so that the sample pretreatment steps would be minimized. Because of the high stability constant of the Hg(II)-dithizone complex, the dithizone formed a much more stable complex with Hg(II). The extraction process of Hg(II) was performed using dithizone with a high selectivity to avoid any interfering elements in the extraction process.

The detection principle for the mercury assay of this test strip method is shown in Figure 1 [24]. Dithizone is one of the foremost extractants that is recognized as a sensitive reagent for the determination of Hg(II) in acidic media. It is capable of forming primary and secondary dithizonates with Hg(II) [25, 26]. Meanwhile, due to a higher absorptivity coefficient and solubility in the organic phase, the primary mercury-dithizonate chelate is preferred in the spectrophotometric determination as well.

The dissolved diatomic mercury ions, Hg(II), are the common form in aqueous solution. The Hg(II) could react with dithizone under acidic condition, and the complex is colored. In addition, the colored complex and the concentration of the Hg(II) ion are in accordance with the Lambert–Beer law:

$$A = \lg\left(\frac{1}{T}\right) = kbc, \tag{2}$$

where A is the absorbency (L/(mol∗cm)), T is the transmissivity (°C), and b and c are the concentration of the light-absorbing material (mol/L) and the thickness of the absorbed layer (cm), respectively.

Figure 2 shows the characteristic absorption of the mercury-dithizone complexes in aqueous solution. The absorbency gradually increased as the wavelength increased to 490 nm. Afterwards, the absorbency gradually decreased as the wavelength continued to increase. The results showed the maximum absorption peak of this Hg(II)-dithizone complex was at 490 nm. Subsequently, the measurement of spectra reflectivity was measured at 490 nm. The apparent molar absorptivity at 490 nm was 0.6 L·mol^{-1}·cm^{-1}.

Figure 3(a) shows the linear relationship between the absorbency value and different reaction times of the test strip and Hg^{2+} via the dipping-in method, which is that the paper-based test strip is dipped directly into the Hg^{2+}-containing solution firstly and then pulled it out from the solution to measure the absorbency. In this process, the test strip reacted rapidly with Hg^{2+} in solution, which led to color formation of the test strip. However, in this process, the Hg^{2+} contained in solution can continuously react with dithizone which loaded in the paper-based test strip, which caused the color absorbency of mercury-dithizone complexes formed on paper-based test strip increase with the prolongation of dipping time, as shown in Figure 3(a). The dipping-in time of the test strip needs to be strictly and accurately controlled if this dipping-in method was used in determining Hg^{2+} concentration, which is very difficult to operate by hand especially in on-site screening analysis. To solve this issue, the dropping-on method was chosen to use. In this method, certain amount of Hg^{2+}-containing solution was dropped on the dithizone-loaded paper-based test strip. Subsequently, the absorbency of the test strip colored by mercury-dithizone complexes was measured using spectrophotometer. During the measurement of using this dropping-on method, it was found that the colored liquid (mercury-dithizone complexes) diffused in indefinite directions due to liquid diffusivity. For the dropping-on method, although the absorbency value does not continued to increase with the reaction time, it was still difficult to accurately determine the result of color forming due to nonuniformity of colored complexes on the paper strip due to the Hg^{2+} or liquid diffusivity.

To overcome this diffusivity phenomenon, a horizontal circle 9 mm in diameter was made on the paper-based test strip [27]. This marker circle served as a hydrophobic barrier to prevent the dropped mercury solution from dispersing, and the uniform color is displayed in Figure 3(b). The test strip shows a light green color at the absence of mercury solution. When a drop of test solution was placed on the test strip, as shown in Figure 3(b), the test strip turned into pink at the presence of mercury solution. According to a series of trials, 3.7 μL was a suitable volume to be placed in the circle for preventing the mercury solution drop from dispersing and forming a uniform color. In addition, it should be noted that the dithizone loaded in this 90 mm circle test strip, 11.5 mmol, is much higher, 10 times higher, than that required for reacting with Hg^{2+}, 1.5–3.5 mmol, in 3.7 μL solution.

After solving this diffusivity phenomenon of mercury solution on the paper-based test strip, this dropping-on method could be used to determine the Hg^{2+}

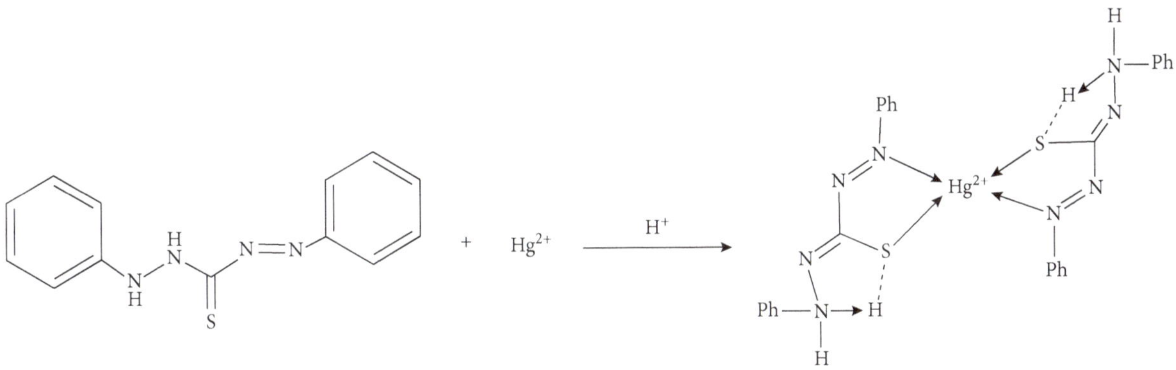

FIGURE 1: The detection principle for the mercury assay.

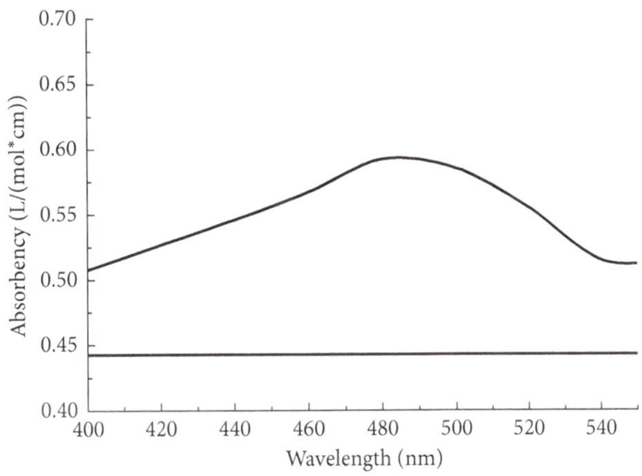

FIGURE 2: Characteristic absorption of the mercury-dithizone complexes in aqueous solution.

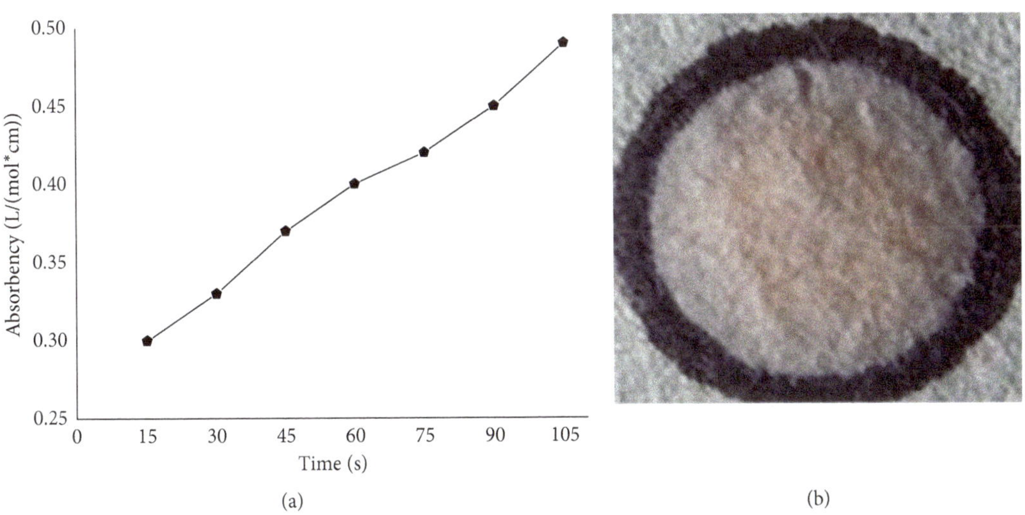

FIGURE 3: Different images of dipping-in method and dropping-on method: (a) the linear relationship between absorbency value and different dipping time of test strip and Hg^{2+} by dipping-in method and (b) test strip with marker circle and mercury-dithizone colored complexes forming by dropping-on method.

concentration in solution. In this method, the Hg^{2+}-containing solution was quantificationally dropped on the surface of circled paper-based test strip, and then the Hg^{2+} in the dropped solution reacts with dithizone loaded in the test strip to form mercury-dithizone complexes which is colored in pink. The color intensity formed on the test strip is mainly affected by the reaction time between Hg^{2+} and dithizone, here, namely, color-forming time. The test strip absorbency (color intensity) served as a standard function of color-forming time. The absorbency, at 490 nm, of colored complexes formed using 30 μg/mL mercury solution to react with dithizone-loaded test strip at different color-forming time (0.5 to 10 min) is presented in Figure 4. As shown in Figure 4, the absorbency of the test strip after dropping of Hg^{2+} solution gradually increased with color-forming time ranging from 0.5 min to 1.5 min, and then, the absorbency value tended to stabilize between 1.5 min and 3.5 min. The absorbency further reduced as the color-forming time prolonged beyond 3.5 min. The absorbency increase at color-forming time from 0.5 min to 1.5 min is due to formation of more mercury-dithizone complexes through the reaction of Hg^{2+} with dithizone. When the color-forming time reached 1.5 min, the absorbency stabilized, illustrating all the Hg^{2+} in the dropped solution reacted completely with dithizone loaded in test strip and formed the colored complexes, and this color intensity kept at a constant value for a while, up to 3.5 min. After 3.5 min, the decrease of absorbency is attributed to the water evaporation of the test strip. The maximum and most stable absorbency value (0.282) appeared at a 2.5 min color-forming time. The results showed that the reliable absorbency measurement of this test strip should be carried out in 1.5–3.5 minutes after dropping of Hg^{2+}-containing solution on the test strip.

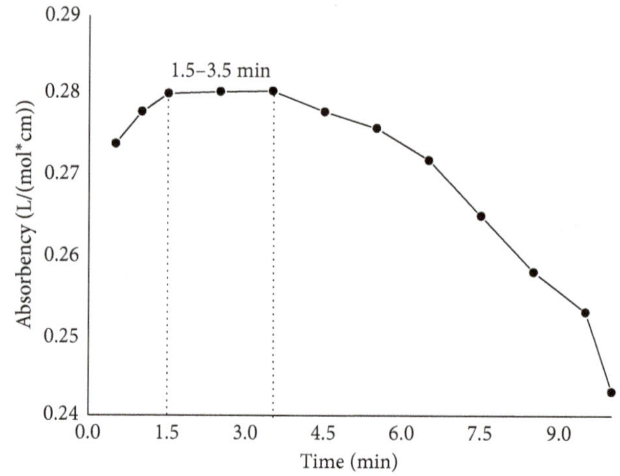

FIGURE 4: Absorbency of colored complexes formed using 3.7 μL 30 μg/mL Hg^{2+} solution and dithizone-loaded test strip at different color-forming time.

4.2. Calibration Curve. To establish the calibration curve, mercury solutions with various concentrations were prepared and used (Figure 5). Under the optimum conditions, pH 2.0 and reaction time 2.5 min, a linear calibration curve was constructed for Hg(II) determination over the range of 0.1 μg/mL to 30 μg/mL. The correlation coefficient (R2) was 0.9971, which showed an acceptable linearity of the calibration curve. Through the data analysis presented in Figure 5, it is found that the upper and lower detection limits of this paper-based test strip method are 30 μg/mL and 0.1 μg/mL, respectively. Although this lower detection limit was not enough for detection of Hg^{2+} ions in drinking waters or lake waters, it could be used for determination of waste waters from industry. The paper-based test strip with much lower detection limit, suitable for detecting lake waters or drinking waters, is our another research work which is going and will be reported in detail in our another paper.

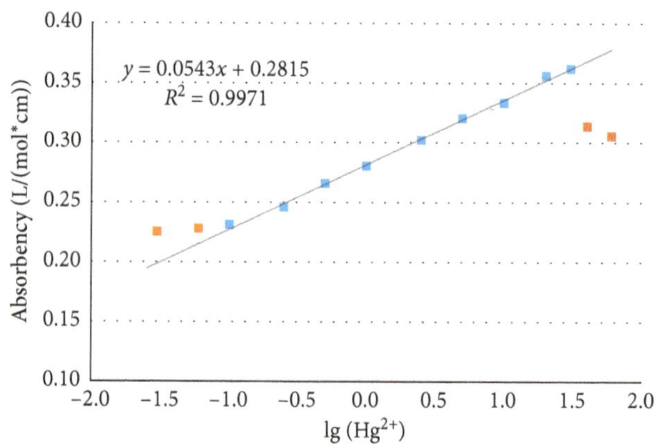

FIGURE 5: Working range of paper-based test strip (pH 2.0, color-forming time: 2.5 min).

4.3. Selectivity of Paper-Based Test Strip. In practice, Hg^{2+} cannot exist alone because large numbers of anions and cations also exist. However, the dithizone is effective on various ions. As shown in Figure 6, the test strips have absorbency for the different ions (Hg^{2+}, K^+, Ca^{2+}, Na^+, Mn^{2+}, Mg^{2+}, Cu^{2+}, Ag^+, Ba^{2+}, Al^{3+}, Fe^{3+}, Pb^{2+}, Zn^{2+}, PO_4^{3-}, Cl^-, and SO_4^{2-}). Furthermore, Hg^{2+} displayed excellent absorbability due to the test strips having better sensitivity and selectivity with the Hg^{2+} in the solution. Occasionally, Mn^{2+}, Mg^{2+}, Pb^{2+}, Cu^{2+}, Ag^+, Cl^-, and SO_4^{2-} interfered a little with mercury [2], and according to

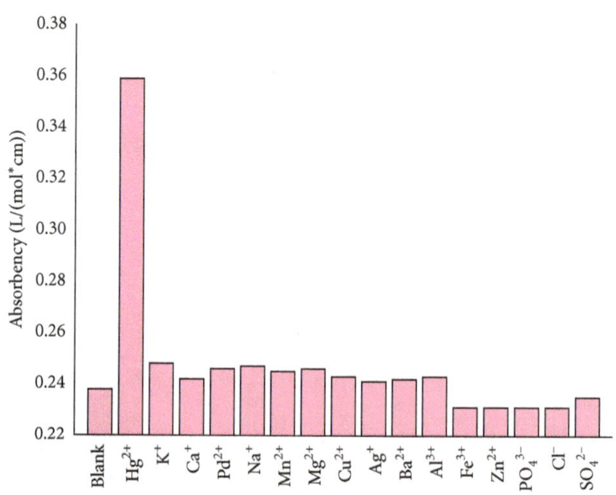

FIGURE 6: Absorbency of various metals in contact with the dithizone-loaded test strip; shown are the responses from blank (MilliQ water), 20 mg/L of Hg^{2+}, K^+, Ca^{2+}, Na^+, Mn^{2+}, Mg^{2+}, Cu^{2+}, Ag^+, Ba^{2+}, Al^{3+}, Fe^{3+}, Zn^{2+}, PO_4^{3-}, Cl^-, and SO_4^{2-}, and the concentration of other ions were one thousand times of Hg^{2+} (color-developing time: 2.5 min).

Table 1: Determination of Hg(II) concentration in a spiked tap water sample based on the test strip method.

Sample ID	Test strip method		Reference method	
	$[Hg^{2+}]$ (μg/mL)*	Standard deviation (μg/mL)	$[Hg^{2+}]$ (μg/mL)*	Standard deviation (μg/mL)
1	2.9	0.05	3.1	0.05
2	6.2	0.21	6.0	0.18
3	9.1	0.25	8.8	0.30
4	13.5	0.19	13.7	0.32

*Average value ($n = 3$).

the calibration curve, the deviations resulted from the interference of these ions are all less than 5%, which is acceptable usually for analysis [1, 9].

4.4. Verification of the Method with Spiked Water Samples. In this study, a high-efficiency test strip was developed in the circle with 3.7 μL of the mercury solution by pipette, the color-forming time was fixed at 2.5 min, and compared with the reference method, UV-Vis spectroscopy method, by determining same mercury concentrations [18]. The mean values from three replicate samples and the standard deviations of these measurements are shown in Table 1. The measured concentrations of Hg(II) using the test strip were similar as these from the reference method (2.9 and 3.1; 6.2 and 6.0; 9.1 and 8.8; and 13.5 and 13.7). The results exhibited a good sensitivity and selectivity, and also the deviation of this test-strip method was similar as that of reference method, even better than the reference method at high Hg (II) concentration (0.05 and 0.05; 0.21 and 0.18; 0.25 and 0.30; and 0.19 and 0.32). The measured data denoted that the proposed method can be satisfactorily applied to the determination of trace Hg(II) in real samples.

5. Conclusions

A cellulosic paper-based test strip for specially determining Hg^{2+} was developed. The color-forming reagent dithizone was physically loaded into the quantitative filter paper by impregnation process, and the mercury-dithizone complexes have a characteristic magenta color at acidic condition. The high-efficiency test strip method was developed by dropping 3.7 μL of the mercury solution via pipette on a circle test strip, and the color-forming time was fixed at 2.5 min. Furthermore, the paper-based test strips have a high selectivity to Hg^{2+} ion.

The developed method offered a good sensitivity and selectivity for the determination of Hg(II) in the concentration range of 0.1 μg/mL to 30 μg/mL. Although this lower detection limit was not enough for detection of Hg^{2+} ions in drinking waters or lake waters, it is enough to be used for detection of Hg^{2+} ion in waste waters from industry.

This paper-based test strip method, when applied to samples spiked with Hg^{2+} ions, gave accurate results in comparison with the conventional method for the determination of Hg^{2+}. Therefore, it can be concluded that the paper-based test strip method developed in this study is a simple, effective, and reliable way of determining the Hg^{2+} ion concentration in aqueous sample.

Conflicts of Interest

The authors declare that they have no conflicts of interest.

Authors' Contributions

Zhen Xu was contributed equally.

Acknowledgments

The authors are grateful to the National Natural Science Foundation of China (Grant Nos. 31570566, 31500489, 31800499, and 31600472), the Natural Science Foundation of Shandong (ZR2017LEM009 and ZR2018BEM026), the Key Research and Development Program of Shandong Province (No. 2017GSF17130), the Foundation of Guangxi Key Laboratory of Clean Pulp and Papermaking and Pollution Control of China (KF201717), the Shandong Taishan Scholar Program and the Foundation of Key Laboratory of Pulp and Paper Science and Technology of Ministry of Education/Shandong Province of China (Nos. ZR201707 and ZR201710), National Key R&D Program of China (No. 2017YFB0308000), and Joint Research Fund for Young Doctor of Qilu University of Technology (Shandong Academy of Sciences) (No. 2017BSH2010).

References

[1] J. Chen, Y. Li, W. Zhong, H. Wang, P. Zhang, and J. Jiang, "A highly selective fluorescent and colorimetric chemosensor for Hg^{2+} based on a new rhodamine derivative," *Analytical Methods*, vol. 8, no. 9, pp. 1964–1967, 2016.

[2] M. Kolb, M. Bahadir, and B. Teichgräber, "Determination of chemical oxygen demand (COD) using an alternative wet chemical method free of mercury and dichromate," *Water Research*, vol. 122, pp. 645–654, 2017.

[3] G. Aragay, H. Montón, J. Pons, M. Fontbardía, and A. Merkoçi, "Rapid and highly sensitive detection of mercury ions using a fluorescence-based paper test strip with an N-alkylaminopyrazole ligand as a receptor," *Journal of Materials Chemistry*, vol. 22, no. 13, pp. 5978–5983, 2012.

[4] J. H. Richard and H. Biester, "Mercury removal from contaminated groundwater: performance and limitations of amalgamation through brass shavings," *Water Research*, vol. 99, pp. 272–280, 2016.

[5] D. A. Deeds, A. Ghoshdastidar, F. Raofie, É. A. Guérette, A. Tessier, and P. A. Ariya, "Development of a particle-trap preconcentration-soft ionization mass spectrometric technique for the quantification of mercury halides in air," *Analytical Chemistry*, vol. 87, no. 10, pp. 5109–5116, 2015.

[6] A. Fashi, M. R. Yaftian, and A. Zamani, "Electromembrane extraction-preconcentration followed by microvolume UV–Vis spectrophotometric determination of mercury in water and fish samples," *Food Chemistry*, vol. 221, pp. 714–720, 2017.

[7] K. Ma, X. Li, B. Xu, and W. Tian, "A sensitive and selective "turn-on" fluorescent probe for Hg^{2+} based on thymine–Hg^{2+}–thymine complex with an aggregation-induced emission feature," *Analytical Methods*, vol. 6, no. 7, pp. 2338–2342, 2014.

[8] H. Tao, Y. Lin, J. Yan, and J. Di, "A plasmonic mercury sensor based on silver–gold alloy nanoparticles electrodeposited on indium tin oxide glass," *Electrochemistry Communications*, vol. 40, pp. 75–79, 2014.

[9] L. Farzin, M. Shamsipur, and M. A. Tabrizi, "Biomagnetic separation and pre-concentration of trace amounts of Hg^{2+} in biological samples based on T-rich oligonucleotide modified magnetic beads," *Analytical Methods*, vol. 7, no. 20, pp. 8947–8953, 2015.

[10] B. Silwana, D. H. C. Van, E. Iwuoha, and V. Somerset, "Amperometric determination of cadmium, lead, and mercury metal ions using a novel polymer immobilised horseradish peroxidase biosensor system," *ournal of Environmental Science and Health, Part A*, vol. 49, no. 13, pp. 1501–1511, 2014.

[11] A. Scheuhammer, B. Braune, H. M. Chan et al., "Recent progress on our understanding of the biological effects of mercury in fish and wildlife in the Canadian Arctic," *Science of The Total Environment*, vol. 509-510, pp. 91–103, 2015.

[12] M. Cegłowski and G. Schroeder, "Removal of heavy metal ions with the use of chelating polymers obtained by grafting pyridine–pyrazole ligands onto polymethylhydrosiloxane," *Chemical Engineering Journal*, vol. 259, pp. 885–893, 2015.

[13] M. Tomás-Gamasa, M. S. S. Serdjukow, M. S. M. Su, M. Müller, and P. T. Carell, "Removal of heavy metal ions with the use of chelating polymers obtained by grafting pyridine–pyrazole ligands onto polymethylhydrosiloxane," *Angewandte Chemie International Edition*, vol. 54, no. 3, pp. 796–800, 2015.

[14] C. Pettinari, A. Tăbăcaru, and S. Galli, "Coordination polymers and metal–organic frameworks based on poly (pyrazole)-containing ligands," *Coordination Chemistry Reviews*, vol. 307, pp. 1–31, 2016.

[15] C. Cano-Raya, M. D. Fernández-Ramos, J. Gómez-Sánchez, and L. F. Capitán-Vallvey, "Irreversible optical sensor for mercury determination based on tetraarylborate decomposition," *Sensors and Actuator B: Chemical*, vol. 117, no. 1, pp. 135–142, 2006.

[16] A. V. Yallouz, D. C. R. Calixto, and S. Paciornik, "A low-cost non instrumental method for semiquantitative determination of mercury in fish," *Fresenius' Journal of Analytical Chemistry*, vol. 366, no. 5, pp. 461–465, 2000.

[17] G. Q. Shi and G. Jiang, "A dip-and-read test strip for the determination of mercury(II) ion in aqueous samples based on urease activity inhibition," *Analytical Sciences*, vol. 18, no. 11, pp. 1215–1219, 2002.

[18] L. F. Capitán-Vallvey, C. C. Raya, E. L. López, and F. Ramos, "Irreversible optical test strip for mercury determination based on neutral ionophore," *Analytica Chimica Acta*, vol. 524, no. 1-2, pp. 365–372, 2004.

[19] Z. Chandio, F. Talpur, H. Khan, H. Afridi, G. Khaskheli, and M. Mughal, "On-line preconcentration and determination of ultra trace amounts of mercury using surfactant coated alumina modified by dithizone with cold vapor atomic absorption spectrometry," *RSC Advances*, vol. 4, no. 7, pp. 3326–3331, 2014.

[20] R. F. Gur'eva, S. B. Savvin, and A. V. Mikhailova, "Sorption and determination of vanadium(IV, V) and Mercury(I, II) as their colored complexes of organic reagents," *Journal of Analytical Chemistry*, vol. 58, pp. 623-624, 2003.

[21] R. Sedghi, S. Kazemi, and B. Heidari, "Novel selective and sensitive dual colorimetric sensor for mercury and lead ions derived from dithizone-polymeric nanocomposite hybrid," *Sensors and Actuators B: Chemical*, vol. 245, pp. 860–867, 2017.

[22] S. S. Yamamura and J. H. Sikes, "Use of Citrate-EDTA masking for selective determination of iron with 1, 10-phenanthroline," *Analytical Chemistry*, vol. 38, no. 6, pp. 793–795, 1966.

[23] S. Carrera, G. Santiago, and M. Vega, "Spectrophotometric determination of dithizone–mercury complex by solid phase microextraction in micropipette tip syringe packed with activated carbon xerogel," *Microchemical Journal*, vol. 129, pp. 133–136, 2016.

[24] Z. Zhang, J. Li, X. Song, J. Ma, and L. Chen, "Hg^{2+} ion-imprinted polymers sorbents based on dithizone–Hg^{2+} chelation for mercury speciation analysis in environmental and biological samples," *RSC Advances*, vol. 4, no. 87, pp. 46444–46453, 2014.

[25] M. Mudasir, K. Karelius, N. H. Aprilita, and E. T. Wahyuni, "Adsorption of mercury(II) on dithizone-immobilized natural zeolite," *Journal of Environmental Chemical Engineering*, vol. 4, no. 2, pp. 1839–1849, 2016.

[26] D. Zhang, M. Sun, and L. Zou, *A Review on Spectrometer of Pb (II) in Water*, Springer, Berlin, Germany, 2016.

[27] A. Y. Satoh, J. E. Trosko, and S. J. Masten, "Methylene blue dye test for rapid qualitative detection of hydroxyl radicals formed in a fenton's reaction aqueous solution," *Environmental Science and Technology*, vol. 41, no. 8, pp. 2881–2887, 2007.

14

Chemical Characterization and Oxidative Stability of Medium- and Long-Chain Fatty Acid Profiles in Tree-Borne Seed Oils

Da-Som Kim,[1] Hoe-Sung Kim,[1] Kyoung-Tae Lee,[2] Dong-Lee Hong,[3] Sung-Rae Cho,[3] Jeong Hoon Pan,[4] Yong Bae Park,[2] Yang-Bong Lee,[3] Jae Kyeom Kim,[4] and Eui-Cheol Shin[1]

[1]*Department of Food Science, Gyeongnam National University of Science and Technology, Jinju 52725, Republic of Korea*
[2]*Southern Forest Resources Research Center, National Institute of Forest Science, Jinju 52817, Republic of Korea*
[3]*Department of Food Science and Technology, Pukyong National University of Science and Technology, Busan 48513, Republic of Korea*
[4]*School of Human Environmental Sciences, University of Arkansas, Fayetteville, AR 72701, USA*

Correspondence should be addressed to Eui-Cheol Shin; eshin@gntech.ac.kr

Academic Editor: Boryana M. Nikolova-Damyanova

This study was undertaken to evaluate chemical characteristics and oxidative stability of tree-borne seed oils. A total of 15 different fatty acids were identified in six tree-borne seed oils, which included seven types of saturated fatty acids, four types of monounsaturated fatty acids, and four types of polyunsaturated fatty acids. Japanese camphor tree (JCT) had a high content of medium-chain fatty acids ($97.94 \pm 0.04\%$), in which fatty acid composition was distinct from those of the other five plant seed oils. Overall, contents of tocopherols, a type of fat-soluble vitamin, ranged between 3.82 ± 0.04 mg/100 g and 101.98 ± 1.34 mg/100 g, respectively. Phytosterol contents ranged from 117.77 ± 1.32 mg/100 g to 479.45 ± 4.27 mg/100 g, respectively. Of all tree-borne seed oils, β-sitosterol was the phytosterol at the highest concentration. Contents of unsaponifiables were between 0.13 ± 0.08 and 2.01 ± 0.02, and values of acid, peroxide, and *p*-anisidine were between 0.79 ± 0.01 and 38.94 ± 0.24 mg KOH/g, 3.53 ± 0.21 and 127.67 ± 1.79 meq/kg, and 2.07 ± 0.51 and 9.67 ± 0.25, respectively. Oxidative stability of tree-borne seed oils was assessed through measurement of oxidation-induction periods. These results should serve as a foundation to identify the potential of tree-borne seed oils in industrial application as well as in providing fundamental data.

1. Introduction

In modern society, forest plants and their products have been increasingly viewed as useful resources, which have prompted research on applications of plant-based products and utilization of associated organic compounds with diverse physiological activities [1]. Given the potential utility of plant-based products, in particular plant-based oils, it is surprising that relatively few studies have been conducted on oil resources of forest trees, which may be an important source of functional chemical substances. In general, vegetable oils are obtained from grains, and their major components are fatty acids, accounting for over 90% of ester forms. Vegetable fats contain more unsaturated fatty acids than animal fats, and unsaturated fatty acids have important nutritional and functional properties. In particular, many seed oils contain high levels of unsaturated fatty acids. Of them, the n-6 series of linoleic acid (C18 : 2) generate conjugated linoleic acid (CLA), through microbial or chemical reactions, which is an isomer known to have useful functions in the human body [2, 3]. In addition, the n-3 series of alpha-linolenic acid (C18 : 3, n-3), which are particularly representative fatty acids of perilla oil, can be converted to eicosapentaenoic acid (EPA, C20 : 5, n-3) and docosahexaenoic acid (DHA, C22 : 6, n-3) by desaturation and chain elongation in human bodies [4, 5]. These n-3 series of fatty acids are essential to synthesize brain phospholipids and are known to improve brain function, including learning ability [6]. Vegetable fats also contain high levels of phytosterols and vitamin E, thus have antioxidant activity; these constituents confer oxidative stability for food processing, storage, and distribution [7, 8].

Among vegetable oils that are known as sources of fat-soluble components, there have been many studies on seed oils of legumes and nuts [9, 10], but there have been few studies on seed oils derived from tree-borne seeds [11]. Therefore, in this study, we aimed at characterizing chemical characteristics of tree-borne seed oils: mastic-leaf prickly ash (MPA, Zanthoxylum schinifolium Siebold & Zucc), Japanese camphor tree (JCT, *Cinnamomum yabunikkei* H.Ohba), Korean evodia (KOE, *Euodia daniellii* (Benn.) Hemsl.), Japanese lacquer tree (JLT, *Rhus verniciflua* Stokes), Tung tree (TUT, *Vernicia fordii* (Hemsl.) Airy Shaw), and oriental raisin tree (ORT, *Hovenia dulcis* Thunb.). Although these are plants commonly found in the wild, there has been little research on fat-soluble components of their seed oils. In fact, most literature on the use of plant-based products tend to report pharmacological effects of beverage or root components using leaves or fruits of trees, whereas reports on oils tended to focus on simple topics, such as fatty acid composition. As a follow-up study of our recent report about physicochemical properties of tree-borne seed oils [11], here we identified physicochemical properties of oils extracted from 6 trees, by which we evaluated their potentials as oil resources, and provided fundamental data for their use as oil resources.

2. Materials and Methods

2.1. Materials. The trees from which we obtained tree-borne seed oils for this study were MPA, JCT, KOE, JLT, TUT, and ORT. Samples were harvested and collected by Southern Forest Resources Research Center, Jinju, Republic of Korea, in 2016, and oils were extracted from seed endosperm, for which seed coats and impurities were removed manually. For solvent extraction, samples were mixed with *n*-hexane (10 times sample weight) and stirred at 4°C for 12 h, followed by concentration using a stream of nitrogen (N_2) gas, resulting in fat-soluble substances, which were used for experiments [12].

2.2. Fatty Acid Composition Analysis. To identify fatty acid composition of the extracted tree-borne seed oils, the seed oils were derivatized to methyl ester using boron trifluoride-(BF3-) methanol before analysis [13]. Approximately 0.1 g of tree-borne seed oils were collected in test tubes, to which 0.5 mL heptadecanoic acid (C17 : 0) (1 mg/mL hexane) was added as an internal reference. After adding 2 mL of 0.5 M NaOH methanol to the test tubes, the tube was heated at 110°C for 10 min using a Reacti-Therm III Heating/Stirring Module (Thermo Fisher Scientific Co., Rockford, IL, USA), which was then cooled to room temperature, and heated again at 110°C after adding 4 mL BF3-methanol for 1 h. After cooling, 2 mL hexane was added to the sample mix and vortexed for 1 min, followed by collection of the hexane layer. Solvent was removed from the hexane layer collected through three repetitions of hexane addition and extraction using nitrogen gas, and then the remaining residue was dissolved with 1 mL hexane again, which was used as the sample for fatty acid analysis. Fatty acids of tree-borne seed oils were analyzed by gas chromatography (GC) using Agilent Technologies 6890N (Agilent Technologies, Santa Clara, CA, USA). An SP-2560 capillary column (100 m × 0.25 mm i.d., 0.25 μm film thickness; Agilent Technologies) was used as the column for analysis, and nitrogen (2.7 mL/min) was used as carrier gas. Temperature of the inlet and the detector was 250°C. The sample ratio to the inlet was set to 10 : 1, and rates of hydrogen and air for flame ionization at the detector were set to 40 mL and 450 mL/min, respectively. The initial oven temperature was 130°C for 5 min and gradually increased to 240°C by 4°C/min, and maintained at 240°C for 15 min. Analyzed results were identified by retention time on the basis of 37 fatty acids reference standards (Supelco 37 FAME, Sigma-Aldrich Co., St. Louis, MO, USA).

2.3. Vitamin E (Tocopherol) Analysis. To identify tocopherol contents in tree-borne seed oils, the solvent extraction method was used [9]. For pretreatment of samples, 1 g oil was mixed with hexane up to 50 mL and homogenized for 30 s, and this mixture was passed through a 0.45 μm nylon membrane filter (GE Osmonics Lab Store, Minnetonka, MN, USA). A normal-phase HPLC system (Agilent 1260, Agilent Technologies) with a fluorescence detector was used for analysis. The column for separation was LiChrosorb Si-60 column (4 mm × 250 mm, 5 μm particle size; Hibar® Fertigsäule RT, Merck, Darmstadt, Germany) in a normal phase condition. Mobile phase was 0.85% isopropanol/hexane, and flow rate was set as 1 mL/min. Wavelengths of the detector for vitamin E analysis were 290 nm for excitation and 330 nm for emission. For quantification of tocopherol, reference standards for α-, β-, γ-, and δ-tocopherol were used. To assess purity of each tocopherol, the reference standards were dissolved in 25 mL ethanol, and then extinction coefficients were measured by a UV-spectrophotometer (DU-62, Beckman Instruments, Inc., Fullerton, CA, USA). Wavelengths for each isomer were 294 (α-T), 297 (β-T), 298 (γ-T), and 298 (δ-T) nm. Purity of α-, β-, γ-, and δ-tocopherol was 99.07%, 82.47%, 98.71%, and 89.16%, respectively. When judged based on purity of the reference standards, concentrations of stock solutions for α-, β-, γ-, and δ-tocopherols were 1.96, 1.65, 3.65, and 1.80 mg/mL, respectively, and the standard solutions were stored at −40°C for further use. For the calibration of each tocopherol, concentrations range for α-, β-, γ-, and δ-tocopherols were 0.25–1.96, 0.21–1.65, 0.46–3.65, and 0.23–1.80 mg/mL, respectively.

2.4. Contents of Unsaponifiables and Phytosterols. Contents of unsaponifiables in tree-borne seed oils, particularly phytosterols, the representative unsaponifiables, were analyzed by an alkaline saponification method [10]. We mixed 0.5 g of extracted seed oils with 0.5 mL 5α-cholestane (1 mg/mL hexanes), an internal reference, followed by addition of 8 mL of 3% (w/v) pyrogallol-ethanol solution for antioxidative capacity and 1 mL saturated potassium hydroxide (KOH) solution, which was then well mixed, saponified at 80°C for 1 h, and cooled in an ice bucket. For extraction, 10 mL distilled water and 7 mL hexanes were added to the sample mix and then the supernatant corresponding to the hexane layer was separated. Extraction of the hexane layer was repeated three times. The solvent was completely evaporated from the separated hexane layer under nitrogen gas to concentrate the residue. Weight of the completely concentrated residue was measured and compared with input sample amount, by which residue weight was calculated as the ratio of unsaponifiables.

For phytosterol analysis, derivatization was performed by using the trimethylsilyl group. Unsaponifiables were mixed with 0.5 mL of BSTFA (N, O-bis(trimethylsilyl)trifluoroacetamide + trimethylchlorosilane) that contained 1% trimethylsilyl chloride and 1 mL pyridine, and dissolved again, followed by induction of reaction at 80°C for 1 h. After completion of the reaction, samples were slowly cooled at room temperature and filtered with a 0.45 μm filter to remove particles, which were used as samples for phytosterol analysis. Phytosterol was analyzed by gas chromatography, for which HP-5 column (30 m × 0.32 mm i.d., 0.25 μm in film thickness; Agilent Technologies) was used, and helium (2.7 mL/min) was used as the carrier gas. The oven temperature was 260°C at the beginning, increased to 300°C by 3°C/min, and then maintained at 300°C for 15 min. Temperature of the inlet and the detector was 300°C and 320°C, respectively, and the split ratio was 10 : 1. Detection was performed by a flame ionization detector. For identification and measurement of phytosterols, each reference standard, campesterol, stigmasterol, and β-sitosterol, was purchased from Sigma-Aldrich Co., and retention time and peak area were used for identification. Phytosterol quantification was performed using 5α-cholestane as the internal standard. To accurately quantify the phytosterols in oils, relative response factors (RRFs) to 5α-cholestane were determined based on three replicate analyses. The RRFs were determined according to the following equation:

$$\mathrm{RRF} = \frac{\mathrm{PA}_{\mathrm{sterol}}}{\mathrm{W}_{\mathrm{sterol}}} \times \frac{\mathrm{W}_{\mathrm{is}}}{\mathrm{PA}_{\mathrm{is}}}, \tag{1}$$

where $\mathrm{PA}_{\mathrm{sterol}}$ is the peak area of the sterol, $\mathrm{W}_{\mathrm{sterol}}$ is the mass (mg) of the sterol, $\mathrm{PA}_{\mathrm{is}}$ is the peak area of the internal standard, and W_{is} is the mass (mg) of the internal standard.

On the basis of the RRFs, phytosterol contents were determined using the following equation:

$$\text{Sterols (mg/100 mg lipid extract)} = \frac{\mathrm{PA}_{\mathrm{sterol}}}{\mathrm{PA}_{\mathrm{is}}} \times \frac{1}{\mathrm{RRF}} \times \frac{\mathrm{W}_{\mathrm{is}}}{\mathrm{W}_{\mathrm{sample}}} \times 100, \tag{2}$$

where $\mathrm{PA}_{\mathrm{sterol}}$ is the peak areas of the sterol, $\mathrm{PA}_{\mathrm{is}}$ is the peak area of the internal standard, W_{is} is the mass (mg) of the internal standard, and $\mathrm{W}_{\mathrm{sample}}$ is the mass (g) of the oil.

Commercial standard of Δ5-avenasterol was not available, so that right after peak of β-sitosterol was considered as Δ5-avenasterol because of its similar molecular weight and close retention time to Δ5-avenasterol and based on the literature [10].

2.5. Acid Value, Peroxide Value, and p-Anisidine Value (p-AV). Acid value of tree-borne seed oils was measured using the American Oil Chemist's Society (AOCS) method [14]. After adding 1 g oil to a 150 mL conical flask, 30 mL of the solution was mixed with ethanol-ether in 1 : 1 (v/v) ratio to dissolve oils. To this solution, we added (and mixed well) 100 μL of 1% phenolphthalein as an indicator and then the mix was titrated with 0.1 M potassium hydroxide (KOH)/ethanol solution until it became light red. Acid value was finally identified through a mock experiment without sample.

Peroxide value of seed oils were measured using the AOCS method [15]. After collecting 1 g oil in a 250 mL conical flask, oil samples were dissolved with 25 mL of solution mixed with acetic acid-chloroform in 3 : 2 (v/v) ratio, to which 1 mL saturated potassium iodide (KI) solution was added. The mix was then shaken for 1 min, incubated in a dark condition for 10 min, and mixed well with 75 mL distilled water. Then, 1 mL of 1% starch solution was added, and the solution was titrated with 0.01 N sodium thiosulfate ($Na_2S_2O_3$) solution until it became transparent. Peroxide value was finally identified through a mock experiment without sample.

To identify amounts of aldehydes and ketones that are secondary oxidation products generated during acidification of seed oils, p-AV was utilized. After mixing 100 mg oil with 25 mL isooctane, optical density (OD) was measured at 350 nm. Following reaction of 2.5 mL of this solution with 0.5 mL of 0.25% (w/v) p-anisidine/acetic acid solution for 15 min, OD was measured at the same wavelength. In this way, OD values before and after reaction with anisidine solution were measured to find p-AV value [16].

2.6. Color Scale and Brown Color Intensity Using Color Difference Meter and Spectrophotometer. Following transfer of 5 g tree-borne seed oils to Petri dishes for color scale analysis, color scale was measured by a color difference meter (CR-300, Konica Minolta Inc., Tokyo, Japan). Each of L^* (0/100 darkness/lightness), a^* (±, redness/greenness), and b^* (±, yellowness/blueness) in the Hunter scale were measured three times, from which mean values were calculated. Values of the standard plate, used as a reference, were $L^* = 93.59$, $a^* = 2.62$, and $b^* = 1.88$. To measure brown color intensity that is related with browning of seed oils, 200 μL seed oils were added to a 96-well plate, and OD was measured at 420 nm using a spectrophotometer (Multiskan Go, Thermo Fisher Scientific, Waltham, MA, USA). A higher OD value was considered as a higher brown color intensity of oil [11].

2.7. Oxidation-Induction Periods. To evaluate oxidative stability depending on physicochemical properties of tree-borne seed oils, a rancimat (892 Professional rancimat, Metrohm AG, Herisau, Switzerland) instrument was used. After 3 g of oil was transferred to a rancimat tube, air was injected to the tube with oil at a rate of 20 L/h and a reaction was performed at 120°C in order to induce forced oxidation.

Table 1: Fatty acid profiles in tree-borne seed oils.

Fatty acid	Composition (%)					
	JCT	KOE	JLT	MPA	ORT	TUT
Capric acid (C10 : 0)	30.22 ± 0.03	nd	nd*	nd	nd	nd
Lauric acid (C12 : 0)	65.83 ± 0.03	nd	nd	nd	nd	nd
Myristic acid (C14 : 0)	1.02 ± 0.01	nd	nd	nd	nd	nd
Palmitic acid (C16 : 0)	0.38 ± 0.01	9.80 ± 0.01	13.82 ± 0.06	11.26 ± 0.03	6.75 ± 0.01	3.84 ± 0.01
Stearic acid (C18 : 0)	nd	2.61 ± 0.01	3.52 ± 0.01	1.85 ± 0.01	3.98 ± 0.01	2.89 ± 0.01
Arachidic acid (C20 : 0)	nd	nd	nd	nd	0.85 ± 0.01	nd
Behenic acid (C22 : 0)	nd	nd	nd	nd	0.43 ± 0.01	nd
ΣSaturated fatty acids	97.94 ± 0.04^{a}	12.35 ± 0.02^{c}	16.24 ± 0.07^{b}	13.24 ± 0.05^{c}	12.14 ± 0.02^{c}	6.76 ± 0.02^{d}
Palmitoleic acid (C16 : 1ω7)	nd	5.04 ± 0.01	0.72 ± 0.01	2.38 ± 0.01	0.53 ± 0.01	nd
Oleic acid (C18 : 1ω9)	1.82 ± 0.01	21.40 ± 0.09	20.54 ± 0.04	34.01 ± 0.01	31.38 ± 0.04	10.33 ± 0.05
Gondoic acid (C20 : 1ω9)	nd	nd	nd	0.50 ± 0.01	nd	nd
Nervonic acid (C24 : 1ω9)	nd	nd	nd	nd	nd	42.78 ± 0.76
ΣMonounsaturated fatty acids	1.82 ± 0.01^{e}	27.17 ± 0.09^{c}	21.08 ± 0.05^{d}	37.19 ± 0.02^{b}	32.10 ± 0.05^{bc}	53.34 ± 1.21^{a}
Linoleic acid (C18 : 2ω6)	0.73 ± 0.01	38.96 ± 0.04	60.41 ± 0.12	27.84 ± 0.01	18.65 ± 0.01	11.96 ± 0.01
Eicosadienoic acid (C20 : 2)	nd	nd	nd	0.40 ± 0.01	nd	nd
Linolenic acid (C18 : 3ω3)	nd	21.95 ± 0.04	0.99 ± 0.01	21.75 ± 0.02	35.13 ± 0.08	nd
Docosahexaenoic acid (C22 : 6ω3)	nd	nd	nd	nd	nd	28.21 ± 1.20
ΣPolyunsaturated fatty acids	0.73 ± 0.01^{f}	60.39 ± 0.06^{b}	62.01 ± 0.14^{a}	49.58 ± 0.03^{d}	55.01 ± 0.09^{c}	40.03 ± 1.32^{e}
Oil content (g/100 g)	20.17 ± 1.24^{c}	33.59 ± 2.01^{a}	24.60 ± 1.73^{c}	9.40 ± 0.72^{d}	29.6 ± 1.96^{b}	4.6 ± 0.28^{e}

*Means with different letters (a–f) within a row are significantly different by Tukey's multiple test ($P < 0.05$); nd = not detected.

Oxidation products such as formic acid, ketones, aldehydes, and carboxylic acids were measured through an oxidation-detecting sensor, based on which oxidation-induction periods were investigated [11].

2.8. Statistical Analysis. All experiments were repeated three times, and the results are expressed as mean ± standard deviation. We compared mean values using Tukey's multiple range test in SAS 9.2 (Statistical Analysis System, Version 9.2, SAS Institute Inc., Cary, NC, USA) program. A P value less than 0.05 was considered as statistically significant.

3. Results and Discussion

3.1. Fatty Acid Composition. Fatty acid analysis results of tree-borne seed oils are shown in Table 1. A total of 15 kinds of fatty acids were identified from six tree-borne seed oils, which included seven types of saturated fatty acids, four types of monounsaturated fatty acids, and four types of polyunsaturated fatty acids. For saturated fatty acids, JCT had 97.94 ± 0.04%, corresponding to the highest level, and the remaining five seed oils contained in the range between 6.76 ± 0.02% and 16.24 ± 0.07%. In particular, JCT showed a high medium-chain fatty acid content (>96%) which is interesting given health implications of these fatty acids. Specially, fatty acids of edible oils are long-chain triglycerides composed of 14–20 carbons, which are absorbed by lymph vessels through a fat-specific digestive process, including emulsification and chylomicron formation. Owing to this long digestion process, long-chain triglycerides are relatively less utilized for energy than are carbohydrates or proteins and are likely to be stored in the body, thus likely causing obesity [17]. In contrast, medium-chain triglycerides that have short fatty acids with 8–12 carbons are absorbed by the liver, through the hepatic portal vein, without formation of chylomicron (the general characteristic of fat digestion) and can be digested without bile or lipase. The medium-chain triglycerides are mostly obtained through hydrolysis of coconut oil, butter, and palm oil and are also found in human breast milk, in which they account for approximately 20% of milk. Because medium-chain triglycerides do not need proteins such as fatty acid binding protein, fatty acid transport protein, and fatty acid translocase for their processing, their oxidation rate is faster than that of long-chain fatty acids, and thus, they tend to be used as an energy source rather than being accumulated in the body [17, 18]. According to previous reports, diets with medium-chain fatty acids had a better thermic effect of food than those with long-chain fatty acids [19–21]. Another study also reported that medium-chain fatty acids had a higher thermic effect of food than long-chain fatty acids, reduced appetite, and increased satiety during a 2-week study, resulting in weight loss [22]. In respect to the practicality of cooking with medium-chain fatty acids, their smoke point is low and bubbles are generated when mixed with other vegetable oils or when they are used for frying. To solve these issues, a study on medium- and long-chain triglycerides that contain a certain amount of long-chain triglycerides with essential fatty acids has been ongoing [23].

For monounsaturated fatty acids, the highest level was 53.34 ± 1.21% found in TUT, and its level ranged from 1.82 ± 0.01% in TUT to 37.19 ± 0.02% in MPA. Oleic acid, although not an essential fatty acid, is known to lower cholesterol levels in the body, contributing to prevention of cardiovascular disease, so that it is a nutritionally significant fatty acid [24]. In addition, monounsaturated fatty acids have possessed a higher oxidative stability than polyunsaturated fatty acids, showing a higher convenience and stability in oil processing [25]. Levels of polyunsaturated fatty acids were

TABLE 2: Tocopherol (T) composition in tree-borne seed oils.

Sample	Tocopherols (mg/100 g oil)			Total-T
	α-T	γ-T	δ-T	
MPA	$6.01 \pm 0.06^{b*}$	nd	nd	6.01 ± 0.06^{e}
JCT	10.88 ± 0.43^{a}	11.13 ± 0.70^{d}	4.78 ± 0.60^{c}	26.79 ± 1.72^{d}
KOE	nd	3.66 ± 0.05^{e}	0.16 ± 0.01^{e}	3.82 ± 0.04^{f}
JLT	nd	42.47 ± 1.28^{b}	33.92 ± 0.47^{a}	76.40 ± 1.73^{b}
TUT	2.81 ± 0.10^{c}	35.52 ± 0.21^{c}	2.45 ± 0.04^{d}	40.78 ± 0.11^{c}
ORT	nd	95.59 ± 1.24^{a}	6.39 ± 0.16^{b}	101.98 ± 1.34^{a}

*Means with different letters (a–f) within a column are significantly different by Tukey's multiple test ($P < 0.05$); nd = not detected.

high in JLT and KOE (62.01 ± 0.14% and 60.39 ± 0.06%). While MPA, ORT, and TUT also had over 40% of polyunsaturated fatty acid contents, their levels in JCT were as low as 0.73 ± 0.01%. Linoleic acid, which is the representative of omega-6 polyunsaturated fatty acids, is mostly found in plant grains, and it lowers blood cholesterols and inhibits arterial thrombus formation. Linoleic acid levels in JLT were as high as 60.41 ± 0.12%. As omega-6 fatty acids are generally unable to be synthesized in humans and mammals, they are classified as essential fatty acids [26].

Omega-3 fatty acids, which are represented by docosahexaenoic acid (DHA) and linolenic acid, have various nutritional functions. ORT was found to have a high linolenic acid content with 35.13 ± 0.05%, and TUT had 28.21 ± 1.20% of DHA (C22 : 6ω3). Omega-3 fatty acids that are represented by DHA and linolenic acid have various nutritional functions, which not only serve as a fuel of metabolism but also protect and maintain cells and play important roles in maintenance and apoptosis of heart cells [27]. In addition, linolenic acid helps blood circulation by lowering blood cholesterol levels, affecting prevention or improvement of vascular diseases such as hypertension, arteriosclerosis, angina pectoris, obesity, diabetes, and atopy [27]. The six tree-borne seed oils in this study showed various fatty acid compositions. Some seed oils had a high level of medium-chain fatty acids, others had a high level of monounsaturated fatty acids, and yet others had a high level of polyunsaturated fatty acids. Although polyunsaturated fatty acids are known to function as various biological membranes and are involved in cell metabolism, the human body often fails to synthesize sufficient amounts; therefore, such deficiency needs to be supplemented by an external source (food) [28]. Most seed oils, except JCT oil, contained high levels of unsaturated fatty acids. While contributing to nutritional functions, such high unsaturation levels are vulnerable to acidification. Thus, prevention of acidification for oil storage and processing needs to be studied in order to enable tree-borne seed oils to be used as food and functional food materials.

3.2. Tocopherol Contents. Tocopherol contents of tree-borne seed oils are presented in Table 2. A total of three tocopherol isomers (α-T, γ-T, and δ-T) were identified, whereas tocotrienols were undetected. Vitamin E contents in the six tree-borne seed oils ranged from 3.82 ± 0.04 mg/100 g to 101.98 ± 1.34 mg/100 g. KOE oil had the lowest level of tocopherols, whereas oil from ORT had the highest content. Vitamins are organic compounds that support nutrient functions in the body, and tocopherols have antiaging effects due to their antioxidant activity [29]. Tocopherols occur as a viscous, pale-yellow, oily substance in nature and are defined as a derivative group of 2-methyl-6-hydroxy-chroman; in this group, there are four types of tocopherols—α, β, γ, and δ—depending on the degree of the methyl group bound to the chroman-ring [9, 29]. For α-T, JCT oil had the highest level (10.88 ± 0.43 mg/100 g) among the six tree-borne seed oils. α-T has the highest vitamin E activity. β-T is present in a trace amount in general vegetable oils but was undetected in all the tree-borne seed oils. γ-T is used as antioxidant in major foods and was present in the highest proportion of all tocopherols in the tree-borne seed oils. For γ-T, ORT (95.59 ± 1.24 mg/100 g) had the highest level. δ-T, which is, like β-T, mostly present as a trace element in plants, was found to be at the highest level in JLT (33.92 ± 0.47 mg/100 g), whereas the other oils had only trace amounts of δ-T. Tocopherol activity reduces in the order α-T > β-T > γ-T > δ-T; tocopherols mainly inhibit oxidation of unsaturated fatty acids, phospholipids, and fat-soluble vitamin A, preventing arteriosclerosis, cataract, and cancer that are caused by lipid peroxide that is generated during lipid acidification. Thus, tocopherols are important vitamins for modern humans [29]. Individual contents in fat-soluble vitamin analysis may vary depending on the extraction method. When Miraliakbari and Shahidi extracted tocopherols from tree nut oils, extraction using chloroform/methanol as solvent showed a significantly higher yield than that of hexane ($P < 0.05$) [12]. It was postulated that such difference in contents could be attributable to a higher solubility in the chloroform/methanol system. While we used hexane in this study, it would vary in tocopherol content depending on the extraction solvent system. As Miraliakbari and Shahidi reported 6.18–19.87 mg/100 g as the range of tocopherol contents in seven tree nut oils [12], it seemed that tocopherol contents of tree-borne seed oils, except that of KOE, should not be less than those of tree nut oils that are used in food.

3.3. Phytosterol Contents. Phytosterol contents within tree-borne seed oils are listed in Table 3. Phytosterol contents ranged from 117.77 ± 1.32 mg/100 g in TUT to 479.45 ± 4.27 mg/100 g in JLT. Of phytosterols, campesterol was found at the highest level (59.43 ± 0.31 mg/100 g) in MPA, and JLT had the highest levels of stigmasterol (20.20 ± 0.17 mg/100 g), β-sitosterol (388.84 ± 3.19 mg/100 g), and Δ5-avenasterol (34.54 ± 0.74 mg/100 g). According to a literature [30],

TABLE 3: Phytosterol contents in tree-borne seed oils.

Sample	Phytosterols (mg/100 g oils)				
	Campesterol	Stigmasterol	β-Sitosterol	Δ5-Avenasterol	Total
MPA	$59.43 \pm 0.31^{a*}$	5.84 ± 0.10^{c}	99.76 ± 0.88^{d}	12.25 ± 0.17^{c}	177.28 ± 1.29^{c}
JCT	12.67 ± 0.19^{e}	2.82 ± 0.14^{d}	114.98 ± 2.30^{c}	10.85 ± 0.58^{d}	141.32 ± 3.21^{d}
KOE	19.17 ± 0.25^{d}	3.14 ± 0.05^{d}	90.43 ± 1.12^{e}	12.19 ± 0.13^{c}	124.93 ± 1.55^{e}
JLT	35.87 ± 0.43^{b}	20.20 ± 0.17^{a}	388.84 ± 3.19^{a}	34.54 ± 0.74^{a}	479.45 ± 4.27^{a}
TUT	5.16 ± 0.06^{f}	13.26 ± 0.11^{b}	90.73 ± 1.04^{e}	8.62 ± 0.14^{e}	117.77 ± 1.32^{f}
ORT	28.42 ± 1.09^{c}	19.31 ± 0.84^{a}	161.18 ± 6.59^{b}	18.15 ± 0.62^{b}	227.06 ± 9.13^{b}

*Means with different letters (a–f) within a column are significantly different by Tukey's multiple test ($P < 0.05$).

comparisons of phytosterols extracted from coriander and seed oils of caraway, anise, nutmeg, and white mustard showed that their phytosterol contents ranged between 283.64 and 859.79 mg/100 oil, based on which it was considered that phytosterol contents in tree-borne seed oils except JLT oil were lower than those reported in the literature. Phytosterols refer to sterols originated from plants, and over 250 phytosterols have been reported to date. A clinical trial confirmed that phytosterols lowered cholesterol [31]. As phytosterols are not synthesized in the human body, they need to be provided through diet. The human body has an extremely low level of phytosterol in the blood, which is because of a low intestinal absorption rate of phytosterols [31].

3.4. Unsaponifiables. Proportions of unsaponifiables within tree-borne seed oils are presented in Table 4. ORT and JLT had the highest level of unsaponifiables in tree-borne seed oils, which seemed to be proportional to the high tocopherol level (101.98 ± 1.34 mg/100 g oil; Table 2) and a high phytosterol content (479.45 ± 4.27 mg/100 g oil; Table 3) of JLT. For industrial application of unsaponifiables (fat-soluble vitamins, fat-soluble pigments, and phytosterols), the study results on total oil contents of seeds will be informative.

3.5. AV, POV, and p-AV. Contents of AV, POV, and *p*-AV in tree-borne seed oils are presented in Table 5. Acid value indicates acidity when fatty acids of oils bound to glycerol were separated to free fatty acids, by the change in quality; a high acid value means a high proportion of free fatty acids or a low quality of oils. Of the oil resources, MPA had the highest acid value with (38.94 ± 0.24), whereas TUT had the lowest value (0.79 ± 0.01). Additional research is needed to determine if the high acid value of MPA was due to acidification or due to preexisting free fatty acids. For POV, JCT oil showed the lowest value (3.53 ± 0.21 meq/kg) among the tree-borne seed oils, whereas MPA oil showed a relatively high value (127.67 ± 1.79 meq/kg). Although POV is a basic method to measure primary oxidation products, it is difficult to use in cases of high temperature or increasing acidification. POV is obtained by measuring lipid hydroperoxide, the primary oxidation product, which is used for detection of acidification in oils or for the measurement of the induction period. At levels of 60 meq/kg or lower (the reference value for POV content of vegetable oils for frying), it is possible that acidification by generation of primary oxidation

TABLE 4: Unsaponifiables in tree-borne seed oils.

Sample	Unsaponifiables (%)
MPA	$0.39 \pm 0.04^{d*}$
JCT	0.52 ± 0.08^{c}
KOE	0.28 ± 0.18^{e}
JLT	0.84 ± 0.05^{b}
TUT	0.13 ± 0.08^{e}
ORT	2.01 ± 0.02^{a}

*Means with different letters (a–e) within a column are significantly different by Tukey's multiple test ($P < 0.05$).

products (e.g., in tree-borne seed oils except MPA oil) might not have been completed yet [32].

Oil acidification leads to elevation of aldehyde levels, and *p*-anisidine forms yellow complexes by reacting with aldehydes in the presence of acetic acid, so that the degree of acidification is assessed through measurement of such complexes [33]: 2-alkenal or 2,4-alkadienal aldehydes. In general, *p*-AV is a measure for the level of secondary oxidation products among primary and secondary metabolites [34]. In this study, *p*-AV ranged from 2.07 ± 0.51 to 9.67 ± 0.25. Except MPA oil, all tree-borne seed oils had stable values from 2.07 ± 0.51 to 4.18 ± 0.50. In a study of Miraliakbari and Shahidi, oxidative stability of seven tree nut oils were measured, in which *p*-AV ranges of tree nut oils after being freshly collected and after 12 days of storage were between 0.24 and 0.55 and from 9.36 to 48.83, respectively [12]. Considering *p*-AV of tree-borne seed oils, it is likely that aldehydes were formed in all the oils to some degree; thus, it seems to be necessary to block oxygen or maintain low temperature during oil extraction in order to prevent oxidation in future.

3.6. Brown Color Intensity and Color Scale. Measured brown color intensities of tree-borne seed oils using as an indicator of oil oxidation are shown in Table 6. To measure browning intensity, we used the absorption of wavelength at 420 nm that specifically absorbs brown colors. Highest browning intensities were observed in JCT and ORT oils (2.84 ± 0.23 and 2.29 ± 0.06), whereas TUT oil had the lowest browning intensity (0.20 ± 0.01). Color scales identified at 420 nm not only included brown color intensity from browning but also reflected changes by each oil's own brown pigment. Further, color scales of oils of tree-borne seed were measured by a color difference meter, and the results are shown in Table 6. Although there was no significant difference in L^* values that

Table 5: Measurements of acid value (AV), peroxide value (POV), and p-anisidine value (p-AV) in tree-borne seed oils.

Sample	Acid value (mg KOH/g)	Peroxide value (meq/kg)	p-Anisidine value
MPA	$38.94 \pm 0.24^{a*}$	127.67 ± 1.79^{a}	9.67 ± 0.25^{a}
JCT	34.92 ± 1.94^{b}	3.53 ± 0.21^{e}	2.48 ± 0.48^{d}
KOE	9.87 ± 0.43^{d}	12.98 ± 0.17^{c}	2.52 ± 0.40^{d}
JLT	11.32 ± 0.14^{c}	8.21 ± 0.18^{d}	2.07 ± 0.51^{e}
TUT	0.79 ± 0.01^{e}	40.86 ± 0.70^{b}	4.18 ± 0.50^{b}
ORT	9.62 ± 0.48^{d}	14.93 ± 0.50^{c}	2.84 ± 0.06^{c}

*Means with different letters (a–e) within a column are significantly different by Tukey's multiple test ($P < 0.05$).

Table 6: Measurements of browning intensity and color scale in tree-borne seed oils.

Sample	Browning	L^*	a^*	b^*
MPA	$0.85 \pm 0.09^{d\#}$	39.04 ± 0.10^{c}	-1.14 ± 0.03^{a}	1.30 ± 0.02^{d}
JCT	2.84 ± 0.23^{a}	37.86 ± 0.01^{d}	-1.26 ± 0.03^{b}	5.91 ± 0.02^{b}
KOE	0.39 ± 0.01^{e}	40.57 ± 0.03^{b}	-1.39 ± 0.05^{b}	1.26 ± 0.03^{d}
JLT	1.61 ± 0.19^{c}	35.19 ± 0.02^{e}	-2.22 ± 0.06^{c}	4.45 ± 0.04^{c}
TUT	0.20 ± 0.01^{f}	41.03 ± 0.02^{a}	-1.08 ± 0.02^{a}	0.04 ± 0.01^{e}
ORT	2.29 ± 0.06^{b}	39.53 ± 0.01^{c}	-4.25 ± 0.01^{d}	10.47 ± 0.01^{a}

#Means with different letters (a–f) within a column are significantly different by Tukey's multiple test ($P < 0.05$).

Table 7: Induction times of tree-borne seed oils.

Sample	Induction time (h)
MPA	$0.02 \pm 0.00^{e*}$
JCT	0.03 ± 0.00^{e}
KOE	0.44 ± 0.20^{b}
JLT	6.12 ± 0.04^{a}
TUT	0.15 ± 0.03^{d}
ORT	0.22 ± 0.03^{c}

*Means with different letters (a–e) within a column are significantly different by Tukey's multiple test ($P < 0.05$).

reflected brightness in the colorimeter, TUT oil had the highest color scale (41.03 ± 0.02). Overall, a^* values of color scale had negative values, indicating greenness. ORT oil showed a relatively high greenness (-4.25 ± 0.01). As for the b^* values in the color scale, a higher positive value indicates an increase in yellowness, and ORT oil (10.47 ± 0.01) had a relatively high yellowness. Such high yellowness was consistent with browning intensity (2.29 ± 0.06). In addition, TUT oil also was consistent in terms of the lowest browning intensity (0.20 ± 0.01) and the lowest yellowness (0.04 ± 0.01).

3.7. Oxidative Stability. Oxidative stability of tree-borne seed oils was measured by the rancimat test, and the results are presented in Table 7 and Figure 1. Lipids can produce oxidative compounds such as volatile organic acids (mainly formic acid and acetic acid) during oxidation processes. Oxidation-induction period refers to the time when oxygen absorption rate reached the maximum immediately before acidification. Volatile dicarboxylic acids are generated from acidification of tree-borne seed oils, and they produce secondary oxidative compounds including alcohols, aldehydes, ketones, and low molecular weight products causing electrical conductivity to increase within the rancimat system, by which the oxidation-induction period was assessed. The oxidation-induction period is known to be an indicator of antioxidative activity and oxidative stability of oils [35]. In general, a longer oxidation-induction period means a higher oxidative stability in oils. Of the six types of tree-borne seed oils in this study, JLT oil had the longest oxidation-induction period of 6.12 ± 0.04 h, whereas the shortest oxidation-induction period was observed in MPA oil (0.02 ± 0.01 h). Fatty acid profiles are an important parameter for oxidative stability in oils, but this study shows somewhat different patterns for the induction time compared to the general principle of rancimat method. JLT oil represented a high content of polyunsaturated fatty acids (68%) as well as excellent oxidative stability; in contrast, MPA oil had 49% of polyunsaturated fatty acid yet showed relatively low oxidative stability, these inconsistent trends indicate that induction time cannot be simply explained by a degree of polyunsaturated fatty acid compositions. In fact, the induction times for MPA (0.02 hr) and JCT (0.03 hr) were too short to produce oxidative compounds from the fatty acid decomposition (Figure 1). These induction times can be caused by other naturally occurring volatile compounds in oils prior to oxidation. Chemical parameters of MPA are consistent with these patterns; high AV (38.94 ± 0.24 mg KOH/g), POV (127.67 ± 1.79 meq/kg), and p-AV (9.67 ± 0.25) values. Moreover, MPA oil had lower tocopherol content (6.01 ± 0.06 mg/100 g) than others. Therefore, the short oxidation-induction period in MPA oil seemed reasonable. Hidalgo and others demonstrated that the addition of phospholipid to edible oils can increase their induction times [35]. The effect of phospholipid was similar to that of BHT (a synthetic antioxidant) when added [35]. For the comparison between low unsaturated fatty acids and high unsaturated fatty acid, edible oils should be refined because crude oils can contain free fatty acids, phospholipid, and various fat-soluble components which can affect the measurement of induction time. One study reported the various oxidation-induction period of olive oil from 65 to 140 h at low temperatures, demonstrating that edible oil with similar fatty acid profiles may represent varying induction times [36]. The report showed a large difference in the induction time (65–140 h) even though identical oil sample was used; hence, subtle differences in nutrients could result in a broad range of oxidation-induction period lengths. Taken together, it seems that the range of oxidation-induction periods in the six types of tree-borne seed oils may be due to both innate differences in oxidative

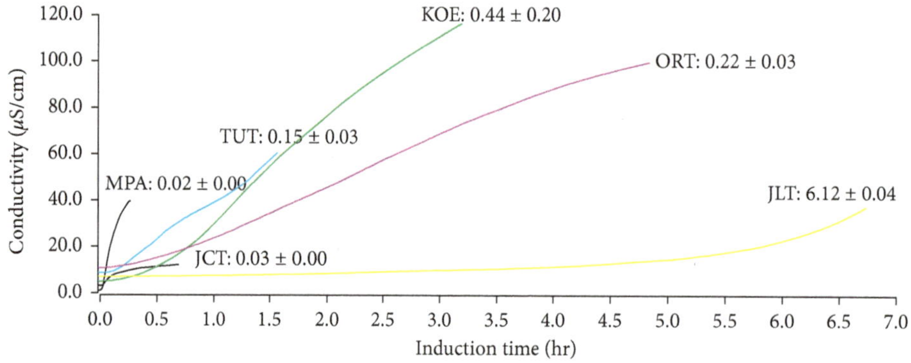

Figure 1: Measurement of induction times of tree-borne oils by the rancimat method.

stability and a reduced quality and stability caused by changes during extraction.

4. Conclusion

This work was to identify potential of oil resources extracted from trees and fat-soluble nutrients, and their physicochemical features were investigated. As for nutrients, all the tree-borne seed oils except JCT oil had high levels of unsaturated fat (>80%) and the tree-borne seed oils had variable contents of tocopherol, fat-soluble vitamins, and phytosterols (plant sterols). For physicochemical features, acid values and POVs for the primary oxidation product were stable in all seed oils except MPA oil, and *p*-AV, the indicator for secondary oxidation products, was also identified. For browning intensity and color scale, it was difficult to conclude freshness because the values from these tests also contained oil's own color scale; however, these tests provided fundamental information about color scale. As for the oxidation-induction period, the indicator of oxidative stability, results varied depending on the level of unsaturated fatty acids and α-T, fat-soluble antioxidant. This study demonstrated that the chain length and rate of unsaturation in fatty acids can be a parameter for the use of edible oils as a nutritional supplement, the content of each tocopherol acting as antioxidant agent, and the phytosterol in tree-borne seed oils as cholesterol-lowering agent in our body. Physicochemical results in this study will be useful parameters for the refining process of tree-borne seed oils in industrial applications.

Conflicts of Interest

The authors declare that they have no conflicts of interest regarding the publication of this research.

Authors' Contributions

Jae Kyeom Kim and Eui-Cheol Shin contributed equally to this work.

Acknowledgments

This work was supported by Gyeongnam National University of Science and Technology Grant in 2017.

References

[1] Y. K. Park, H. J. Roh, J. H. Jeon, and H. H. Kim, "Analyzing the type and priority order of forest functions for private forests," *Journal of Agricultural & Life Science*, vol. 44, pp. 51–59, 2010.

[2] Y. L. Ha and M. W. Pariza, "Naturally-occurring novel anticarcinogens: conjugated linoleic acid (CLA)," *Journal of the Korean Society of Food Science and Nutrition*, vol. 20, pp. 401–407, 1991.

[3] S. J. Hur, J. I. Lee, Y. L. Ha, G. B. Park, and S. T. Joo, "Biological activities of conjugated linoleic acid (CLA) and animal products," *Journal of the Korean Society of Animal Sciences and Technology*, vol. 44, no. 4, pp. 427–442, 2002.

[4] J. Dyerberg, H. O. Bang, E. Stoffersen, S. Moncada, and J. R. Vane, "Eicosapentaenoic acid and prevention of thrombosis and atherosclerosis?," *The Lancet*, vol. 1, pp. 117–126, 1978.

[5] A. Hirai, A. Hamazaki, and J. Jajili, "Eicosapentaenoic acid platelet function in Japanese," *The Lancet*, vol. 22, pp. 1132–1135, 1980.

[6] M. Neuringer and W. E. Connor, "n-3 fatty acids in the brain and retina: evidence for their essentiality," *Nutrition Reviews*, vol. 44, no. 9, pp. 285–294, 1986.

[7] M. I. Gurr, J. L. Harwood, and K. N. Frayn, *Lipid Biochemistry*, Blackwell Science Ltd., Oxford, UK, 5th edition, 2002.

[8] I. B. Lee, K. J. Choi, K. K. Yu, and K. W. Chang, "Tocopherols and fatty acids in plant seeds from Korea," *Journal of the Korean Chemical Society*, vol. 35, pp. 1–5, 1992.

[9] E. C. Shin, Y. Z. Huang, R. B. Pegg, R. D. Phillips, and R. R. Eitenmiller, "Commercial runner peanut cultivars in the United States: tocopherol composition," *Journal of Agricultural and Food Chemistry*, vol. 57, no. 21, pp. 10289–10295, 2009.

[10] E. C. Shin, R. B. Pegg, R. D. Phillips, and R. R. Eitenmiller, "Commercial peanut (*Arachis hypogaea* L.) cultivars in the United States: phytosterol composition," *Journal of Agricultural and Food Chemistry*, vol. 58, no. 16, pp. 9137–9146, 2010.

[11] M. S. Kim, J. H. Park, H. J. Lim et al., "Nutritional components and physicochemical properties of lipids extracted from forest resources," *Journal of the Korean Society of Food Science and Nutrition*, vol. 46, no. 4, pp. 529–536, 2017.

[12] H. Miraliakbari and F. Shahidi, "Oxidative stability of tree nut oils," *Journal of Agricultural and Food Chemistry*, vol. 56, no. 12, pp. 4751–4759, 2008.

[13] E. Prato and F. Biandolino, "Total lipid content and fatty acid composition of commercially important fish species from the Mediterranean, Mar Grande Sea," *Food Chemistry*, vol. 131, no. 4, pp. 1233–1239, 2012.

[14] AOCS, "Acid value of fats and oils," in *Official Methods and Recommended Practices of the AOCS*, American Oil Chemists' Society, Urbana, IL, USA, 7th edition, 2017.
[15] AOCS, "Peroxide value, acetic acid, isooctane method," in *Official Methods and Recommended Practices of the AOCS*, American Oil Chemists' Society, Urbana, IL, USA, 7th edition, 2017.
[16] AOCS, "p-anisidine value," in *Official Methods and Recommended Practices of the AOCS*, American Oil Chemists' Society Press, Urbana, IL, USA, 7th edition, 2017.
[17] A. Bach, G. Debry, and P. Metais, "Hepatic metabolism of medium-chain triglyceride," *Bibliotheca Nutritio Et Dieta*, vol. 25, pp. 24–35, 1977.
[18] A. C. Bach and V. K. Babayan, "Medium-chain triglycerides: an update," *American Journal of Clinical Nutrition*, vol. 36, no. 5, pp. 950–962, 1982.
[19] T. B. Seaton, S. L. Welle, and M. K. Warenko, "Thermic effect of medium-chain and long-chain triglycerides in man," *American Journal of Clinical Nutrition*, vol. 44, no. 5, pp. 630–634, 1986.
[20] J. O. Hill, J. C. Peters, D. Yang et al., "Thermogenesis in humans during overfeeding with medium-chain triglycerides," *Metabolism*, vol. 38, no. 7, pp. 641–648, 1989.
[21] A. G. Dulloo, M. Fathi, N. Mensi, and L. Cirardier, "Twenty-four-hour energy expenditure and urinary catechol amines of humans consuming low-to-moderate amounts of medium-chain triglycerides: a dose-response study in human respiratory chamber," *European Journal of Clinical Nutrition*, vol. 50, pp. 152–158, 1996.
[22] V. V. Wymelbeke, J. Louis-Sylvestre, and M. Fantino, "Substrate oxidation and control of food intake in men after a fat-substitute meal compared with meals supplemented with an isoenergetic load of carbohydrate, long-chain triacylglycerols, or medium-chain triacylglycerols," *American Journal of Clinical Nutrition*, vol. 74, no. 5, pp. 620–630, 2001.
[23] S. P. Koh, N. Arifin, O. M. Lai, M. S. A. Yusoff, K. Long, and C. P. Tan, "Medium-chain fatty acids–nutritional function and application to cooking oil," *Journal of the Science of Food and Agriculture*, vol. 89, pp. 455–462, 2009.
[24] S. Noh and S. H. Yoon, "Stereospecific positional distribution of fatty acids of Camellia (*Camellia japonica* L.) seed oil," *Journal of Food Science*, vol. 77, no. 10, pp. C1055–C1057, 2012.
[25] S. F. O'Keefe, V. A. Wiley, and D. A. Knauft, "Comparison of oxidative stability of high-and normal-oleic peanut oils," *Journal of the American Oil Chemists' Society*, vol. 70, no. 5, pp. 489–492, 1993.
[26] S. M. Watkins and J. B. German, "Unsaturated fatty acids," in *Food Lipids*, C. C. Akoh and D. B. Min, Eds., Marcel Dekker Inc., New York, NY, USA, 2002.
[27] G. J. Nelson, "Dietary fatty acids and lipid metabolism," in *Fatty acids in Foods and Their Health Implications*, C. K. Chow, Ed., Marcel Dekker Inc., New York, NY, USA, 1992.
[28] Y. P. Timilsena, J. Vongsvivut, R. Adhikari, and B. Adhikari, "Physicochemical and thermal characteristics of Australian chia seed oil," *Food Chemistry*, vol. 228, pp. 394–402, 2017.
[29] R. R. Eitenmiller and J. Lee, *Vitamin E: Food Chemistry, Composition, and Analysis*, Marcel Dekker Inc., New York, NY, USA, 2004.
[30] M. Kozłowska, E. Gruczyńska, I. Ścibisz, and M. Rudzińska, "Fatty acids and sterols composition, and antioxidant activity of oils extracted from plant seeds," *Food Chemistry*, vol. 213, pp. 450–456, 2016.
[31] R. A. Moreau, B. D. Whitaker, and K. B. Hicks, "Phytosterols, phytostanols, and their conjugates in foods: structural diversity, quantitative analysis, and health-promoting uses," *Progress in Lipid Research*, vol. 41, no. 6, pp. 457–500, 2002.
[32] K. S. Lee, G. H. Kim, H. H. Kim et al., "Physicochemical properties of frying ginseng and oils derived from deep-frying ginseng," *Journal of the Korean Society of Food Science and Nutrition*, vol. 42, no. 6, pp. 941–947, 2013.
[33] C. C. Akoh and D. B. Min, *Food Lipids-Chemistry, Nutrition, and Biotechnology*, Marcel Dekker Inc., New York, NY, USA, 2002.
[34] S. Naz, H. Sheikh, R. Siddiqi, and S. A. Sayeed, "Oxidative stability of olive, corn, and soybean oil under different conditions," *Food Chemistry*, vol. 88, no. 2, pp. 253–259, 2004.
[35] F. J. Hidalgo, M. M. Leon, and R. Zamora, "Antioxidative activity of amino phospholipids and phospholipid/amino acid mixtures in edible oils as determined by the rancimat method," *Journal of Agricultural and Food Chemistry*, vol. 54, no. 15, pp. 5461–5467, 2006.
[36] V. Mancebo-Campos, M. D. Salvador, and G. Fregapane, "Comparative study of virgin olive oil behavior under rancimat accelerated oxidation conditions and long-term room temperature storage," *Journal of Agricultural and Food Chemistry*, vol. 55, no. 20, pp. 8231–8236, 2007.

15

Sensing of Vascular Permeability in Inflamed Vessel of Live Animal

Sang A Park, Soi Jeong, Young Ho Choe, and Young-Min Hyun

Department of Anatomy and Brain Korea 21 PLUS Project for Medical Science, Yonsei University College of Medicine, Seoul 03722, Republic of Korea

Correspondence should be addressed to Young-Min Hyun; ymhyun@yuhs.ac

Academic Editor: Dokyoung Kim

Increase in vascular permeability is a conclusive response in the progress of inflammation. Under controlled conditions, leukocytes are known to migrate across the vascular barriers to the sites of inflammation without severe vascular rupture. However, when inflammatory state becomes excessive, the leakage of blood components may occur and can be lethal. Basically, vascular permeability can be analyzed based on the intensity of blood outflow. To evaluate the amount and rate of leakage in live mice, we performed cremaster muscle exteriorization to visualize blood flow and neutrophil migration. Using two-photon intravital microscopy of the exteriorized cremaster muscle venules, we found that vascular barrier function is transiently and locally disrupted in the early stage of inflammatory condition induced by N-formylmethionyl-leucyl-phenylalanine (fMLP). Measurement of the concentration of intravenously (i.v.) injected Texas Red dextran inside and outside the vessels resulted in clear visualization of real-time increases in transient and local vascular permeability increase in real-time manner. We successfully demonstrated repeated leakage from a target site on a blood vessel in association with increasing severity of inflammation. Therefore, compared to other methods, two-photon intravital microscopy more accurately visualizes and quantifies vascular permeability even in a small part of blood vessels in live animals in real time.

1. Introduction

Control of vascular permeability allows leukocyte migration, provides nutrition, and maintains homeostasis of the body [1]. Due to its importance, therefore, vascular permeability has been actively investigated in the field of medical science [2–4]. Various stimuli act on endothelial cells and adjust the tightness of the vascular barrier by remodeling actins [3]. VE-cadherin, which is only expressed in the epithelial cell, and JAM-1 are key factors in vascular hemostasis and angiogenesis [5, 6]. Metastasis of cancer cells is strongly associated with vascular integrity since a highly permeable endothelial barrier induces angiogenesis, which eventually supports the dissemination of tumor cells [7–9]. Despite numerous studies, detailed steps of controlling mechanisms about vascular permeability have not been uncovered. Therefore, we prepared cremaster muscle of mice in order to observe the changes in vascular integrity in acute inflammatory conditions. The cremaster muscle was treated with bacterial chemoattractant, fMLP, and two-photon intravital imaging of cremaster muscle venules was conducted for more than 3 hours. fMLP was used to recruit neutrophils along the cremaster vessels [10, 11]. The major premise of this study was that neutrophils activated by fMLP would increase the secretion of vascular endothelial growth factor A (VEGF-A); therefore, vascular integrity would be decreased by cytoskeletal contraction [2, 12]. We were able to detect increased vascular permeability with this methodology; we analyzed the characteristics of this phenomenon by comparing several targeted locations on inflamed vessels. Morphological changes in leukocytes and the intensity of blood flow were also evaluated.

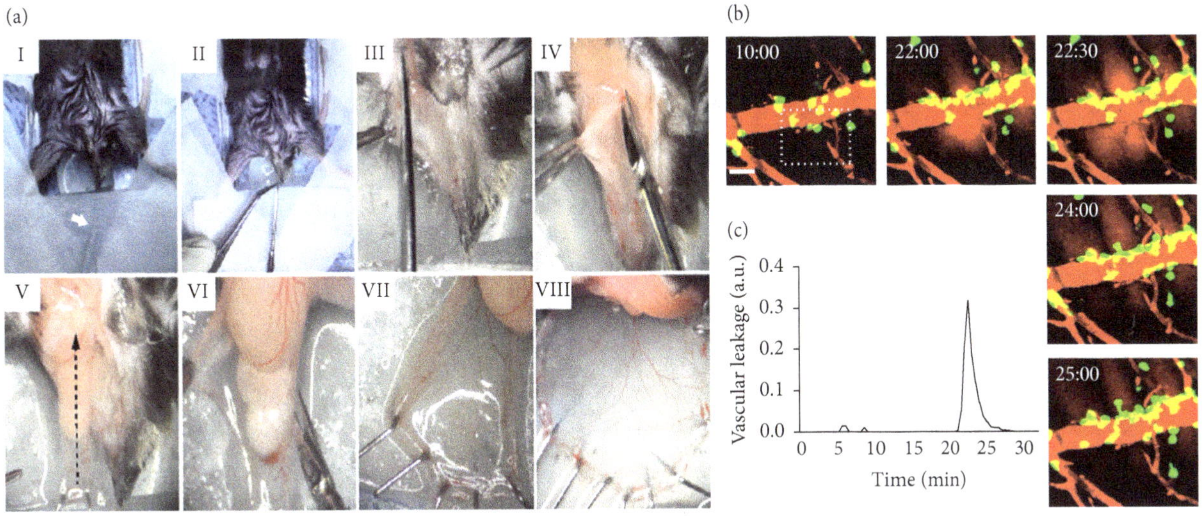

FIGURE 1: Cremaster surgery and intravital imaging of inflamed cremaster vessel. (a) (I) The body of the mouse was fixed in a chamber, using surgical tape, with traction on the scrotum using suture thread. (II) The scrotum was cut with scissors after filling the chamber with sterile 1x PBS. (III) The epidermis in the scrotum was cut along its edge. (b, c) Changes in vessel permeability with time (min) were observed using intravital imaging. Vascular leakage was measured as an arbitrary unit. An intact vessel with zero leakage was set as the baseline for this unit. A graph of vascular leakage shows local and transient increase in blood permeability. Scale bar: 30 μm. See Video 1.

2. Materials and Methods

2.1. Mice. LysM-GFP mice were used for two-photon intravital imaging [13, 14]. The animal and experimental facilities were located at Yonsei University College of Medicine's specific pathogen-free zone. Only heterogenetic mice (GFP/+) were used for experiments. The institutional review board of Yonsei University College of Medicine approved the study.

2.2. Cremaster Exteriorization for Intravital Imaging. The mouse cremaster muscle was chosen for intravital imaging, as it can be locally stimulated by exposure to fMLP after exteriorization [10]. Mice were anesthetized with a Zoletil and Rompun mixture (3 : 2) via intramuscular injection (10 mg/kg). Texas Red dextran was injected via i.v. injection in order to label the blood flow. After a few seconds, mice were placed on the cremaster imaging chamber and firmly fixed with surgical tape. Body hair was removed with a razor and shaving cream along the scrotum and testes. The upper skin of the scrotal sac was retracted with forceps and secured with insect pins. After this procedure, the mouse skin was ready for exposure. From this point, 1x phosphate-buffered saline (PBS) preheated to around 34–35°C was always running inside the chamber sink in order to maintain the temperature and humidity of the cremaster tissue. Attachment of a yellow tip to the PBS tube was useful for continuous water irrigation. Under stereoscopic microscopy, the ventral surface of the pinned scrotal sac was incised. The testicle was removed through the sac with mild pressure on the abdominal cavity. Connective tissues between the scrotal skin and cremaster muscle were gently detached, and the separated scrotal skin was everted to avoid disturbing the cremaster muscle preparation. Finally, the cremaster muscle was gently spread on the silicone bed with several pins. Before imaging, the prepared cremaster muscle was washed once again with preheated 1x PBS in order to obtain a clear view.

2.3. Two-Photon Intravital Imaging of Inflamed Cremaster Muscle Venules. After cremaster surgery, vessel inflammation was induced with drops of fMLP (1 μM). Multiphoton microscopy (LSM 7 MP; Zeiss) with two-photon excitation (3 colors: blue, green, and red. Each of the three channels has an emission wavelength of 420–475 nm, 500–550 nm, and 565–610 nm.) was used at a wavelength of 880–900 nm. Second harmonic generation was represented by blue, and Texas Red dextran and neutrophils were visualized with red and green, respectively. MaiTai Deep See laser system (Spectra-Physics) which detects wavelengths from 690 to 1040 nm was used for excitation. Video resolution was fixed at 512 × 512 pixels.

2.4. Vascular Permeability Analysis. Vascular permeability or leakage was measured using an arbitrary unit, a.u. An intact vessel with zero leakage was set as the baseline for this unit (Figure 1). Absolute values were used to represent intensity in % and blood leakage volumes (Figures 2 and 3).

2.5. Imaging Data Analysis. ZEN operating software (Carl Zeiss) was used for operating software of two-photon microscope and data gathering. Volocity (PerkinElmer) was used for tracking of neutrophils and vascular permeability. Graphs were drawn using Prism Software (GraphPad).

3. Results

3.1. Angiorrhexis Is a Local and Transient Event throughout a Blood Vessel during Inflammation. fMLP is a bacterial chemoattractant used to induce an inflammatory state, in which neutrophils infiltrate across blood vessels in damaged or infected interstitium; this may induce vascular disruption

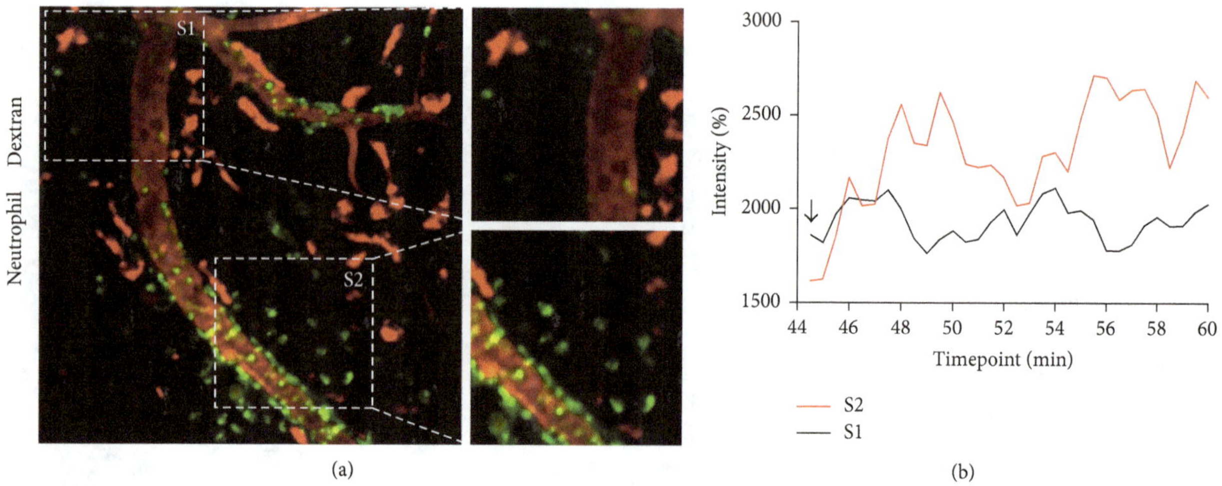

FIGURE 2: Vascular leakage occurs at specific sites on a vessel despite uniform fMLP treatment. (a) Blood vessels are visualized with dextran, and neutrophils appear green on the image. When fMLP was used to activate neutrophils, the site of leakage was designated as S2 and the area where leakage did not occur was designated as S1 in the same blood vessel (blood leakage occurred at S2, with no apparent change in S1). Scale bar: 30 μm. See Video 2. (b) When leakage occurred at S1 and S2, blood was extruded and the intensity of red was increased. The graphical display compares the intensity at two areas at 44 to 60 min in Video 2. At the time of initial imaging (arrow), the intensity of S1 is higher than that of S2 because the background becomes reddish as overlap occurs due to motion during in vivo imaging. Graph B shows the percent intensity (%) of blood flow in the two areas, as measured by the leakage of dextran.

[15–17]. The mouse cremaster muscle was prepared for this experiment since a few drops of fMLP can be directly applied to induce vascular inflammation (Figure 1(a)). We hypothesized that fMLP-induced inflammation would lead to neutrophil infiltration and disruption of the vascular barrier in whole blood vessels. However, when fMLP was evenly applied to the exteriorized cremaster muscle, vascular leakage only increased in part of the blood vessel (Figure 1(b) and Video 1). Thus, disruption did not develop simultaneously in the entire inflamed vessel. On the contrary, vascular leakage occurred locally, and the intensity varied significantly between intact and damaged vessel walls (Figure 1(c)). In addition, neutrophil recruitment was only observed near the site of vessel wall disruption (S2 in Figure 2(a) and Video 2), and only a few neutrophils were detected near the intact vessel structures (S1 in Figure 2(a) and Video 2). Interestingly, neutrophils in S2 were attached in a row at the outer surface of the weakened vessel, as if supporting the unstable membrane. Comparison of S1 and S2 clearly showed differences in inflammation progression. Indeed, the change in background intensity, namely, the black space in Figure 2(a), showed that blood leakage was more intense at S2 than at S1 (Figure 2(b)). Therefore, even though it is known that neutrophil infiltration does not coincide with vascular leakage at the same site in a blood vessel [18, 19], neutrophil adhesion to the abluminal side could induce vascular leakage at a delayed time point following infiltration at the same site. As intravital time-lapse imaging was performed in live animals, shaking occasionally occurred during image capture. This could cause cremaster muscle contraction and negligible variation in Texas Red intensity, as indicated by the arrow in S1 in Figure 2(b).

3.2. Repetitive and Transient Leakage Occurs at a Consistent Point on the Inflamed Blood Vessel. Vascular leakage is a transient and local event in the blood vessel under inflammation. Thus, a point on the stimulated vessel can be more rapidly disrupted than another site on the same stimulated vessel, as shown in Figures 1(b) and 2(a). After detection of vascular leakage at a vulnerable point on the inflamed vessel (S2 in Figure 2(a)), we further tracked the target area to quantify blood outflow using a rainbow scale. The principle of the rainbow scale is similar to that of the contour line; when the rate of leakage is high, the area appears red. Areas with relatively low rates of leakage appear blue to green. From the rainbow scale-based tracking of S2, we observed that there were two separate disruptive events at the same site on the vessel (Figure 3(a)). A period of mild vascular leakage was observed at one site, with outflow of about 75 μm^3 of dextran at peak for 48–50 min. Once outflow began, the Texas Red dextran quickly and widely spread across the interstitial area and became undetectable on two-photon intravital imaging. Interestingly, after an interval of 5 min, another more significant leakage event occurred at the same site on the blood vessel. The peak volume of the second leakage event reached 180 μm^3 (Figure 3(b)). Based on the graphical display of vascular leakage, blood was suddenly extruded from this particular site on the vessel, and the second extrusion phase was more significant than the first. Thus, leakage from inflamed vasculature is a discontinuous event and may repeatedly occur depending on the progress of inflammation. Therefore, our data verified that two-photon intravital microscopy can be used as a precise tool for research on leakage at an individual vascular level.

4. Discussion

Traditional *in vitro* methodologies used to study vascular permeability include measurement of hydraulic

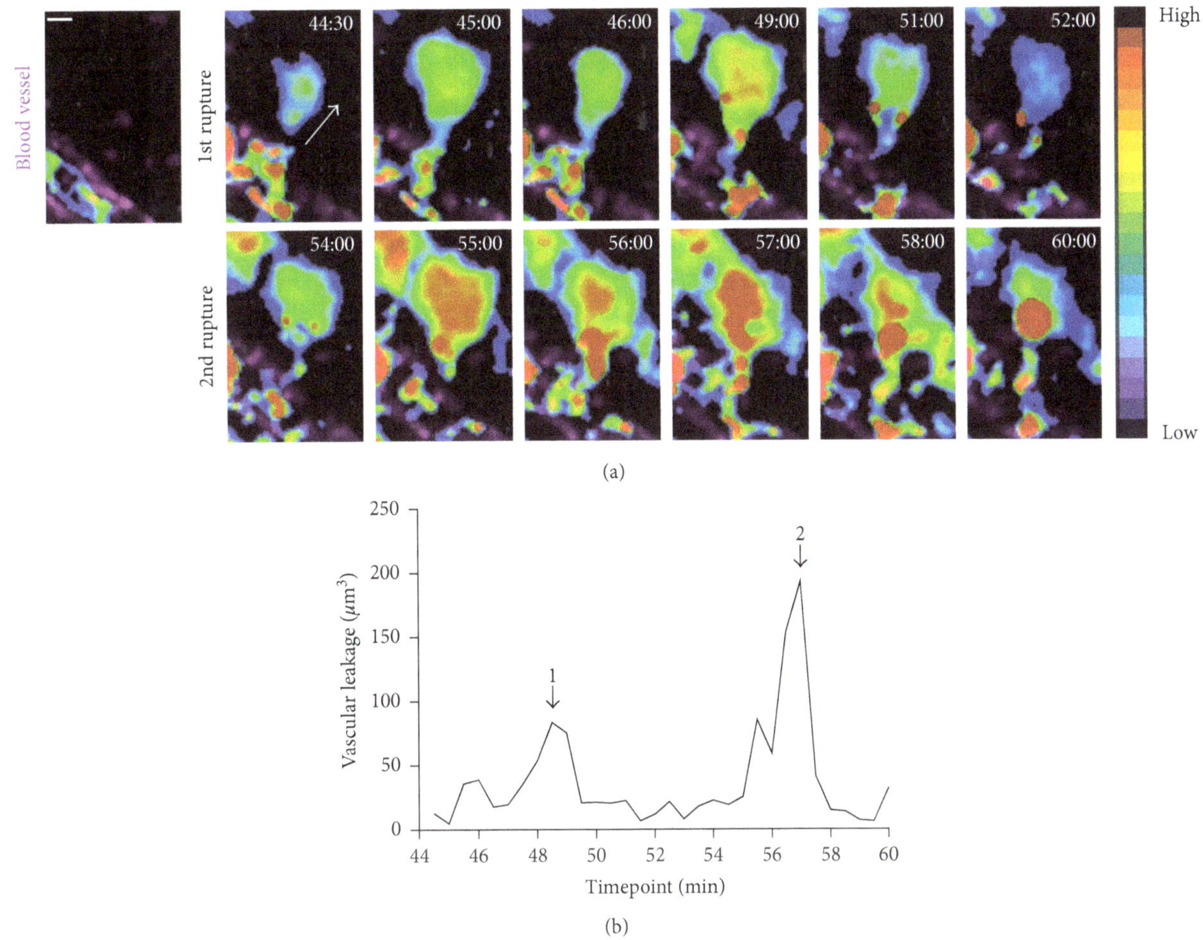

FIGURE 3: Leakage occurs in a repetitive and consistent manner. (a) The first image without blood leakage in S2 was set as a control condition. Color change shows the amount of leakage. Vessels appear magenta and blood flow is seen with rainbow colors. It is apparent that two disruptive events occurred within 16 min. A much greater amount of leakage was observed during the second disruptive event. (b) Vascular permeability was visually demonstrated, and the graph quantitatively shows about twice as much leakage. The graphical display is consistent with transient leakage from the same site. Endothelial barrier disruption did not always result in leakage. Two arrows in the figure indicate that leakage has occurred. Graph B shows the volume of leakage measured according to site. Scale bar: 10 μm.

conductivity, transendothelial electrical resistance, and albumin transport [20]. However, these techniques are inadequate for replication of actual vascular condition and disease models. In addition to the methods described above, a human microvascular endothelial cell (hMVEC) assay is used to reproduce an actual microvascular environment. With this hMVEC assay system, scientists can directly apply cytokines and growth factors such as interleukin-8 and VEGF to induce inflammatory conditions that are similar to an *in vivo* setting [21]. However, it is important to note that almost every *in vitro* experiment shows greater permeability levels compared to *in vivo* trials [20]. *In vitro* studies are performed under precisely controlled conditions, whereas *in vivo* studies are always affected by complicated and confounding factors.

Vascular permeability assays *in vivo* using Evans blue dye can be useful in overcoming the limitation of cellular and membrane-based experiments. Intravenous injection of Evans blue, a dye that binds to albumin, will circulate throughout the animal body. Due to the impermeability of the endothelium to albumin, Evans blue should remain inside the liver in the control animal [22]. However, when vascular permeability is increased with inflammatory stimulation, leakage of Evans blue will be detected around affected vessels and organs. For example, Han et al. [23] were able to prove that angioedema is mediated by bradykinin via Bk2R by comparing the intensity of Evans blue concentration in the digested mouse ear and intestinal tissues. A vascular permeability assay, however, cannot directly show real-time leakage of blood. Combining our cremaster fluorescence analysis and Evans blue dye assay may compensate for the shortcomings of these methods.

In this study, we performed two-photon intravital microscopy to measure how vascular leakage occurs at the subvascular level under fMLP-induced inflammatory conditions. The cremaster muscle is useful for intravital fluorescent imaging since it can be easily exteriorized and is transparent. We investigated how blood vessels are disrupted and vascular leakage occurs during acute inflammation in correlation with neutrophil infiltration. Therefore, we used LysM-GFP mice, in which neutrophils were clearly visualized by detection of green fluorescence following intravenous injection of Texas Red dextran (70 kDa). Texas Red dextran demonstrates a green color in the absence of bleeding, with use of a long wavelength laser (880–900 nm). This wavelength rarely induces tissue phototoxicity or photobleaching. Thus, it was possible to perform intravital imaging at the same site for up to several hours without severe photo-induced damage. When injected, dextran circulates within and clearly outlines the vasculature. Therefore, Texas Red leakage is a marker of vascular disruption. We locally induced an acute inflammatory state by treatment of the exteriorized mouse cremaster muscle with fMLP. As precise two-photon intravital imaging detects the concentration of Texas Red dextran inside and outside blood vessels, we were able to visualize how leakage occurs from any site on a blood vessel in a living animal. Therefore, this intravital imaging method enables more accurate assessment of vascular leakage in terms of time and location. As shown in our data, vascular leakage is a local and transient event in a blood vessel during acute inflammation. Use of intravital imaging to visualize leakage of intravascular components under inflammatory conditions makes it possible to detect subvascular level disruption of the blood vessel in the target organ. In combination with further research on signaling mechanisms such as cytokine interactions among cells and receptors, this fluorescent intravital approach can provide essential information for development of therapeutic methods for diseases associated with blood vessel malfunction.

Conflicts of Interest

The authors declare that they have no conflicts of interest.

Authors' Contributions

Sang A Park and Soi Jeong contributed equally. Sang A Park and Soi Jeong performed the experiments and wrote the manuscript. Young Ho Choe performed the experiments. Young-Min Hyun conceived the study, performed the experiments, and wrote the manuscript.

Acknowledgments

This study was supported by the faculty research grant of Yonsei University College of Medicine (6-2016-0132) and the National Research Foundation of Korea (NRF) grant funded by the Korean government (MSIP) (2016R1A2B4008199) (Young-Min Hyun) and the Ministry of Health & Welfare, Republic of Korea (Grant no. HI14C1324).

References

[1] G. Egawa, S. Nakamizo, Y. Natsuaki, H. Doi, Y. Miyachi, and K. Kabashima, "Intravital analysis of vascular permeability in mice using two-photon microscopy," *Scientific Reports*, vol. 3, no. 1, p. 1932, 2013.

[2] D. Mehta and A. B Malik, "Signaling mechanisms regulating endothelial permeability," *Physiological Reviews*, vol. 86, no. 1, pp. 279–367, 2006.

[3] E. Vandenbroucke, D. Mehta, R. Minshall, and A. B. Malik, "Regulation of endothelial junctional permeability," *Annals of the New York Academy of Sciences*, vol. 1123, no. 1, pp. 134–145, 2008.

[4] J. A. Nagy, L. Benjamin, H. Zeng, A. M. Dvorak, and H. F. Dvorak, "Vascular permeability, vascular hyperpermeability and angiogenesis," *Angiogenesis*, vol. 11, no. 2, pp. 109–119, 2008.

[5] E. Dejana, R. Spagnuolo, and G. Bazzoni, "Interendothelial junctions and their role in the control of angiogenesis, vascular permeability and leukocyte transmigration," *Thrombosis and Haemostasis*, vol. 86, no. 7, pp. 308–315, 2001.

[6] J. Gavard and J. S. Gutkind, "VEGF controls endothelial-cell permeability by promoting the beta-arrestin-dependent endocytosis of VE-cadherin," *Nature Cell Biology*, vol. 8, no. 11, pp. 1223–1234, 2006.

[7] D. M. McDonald and P. Baluk, "Significance of blood vessel leakiness in cancer," *Cancer Research*, vol. 62, no. 18, pp. 5381–5315, 2002.

[8] H. F. Dvorak, "Vascular permeability factor/vascular endothelial growth factor: a critical cytokine in tumor angiogenesis and a potential target for diagnosis and therapy," *Journal of Clinical Oncology*, vol. 20, no. 21, pp. 4368–4380, 2002.

[9] H. Maeda, J. Wu, T. Sawa, Y. Matsumura, and K. Hori, "Tumor vascular permeability and the EPR effect in macromolecular therapeutics: a review," *Journal of Controlled Release*, vol. 65, no. 1-2, pp. 271–284, 2000.

[10] Y. M. Hyun, R. Sumagin, P. P. Sarangi et al., "Uropod elongation is a common final step in leukocyte extravasation through inflamed vessels," *Journal of Experimental Medicine*, vol. 209, no. 7, pp. 1349–1362, 2012.

[11] S. Baez, "An open cremaster muscle preparation for the study of blood vessels by in vivo microscopy," *Microvascular Research*, vol. 5, no. 3, pp. 384–394, 1973.

[12] P. Kumar, Q. Shen, C. D. Pivetti, E. S. Lee, M. H. Wu, and S. Y. Yuan, "Molecular mechanisms of endothelial hyperpermeability: implications in inflammation," *Expert Reviews in Molecular Medicine*, vol. 11, 2009.

[13] N. Faust, F. Varas, L. M. Kelly, S. Heck, and T. Graf, "Insertion of enhanced green fluorescent protein into the lysozyme gene creates mice with green fluorescent granulocytes and macrophages," *Blood*, vol. 96, no. 2, pp. 719–726, 2000.

[14] D. Kreisel, R. G. Nava, W. Li et al., "In vivo two-photon imaging reveals monocyte-dependent neutrophil extravasation during pulmonary inflammation," *Proceedings of the National Academy of Sciences*, vol. 107, no. 42, pp. 18073–18078, 2010.

[15] D. Feng, J. A. Nagy, K. Pyne, H. F. Dvorak, and A. M. Dvorak, "Neutrophils emigrate from venules by a transendothelial cell pathway in response to FMLP," *Journal of Experimental Medicine*, vol. 187, no. 6, pp. 903–915, 1998.

[16] G. F. Elphick, P. P. Sarangi, Y.-M. Hyun et al., "Recombinant human activated protein C inhibits integrin-mediated neutrophil migration," *Blood*, vol. 113, no. 17, pp. 4078–4085, 2009.

[17] L. Zhu and P. He, "fMLP-stimulated release of reactive oxygen species from adherent leukocytes increases microvessel permeability," *American Journal of Physiology-Heart and Circulatory Physiology*, vol. 290, no. 1, pp. 365–372, 2006.

[18] P. He, "Leucocyte/endothelium interactions and microvessel permeability: coupled or uncoupled?," *Cardiovascular Research*, vol. 87, no. 2, pp. 281–290, 2010.

[19] P. Baluk, P. Bolton, A. Hirata, G. Thurston, and D. M. McDonald, "Endothelial gaps and adherent leukocytes in allergen-induced early- and late-phase plasma leakage in rat airways," *American Journal of Pathology*, vol. 152, no. 6, pp. 1463–1476, 1998.

[20] D. O. Bates, "Vascular endothelial growth factors and vascular permeability," *Cardiovascular Research*, vol. 87, no. 2, pp. 262–271, 2010.

[21] M. Martins-Green, M. Petreaca, and M. Yao, "An assay system for in vitro detection of permeability in human "endothelium"," *Methods in Enzymology*, vol. 443, pp. 137–153, 2008.

[22] M. Radu and J. Chernoff, "An in vivo assay to test blood vessel permeability," *Journal of Visualized Experiments*, vol. 73, p. e50062, 2013.

[23] E. D. Han, R. C. MacFarlane, A. N. Mulligan, J. Scafidi, and A. E. Davis, "Increased vascular permeability in C1 inhibitor-deficient mice mediated by the bradykinin type 2 receptor," *Journal of Clinical Investigation*, vol. 109, pp. 1057–1063, 2002.

16

Feasibility of using an Electrolysis Cell for Quantification of the Electrolytic Products of Water from Gravimetric Measurement

Samuel Melaku, Zewdu Gebeyehu, and Rajeev B. Dabke

Department of Chemistry, Columbus State University, Columbus, GA 31907, USA

Correspondence should be addressed to Rajeev B. Dabke; dabke_rajeev@columbusstate.edu

Academic Editor: Pablo Richter

A gravimetric method for the quantitative assessment of the products of electrolysis of water is presented. In this approach, the electrolysis cell was directly powered by 9 V batteries. Prior to electrolysis, a known amount of potassium hydrogen phthalate (KHP) was added to the cathode compartment, and an excess amount of $KHCO_3$ was added to the anode compartment electrolyte. During electrolysis, cathode and anode compartments produced OH^-(aq) and H^+(aq) ions, respectively. Electrolytically produced OH^-(aq) neutralized the KHP, and the completion of this neutralization was detected by a visual indicator color change. Electrolytically produced H^+(aq) reacted with HCO_3^-(aq) liberating CO_2(g) from the anode compartment. Concurrent liberation of H_2(g) and O_2(g) at the cathode and anode, respectively, resulted in a decrease in the mass of the cell. Mass of the electrolysis cell was monitored. Liberation of CO_2(g) resulted in a pronounced effect of a decrease in mass. Experimentally determined decrease in mass (53.7 g/Faraday) agreed with that predicted from Faraday's laws of electrolysis (53.0 g/Faraday). The efficacy of the cell was tested to quantify the acid content in household vinegar samples. Accurate results were obtained for vinegar analysis with a precision better than 5% in most cases. The cell offers the advantages of coulometric method and additionally simplifies the circuitry by eliminating the use of a constant current power source or a coulometer.

1. Introduction

The quantity of a substance produced at the electrode and the quantity of electric charge passed are linked with Faraday's laws of electrolysis. Many reports present coulometric methods of analysis of a variety of reagents [1–17]. A number of original reports [3, 4, 11, 15–17] and reviews [1, 2, 5–10, 12–14] on coulometric methods were published over the past decades. Coulometric method has been applied to analyze a large variety of inorganic and organic compounds [1, 6, 13]. Standardization of reagents is not required in coulometric analysis as the quantity of reagent prepared is directly determined from the charge passing through the cell.

Alternatively, the quantification of a reagent produced in an electrolysis cell without direct monitoring of charge is also well known [5]. This work includes early research on estimating the charge passing through the cell from measuring the mass of silver deposited on the platinum electrode [18, 19]. Similarly, the measurement of color change has been applied to quantify the charge passing through the cell [20]. This method was employed for monitoring the reduction of permanganate ions at platinum electrode and the oxidation of copper to blue cupric triethanolamine complex. Although these methods sidestep the use of a coulometer or a constant current source, they had practical constraints for routine analytical purposes due to the involvement of large-size precious metal electrodes or a rotating electrode placed in a special type of electrochemical cell.

Simple measurement of the volumes of hydrogen and oxygen gases produced at the cathode and anode in an electrolysis cell offers advantages of coulometric method without actually using a coulometer or any electronic equipment. In fact, early studies indicated that hydrogen-oxygen coulometer was easy to assemble and capable of measuring the quantity of electric charge with an accuracy of ±0.1% or better [4]. By choosing a proper size buret for monitoring the volume of gases, the instrument was readily adapted to measuring diverse quantities of electricity down to about 10 coulombs [5]. Enhanced current efficiency was noted for the hydrogen–nitrogen coulometer [21] in which hydrazine sulfate was used as an electrolyte responsible for producing nitrogen gas instead of oxygen gas at the platinum anode. An electrolysis cell was used to measure the volume of O_2(g) and monitor an acid-base titration [22].

As stated in (1), electrolysis of water produces OH^-(aq) in the cathode compartment. In a 1 : 1 stoichiometric ratio (2), electrolytically produced OH^-(aq) neutralizes the weak acid (e.g., potassium hydrogen phthalate ($KC_8H_4O_4H$) or $C_8H_4O_4H^-$ in an anionic form) added to this compartment:

$$2H_2O\,(l) + 2e \rightarrow 2OH^-\,(aq) + H_2\,(g) \qquad (1)$$

$$C_8H_4O_4H^-\,(aq) + OH^-\,(aq) \rightarrow C_8H_4O_4^{2-}\,(aq) + H_2O\,(l) \qquad (2)$$

Concurrently, anode produces H^+(aq) the following equation:

$$H_2O\,(l) \rightarrow 2H^+\,(aq) + \frac{1}{2}O_2\,(g) + 2e \qquad (3)$$

Electrolysis of water in cathode and anode compartments (1) and (3) also produces H_2(g) and O_2(g) at the respective electrodes. A chemical reaction (4) between electrolytically produced H^+(aq) and externally added $KHCO_3$(aq) produces CO_2(g):

$$H^+\,(aq) + HCO_3^-\,(aq) \rightarrow H_2O\,(l) + CO_2\,(g) \qquad (4)$$

When a known quantity of KHP (a primary standard substance) is added to the cathode compartment, the endpoint of the neutralization reaction between electrolytically produced OH^-(aq) and KHP can be determined visually with use of phenolphthalein as an indicator. Electrolysis is promptly stopped at the endpoint. Prior to electrolysis, an excess amount of $KHCO_3$ is added to the anode compartment.

Concurrent liberation of chemically produced CO_2(g) in addition to H_2(g) and O_2(g) amplified the effect of decreasing in the mass. Mass of the cell is continuously monitored before, during, and after electrolysis. From the mass change experienced by the cell, the amount of KHP can be determined to assess the stoichiometric relations between the electrolytic products of water. In this study, the feasibility of monitoring the mass change experienced by an electrolysis cell due to the liberation of gases and quantifying the electrolytic products of water was assessed.

2. Materials and Methods

2.1. Chemicals and Materials. Reagent-grade potassium nitrate, KHP, potassium hydrogen çarbonate, and agar were obtained from Fisher. Millipore deionized water was used for preparation of reagents. Alfa Aesar platinum wire (0.762 mm diameter) was used as purchased. 0.1% phenolphthalein was used as a visual indicator to detect the completion of neutralization. Ohaus Pioneer balance (Model: PA114, 0.0001 g readability, and ±0.0002 g linearity) was used for mass measurements. Vinegar samples were locally purchased from a grocery store.

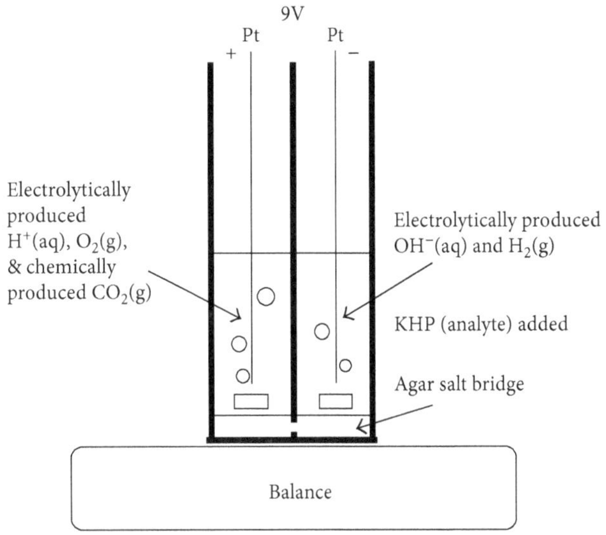

FIGURE 1: Schematic diagram of the water electrolysis cell. KNO_3(aq) was used as an electrolyte in both compartments. $KHCO_3$(aq) was added to the anode compartment. Effective concentrations of KNO_3 and $KHCO_3$ were 1 M each. Desired number of moles of KHP(aq) and a drop of 0.1% phenolphthalein were added to the cathode compartment. The volumes of electrolyte in each compartment were approximately 3.2 mL. Platinum wires were immersed in the electrolytes and directly connected to two 9 V batteries connected in parallel. The cell was placed on a digital balance housed in a glass cabinet.

2.2. Electrolysis Cell. Electrolysis cell was made of spectrophotometer plastic cuvettes. A 3 mm hole was drilled on a side near the base of each cuvette, and the cuvettes were glued together using an epoxy. Bottom portions of two additional cuvettes were cut and glued on top (Figure 1). The extra depth offered by these cuvettes prevented the mass loss due to the spattering of electrolyte mist formed due to liberation of gas bubbles during electrolysis. A mixture of agar and potassium nitrate (3% by mass and 1 M, resp.) was added to water, and the mixture was heated to 90°C. A few drops of this mixture were placed on the bottom of the cuvettes. A gel was formed on the bottom of the cuvettes, serving as a salt bridge. KNO_3(aq) was used as an electrolyte in both compartments. $KHCO_3$ was added to the anode compartment. Effective concentrations of KNO_3 and $KHCO_3$ were 1 M each. Platinum wires served as cathode and anode and were directly powered by two 9–V batteries connected in parallel. Approximately 3.5 cm platinum wire was dipped in the electrolyte in each compartment. The current passing through the cell was in the range of 55 mA to 90 mA. A multimeter was used to display the current passing through the electrolysis cell. Two miniature magnetic stir bars were placed in the compartments, and an external stir bar retriever was used to maneuver the internal stir bars in up and down directions. The movement of stir bars helped dislodge the gas bubbles adhering to the

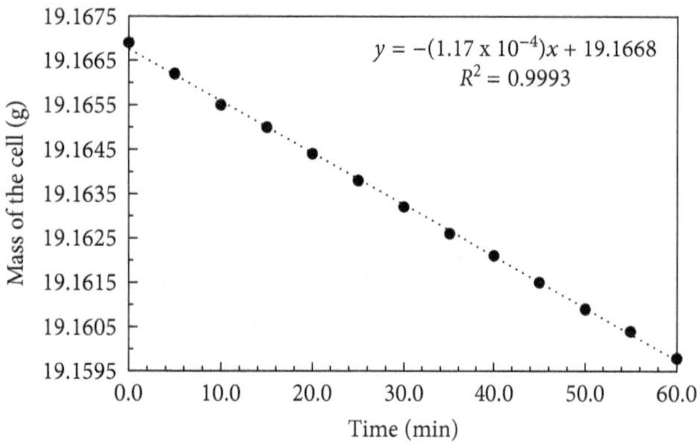

FIGURE 2: Monitoring the mass of an idle cell (no electrolysis).

electrodes and sidewalls and stir the electrolytes in both compartments. Additionally, electrolyte stirring assisted uniform mixing of the contents of cathode compartment. This action was periodically performed, particularly before mass measurement.

2.3. Determination of Mass Correction Factor. Mass correction factor was determined to quantify an error associated with the mass loss due to the evaporation of electrolytes. In this experiment, mass of an idle cell (no electrolysis) was monitored at an interval of 5 minutes. Mass of the cell was monitored for 60 minutes (Figure 2).

2.4. Electrolytic Titrations of KHP. Desired moles of KHP were added to the cathode compartment. Mass of the cell was monitored at a time interval of one minute for the first five minutes before electrolysis. After this initial rest period, electrolysis was started. Electrolysis was paused at a two-minute time interval, stir bars were maneuvered in up and down position, and the mass was recorded after a two-minute rest period. Electrolysis was promptly stopped when the cathode compartment electrolyte turned light pink indicating an endpoint of the titration. Mass measurement was continued for additional eight minutes. The results of this experiment are presented in Figure 3. Electrolytic titration of KHP was repeated to determine the mass change experienced by the cell for various concentrations (0 to 6.0×10^{-4}) of KHP (Figure 4).

2.5. Coulometric Measurements. In view of confirming the stoichiometric relations between the moles of $OH^-(aq)$ and the mass loss due to the liberation of electrolytically produced $H_2(g)$ and $O_2(g)$ and chemically produced $CO_2(g)$, we monitored the charge passing through the electrolysis cell on Faraday MP potentiostat. Electrolysis was paused at 10 coulombs interval, stir bars were maneuvered in up and down position, and the mass was recorded after one-minute rest period (Figure 5).

2.6. Electrolytic Titrations of Real Sample. A known volume of commercial vinegar sample was added to the cathode compartment and titrated against electrolytically produced $OH^-(aq)$. The quantity of acid in a sample was determined from the mass change experienced by the cell.

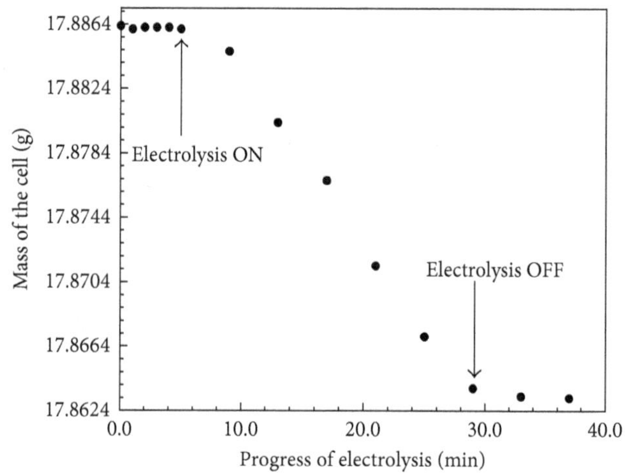

FIGURE 3: Monitoring the mass of the electrolysis cell before, during, and after electrolysis.

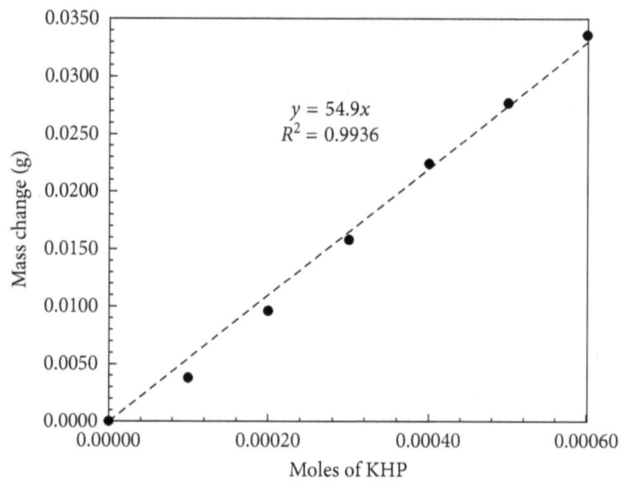

FIGURE 4: A plot of the mass change experienced by the cell versus number of moles of KHP neutralized in the cathode compartment.

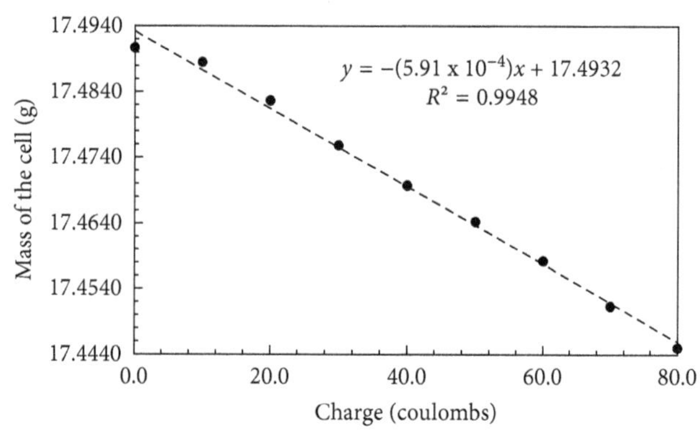

FIGURE 5: A plot of the mass of the electrolysis cell versus the charge passing through the cell.

Table 1: Results of titration of an acid in commercial vinegar samples by electrolytically produced $OH^-(aq)$.

Percent acid content stated on manufacturer's label	Mass change in electrolysis cell (g)	Moles of acid in (mmol)	Average molarity	Percent acid content determined from this study (95% CI, $n = 3$)	Percent RSD
4	0.0165	0.311	0.623	3.74% (±1.32)	14.3
5	0.0222	0.419	0.838	5.03% (±0.53)	4.3
7	0.0305	0.575	1.15	6.91% (±0.52)	3.0

3. Results and Discussion

3.1. Mass Correction Factor. The mass of an idle cell linearly decreased with time (Figure 2). A linear decrease in mass indicated uniform mass loss resulting from the evaporation of electrolyte. The slope of this plot quantified the mass correction factor of 1.17×10^{-4} g/min or 1.95×10^{-6} g/s. This factor was used to determine the mass loss due to evaporation during the period for each trial of the experiment.

3.2. Quantitative Assessment of the Electrolytic Products. Faraday's laws of electrolysis and (1) and (3) indicate, on passing of one Faraday charge through water, the electrolysis cell produces one mole of $OH^-(aq)$ and $H^+(aq)$ in their respective compartments. Additionally, 0.5 moles of $H_2(g)$, 0.25 moles of $O_2(g)$, and 1 mole of $CO_2(g)$ are liberated. Considering molar masses of these gases, a decrease in mass by 53.0 g was expected to result from passing one Faraday of charge.

Figure 3 presents data on the mass of the cell before, during, and after electrolytic titration. Desired moles (4.0×10^{-4} moles) of aqueous KHP were added to the cathode compartment, and the charge was passed through the cell. Mass of the cell significantly decreased during the course of electrolysis. Mass of the cell slightly decreased prior to electrolysis, resulting from the evaporation of electrolyte, and it continued to decrease after electrolysis. Mass correction factor was applied to minimize the error associated with the evaporative mass losses during the course of electrolysis.

The electrolytic titration experiment presented in Figure 3 was repeated for various concentrations of KHP, and the mass correction factor was independently applied to each trial of experiment. A linear response of mass change versus moles of KHP confirmed the quantitative relationship between the moles of KHP and the drop in the mass due to the liberation of gases (Figure 4). Slope of the plot indicated a mass drop of 54.9 g due to the liberation of gases. This mass accounted for the electrolytic neutralization of one mole of KHP after passing one Faraday charge. This quantity of mass obtained from the slope matched with the estimated mass loss of 53.0 g/Faraday. A blank trial (a data point corresponding to zero moles of KHP in Figure 4) confirmed minimal contribution of acidic impurities and dissolved $CO_2(g)$ present in the electrolyte. Literature values [23] of solubility of $H_2(g)$, $O_2(g)$, and $CO_2(g)$ at 1 atm pressure and at 25°C temperature also confirm a minor error in mass measurement (<0.15%) associated with the dissolution of gases produced in the cell. The trials presented in Figure 4 were repeated to determine the relative standard deviation in the slope. The RSD value (0.93%, $n = 3$) signified a reliable quantitative relationship between the mass change and the moles of KHP.

A linear response of the plot (Figure 5) confirmed the mole stoichiometry presented in (1), (3), and (4). A slope of this plot indicated that 5.91×10^{-4} g mass drop occurred per coulomb charge passed through the cell. Accounting for the mass loss due to evaporation of the electrolyte, a correction factor (1.95×10^{-6} g/s) was applied to the slope. Considering an estimated time (17.5 s) required to pass one coulomb charge, mass loss due to evaporation during the passage of one coulomb was 3.41×10^{-5} g/C. The mass-corrected slope after deducting this factor was 5.57×10^{-4} g/C or 53.7 g/Faraday. This mass drop was in agreement with the estimated mass drop of 53.0 g/Faraday determined from the stoichiometric relation presented in (1–4). A linear relation between the mass loss due to liberation of gases and the charge passed through the cell (Figure 5) indicated negligible interference of dissolution of gases produced during electrolysis. The slope of the plot presented in Figure 5 was confirmed from three independent experiments (RDS 0.49%).

3.3. Analysis of Real Sample. The efficacy of the electrolytic titration method presented in this paper was tested for the analysis of acid content in household vinegar samples. A known volume (500 μL) of vinegar was added to the cathode compartment and titrated against electrolytically produced $OH^-(aq)$ as presented in the experimental section. The molarity and percent acid content in the sample were determined from the mass change experienced by the cell (Table 1). The mass correction factor was independently applied to each trial. Quantity of acid determined from electrolytic titration was in agreement with the manufacturer's label, and the quantities were consistent with three household vinegar samples.

4. Conclusions

In this work, the feasibility of using an electrolysis cell for quantification of the electrolytic products of water from gravimetric measurement was tested. The cell presented in this paper enables in situ production of reagents and their direct quantification and does not require standardization of reagents. The electrolysis cell directly powered by 9 V batteries eliminates the requirement of a constant current source or a coulometer, yet offers advantages of coulometric method of titration. The cell utilizes minimal volume of reagents (3.2 mL or less in each compartment). The electrolysis cell is simple, transparent, and easy to fabricate. The cell eliminates the requirement of an external salt bridge or a fritted glass membrane. Linear response of the decrease in the mass of the cell to the moles of KHP added and to the charge passing through the cell validates the applicability of the method. An agreement between the estimated drop in mass (determined from Faraday's laws of electrolysis) and the experimentally determined drop in mass underlines the feasibility of using the electrolysis cell for quantification of the electrolytic products. With acceptable values of relative standard deviation (better than 5% in most cases), the mass drop experienced by the cell quantitatively responds to the acid content in household vinegar samples.

Conflicts of Interest

The authors declare that there are no conflicts of interest regarding the publication of this paper.

Acknowledgments

The authors thank James O. Schreck and reviewers of the manuscript for their helpful comments and suggestions.

References

[1] D. T. Sawyer, A. Sobkowiak, and J. L. Roberts, *Electrochemistry for Chemists*, Wiley, New York, NY, USA, 2nd edition, 1995.

[2] A. M. Bond, *Broadening Electrochemical Horizons*, Oxford Science Publications, New York, NY, USA, 2002.

[3] L. Szebelledy and Z. Somogyi, "Coulometric analysis as a precision method," *Analytical Chemistry*, vol. 112, pp. 313–400, 1938.

[4] J. J. Lingane, "Coulometric analysis," *Journal of the American Chemical Society*, vol. 67, no. 11, pp. 1916–1922, 1945.

[5] J. J. Lingane, *Electroanalytical Chemistry*, Interscience Publishers, Geneva, Switzerland, 2nd edition, 1958.

[6] A. J. Bard, "Electroanalysis and coulometric analysis," *Analytical Chemistry*, vol. 40, no. 5, pp. 64–74, 1968.

[7] D. G. Davis, "Electroanalysis and coulometric analysis," *Analytical Chemistry*, vol. 46, no. 5, pp. 21–28, 1974.

[8] J. T. Stock, "Amperometric, bipotentiometric, and coulometric titration," *Analytical Chemistry*, vol. 54, no. 5, pp. 1–9, 1982.

[9] T. Yoshimori, "Precise coulometric titration-an approach to the highest accuracy," *Reviews in Analytical Chemistry*, vol. 6, no. 1, pp. 13–48, 1982.

[10] G. W. Ewing, "Coulometric Titrimetry," in *Electrochemistry, Past and Present, ACS Symposium Series*, Vol. 390, pp. 402–416, American Chemical Society, Washington, DC, USA, 1989.

[11] A. Cedergren and C. Oraedd, "Coulometric study of the influence of aldehydes and ketones on the karl fischer titration of water using standard pyridine/methanol reagents," *Analytical Chemistry*, vol. 66, no. 13, pp. 2010–2016, 1994.

[12] W. Olthuis and P. Bergveld, "Integrated coulometric sensor-actuator devices," *Mikrochimica Acta*, vol. 121, no. 1–4, pp. 191–223, 1995.

[13] G. H. Jeffery, J. Bassett, J. Mendham, and R. C. Denney, *Vogel's Textbook of Quantitative Chemical Analysis*, Longman, Essex, 5th edition, 1994.

[14] R. P. Mihajlovic, L. N. Jaksic, and R. M. Dzudovic, "Coulometric generation of acids and bases for acid-base titrations in non-aqueous solvents," *Analytica Chimica Acta*, vol. 557, no. 1-2, pp. 37–44, 2006.

[15] R. B. Dabke, Z. Gebeyehu, and R. Thor, "Coulometric analysis experiment for the undergraduate chemistry laboratory," *Journal of Chemical Education*, vol. 88, no. 12, pp. 1707–1710, 2011.

[16] M. Ghahraman, G. A. Crespo, and C. E. Bakker, "Coulometric calcium pump for thin layer sample titrations," *Analytical Chemistry*, vol. 87, no. 19, pp. 10125–10130, 2015.

[17] K. Ikemoto, T. Seki, S. Kimura et al., "Microfluidic separation of redox reactions for coulometry based on metallization at the mixed potential," *Analytical Chemistry*, vol. 88, no. 19, pp. 9427–9434, 2016.

[18] T. W. Richards and F. O. Anderegg, "Complications at the anode in the silver coulometer," *Journal of the American Chemical Society*, vol. 37, no. 4, pp. 675–693, 1915.

[19] E. B. Rosa and G. W. Vinal, "Summary of experiments on the silver voltmeter at the Bureau of Standards and proposed specifications," *Bureau of Standards (U. S.) Bulletin*, vol. 13, no. 3, pp. 447–479, 1916.

[20] T. C. Franklin and C. C. Roth, "A colorimetric coulometer," *Analytical Chemistry*, vol. 27, no. 7, pp. 1197–1999, 1955.

[21] J. A. Page and J. J. Lingane, "A hydrogen-nitrogen gas coulometer," *Analytica Chimica Acta*, vol. 16, pp. 175–179, 1957.

[22] R. B. Dabke and Z. Gebeyehu, "Using mole ratios of electrolytic products of water for analysis of household vinegar: an experiment for the undergraduate physical chemistry laboratory," *Journal of Chemical Education*, vol. 89, no. 9, pp. 1198–1200, 2012.

[23] J. A. Dean, *Lange's Handbook of Chemistry*, McGraw-Hill, New York, NY, USA, 15th edition, 1999.

17

Geographical Characterization of Tunisian Olive Tree Leaves (cv. Chemlali) using HPLC-ESI-TOF and IT/MS Fingerprinting with Hierarchical Cluster Analysis

Amani Taamalli,[1] David Arráez Román,[2,3] Ana María Gómez Caravaca,[2,3] Mokhtar Zarrouk,[1] and Antonio Segura Carretero[2,3]

[1]*Laboratoire de Biotechnologie de l'Olivier, Centre de Biotechnologie de Borj-Cedria, Hammam-Lif, Tunisia*
[2]*Center of Research and Development of Functional Food, Health Science Technological Park, Avda. del Conocimiento s/n, 18100 Granada, Spain*
[3]*Department of Analytical Chemistry, Faculty of Sciences, University of Granada, 18071 Granada, Spain*

Correspondence should be addressed to Amani Taamalli; taamalli_ameni@yahoo.fr

Academic Editor: Antony C. Calokerinos

The olive plant has been extensively studied for its nutritional value, whereas its leaves have been specifically recognized as a processing by-product. Leaves are considered by-products of olive farming, representing a significant material arriving to the olive mill. They have been considered for centuries as an important herbal remedy in Mediterranean countries. Their beneficial properties are generally attributed to the presence of a range of phytochemicals such as secoiridoids, triterpenes, lignans, and flavonoids. With the aim to study the impact of geographical location on the phenolic compounds, *Olea europaea* leaves were handpicked from the Tunisian cultivar "Chemlali" from nine regions in the north, center, and south of Tunisia. The ground leaves were then extracted with methanol : water 80% (*v/v*) and analyzed by using high-performance liquid chromatography coupled to electrospray time of flight and ion trap mass spectrometry analyzers. A total of 38 compounds could be identified. Their contents showed significant variation among samples from different regions. Hierarchical cluster analysis was applied to highlight similarities in the phytochemical composition observed between the samples of different regions.

1. Introduction

Natural antioxidants present in the diet have evidenced the increase of the resistance toward oxidative damages, and they may have a substantial impact on human health [1]. Furthermore, the use of natural extracts from plants as functional ingredients in food, beverage, and cosmetic applications is receiving a great deal of attention among scientists, as well as among consumers and food manufacturers [2]. Recently, there has been a renewed interest in natural product research due to the failure of alternative drug discovery methods to deliver many lead compounds in key therapeutic areas such as immune suppression, anti-infectives, and metabolic diseases [3]. Despite competition from other drug discovery methods, natural products are still providing their fair share of new clinical candidates and drugs [3]. We can find several natural products as sources of new drugs that have been recently reviewed in literature [4–7].

The olive plant has been extensively studied for its nutritional value, whereas its leaves have been specifically recognized as a processing by-product [8]. Considered as by-products of olive farming, olive tree leaves represent a significant material arriving to the olive mill [9]. They have been considered for centuries as an important herbal remedy in Mediterranean countries. Their positive effects on human health are multiple: antioxidant [10–16], hypoglycemic [17], antimicrobial [18], and anticancerous [19–22] among others. These properties are generally attributed to the presence of

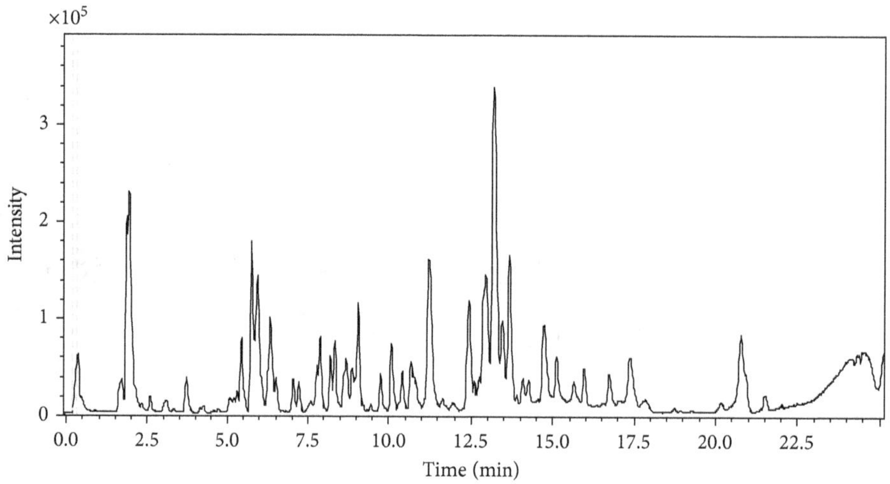

FIGURE 1: Base peak chromatogram of "Chemlali" olive leaf extract from Zaghouan region.

a range of compounds, such as secoiridoids, triterpenes, lignans, and flavonoids. It has been reported that by-products can have a similar or even higher proportion of bioactive compounds than the usable parts of produce [23]. For instance, the characterization and quantification of bioactive compounds is one of the first steps to be taken in any evaluation of the putative contribution of the olive and its derivatives to human health [24] since the chemical composition of olive tree leaves varies depending on several conditions such as origin, proportion of branches on the tree, storage conditions, climatic conditions, moisture content, and degree of contamination with soil and oils [25]. Recently, the effects of abiotic and biotic factors on the phenolic composition in olive tree leaves have been reviewed in literature [26]. Biotic factors such as genotype, load bearing, leaves age, and fungi and bacteria attacks as well as abiotic factors such as hydric deficiency, salinity, fertilization, sampling time, geographical zone, and climatic conditions influence the contents of phenolic compounds [26]. Recent researches revealed the impact of the geographical location on the total phenolic composition of olive leaves [27–30]. However, the data on the variation of the individual phenolic compounds in olive leaves are scarce. According to our knowledge, this is the first study on the phenolic compounds in olive tree leaves from Tunisian cultivars according to the geographical region.

Considering the importance of phenolic compounds in olive leaves, their determination and the scarce literature related to their behavior according to geographical origin, the aim of this study was to explore the phenolic composition of the leaves from the Tunisian olive cultivar "Chemlali" from different localities. This cultivar was chosen since it is widespread throughout Tunisia from the north to the south.

2. Materials and Methods

2.1. Sample Collection and Preparation of Extracts. Fresh olive leaves from the Tunisian "Chemlali" cultivar were obtained at the time of olive pruning from different regions from Tunisia (north, center, and south from coastal and mountainous areas) (available here). Leaves were dried at 25°C, ground, and then 1 g of the obtained homogenized powder was dissolved in 10 mL of methanol : water 80% (*v/v*), and the mixture was allowed to stand in the dark for 24 h [31]. The extract was centrifuged at 5000 g for 10 min at room temperature, and the supernatants were then filtered using a 0.45 μm syringe filter to be analyzed by high-performance liquid chromatography (HPLC).

2.2. Chromatographic Separation of Phenolic Compounds. An Agilent 1200 HPLC system (Agilent Technologies, Waldbronn, Germany) equipped with a vacuum degasser, autosampler, a binary pump, and a DAD detector was used for the chromatographic determination. Phenolic compounds were separated by using an Agilent Eclipse Plus C18-column (4.6 × 150 mm, 1.8 mm particle size) operating at 25°C and a flow rate of 0.8 mL/min. The mobile phases used were water with acetic acid (0.5%) (phase A) and acetonitrile (phase B), and the solvent gradient changed according to the following conditions: 0 to 10 min, 5–30% B; 10 to 12 min, 30–33% B; 12 to 17 min, 33–38% B; 17 to 20 min, 38–50% B; and 20 to 23 min, 50–95% B. Finally, the B content was decreased to the initial conditions (5%) in 2 min, and the column reequilibrated for 10 min. A volume of 10 mL of the extracts was injected. The separated compounds were monitored in sequence first with a diode-array detector (DAD) at 240 and 280 nm and then with a mass spectrometry (MS) detector.

2.3. Time of Flight and Ion Trap Mass Spectrometry Detection. The HPLC system was firstly coupled to a Bruker Daltonics micrOTOF mass spectrometer (Bruker Daltonics, Bremen, Germany) using an orthogonal electrospray interface. The mass spectrometer was operated in the negative ionization mode and acquired data in the mass range from *m/z* 50 to 1000 with a spectra rate of 1 Hz. The capillary was set at +4 kV, the end plate offset at −500 V, the nebulizer gas

Table 1: Identification data for the compounds detected in olive leaf extracts.

m/z [M-H]$^-$	RT (min)	Formula	Compound	IT/MS^2 fragments	Class
191.0561	2.0	$C_7H_{12}O_6$	Quinic acid	103	Other polar compounds
315.1085	5.8	$C_{14}H_{20}O_8$	Hydroxytyrosol-hexose	123, 153	Simple phenols
153.0557	6.6	$C_8H_{10}O_3$	Hydroxytyrosol	123	Simple phenols
299.1136	7.0	$C_{14}H_{20}O_7$	Tyrosol glucoside		Simple phenols
403.1246	7.7	$C_{17}H_{24}O_{11}$	Elenolic acid glucoside isomer 1		Secoiridoids and related compounds
389.1089	7.8	$C_{16}H_{22}O_{11}$	Secologanoside	183, 209, 227	Secoiridoids and related compounds
353.0878	7.9	$C_{16}H_{18}O_9$	Caffeoylquinic acid isomer 1	191	Other polar compounds
403.1246	8.3	$C_{17}H_{24}O_{11}$	Elenolic acid glucoside isomer 2		Secoiridoids and related compounds
305.0667	8.7	$C_{15}H_{14}O_7$	Gallocatechin	225	Flavanol
401.1453	8.7	$C_{18}H_{26}O_{10}$	Benzyl alcohol hexose-pentose	161	Other polar compounds
403.1246	8.9	$C_{17}H_{24}O_{11}$	Elenolic acid glucoside isomer 3	179, 223, 371	Secoiridoids and related compounds
353.1606	9.0	$C_{16}H_{18}O_9$	Caffeoylquinic acid isomer 2	191	Other polar compounds
377.1453	9.1	$C_{16}H_{26}O_{10}$	Oleuropein aglycon derivative		Secoiridoids and related compounds
403.1246	9.8	$C_{17}H_{24}O_{11}$	Elenolic acid glucoside isomer c		Secoiridoids and related compounds related compounds
525.1614	10.5	$C_{24}H_{30}O_{13}$	Demethyloleuropein		Secoiridoids and related compounds
609.1461	10.7	$C_{27}H_{30}O_{16}$	Rutin	300	Flavonol
593.1512	11.0	$C_{27}H_{30}O_{15}$	Luteolin rutinoside	285	Flavone
623.2075	11.0	$C_{29}H_{36}O_{15}$	Verbascoside	461, 315	Secoiridoids and related compounds
447.0933	11.2	$C_{21}H_{20}O_{11}$	Luteolin glucoside isomer 1	285	Flavone
701.2298	11.6	$C_{31}H_{42}O_{18}$	Oleuropein glucoside	539, 377, 307	Secoiridoids and related compounds
607.1668	11.8	$C_{28}H_{32}O_{15}$	Diosmin isomer 1	299	Flavone
577.1563	11.7	$C_{27}H_{30}O_{14}$	Apigenin rutinoside	269	Flavone
607.1668	12.1	$C_{28}H_{32}O_{15}$	Diosmin isomer 2	299	Flavone
303.051	12.2	$C_{15}H_{12}O_7$	Taxifolin		Flavanonol
431.0984	12.4	$C_{21}H_{20}O_{10}$	Apigenin glucoside	268	Flavone
447.0933	12.4	$C_{21}H_{20}O_{11}$	Luteolin glucoside isomer 2	285	Flavone
461.1089	12.6	$C_{22}H_{22}O_{11}$	Chrysoeriol glucoside		Flavone
569.1876	12.9	$C_{26}H_{34}O_{14}$	Methoxyoleuropein	537, 403, 337	Secoiridoids and related compounds
539.1770	13.2	$C_{25}H_{32}O_{13}$	Oleuropein isomer 1	275, 377, 307	Secoiridoids and related compounds
539.1770	13.4	$C_{25}H_{32}O_{13}$	Oleuropein isomer 2	275, 377, 307	Secoiridoids and related compounds
539.177	13.6	$C_{25}H_{32}O_{13}$	Oleuropein isomer 3	275, 377, 307	Secoiridoids and related compounds
523.1820	14.5	$C_{25}H_{32}O_{12}$	Ligstroside	291, 259	Secoiridoids and related compounds
377.1242	14.7	$C_{19}H_{22}O_8$	Oleuropein aglycon isomer 1		Secoiridoids and related compounds
285.0405	15.9	$C_{15}H_{10}O_6$	Luteolin	175, 151	Flavone
301.0354	16.2	$C_{15}H_{10}O_7$	Quercetin		Flavonol
377.1242	17.3	$C_{19}H_{22}O_8$	Oleuropein aglycon isomer 2	275	Secoiridoids and related compounds
269.0455	18.6	$C_{15}H_{10}O_5$	Apigenin	225, 151	Flavone
299.0561	19.2	$C_{16}H_{12}O_6$	Diosmetin		Flavone

at 2 bar, and the dry gas at 9 L/min at 250°C. The accurate mass data of the molecular ions were processed through the software data analysis and target analysis 4.0 (Bruker Daltonics). External mass spectrometer calibration was performed using a Cole-Parmer syringe pump (Vernon Hills, Illinois, USA) directly connected to the interface, equipped with a Hamilton syringe (Reno, Nevada, USA). The standard solution was injected at the beginning of the run, and all the spectra were calibrated prior to phenolic compounds' identification. Then, the same HPLC system was coupled to a Bruker Daltonics Esquire 2000 ion trap mass spectrometer (Bruker Daltonics, Bremen, Germany) equipped with an electrospray interface (Agilent Technologies, CA, USA) operating in the negative ionization mode. The ion trap scanned at the 50–1000 m/z range at 13000 μ/s during the separation and detection. The maximum accumulation time for the ion trap was set at 200 ms, the target count at 20000, and compound stability was set at 50%. The optimum values of the ESI-MS parameters were as follows: capillary voltage, +3.0 kV; drying gas temperature, 300°C; drying gas flow, 7.0 L/min; and nebulizing gas pressure, 21.7 psi. The instrument was controlled by Esquire NT software from Bruker Daltonics. Peak identification was performed on the basis of their relative retention time values, TOF-MS and IT-MS/MS data, comparison with authentic standard solutions, and using the information previously reported in the literature [32]. Limits of detection (LOD) and quantification (LOQ) were respectively as follow: 0.09 and 0.3 ppm for hydroxytyrosol, 0.31 and 1.03 ppm for tyrosol, 0.02 and 0.06 ppm for luteolin, and 0.02 and 0.06 ppm for apigenin.

2.4. Radical-Scavenging Capacity (RSC). The olive leaf samples were analyzed for their capacity to scavenge the stable DPPH$^\bullet$ radical [33]. A volume of 3 mL of a freshly prepared DPPH solution (10^{-4} M in methanol) was added to 0.5 mL of

TABLE 2: Phenolic composition (in % of total phenolics) of "Chemlali" olive leaves from different regions.

	Gabes	Ghraiba	El Hajeb	Jelma	Kondar	Lamta	Sidi Bou Ali	Teboulba	Zaghouan
Quinic acid	6.96 ± 0.05	5.21 ± 0.01	8.65 ± 0.45	7.86 ± 0.41	7.82 ± 0.51	9.26 ± 0.36	8.83 ± 0.79	6.58 ± 0.24	7.55 ± 1.02
Hyty-Gl	8.59 ± 0.63	7.05 ± 0.16	6.98 ± 0.21	3.02 ± 0.15	5.13 ±	7.52 ± 0.21	12.96 ± 0.80	9.06 ± 0.37	1.75 ± 0.15
Hyty	1.65 ± 0.02	2.24 ± 0.02	1.01 ± 0.02	1.16 ± 0.01	1.83 ± 0.13	3.90 ± 0.06	1.56 ± 0.08	2.00 ± 0.09	1.92 ± 0.01
Ty-Gl	0.44 ± 0.03	1.24 ± 0.11	0.86 ± 0.07	0.82 ± 0.02	0.45 ± 0.13	0.36 ± 0.15	0.79 ± 0.21	1.39 ± 0.1	3.74 ± 0.33
EA-Gl isomer 1	nd	0.58 ± 0.03	nd	1.39 ± 0.01	0.49 ± 0.1	nq	nd	0.70 ± 0.02	1.34 ± 0.11
Secologanoside	nd	15.78 ± 0.21	9.73 ± 0.29	18.27 ± 0.87	14.36 ± 0.66	15.32 ± 0.81	9.72 ± 0.57	12.77 ± 0.78	14.17 ± 0.53
Caffeoylquinic acid isomer 1	0.30 ± 0.01	0.40 ± 0.08	0.18 ± 0.03	0.60 ± 0.1	0.32 ± 0.08	0.82 ± 0.06	0.42 ± 0.03	0.38 ± 0.02	0.74 ± 0.02
EA-Gl isomer 2	nd	nd	nd	0.95 ± 0.13	nd	nd	nd	0.07 ± 0.01	0.35 ± 0.07
Gallocatechin	0.16 ± 0.01	2.58 ± 0.12	2.39 ± 0.1	2.54 ± 0.01	3.78 ± 0.4	8.89 ± 0.3	4.35 ± 0.24	2.39 ± 0.35	1.24 ± 0.43
Benzyl alcohol hexose-pentose	1.90 ± 0.04	0.70 ± 0.02	0.90 ± 0.01	0.54 ± 0.11	1.03 ± 0.31	1.23 ± 0.27	1.35 ± 0.45	1.26 ± 0.2	0.86 ± 0.11
EA-Gl isomer 3	2.13 ± 0.03	1.90 ± 0.00	1.44 ± 0.01	1.76 ± 0.07	2.49 ± 0.38	1.14 ± 0.13	1.77 ± 0.47	1.42 ± 0.29	1.70 ± 0.19
Caffeoylquinic acid isomer 2	nd	0.20 ± 0.01	0.17 ± 0.01	0.24 ± 0.01	0.19 ± 0.05	0.33 ± 0.01	0.21 ± 0.00	0.13 ± 0.02	0.25 ± 0.02
Oleuropein aglycon derivative	5.29 ± 0.06	7.96 ± 0.03	7.95 ± 0.39	6.83 ± 0.13	10.42 ± 0.81	13.30 ± 0.59	8.85 ± 0.77	7.42 ± 0.94	7.16 ± 0.43
EA-Gl isomer 4	nd	2.98 ± 0.14	3.85 ± 0.14	2.51 ± 0.02	3.41 ± 0.23	0.29 ± 0.07	2.51 ± 0.09	1.36 ± 0.15	1.14 ± 0.21
Demethyloleuropein	nd	0.03 ± 0.00	0.20 ± 0.01	0.04 ± 0.01	0.54 ± 0.24	nd	0.18 ± 0.02	0.19 ± 0.01	nq
Rutin	nd	0.18 ± 0.01	0.29 ± 0.04	0.20 ± 0.01	0.28 ± 0.03	0.27 ± 0.02	0.47 ± 0.17	0.53 ± 0.10	0.42 ± 0.05
Luteolin rutinoside	nq	0.05 ± 0.01	nd	0.06 ± 0.01	nq	0.19 ± 0.03	nd	nq	0.10 ± 0.09
Verbascoside	nq	nq	0.23 ± 0.04	nq	nq	nq	0.20 ± 0.11	1.38 ± 0.27	0.62 ± 0.18
Luteolin glucoside isomer 1	3.63 ± 0.01	2.50 ± 0.03	1.74 ± 0.54	1.79 ± 0.32	1.91 ± 0.01	2.53 ± 0.14	2.12 ± 0.65	1.48 ± 0.24	2.57 ± 0.22
Oleuropein glucoside	0.37 ± 0.02	0.20 ± 0.01	0.13 ± 0.08	0.85 ± 0.05	0.02 ± 0.01	nq	0.10 ± 0.01	0.18 ± 0.02	0.57 ± 0.6
Diosmin isomer 1	nd	nq	nd	nq	nq	nq	nq	nd	nq
Apigenin rutinoside	nq	0.04 ± 0.01	0.02 ± 0.00	0.02 ± 0.01	0.04 ± 0.012	0.15 ± 0.01	0.05 ± 0.01	nq	0.07 ± 0.02
Diosmin isomer 2	nq	nd	nd	nq	nq	nq	nq	nd	0.08
Taxifolin	0.13 ± 0.01	0.14 ± 0.02	0.13 ± 0.01	0.77 ± 0.05	0.12 ± 0.02	1.35 ± 0.31	0.29 ± 0.14	0.60 ± 0.01	2.09 ± 0.31
Apigenin glucoside	0.07 ± 0.01	0.22 ± 0.11	0.12 ± 0.06	0.36 ± 0.08	0.07 ± 0.01	0.58 ± 0.05	0.15 ± 0.02	0.12 ± 0.01	0.41 ± 0.04
Luteolin glucoside isomer 2	1.81 ± 0.02	1.21 ± 0.13	0.82 ± 0.04	0.75 ± 0.15	0.68 ± 0.03	1.08 ± 0.12	1.16 ± 0.04	0.77 ± 0.011	1.11 ± 0.01
Chryseriol glucoside	0.47 ± 0.03	0.33 ± 0.02	0.31 ± 0.08	nd	0.28 ± 0.01	0.34 ± 0.03	0.33 ± 0.01	nq	0.17 ± 0.02
Methoxyoleuropein	13.46 ± 0.15	11.27 ± 0.23	9.54 ± 0.12	7.55 ± 0.10	11.56 ± 0.26	9.20 ± 0.41	12.01 ± 0.10	4.60 ± 0.31	7.52 ± 0.16
Oleuropein isomer 1	25.39 ± 1.68	17.87 ± 0.15	20.25 ± 0.15	23.77 ± 0.22	15.76 ± 0.35	12.41 ± 0.27	15.20 ± 0.19	18.96 ± 0.38	22.87 ± 0.47
Oleuropein isomer 2	5.31 ± 0.31	3.39 ± 0.08	4.92 ± 0.37	nd	3.00 ± 0.28	1.15 ± 0.10	1.99 ± 0.09	4.25 ± 0.24	nd
Oleuropein isomer 3	11.55 ± 0.45	6.62 ± 0.11	12.03 ± 0.14	9.79 ± 0.07	5.56 ± 0.02	2.24 ± 0.15	4.49 ± 0.07	11.00 ± 0.12	5.91 ± 0.07
Ligstroside	0.55 ± 0.02	2.15 ± 0.03	2.00 ± 0.10	5.24 ± 0.07	1.57 ± 0.12	0.21 ± 0.01	0.90 ± 0.02	1.72 ± 0.08	6.83 ± 0.03
Oleuropein aglycon isomer 1	6.84 ± 0.05	2.98 ± 0.05	2.77 ± 0.04	nd	4.10 ± 0.09	2.38 ± 0.10	3.73 ± 0.14	4.02 ± 0.07	3.58 ± 0.04
Luteolin	0.72 ± 0.02	1.60 ± 0.01	0.45 ± 0.03	nd	2.19 ± 0.015	2.23 ± 0.34	2.82 ± 0.09	1.22 ± 0.084	1.08 ± 0.01
Quercetin	0.01 ± 0.00	0.08 ± 0.01	0.06 ± 0.01	0.30 ± 0.02	0.06 ± 0.00	0.74 ± 0.01	0.17 ± 0.03	0.40 ± 0.01	0.20 ± 0.02
Oleuropein aglycon isomer 2	2.57 ± 0.01	nd	nd	nd	nq	nd	nd	1.56 ± 0.12	nd
Apigenin	nq	0.13 ± 0.01	nq	nd	0.25 ± 0.05	0.61 ± 0.012	0.32 ± 0.03	0.08 ± 0.01	0.04 ± 0.01
Diosmetin	nq	0.18 ± 0.01	nd	nd	0.29 ± 0.02	nd	nd	0.04 ± 0.01	nd

nd: not detected; nq: not quantified (under the limit of quantification).

the polar extract. The reaction mixture was shaken vigorously for 10 s in a vortex apparatus and then maintained in the dark for 20 min, after which a steady state was reached. The absorbance of the mixture was measured at 515 nm against a blank solution (without radical). A control sample (without olive leaf extract) was prepared and measured. The RSC toward $DPPH^{\bullet}$ was expressed as the % reduction in $DPPH^{\bullet}$ concentration by the constituents of the oils: % $[DPPH^{\bullet}]$ red $= 100*(1 - [DPPH^{\bullet}]_{20}/[DPPH^{\bullet}]_0)$ where $[DPPH^{\bullet}]_0$ and $[DPPH^{\bullet}]_{20}$ were the concentrations of

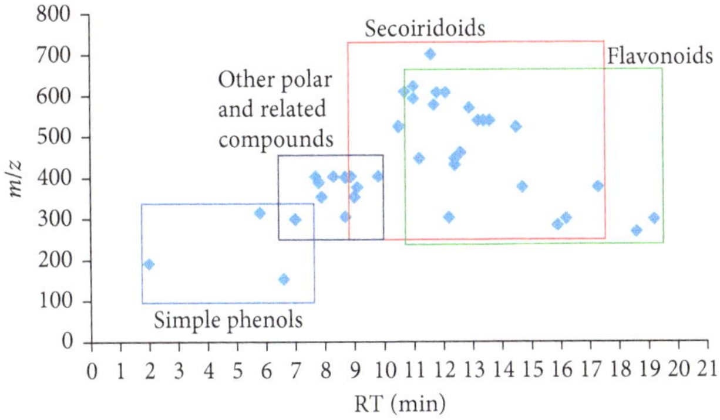

FIGURE 2: Distribution of the identified compounds according to their retention time and m/z.

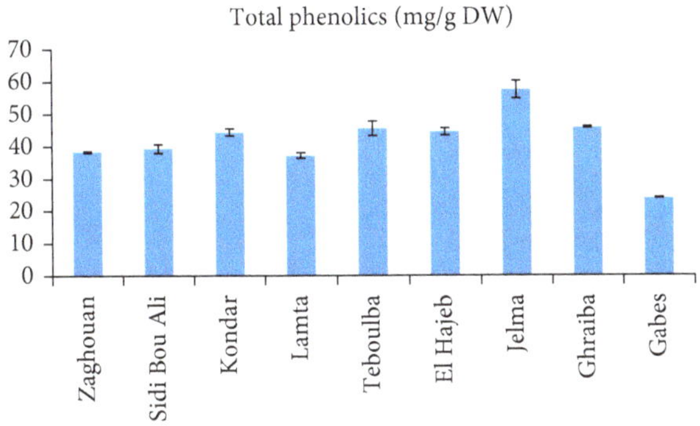

FIGURE 3: Total phenolic contents in "Chemlali" olive leaf extracts according to the cultivation regions.

DPPH$^{\bullet}$ in the control sample (t = 0) and in the test mixture after the 20-minute reaction, respectively.

2.5. Statistical Analysis. Statistical analysis was performed by means of XLSTAT for Windows. Data are given as means of triplicates. Hierarchical cluster analysis (HCA) was performed using Ward's method based on Euclidean distance.

3. Results and Discussion

3.1. HPLC-MS Analysis of the Olive Leaf Extract. The base peak chromatogram (BPC) of "Chemlali" olive leaf extract and the compounds characterized in the negative ion mode are shown in Figure 1. Peak identification was carried out using Target Analysis software (Bruker Daltonics), by comparing both migration time and accurate MS and MS/MS spectral data obtained from olive leaf samples and commercial standards, together with the information previously reported in the literature.

Table 1 summarizes the information about the identified phenolic compounds, in "Chemlali" olive leaf extracts, together with their corresponding retention time, theoretical mass to charge ratio (m/z), molecular formula, and IT/MS^2 fragments. A total of 38 compounds could be identified. Three main phenolic groups could be clearly distinguished: phenolic alcohols (hydroxytyrosol, tyrosol, and tyrosol glucoside), flavonoids (luteolin rutinoside, 2 isomers of luteolin glucoside, 2

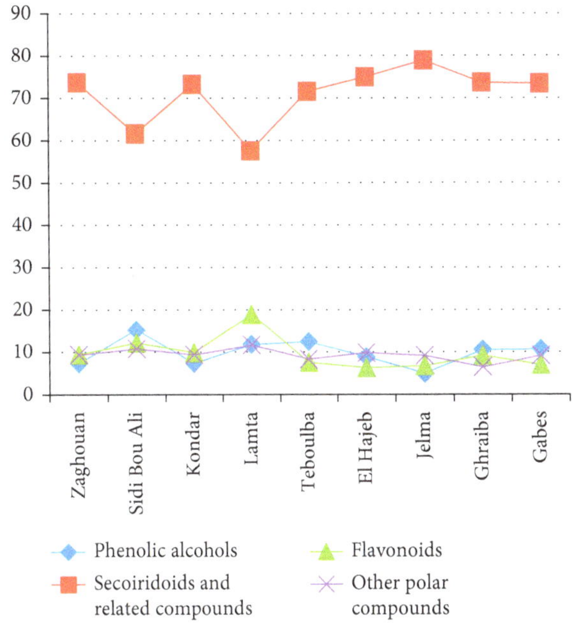

FIGURE 4: Variation in the distribution of the phenolic compound classes in "Chemlali" olive leaf extracts according to the cultivation regions.

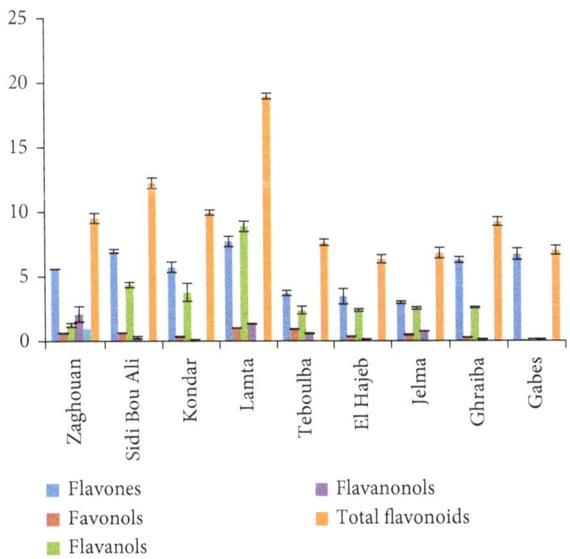

FIGURE 5: Variation in the distribution of flavonoids in "Chemlali" olive leaf extracts according to the cultivation regions.

isomers of diosmin, apigenin rutinoside, apigenin glucoside, chryseriol glucoside, luteolin, apigenin, rutin, quercetin, gallocatechin, taxifolin, and diosmetin), and various secoiridoids (particularly, oleuropein with 3 isomers, methoxyoleuropein, ligstroside, oleuropein aglycon derivative and related compounds such as elenolic acid glucoside isomers, and secologanoside). Apart from them, other polar compounds were also found; they are quinic acid, caffeoylquinic acid isomers, and benzyl alcohol hexose-pentose. A certain tendency in the elution order of the compounds related to their chemical structure class was observed, appearing in the following order

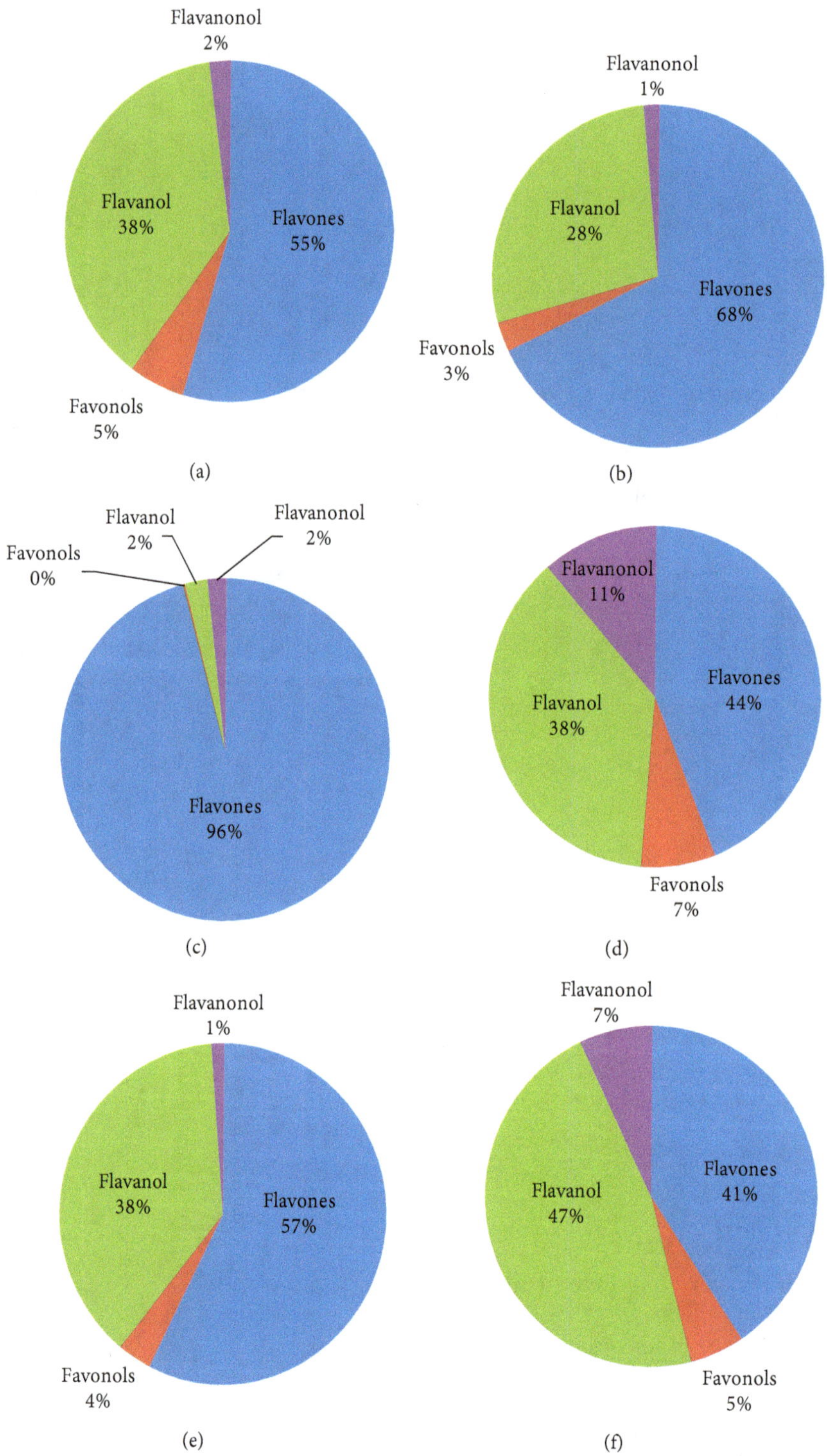

FIGURE 6: Continued.

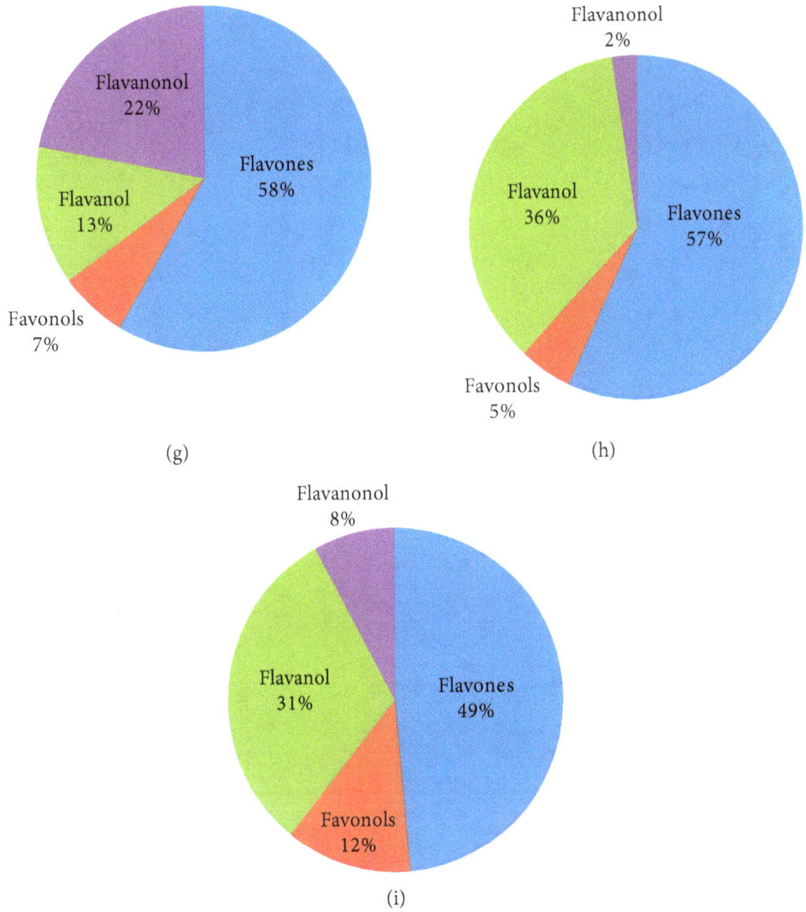

FIGURE 6: Flavonoid subclasses distribution. (a) El Hajeb. (b) Ghraiba. (c) Gabes. (d) Jelma. (e) Kondar. (f) Lamta. (g) Zaghouan. (h) Sidi Bou Ali. (i) Teboulba.

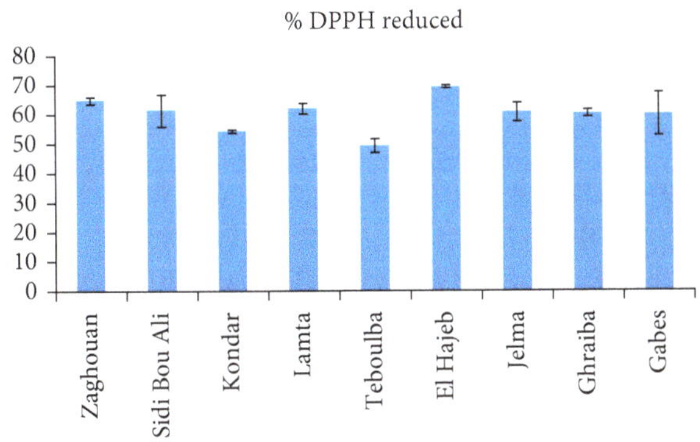

FIGURE 7: Radical-scavenging capacity of olive leaf extracts according to the cultivation regions.

of increasing retention time (RT), and thus hydrophobicity: simple phenols, secoiridoids and related compounds, and flavonoids (Figure 2).

As can be observed in Table 2, some qualitative differences were registered among olive leaf samples from the geographical studied regions. Quantitatively, significant differences were found in a wide number of phenolic

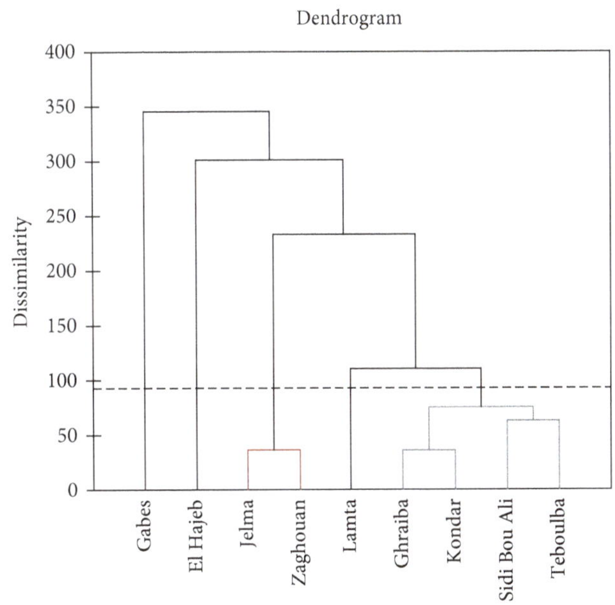

FIGURE 8: Hierarchical cluster analysis dendrogram. Groups: (1) Gabes; (2) Ghraiba, Kondar, Sidi Bou Ali, Teboulba; (3) El Hajeb; (4) Jelma, Zaghouan; (5) Lamta.

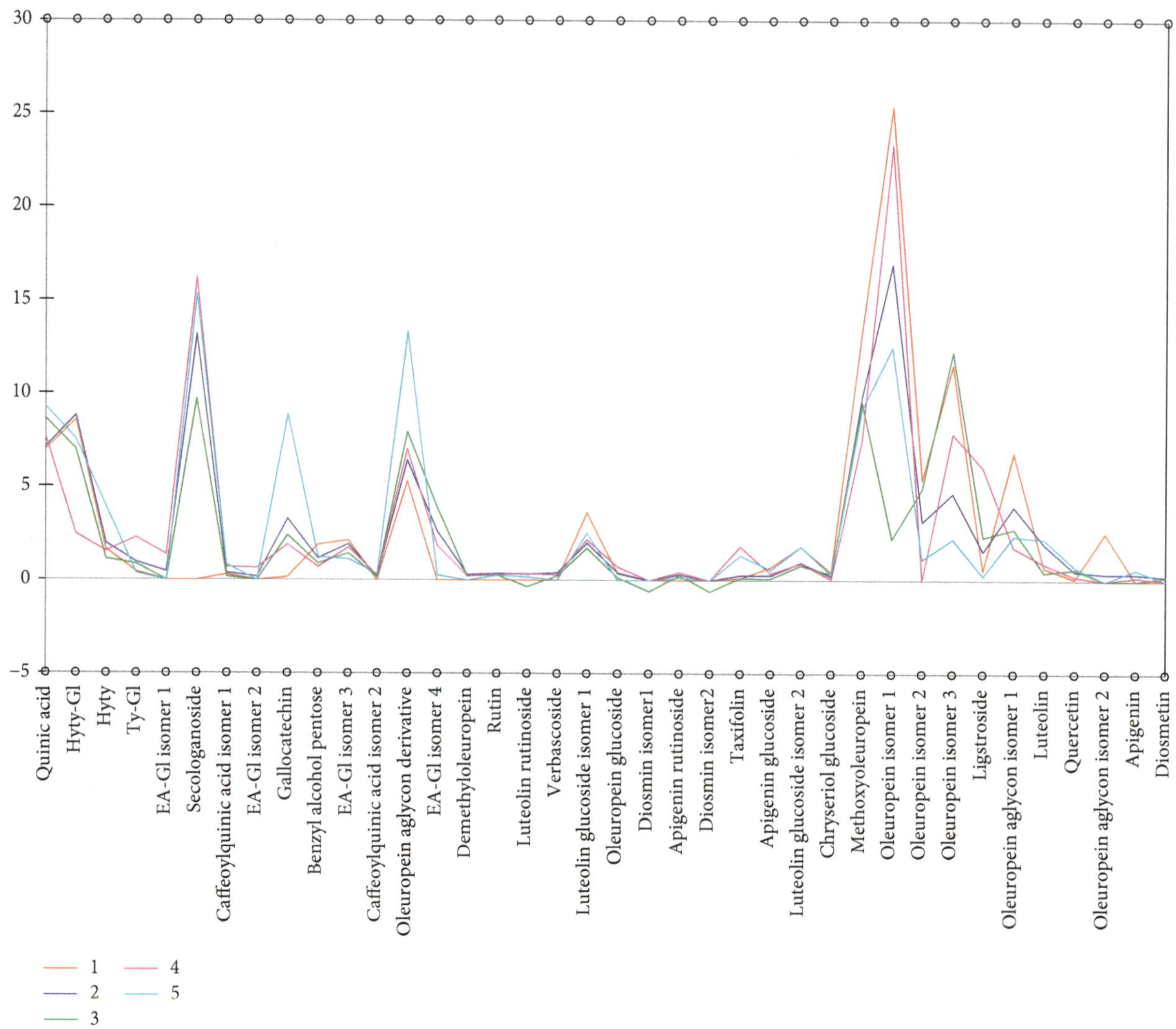

FIGURE 9: HCA groups' profiles according to the phenolic composition. Groups: (1) Gabes; (2) Ghraiba, Kondar, Sidi Bou Ali, Teboulba; (3) El Hajeb; (4) Jelma, Zaghouan; (5) Lamta.

compounds according to the geographical origin of the samples. Our results are in agreement with those reported in literature. In deed comparing Tunisian cultivars located in southern and northern Tunisia, significant differences were observed for phenolic composition as well as antioxidant activity where higher levels were registered for samples from the north [29].

Bilgin and Şahin [28] demonstrated that total phenolic content and extract yield varied significantly among Turkish olive cultivars from six sites in Anatolia. The authors observed that the total phenolic contents in olive leaves decreased as the geographical altitude decreases [28]. In another study on Greek olive cultivars, leaves collected from three different locations showed differences in total phenolic contents and antioxidant activity [27]. Similarly, the geographical zone has influenced the phenolic composition of Arbequina cultivar samples collected from six regions in Spain [30]. Using hierarchical cluster analysis, the latter study showed high similarities between the olive leaf samples from Cordoba and Navarra and both samples from Mallorca, respectively.

In our study, significant differences were observed for total phenolic contents in "Chemlali" olive leaf extracts from different regions (Figure 3). These differences could be explained by the most marked variation that was observed for flavonoids and secoiridoids and related compounds as well as for phenolic alcohols classes especially for the leaves collected from Lamta region. These classes presented amounts that reached 18, 78 and 15% for Lamta, Jelma, and Sidi Bou Ali, respectively. Furthermore, we can observe that secoiridoids' tendency was the inverse of that of flavonoids (Figure 4).

Being the major one among the identified compounds, oleuropein amounts varied significantly among olive leaf samples. The registered values were between 15.8 and 42.25% for samples from Gabes and Lamta regions, respectively.

Considering the subclasses of flavonoids, the most pronounced variation among samples from different regions was registered for flavones and flavanols. Flavones presented the highest percentage among the rest of flavonoids followed by flavanols. Among analyzed samples, olive leaves from

Gabes showed the highest percentage in flavones, whereas leaves from Lamta showed the highest levels in flavanols presented by gallocatechin (Figures 5 and 6).

Radical-scavenging activity is very important, due to the deleterious role of free radicals in foods and in biological systems. This test is a standard assay in antioxidant activity studies and offers a rapid technique for screening the radical-scavenging activity of specific compounds. The stable free DPPH radical is a useful reagent to investigate the scavenger properties of phenolics [34]. The radical-scavenging capacity values varied significantly between studied samples: the leaves from Zaghouan presented the highest radical-scavenging capacity, whilst leaves from Teboulba showed a radical-scavenging capacity of 49% (Figure 7).

3.2. Chemometrics. Data clustering is a common technique for statistical data analysis, which is used in many fields. It is known as the classification of objects into groups (clusters), so that the data in each subset share some common trait—often proximity according to some defined distance measure [35]. The hierarchical analysis is a kind of very important clustering methods. It is an unsupervised technique that examines the interpoint distances between all of the samples and represents that information in the form of a two dimensional plot called a hierarchical tree or dendrogram.

In our study, an overall cluster analysis based on the determined metabolites and radical-scavenging capacity was performed to study the general differences among samples. To display the similarities between olive leaf samples from different provenance, a dendrogram was produced by applying a hierarchical clustering algorithm to the data.

As can be seen in Figure 8, there is a clear separation among the clusters established for different samples. Applying Ward's method based on Euclidean distance, olive leaf samples could be gathered into five different groups with respect to their phenolic composition: the first presented by Gabes region; the second by Ghraiba, Kondar, Sidi Bou Ali, and Teboulba; and the third by El Hajeb. The fourth group is composed by Jelma and Zaghouan and the fifth by Lamta.

It is useful to mention that HCA could be successfully applied for the classification and evaluation of the regional growing effects on the specification of the olive leaves based on their phenolic profile. In Figure 9, the profiles of the distinguished clusters or groups are illustrated. As can be seen, the leaves from Lamta region located in the coastal center of Tunisia are richer in gallocatechin and oleuropein aglycone derivative. Groups 1 and 4 presented the leaves richer in oleuropein isomer 1. Moreover, leaves from Gabes (group 1) located in the coastal south of Tunisia showed their richness in oleuropein aglycone.

These results are important since they may be useful for further researches and give an insight into the behavior of phenolic compounds in olive leaves from different environments. Such compounds have shown to be sensible to the edaphoclimatic variation between the considered regions of cultivation of "Chemlali" cultivar.

Conflicts of Interest

The authors declare no conflicts of interest.

Acknowledgments

The authors are grateful to the Tunisian Ministry of Higher Education and Scientific Research, the Spanish Ministry of Economy and Competitiveness (MINECO) (AGL2015-67995-C3-2-R), and the Andalusian Regional Government Council (P11-CTS-7625).

References

[1] B. Dimitrios, "Sources of natural phenolic antioxidants," *Trends in Food Science & Technology*, vol. 17, no. 9, pp. 505–512, 2006.

[2] D. Komes, A. Belščak-Cvitanović, D. Horžić, G. Rusak, S. Likić, and M. Berendika, "Phenolic composition and antioxidant properties of some traditionally used medicinal plants affected by the extraction time and hydrolysis," *Phytochemical Analysis*, vol. 22, no. 2, pp. 172–180, 2011.

[3] M. S. Butler, "The role of natural product chemistry in drug discovery," *Journal of Natural Products*, vol. 67, no. 12, pp. 2141–2153, 2004.

[4] R. Gyawali and S. A. Ibrahim, "Natural products as antimicrobial agents," *Food Control*, vol. 46, pp. 412–429, 2014.

[5] D. J. Newman and G. M. Cragg, "Natural products as sources of new drugs from 1981 to 2014," *Journal of Natural Products*, vol. 79, no. 3, pp. 629–661, 2016.

[6] G. M. Cragg and D. J. Newman, "Natural products: a continuing source of novel drug leads," *Biochimica et Biophysica Acta*, vol. 1830, no. 6, pp. 3670–3695, 2013.

[7] L. M. Blair and J. Sperry, "Natural products containing a nitrogen–nitrogen bond," *Journal of Natural Products*, vol. 76, no. 4, pp. 794–812, 2013.

[8] J.-H. Park, J.-H. Jung, J.-Y. Yang, and H.-S. Kim, "Olive leaf down-regulates the oxidative stress and immune dysregulation in streptozotocin-induced diabetic mice," *Nutrition Research*, vol. 33, no. 11, pp. 942–951, 2013.

[9] N. Xynos, G. Papaefstathiou, M. Psychis, A. Argyropoulou, N. Aligiannis, and A.-L. Skaltsounis, "Development of a green extraction procedure with super/subcritical fluids to produce extracts enriched in oleuropein from olive leaves," *Journal of Supercritical Fluids*, vol. 67, pp. 89–93, 2012.

[10] M. Ben Salah, H. Abdelmelek, and M. Abderraba, "Study of phenolic composition and biological activities assessment of olive leaves from different varieties grown in Tunisia," *Medicinal Chemistry*, vol. 2, no. 5, pp. 107–111, 2012.

[11] J. E. Hayes, P. Allen, N. Brunton, M. N. O'Grady, and J. P. Kerry, "Phenolic composition and in vitro antioxidant capacity of four commercial phytochemical products: Olive leaf extract (*Olea europaea* L.), lutein, sesamol and ellagic acid," *Food Chemistry*, vol. 126, no. 3, pp. 948–955, 2011.

[12] K. Kiritsakis, M. G. Kontominas, C. Kontogiorgis, D. Hadjipavlou-Litina, A. Moustakas, and A. Kiritsakis, "Composition and antioxidant activity of olive leaf extracts from Greek olive cultivars," *Journal of the American Oil Chemists' Society*, vol. 87, no. 4, pp. 369–376, 2009.

[13] O.-H. Lee, B.-Y. Lee, J. Lee et al., "Assessment of phenolics-enriched extract and fractions of olive leaves and their antioxidant activities," *Bioresource Technology*, vol. 100, no. 23, pp. 6107–6113, 2009.

[14] K. Hassanzadeh, K. Akhtari, H. Hassanzadeh, S. A. Zarei, N. Fakhraei, and K. Hassanzadeh, "The role of structural C–H compared with phenolic OH sites on the antioxidant activity of oleuropein and its derivatives as a great non-flavonoid family of the olive components: a DFT study," *Food Chemistry*, vol. 164, pp. 251–258, 2014.

[15] D. Raederstorff, "Antioxidant activity of olive polyphenols in humans: a review," *International Journal for Vitamin and Nutrition Research*, vol. 79, no. 3, pp. 152–165, 2009.

[16] A. M. Al-attar and N. A. Shawush, "Physiological investigations on the effect of olive and rosemary leaves extracts in male rats exposed to thioacetamide," *Saudi Journal of Biological Sciences*, vol. 21, no. 5, pp. 473–480, 2014.

[17] A. Kaeidi, S. Esmaeili-Mahani, V. Sheibani et al., "Olive (*Olea europaea* L.) leaf extract attenuates early diabetic neuropathic pain through prevention of high glucose-induced apoptosis: in vitro and in vivo studies," *Journal of Ethnopharmacology*, vol. 136, no. 1, pp. 188–196, 2011.

[18] F. Brahmi, G. Flamini, M. Issaoui et al., "Chemical composition and biological activities of volatile fractions from three Tunisian cultivars of olive leaves," *Medicinal Chemistry Research*, vol. 21, no. 10, pp. 2863–2872, 2011.

[19] R. Fabiani, P. Rosignoli, A. De Bartolomeo et al., "Oxidative DNA damage is prevented by extracts of olive oil, hydroxytyrosol, and other olive phenolic compounds in human blood mononuclear cells and HL60 cells," *Journal of Nutrition*, vol. 138, no. 8, pp. 1411–1416, 2008.

[20] A. Taamalli, D. Arráez-Román, E. Barrajón-Catalán et al., "Use of advanced techniques for the extraction of phenolic compounds from Tunisian olive leaves: phenolic composition and cytotoxicity against human breast cancer cells," *Food and Chemical Toxicology*, vol. 50, no. 6, pp. 1817–1825, 2012.

[21] E. Barrajón-Catalán, A. Taamalli, R. Quirantes-Piné et al., "Differential metabolomic analysis of the potential antiproliferative mechanism of olive leaf extract on the JIMT-1 breast cancer cell line," *Journal of Pharmaceutical and Biomedical Analysis*, vol. 105, pp. 156–162, 2015.

[22] L. Abaza, T. P. N. Talorete, P. Yamada, Y. Kurita, M. Zarrouk, and H. Isoda, "Induction of growth inhibition and differentiation of human leukemia HL-60 cells by a Tunisian Gerboui olive leaf extract," *Bioscience, Biotechnology, and Biochemistry*, vol. 71, no. 5, pp. 1306–1312, 2007.

[23] J. F. Ayala-Zavala, V. Vega-Vega, C. Rosas-Domínguez et al., "Agro-industrial potential of exotic fruit byproducts as a source of food additives," *Food Research International*, vol. 44, no. 7, pp. 1866–1874, 2011.

[24] A. Taamalli, D. Arráez-Román, M. Zarrouk, J. Valverde, A. Segura-Carretero, and A. Fernández-Gutiérrez, "The occurrence and bioactivity of polyphenols in Tunisian olive products and by-products: a review," *Journal of Food Science*, vol. 77, no. 4, pp. R83–R92, 2012.

[25] S. N. El and S. Karakaya, "Olive tree (*Olea europaea*) leaves potential beneficial effects on human health," *Nutrition Reviews*, vol. 67, no. 11, pp. 632–638, 2009.

[26] N. Talhaoui, A. Taamalli, A. M. Gómez-Caravaca, A. Fernández-Gutiérrez, and A. Segura-Carretero, "Phenolic compounds in olive leaves: analytical determination, biotic and abiotic influence, and health benefits," *Food Research International*, vol. 77, pp. 92–108, 2015.

[27] E. Giannakopoulou, G. Mitsopoulos, M. Hagidimitriou, V. Papageorgiou, and M. Komaitis, "Influence of cultivar, harvesting season and geographical origin on phenolic content in leaves of Greek olive cultivars," *Acta Horticulturae*, vol. 924, pp. 437–444, 2011.

[28] M. Bilgin and S. Şahin, "Effects of geographical origin and extraction methods on total phenolic yield of olive tree (*Olea europaea*) leaves," *Journal of the Taiwan Institute of Chemical Engineers*, vol. 44, no. 1, pp. 8–12, 2013.

[29] F. Brahmi, B. Mechri, M. Dhibi, and M. Hammami, "Variation in antioxidant activity and phenolic content in different organs of two Tunisian cultivars of *Olea europaea* L.," *Acta Physiologiae Plantarum*, vol. 36, no. 1, pp. 169–178, 2013.

[30] R. Japón-Lujan, J. Ruiz-Jiménez, and M. D. L. de Castro, "Discrimination and classification of olive tree varieties and cultivation zones by biophenol contents," *Journal of Agricultural and Food Chemistry*, vol. 54, no. 26, pp. 9706–9712, 2006.

[31] L. Abaza, N. Ben Youssef, H. Manai, F. M. Haddada, K. Methenni, and M. Zarrouk, "Chétoui olive leaf extracts: influence of the solvent type on phenolics and antioxidant activities," *Grasas y Aceites*, vol. 62, no. 1, pp. 96–104, 2007.

[32] A. Taamalli, D. Arráez-Román, E. Ibañez, M. Zarrouk, A. Segura-Carretero, and A. Fernández-Gutiérrez, "Optimization of microwave-assisted extraction for the characterization of olive leaf phenolic compounds by using HPLC-ESI-TOF-MS/IT-MS2," *Journal of Agricultural and Food Chemistry*, vol. 60, no. 3, pp. 791–798, 2012.

[33] G. Kalantzakis, G. Blekas, K. Pegklidou, and D. Boskou, "Stability and radical-scavenging activity of heated olive oil and other vegetable oils," *European Journal of Lipid Science and Technology*, vol. 108, no. 4, pp. 329–335, 2006.

[34] İ. Gülçin, "Antioxidant activity of food constituents: an overview," *Archives of Toxicology*, vol. 86, no. 3, pp. 345–391, 2012.

[35] Z. Piravi-Vanak, J. B. Ghasemi, M. Ghavami, H. Ezzatpanah, and E. Zolfonoun, "The influence of growing region on fatty acids and sterol composition of Iranian olive oils by unsupervised clustering methods," *Journal of the American Oil Chemists' Society*, vol. 89, no. 3, pp. 371–378, 2012.

18

Development of QuEChERS Method for the Determination of Polycyclic Aromatic Hydrocarbons in Smoked Meat Products using GC-MS from Qatar

Hussain Al-Thaiban, Nada Al-Tamimi, and Murad Helaleh

Anti Doping Lab-Qatar, Toxicology Multipurpose laboratory, P.O. Box 27775, Doha, Qatar

Correspondence should be addressed to Murad Helaleh; mrhelala@yahoo.com

Academic Editor: Valdemar Esteves

A simple and fast method for the determination of PAHs in smoked meat samples was described. The QuEChERS (Z-Sep) procedure was used for sample preparation. Gas chromatograph-mass spectrometer with electron ionization (EI) was used to separate and detect the PAHs. All 16 common PAHs were analyzed successfully. Matrix-matched calibration was applied. Spiked samples were performed at 1 ng/g (n = 10) and 10 ng/g (n = 10) for two days. Overall recoveries of PAHs were within 74 to 117%, with RSDs within 1.15 to 37.57% and 1 and 10 ng/g wet weight for first and second day, respectively. In most of the analyzed smoked meat samples, there were no exceeded levels compared to the maximum levels declared by Commission Regulation (EU) number 835/2011. The method can be recommended for routine analysis for laboratories having a large number of samples.

1. Introduction

PAHs represents a group of organic compounds (hydrophobic) consisting of two or more fused aromatic rings that are ubiquitous pollutants and those can be generated during food processing [1]. PAHs containing five or more rings are considered having potential carcinogenic and genotoxic to humans and regarded as concerned to public health [2–4]. Therefore, exposure to PAHs is recommended to be as low as possible [5]. Diet is considered to be one of the highest exposures to PAHs to the most population in different countries and counts to be more than 90% [2, 6, 7]. However, high level of PAHs is not usually detected in raw foods [7].

PAHs are widely distributed in the environment and remains for a long time [8]. Benzo[a]pyrene (B[a]P) is carcinogenic and mutagenic [9] and may cause a substantial contributed to burden cancer in human [10]. However, according to scientific committee on food (SCF) [11], B[a]P was considered as a marker for the occurrence and impact of PAHs carcinogenic in food.

PAH compounds those containing five or more aromatic rings are known as "heavy molecular weight (HMW)" PAHs and those with less than five rings are known as "light molecular weight" (LMW) PAHs. Heavy and light PAHs are nonpolar compounds, showing high lipophilic nature. However, HMW PAHs is more toxic than LMW PAHs [12]. Other groups of PAHs which are not defined as carcinogens may act as synergists. As a general concept, PAHs are present in a mixture and are regarded the priority organic pollutants [12]. The fact that PAHs are present in food insists their control in food is necessary [13].

A method for determining the PAHs levels in commercial smoked meat and fish products from polish market was reported [14], PAHs were determined in grilled meat [15], in smoked fish [16] and B[a]P in smoked meat products [17, 18],in meat products smoked with different types of wood and smoking spices [19], and in Swedish smoked meat and fish [20]. PAHs in smoked ham produced at household in Serbia were reported [21]. A study was reported for the evaluation of the health risk of long-term exposure to PAHs in daily intake of smoked meat in Southwest China [22].

There are different routes of contamination on food, which include and not limited to (i) packing materials and thermal processing of food from animal origins [23] and even during home preparation and (ii) direct deposition of PAHs from weathering of fruits and vegetables [24].

The principle of sample preparation is to remove the complexities of the matrix from the sample and to make it suitable for qualitative and quantitative analysis [25]. Extraction techniques such as Soxhlet can be simple , which consumes large solvent volumes and takes long extraction times. However, the cleanup techniques to remove residues such as lipids are challenging. Standard methods for cleanup such as GPC [26], florisil [27], and silica gel [28], which required an extensive quantity of reagents, solvents, and materials, are available in the literature. A challenge for rapid extraction and clean-up method has become essential, in order to minimize the sample preparation costs and minimize consumable costs. Quenchers method was first developed by Anastassiades et al., to provide a rapid, inexpensive, and fast method for analyzing pesticides residues in fruits and vegetables [29]. The method was extended to extract pesticide and cleanup the extract from fatty foods (e.g., avocados, milk, and eggs) [30]. Quenchers method has been shown to efficiently analyze fish fillets [31–33] and shrimp [34]. The nature of Quenchers sample preparation is promising and permits its use for extraction and cleanup, and it is efficiently applicable for the analysis of wide matrices.

Quantitative analysis of PAHs from a complex matrices such as smoked meat faces three major challenges: (i) the occurrence of most PAHs at trace levels (i.e., ppb levels), (ii) the coextract of many other components from the matrices with PAHs, which leads to difficult identification in chromatographic spectrum, and (iii) the structural similarity and many of PAHs occur as isomers, which again leads to identification problem in individual PAHs [35].

Quenchers techniques replaced the complications of common, standard extraction techniques, and they provide a simple protocol for sample handling [36]. Acetonitrile was used as an extraction solvent in QuEChERS instead of acetone, which is thought to be an excellent separator from water, after the addition of salts [37, 38]. Moreover, the cleanup protocols are adopted to effectively remove the matrix-matched materials such as polar pigment, fatty acids, organic compounds, and sugar [29].

There was a growing interest in QuEChERS among researchers all over the world; the need for further investigation is mandatory for researchers in order to facilitate and develop rapid, efficient, and effective methods for different complicated matrices.

Quenchers method is basically based on extraction with acetonitrile partitioned from an aqueous matrix using $MgSO_4$ and NaCl followed by cleanup using (d-SPE) with $MgSO_4$ and analysis by GC-MS. Labelled d-PAHs can be used as an internal standard to compensate the analyte loss and matrix effect on chromatographic response.

One of the main obstacles in the determination of fatty food is the high-fat content (e.g., lipids, triglycerides, and fatty acids). However, the removal of lipids is important to maintain the GC system and also to allow the low detection limits (LOD). The sensitivity of the method was confirmed by the ability to detect low PAHs concentrations at the allowable permitted levels. Lipids may have severe effects, such as reproducibility, robustness, and recovery, on analyzing PAHs by GC-MS[39]. Therefore, it is essential to remove lipids prior to analyses.

The aim of the present study is to optimize and validate the applicability of Quenchers for the extraction and cleanup and finally analyzing 16 PAHs in smoked meat samples at low LOQ levels. A few studies in the literature were reported for developing an excellent analytical method utilizing QuEChERS for determining PAHs concentrations in smoked meat samples. However, both levels and distribution patterns in smoked meat product are of interest. The method was developed to cover the analysis of 2–6 rings PAHs. Precision and accuracy of the method were validated.

2. Materials and Methods

2.1. Chemicals. All 16 PAH standards, naphthalene (NAP), acenaphthylene (ACNY), acenaphthene (ACN), fluorene (FL), phenanthrene (PHEN), anthracene (ANTH), fluoranthene (FLU), pyrene (PYR), benz[a]anthracene (B[a]A), chrysene (CHRY), benzo[b]fluoranthene (B[b]F), benzo[k]fluoranthene (B[k]F), benzo[a]pyrene (B[a]P), indeno[1,2,3-cd]pyrene (I-123-cd_P), dibenz[a,h]anthracene (D[ah]A), and benzo[g,h,i]perylene (B[ghi]P), were purchased from Cambridge isotope laboratories (Frontage Road, Andover, MA, USA). The internal standard mixture: naphthalene_d8 (NAPH_d8), acenaphthylene_d8 (ACNY_d8), phenanthrene_d10 (PHEN_d10), fluoranthene_d10 (FLU_d10), pyrene_d10 (PYR_d10), benzo [a]pyrene_d12 (B[a]P_d12), and benzo[g,h,i]perylene_d12 (B[ghi]P_d12) were obtained from Cambridge isotope laboratories (Frontage Road, Andover, MA, USA).

Hexane was obtained from Sigma-Aldrich and dichloromethane (DCM) from fluka. 50 ml polypropylene conical tube (Falcon) was used for the extraction. Acetonitrile (99.9%) was obtained from Merck. Sodium chloride (NaCl) and Magnesium Sulfate ($MgSO_4$) were obtained from Sigma-Aldrich. Fixed speed vortex mixer (Fisher Scientific, USA) was used for the shaking of the tubes. The centrifuge was from Thermo Scientific-SL 16 R. The QuEChERS kits used for the extraction and cleanup of meat samples were purchased from Supel TM QuE Z-Sep (12 ml tubes) (55403-U). Heating block was from Thermo Scientific-Reacti-Therm III no. TS-18824 and Turbo Vap® LV from Biotage.

2.2. Meat Samples. All smoked meat samples ($n = 30$) were purchased from Qatar local market (Qatar, Doha). All samples were weighed before being grinded, followed by dividing them into small portions by blender to homogenize them in order to (i) increase the surface area, (ii) break down the cell structure, and (iii) ensure analytical test portion representing whole samples. However, the meat samples have to be kept cold to avoid analyte break down. Finally, ground meat was stored by placing them in aluminum foil and storing at −20°C.

2.3. PAHs Standard and PAH-d8. A standard mixture containing 16 PAHs solution (2000 μg/ml). Two working solutions were prepared (1000 ng/ml and 50 ng/ml) with DCM, capped using crimper cap and stored in the refrigerator until it is used. Deuterated PAHs concentration of 1000 ng/ml was prepared with DCM from the original PAHs surrogate cocktail (2000 μg/ml), and the vials were capped using crimper cap and stored in the moisture cabinet at room temperature until it is used.

2.4. Matrix-Matched Calibration. The matrix-matched calibration was used to prepare the calibration standards. It is stored in a refrigerator at 4°C. All the standards used to prepare the matrix-spiked calibration should be taken out from the refrigerator and allowed to come at room temperature prior to use, sonicated as per the manufacturer's instructions. The matrix-matched calibration was prepared by spiking the meat sample wet weight (2.0 g) with standard PAHs to obtain seven calibration points (0.5, 1.0, 5.0, 10.0, 20.0, 25.0, and 50.0 ng/g) and with the d-PAHs of 20 ng/g.

2.5. Preparation of PAHs QC Samples. The QC samples must be prepared from a spiking solution with the analytes of interest. The spiking should be made using standards prepared separately from those used for calibration. The QC samples were handled exactly in the same manner as the actual samples. The QC samples were analyzed by applying the same criteria for the method being evaluated. The two QC levels were at 1.0 ng/g and 10 ng/g, with 10 replicates for each concentration level.

2.6. Extraction and Purification of Smoked Meat Samples. Chopped and stored smoked meat was taken out from the freezer, thawed at 4°C before extraction, and purified by QuEChERS method. The QuEChERS purification extract offer a fast, efficient, and accurate method for the determination of PAHs in meat samples. Two grams of smoked meat sample was added into 50 centrifuge tube and spiked with d-PAHs, mixed well, and left for 30 min at room temperature. Water was added (5 ml) and homogenized, 5 ml of ACN was added to the tube, and mixed vigorously for 1 min. Sodium chloride (0.5 g) and magnesium sulfate (3.0 g) were added to the tube; the tube was shaken immediately for 1 min after adding the salts. The content was centrifuged for 10 min at 3400 rpm (Temperature = 20°C). The supernatant was transferred into a 15 ml tube containing Quenchers (Z-Sep) + 500 mg $MgSO_4$ and shaken for 1 min and centrifuged for 10 min at 3400 rpm (Temperature = 20°C). Finally, the extract was transferred into appropriate tubes and dried further using the heating block (45°C) until the volume reaches approximately 100 μl. The schematic diagram of the extraction process is shown in Figure 1. Background reduction was evaluated by analysis of the extract cleaned by Z-Sep, and it shows the lowest background. The large peak eluting 7.5 to 7.9 and 19.8 to 20.5 minutes was identified or unidentified, and it did not interfere with the ions used for quantitation of PAHs.

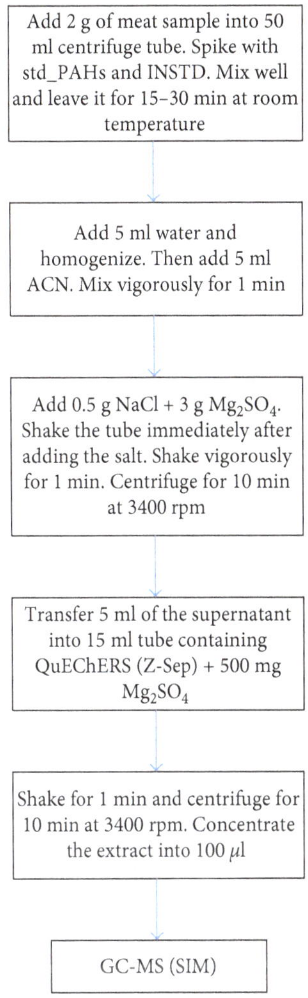

Figure 1: Schematic diagram for the extraction and cleanup of meat using QuEhERS method.

2.7. GC-MS Analysis. The extract was analyzed on Agilent 6890N gas chromatography interfaced to Agilent 5975B mass spectrometer with electron ionization. The gas chromatography was connected with a 30 m HP-5MS capillary column with 0.25 mm × 0.25 μm film. Helium was used as the carrier gas. The column was maintained at a constant flow rate of 1.0 ml/min, and 1 μl was injected into a splitless mode and purge flow to split vent was 50 ml/min at 0.75 min.

Ion source and quadrupole temperature were set at 280°C and 180°C, respectively. Injector and transfer line were maintained at 280°C and 310°C, respectively. The column temperature was initially kept at 35°C for 1 min, ramped to 200°C at a rate of 25°C/min, ramped to 310°C at a rate of 8°C/min, and kept for 3.5 min. Selected-ion monitoring (SIM) acquisition was carried out to analyze the PAHs (m/z) and internal standard (m/z), and comparison of the base peak of each targeted PAH and d-PAH are shown in Table 1. The target ions are the quantifier ions (m/z) selected for each PAHs compound and are listed in Table 1. The qualifier ions were used for confirmation purpose. The peak was identified if the retention time varied within ±0.05 min compared with the calibration standard and quantified if the S/N was ≥3,

TABLE 1: Retention time, quantifier ions, and qualifier ions (*m/z*) for target PAHs and d-PAHs.

Number	Analyte	RT (min)	Quantifier ion (*m/z*)	Qualifier ion (*m/z*)
1	**Naphthalene-d8**	6.192	136	128/137
2	Naphthalene	6.208	128	102/78
3	**Acenaphthylene-d8**	7.746	160	152/161
4	Acenaphthylene	7.746	152	126/76
5	Acenaphthene	7.946	154	152/76
6	Fluorene	8.501	166	115/82
7	**Phenanthrene-d10**	9.741	188	158/189
8	Phenanthrene	9.779	178	152/176
9	Anthracene	9.848	178	152/176
10	**Fluoranthene_D10**	11.904	212	202/213
11	Fluoranthene	11.950	202	200/101
12	**Pyrene_D10**	12.375	212	106/211
13	Pyrene	12.413	202	200/101
14	Benz[a]anthracene	15.349	228	226/114
15	Chrysene	15.453	228	226/114
16	Benzo[b]fluoranthene	18.117	252	126/113
17	Benzo[k]fluoranthene	18.178	252	126/113
18	**Benzo[a]pyrene_D12**	18.814	264	132/252
19	Benzo[a]pyrene	18.883	252	126/113
20	Indeno[1,2,3-cd]pyrene	21.409	276	138/274
21	Dibenz[a,h]anthracene	21.486	278	139/276
22	**Benzo[g,h,i]perylene_D12**	21.880	288	276/125
223	Benzo[g,h,i]perylene	21.945	276	138/274

TABLE 2: Retention time (Rt), regression coefficient (r2), slope, intercept of the calibration curve, limit of detection (LOD), and limit of quantitation (LOQ).

Compound	RT (min)	R2	Calibration range (ng/g)	Intercept	Slope	LOD (ng/g)	LOQ (ng/g)
Naphthalene	5.208	0.9998	0.5–50	0.3217	0.0877	2.74	6.64
Acenaphthylene	7.746	0.9996	0.5–50	0.4983	0.0221	7.60	20.01
Acenaphthene	7.946	0.9998	0.5–50	0.0193	0.0078	2.10	2.77
Fluorene	8.501	0.9981	0.5–50	0.0335	0.0662	1.44	2.48
Phenanthrene	9.779	1.0000	0.5–50	0.0345	0.1582	1.60	4.21
Anthracene	9.848	0.9998	0.5–50	0.0495	0.1039	0.36	0.87
Fluoranthene	11.950	0.9999	0.5–50	−0.0078	0.115	0.48	1.07
Pyrene	12.413	1.0000	0.5–50	−0.0114	0.0707	0.50	0.83
Benz[a]anthracene	15.349	1.0000	0.5–50	−0.0034	0.112	0.69	1.13
Chrysene	15.453	0.9999	0.5–50	−0.0173	0.1231	0.44	0.62
Benzo[b]fluoranthene	18.117	0.9995	0.5–50	0.0841	0.2466	0.24	0.41
Benzo[k]fluoranthene	18.178	0.9999	0.5–50	0.1379	0.1314	0.53	1.05
Benzo[a]pyrene	18.883	0.9996	0.5–50	−0.0163	0.1667	0.34	0.63
Indeno[1,2,3-cd]pyrene	21.409	1.0000	0.5–50	0.0183	0.1884	0.40	0.75
Dibenz[a,h]anthracene	21.486	0.9998	0.5–50	−0.0461	0.1885	0.63	1.22
Benzo[g,h,i]perylene	21.945	0.9998	0.5–50	0.1116	0.1448	1.71	4.41

LOD = average concentration of the matrix blank + 3 ∗ standard deviation of the matrix blank (n = 10); LOQ = average concentration of the matrix blank + 10 ∗ standard deviation of the matrix blank (n = 10).

and the ratio of the quantifier ion to qualifier ion was within ±20%. The sample that shows low response or less than the quantification limit was considered to be nondetected.

2.8. Method Linearity. The internal standard method was applied with seven matrix-matched calibrations ranging from 0.5 to 50 ng/g. The method linearity correlation coefficient (r^2) was higher than 0.995. The retention times (R_t), regression coefficient (r^2), slope, and intercept of the calibration curves are summarized in Table 2. The PAH recoveries were generally higher than 70% for all PAHs. Both recoveries and precisions are acceptable and within the range for all PAH compounds. However, the proposed method meets the EU criteria that set a maximum LOD and LOQ which is equal to 0.3 μg/kg and 0.9 μg/kg for benzo[a]pyrene in food [40, 41]. Detection and quantification limits (LODSs and LOQs) were calculated based on the analysis of the blank response. LODS was defined as the lowest concentration of the analyte in the sample that can be measured and present at a concentration above that in the blank samples (LODS = 3 ∗ SD). LOQ was defined as the lowest concentration of the analyte that can be determined at an acceptable level of uncertainty, and usually, it was the lowest point in the calibration graph (European guideline). Analytes

TABLE 3: : Recovery percentage, relative standard deviation (RSDs, %) and RSDs pooled (%) for the number of determination (n = 10) for each spiking level.

Compound	Recovery ± RSDs (%)				Pooled RSD (%)
	First day 1 ng/g Recovery ± RSD%	Second day 1 ng/g Recovery ± RSD%	First day 10 ng/g Recovery ± RSD%	Second day 10 ng/g Recovery ± RSD%	
Naphthalene	99 ± 6	104 ± 14	101 ± 6	101 ± 6	9
Acenaphthylene	117 ± 24	95 ± 18	99 ± 4	105 ± 7	15
Acenaphthene	111 ± 5	117 ± 16	99 ± 3	103 ± 6	9
Fluorene	101 ± 19	99 ± 16	96 ± 4	106 ± 11	15
Phenanthrene	88 ± 16	107 ± 13	95 ± 5	113 ± 17	14
Anthracene	83 ± 17	106 ± 5	99 ± 2	108 ± 12	11
Fluoranthene	93 ± 11	102 ± 10	98 ± 4	110 ± 15	10
Pyrene	98 ± 7	97 ± 9	95 ± 5	111 ± 16	10
Benz[a]anthracene	100 ± 9	97 ± 9	93 ± 5	110 ± 12	9
Chrysene	101 ± 8	92 ± 11	98 ± 3	110 ± 13	10
Benzo[b]fluoranthene	103 ± 14	96 ± 6	98 ± 3	112 ± 20	13
Benzo[k]fluoranthene	79 ± 38	101 ± 6	100 ± 1	100 ± 11	20
Benzo[a]pyrene	106 ± 10	99 ± 9	98 ± 2	105 ± 18	11
Indeno[1,2,3-cd]pyrene	109 ± 10	74 ± 10	96 ± 2	104 ± 18	12
Dibenz[a,h]anthracene	108 ± 11	92 ± 8	100 ± 1	107 ± 19	12
Benzo[g,h,i]perylene	107 ± 14	97 ± 11	90 ± 4	111 ± 17	12
Recovery range (%)	79–117	74–117	90–101	101–113	
RSDs range (%)	5–38	6–18	1–6	6–20	

are considered to be quantitative when ion signal-to-noise (S/N) ≥ 3 with ±20%. All samples met the criteria but having S/N < 3 were considered to be <LOQ, while those failed the criteria were considered as not detected (nd). Method detection limits (LODSs) and limit of quantitation (LOQs) are listed in Table 2.

2.9. Recovery and RSDs. The recovery of (n = 10) replicates at two levels (1 ng/g and 10 ng/g) is calculated and reported in Table 3. The recovery test was repeated on two different days. The calculated recovery and RSDs (%) are summarized in Table 3. The result shows very good recovery and excellent RSDs (%). The recovery for all sets of 10 replicates was in the average range of 74–117%. The lower spiking level was selected in the test in order to define and to check the capability of the method to detect the PAHs in meat sample at 1 ng/g. The spiking levels covered the entire range, and the recovery did not differ at the lowest and the highest concentrations. Three types of smoked meat were used to perform the recovery test, including beef roasted smoked [1] ; smoked turkey breast [29], and Al-Tag beef M smoked [42]. However, there was no influence on the PAHs determination when using different types of smoked meat.

However, according to Commission Regulation (EC) number 1881/2006 and (EC) 333/2007 [40, 41, 43], the LOD for PAHs in meat is 0.3 ng/g and the LOQ was 0.9 ng/g wet weight and the recovery range of the method should range from 50 to 120%, which clearly indicates that the method was validated according to the regulation and complies with the recommendation criteria. The validated QuEChERS method selected for PAHs gives excellent recoveries, repeatability, reproducibility, and sensitivity.

3. Results and Discussion

In order to obtain the best recovery of the QuEChERS method. All related parameters were optimized. Results of the optimized parameters are discussed. Quenchers is a method of choice for processing samples because it is quick, easy, and inexpensive. The first step was by hydrating the meat samples, and then extracting with acetonitrile followed by partitioning with salts. Hydration steps are important in QuEChERS extraction in order to make partitioning effective [37, 44]. Acetonitrile is an effective solvent and can result in low coextracted substances for some matrices [44, 45].

Validation of the method was performed, by checking the spike recoveries for meat samples, and determined in 10 replicates for each of two levels for two different days. Method detection limits and limits of quantitation for all 16 PAHs were determined using 10 blank replicates with calculating the standard deviations (SDs).

The GC-MS in SIM mode was shown for monitoring PAHs at a very low concentration in meat samples. Figure 2 shows the chromatogram obtained in GC-MS (SIM) mode for meat samples (Figure 2(a)) when fortified with 16 PAHs at 50 ng/g level and for nonfortified sample (Figure 2(b)). The chromatogram separation was clean, excellent, and can be achieved without the effect of the sample matrix interferences.

3.1. Effect of Salts in the Extraction. Magnesium sulfate ($MgSO_4$) was used as a drying agent to ensure a phase separation between organic solvent and water. Z-Sep QUE reduces concentration of fat, proteins, and other matrix components. Combination of Z-Sep and $MgSO_4$ effectively removes polar matrix and water. Acetonitrile liquid-liquid partitioning is

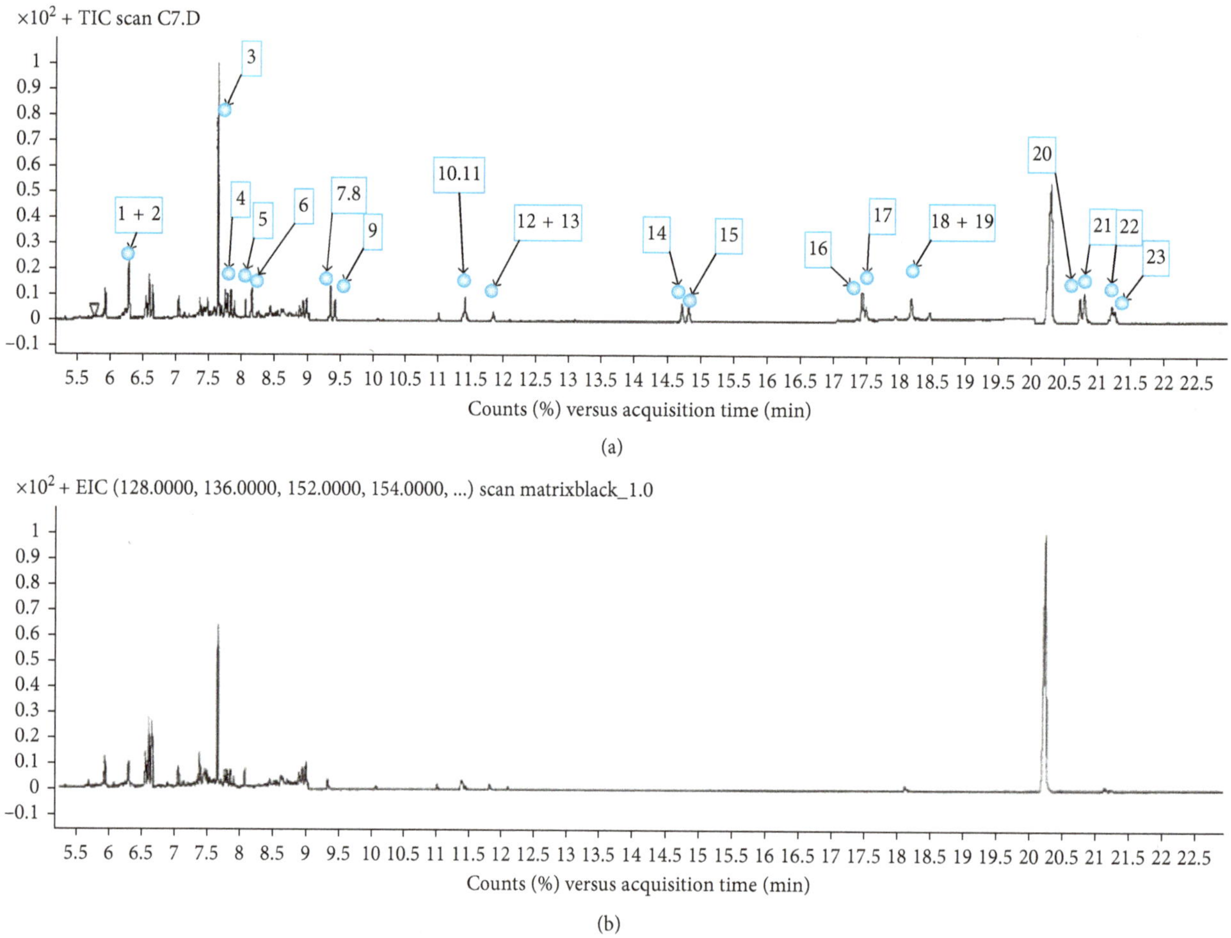

FIGURE 2: The selected-ion monitoring (SIM) obtained by Z-Sep QUE for meat samples under the selected parameters: (a) spike with 50 ng/g of 16 PAHs and I.S. (1 + 2) Naphthalene + Naph_d8, (3) acenaphylene_d8, (4) acenaphthylene, (5) acenaphthene, (6) fluorine (7 + 8) phenanthrene + Phen_d10, (9) anthracene, (10 + 11) fluoranthene + Fluro_d10, (12 + 13) pyrene + Pyr_d10, (14) benzo[a]anthracene, (15) chrysene, (16) benzo[b]fluoranthenem, (17) benzo[k]fluoranthene, (18+19) benzo[a]pyrene + B[a]P_d12, (20) indeno[1,2,3-cd]pyrene, (21) dibenz[a,h]anthracene, (22) benzo[g,h,i]perylene_d12, and (23) benzo[g,h,i]perylene. (b) Nonspike meat sample.

done by adding $MgSO_4$ and NaCl, however, $MgSO_4$ and NaCl generate sample extraction temperature of 45–50°C that persisted for the duration of the extraction. NaCl control the solvent to be removed in contact with the sample, making it to be more effective in the dissolution of analytes and facilitate the partitioning of the analytes from aqueous to the organic layer. The nonpolar PAHs with hydrophobic interaction, with pi-bond being involved, when extracted with relatively polar solvent (i.e., ACN) pi-bond and linear in geometry gave slightly better extraction. The geometry of the solvent should allow maximum interaction with the analyte besides its polarity.

Increase in salt allows greater phase separation. However, amount of salts used can also have effectiveness on the extraction system. Therefore, the role of the salt is to regulate the polarity of the matrix. Anastassiades et al. [29] showed that the use of $MgSO_4$ to remove the excess of water and to provide an exothermic reaction improves the extraction process.

Furthermore, the addition of salt increases the temperature, lowering the activation energy, decreasing the viscosity of the solvent, and finally increasing the interaction of the solvent matrix [46]. Different amounts of salt were used and the optimal amount of 3.0 g of $MgSO_4$ and 0.5 g of NaCl were selected throughout the experiment which gives the highest intensity response for the extracted PAHs from meat samples.

3.2. Effect of Solvent. As advisable, the solvent must be less expensive, compatible with analytical instrument and environment-friendly [29]. However, acetonitrile (ACN) and ethyl acetate have been largely used to extract polar to nonpolar compounds [42]. The solvent volume can play an essential role in recovery and must be in sufficient quantity to allow the full immersion of the sample into maximum solvent-analyte interaction. Different amounts of ACN were tested: 2.5, 5.0, 7.5, 10, and 15 ml. It was found that the highest peak intensity and the maximum recovery were obtained at 5 ml ACN. ACN provides a cleaner chromatogram and is considered to be one of the most selective solvents, and it has advantage over most other solvents used in QuEChERS technique [29]. However, ACN provide easier separators from water compared with other solvents used in QuEChERS in the presence of salts, providing a good phase separation which prevents interaction of polar matrix [36].

3.3. Effect of Centrifuge Time and Speed. The results obtained shows that excellent recovery of PAHs at 10 min, which was chosen as the optimal time for centrifuge. A centrifuge of 3400 rpm was found to be sufficient to obtain a good recovery of PAHs. The centrifuge facilitates the solvent to be more in contact with meat sample, provides more effectiveness in dissolution of the analyte [47], and hence reduces the time required for extraction.

3.4. Effect of Water. For ACN salting out or partitioning to occur, we must have percentage of water associated with the sample. Addition of water creating aqueous environment within the sample reduces the potential for lipids to impact extraction efficiency and minimize the fat extract. Different amounts of water were tested (2.5, 5.0, 7.5, 10, and 15 ml). It was found that the highest peak intensity and the maximum recovery were obtained with 5 ml and 7.5 ml of water, and by increasing the volume, peak intensity starts decreasing.

3.5. Types of Sorbent Used. Three types of cleaning sorbents were used in the extraction of meat samples. The sorbent has different effect on recovery and selectivity. The effect of sorbent types was evaluated based on the recovery achieved. Z-Sep/C18 (combination of Z-Sep and Discovery® DSC-18 particles) was used, and the recovery varied from 14 to 104%; lower recovery was observed for the PAHs having 5 and 6 rings. Another sorbent that is Z-Sep+ (dual bonded C18 and zirconium silica) was tested, and the recovery ranged from 46 to 126%, and the lower recovery was observed for the PAHs having 6 rings. The lower recoveries of PAHs having 5 and 6 rings could be due to limited solubility of these compounds in acetonitrile. However, the Z-Sep is a zirconium-coated silica (recommended for highly hydrophobic analytes such as PAHs and PCBs) yielded the highest average recoveries, and a few matrix interferences were observed in the first few minutes of the GC-MS run. Therefore, Z-Sep QUE was used throughout the experiment.

3.6. Analysis of Real Samples. Seven out of 16 PAHs have been categorized as probable human carcinogens. These are benzo[a]anthracene, benzo[b]fluoranthene, benzo[k]fluoranthene, chrysene, benzo[a]pyrene, dibenzo[a,h]anthracene, and indeno[1,2,3-cd]pyrene [26, 48]. The European Union has setup maximum levels of 2 ng/g wet weight for benzo[a]pyrene and considered to be a marker for carcinogenic risk [49]. However, the European Food Safety Authority [50] declared that benzo[a]pyrene is not a suitable marker for the occurrence and toxicity of PAHs in food. In 2012, four groups of PAHs were considered to provide many values to carcinogenicity, and they are benzo[a]anthracene, benzo[b]fluoranthene, chrysene, and benzo[a]pyrene [51].

In order to verify the effectiveness of the method, different types of smoked meat samples ($n = 30$) were analyzed using the optimal parameters. The results of the smoked meat samples are presented in Table 4.

Noncarcinogenic PAHs (NAPH, ACNY, CAN, FL, PHEN, ANTH, FLUO, and PYR) are accounted to be at proportion more than 70% of the total PAHs detected in meat samples. It is obvious that the noncarcinogenic PAHs in smoked meat samples are predominant. The genotoxic PAH8 was found to be more than 20% relative percentage of the total PAHs detected and the most predominant PAHs was B[a]P and found to be less than 5% relative percentage.

From Table 5, naphthalene, acenaphthene, fluorine, and phenanthrene (noncarcinogenic PAHs) shows the highest mean concentration of 4.78 ng/g, 25.17 ng/g, 1.97 ng/g, and 1.45 ng/g, respectively, whereas the lowest concentration was found for chrysene, I[cd]P and D[ah]A, comprising less than 5.0% of the total PAH concentration with respect to PAH8. B[a]P (0.90 ng/g; <5.0%) with a frequent detection more than 20%, and I[cd]P was found to be the second highest detected PAHs with more than 15%. From the values reported by EFSA, the highest mean concentration for individual PAH8 was 0.47 ng/g and 0.29 ng/g for CHRY and B[a]P, respectively, in grilled meat samples in EU countries [50]. However, the mean concentration for the sum of PAH8 detected in this method was 1.82 ng/g, which is close to the reported mean concentration (1.77 ng/g) by EFSA [50]. B[a]P was detected in 23% of the total sample analyzed, and the mean concentration was 0.90 ng/g, which is below the maximum allowable limit adopted by the European Union (EU) for B[a]P (5 ng/g) in smoked meat [43, 52]. The maximum limits for the rest of carcinogenic PAHs were not yet established. The mean concentration of PAH8 and ∑B[a]Peq, shows relatively low levels of genotoxic PAHs in smoked meat. The European countries set a maximum level of 2 ng/g for individual PAH and 5 ng/g for the sum of PAH8 in olive residual oil [53].

B[a]P content was found to be with a maximum of 3.63 ng/g, which represents a factor below the level permitted by EU (5 ng/g). However, the mean BaP for all analyzed samples was 0.90 ng/g, which is also far below the permitted limits.

EFSA declared that the PAH4 or PAH8 are significantly more effective indicator of PAHs occurrence in food than the concentration of B[a]P alone [50]. However, in relation with PAH4, the total concentration calculated for different smoked meat was 10.84 ng/g, and for PAH8, the total concentration was 14.54 ng/g (Table 5). B[b]F and B[ghi]P was not found in all tested smoked samples.

3.7. Method Uncertainty. Repeatability of the analysis was performed for two levels on different days: ten replicates for each level and the standard deviation were defined. The uncertainty derived from repeatability is calculated by the following equation:

$$U(\text{rep}) = \frac{\text{RSD}}{V_n}, \quad (1)$$

where RSD is the relative standard deviation and n is the number of repetition ($n = 10$). $U(\text{rec})$ is calculated and found to be equal 3.32%. $U(p)$ is the relative standard

TABLE 4: : Levels of individual PAHs detected in different samples (ng/g) (n = 2).

PAHs	Nap	ACN	FL	PHEN	ANTH	FLUO	PYR	B[a]A	CHR	B[a]P	I_cd P	DB[a,h] A	∑PAHs
Al-Tag tky smoked breast	nd	nd	1.58	nd	1.00	nd	nd	nd	nd	0.39	nd	nd	2.97
Prime tky smkd breast strip	nd	nd	1.73	1.45	nd	0.77	1.05	nd	nd	0.36	nd	0.63	6.00
Prime chk smoked breast	nd	nd	1.47	nd	nd	nd	0.51	nd	nd	nd	nd	nd	1.98
Gourmet smoked chicken roll	2.86	nd	nd	nd	nd	0.77	1.13	nd	nd	nd	nd	nd	4.76
Gourmet smoked turkey	nd	nd	nd	nd	nd	nd	nd	nd	nd	0.35	nd	nd	0.35
5 yildiz smoked turkey breast	nd	nd	1.22	nd	nd	nd	nd	nd	nd	nd	nd	nd	1.22
Volys tky smoked breast	5.73	25.17	2.26	nd	nd	nd	nd	nd	nd	nd	0.75	nd	33.90
Siniora tky smoked breast	nd	nd	nd	nd	nd	0.60	nd	nd	nd	nd	0.87	nd	1.47
Euro gourmet smoked chicken	nd	nd	3.34	nd	nd	nd	nd	nd	nd	nd	0.63	nd	3.97
Volys tky smoked strips	4.70	nd	1.63	nd	nd	0.59	0.84	nd	nd	nd	nd	nd	7.76
Al tag chk m plain	nd	nd	1.91	nd	nd	nd	nd	nd	nd	nd	nd	nd	1.91
Al-Tag tky roast	7.86	nd	3.84	nd	nd	nd	nd	nd	nd	nd	0.41	nd	12.11
Volys chk breast fillet	nd	nd	nd	nd	nd	nd	nd	nd	0.53	nd	nd	nd	0.53
Gourmet smkd ckn breast blk	nd	nd	nd	nd	nd	nd	nd	nd	nd	3.63	nd	nd	3.63
Gourmet ckn plain mortadella	nd	nd	nd	nd	nd	nd	nd	nd	nd	nd	nd	nd	nd
Aia tky oven breast low fat	nd	nd	nd	nd	nd	nd	nd	nd	nd	nd	nd	nd	nd
Al-Tag tky breast	3.49	nd	nd	nd	nd	nd	nd	nd	nd	nd	nd	nd	3.49
Aia chk breast low fat	nd	nd	nd	nd	nd	nd	nd	nd	nd	nd	nd	nd	nd
Euro gourmet rstd chicken fi	3.22	nd	nd	nd	nd	nd	nd	nd	nd	nd	nd	nd	3.22
Aia tky roast thigh	3.65	nd	nd	nd	nd	nd	nd	nd	nd	nd	nd	nd	3.65
Al-Tag beef m smoked	nd	nd	nd	nd	nd	nd	nd	nd	nd	nd	0.42	nd	0.42
Al-Tag beef m roasted smoked	nd	nd	nd	nd	0.36	nd	0.52	1.15	0.94	0.48	nd	nd	3.44
Siniora pastrami w/pe	4.07	nd	nd	nd	nd	nd	nd	nd	nd	nd	nd	nd	4.07
Bibi smoke turkey bre	nd	nd	nd	nd	nd	nd	nd	nd	nd	nd	nd	nd	nd
Sams finest tky bst r		nd	1.38	nd	nd	nd	nd	nd	nd	nd	nd	nd	1.38
Sams finest thy bst g	7.42	nd	1.54	nd	nd	nd	nd	nd	nd	0.37	nd	nd	9.32
Bibi chicken mortadel	nd	nd	1.71	nd	nd	nd	nd	nd		nd	nd	nd	1.71
5 yildiz turkey mortad	nd	nd	nd	nd	nd	nd	nd	0.91	0.58	0.71	nd	nd	2.20
5 yildiz beef mortadel	nd	nd	nd	nd	nd	nd	nd	nd	nd	nd	nd	nd	nd
Gourmet beed pastrami		nd	nd	nd	nd	nd	nd	nd	0.45	nd	nd	nd	0.45
∑PAHs	43.00	25.17	23.61	1.45	1.36	2.73	4.05	2.06	2.49	6.29	3.07	0.63	—
Averg. concentration	4.78	25.17	1.97	1.45	0.68	0.68	0.81	1.03	0.62	0.90	0.61	0.63	—
SD	1.83	—	0.84	—	—	0.1	0.29	0.16	0.22	1.3	0.2	—	—
Frequently detected (%)	30.00	3.33	40.00	3.33	6.67	13.33	16.67	6.67	13.33	23.33	16.67	3.33	—

Note: nd refers to values below LODs.

TABLE 5: Mean concentration of PAHs, ∑PAHs, TEF, TEQ (∑B[a]Peq), and relative % to ∑PAHs (ng/g wet/wt).

Compounds	∑ Indiv. PAHs	min	max	Mean ∑PAHs	TEF	TEQ	Relative % to ∑PAHs	Frequent detection (%)
Naphthalene	43.00	2.86	7.86	4.78	0.000	0.000	32.82	30.00
Acenaphthene	25.17	25.17	25.17	25.17	0.001	0.025	19.22	3.33
Fluorene	23.61	1.22	3.84	1.97	0.001	0.024	18.02	40.00
Phenanthrene	1.45	1.45	1.45	1.45	0.001	0.001	1.11	3.33
Anthracene	1.36	0.36	0.36	0.68	0.010	0.014	1.04	6.67
Fluoranthene	2.73	0.59	0.77	0.68	0.010	0.027	2.08	13.33
Pyrene	4.05	0.51	1.13	0.81	0.001	0.004	3.09	16.67
Benz[a]anthracene	2.06	0.91	1.15	1.03	0.100	0.206	1.57	6.67
Chrysene	2.49	0.45	0.94	0.62	0.010	0.025	1.90	13.33
Benzo[a]pyrene	6.29	0.35	3.63	0.90	1.000	6.289	4.80	23.33
Indeno[1,2,3-cd]pyrene	3.07	0.41	0.87	0.61	0.100	0.307	2.35	16.67
Dibenz[a,h]anthracene	0.63	0.63	0.63	0.63	1.000	0.625	0.48	3.33
∑PAHs	115.92							
∑PAH8*	14.54	1.82						
∑PAH4**	10.84	2.71						
∑LMW***	94.59	15.77						
∑HMW****	21.32	2.13						
Mean of the ∑PAHs		3.71						

*B[a]A, Chry, B[b]F, B[k]F, B[a]P, Ind[1,2,3-cd]pyrene, Dibenz[a,h]A, and Benzo[ghi]P. **B[a]A, Chry, B[b]F and B[a]P. ***Low molecular weight (LMW): Naph, ACNY, ACN, Fl, Phen and Anth. ****High molecular weight (HMW): Fluor, Pyr, B[a]A, Chry, B[b]F, B[k]F, B[a]P, Ind[1,2,3-cd]pyrene, Dibenz[a,h]A, and Benzo[ghi]P.

uncertainty due to precision and expressed as RSDs (%) and found to be 11 %.

Standard uncertainty of the stock standard purity (%) (U_{STD}) is 0.00577 and the standard uncertainty of the internal standard purity (%) (U_{INSTD}) is 0.00576.

The combined uncertainty ($U(c)$) is calculated by the following equation:

$$U(c) = \text{square root}(U(p))^2 + (U(\text{rec}))^2 + (U(\text{STD})) + (U(\text{INSTD})) = 11.49\%. \tag{2}$$

Expanded uncertainty is calculated by multiplying the combined uncertainty by a coverage factor k ($k = 2$ at 95% confidence level) as follows:

$$U_{\text{exp}} = U_c * K, \quad U_{\text{exp}} = 22.98\%. \tag{3}$$

4. Conclusions

This study reports for the first time the application of Quenchers for the determination of PAHs in smoked meat samples in Qatar. However, this study reports the levels of BaP, PAH8, ∑LMW, ∑B[a]Peq, ∑PAHs, and ∑HMW in smoked meat items commonly consumed in the Gulf countries. The levels of PAHs in this study found to be not exceeding the maximum levels (according to a regulation of European Commission (EU) no. 835/2011) for PAHs determined in samples [51].

The developed method and performance characteristics are in accordance with the required in legislation. The proposed EU method which defined a range of acceptable recovery between 50 and 120%, with a maximum LODS and LOQ of 0.3 ng/g and 0.9 ng/g, respectively, for B[a]P in food [40, 41].

Conflicts of Interest

The authors declare that they have no conflicts of interest.

References

[1] M. J. Agerstad and K. Skog, "Review genotoxicity of heat-processed foods," *Mutation Research*, vol. 574, pp. 156–172, 2005.

[2] European Commission Scientific Committee on Foods, *Opinion of the Scientific Committee on Food on the Risks to Human Health of Polycyclic Aromatic Hydrocarbons in Food*, European Commission Scientific Committee on Foods, Brussels, Belgium, 2002.

[3] FAO/WHO Expert Committee on Food Additives, *Meeting & World Health Organization, Evaluation of Certain Food Additives: Sixty-Fourth Meeting*, 2005.

[4] I. C. Nisbet and P. K. LaGoy, "Toxic equivalency factors (TEFs) for polycyclic aromatic hydrocarbons (PAHs)," *Regulatory Toxicology and Pharmacology*, vol. 16, no. 3, pp. 290–300, 1992.

[5] Joint FAO/WHO Expert Committee on Food Additives and contaminants in WHO Food Additive Report Series, *International Program on Chemical Safety*, Vol. 28, World Health Organization, Geneva, Switzerland, 1991.

[6] J. M. Llobet, G. Falco, A. Bocio, and J. L. Domingo, "Exposure to polycyclic aromatic hydrocarbons through consumption of edible marine species in Catalonia, Spain," *Journal of Food Protection*, vol. 69, no. 10, pp. 2493–2499, 2006.

[7] World Health Organization, *Selected Non-Heterocyclic Polycyclic Aromatic Hydrocarbons. Environmental Health Criteia No. 202*, WHO, Geneva, Switzerland, 1998.

[8] M. Howsam, K. C. Jones, and P. Ineson, "PAHs associated with the leaves of three deciduous tree species. I—concentrations and profiles," *Environmental Pollution*, vol. 108, no. 3, pp. 413–424, 2000.

[9] V. A. Du Four, N. Van Larebeke, and C. R. Janssen, "Genotoxic and mutagenic activity of environmental air samples in Flanders, Belgium," *Mutation Research/Genetic Toxicology and Environmental Mutagenesis*, vol. 558, no. 1-2, pp. 155–167, 2004.

[10] D. H. Phillips, "Polycyclic aromatic hydrocarbons in the diet," *Mutation Research/Genetic Toxicology and Environmental Mutagenesis*, vol. 443, no. 1, pp. 139–147, 1999.

[11] European Commission, "Commission recommendation 2005/108/EC of 4 February 2005 on the further investigation into the levels of polycyclic aromatic hydrocarbons in certain foods," *Official Journal of the European Union*, vol. L34, pp. 43–45, 2005.

[12] P. Plaza-Bolaños, A. G. Frenich, and J. L. M. Vidal, "Polycyclic aromatic hydrocarbons in food and beverages. Analytical methods and trends," *Journal of Chromatography A*, vol. 1217, no. 41, pp. 6303–6326, 2010.

[13] B. Veyrand, A. Brosseaud, L. Sarcher et al., "Innovative method for determination of 19 polycyclic aromatic hydrocarbons in food and oil samples using gas chromatography coupled to tandem mass spectrometry based on an isotope dilution approach," *Journal of Chromatography A*, vol. 1149, no. 2, pp. 333–344, 2007.

[14] A. Zachara, D. Gałkowska, and L. Juszczak, "Contamination of smoked meat and fish products from Polish market with polycyclic aromatic hydrocarbons," *Food Control*, vol. 80, pp. 45–51, 2017.

[15] M. Kamankesh, A. Mohammadi, H. Hosseini, and Z. M. Tehrani, "Rapid determination of polycyclic aromatic hydrocarbons in grilled meat using microwave-assisted extraction and dispersive liquid–liquid microextraction coupled to gas chromatography–mass spectrometry," *Meat Science*, vol. 103, pp. 61–67, 2015.

[16] V. Ghasemzadeh-Mohammadi, A. Mohammadi, M. Hashemi, R. Khaksar, and P. Haratian, "Microwave-assisted extraction and dispersive liquid–liquid microextraction followed by gas chromatography–mass spectrometry for isolation and determination of polycyclic aromatic hydrocarbons in smoked fish," *Journal of Chromatography A*, vol. 1237, pp. 30–36, 2012.

[17] E. Ledesma, M. Rendueles, and M. Díaz, "Spanish smoked meat products: benzo (a) pyrene (BaP) contamination and moisture," *Journal of Food Composition and Analysis*, vol. 37, pp. 87–94, 2015.

[18] C. Santos, A. Gomes, and L. C. Roseiro, "Polycyclic aromatic hydrocarbons incidence in Portuguese traditional smoked meat products," *Food and Chemical Toxicology*, vol. 49, no. 9, pp. 2343–2347, 2011.

[19] A. Hitzel, M. Pöhlmann, F. Schwägele, K. Speer, and W. Jira, "Polycyclic aromatic hydrocarbons (PAH) and phenolic substances in meat products smoked with different types of wood and smoking spices," *Food Chemistry*, vol. 139, no. 1, pp. 955–962, 2013.

[20] S. Wretling, A. Eriksson, G. A. Eskhult, and B. Larsson, "Polycyclic aromatic hydrocarbons (PAHs) in Swedish smoked meat and fish," *Journal of Food Composition and Analysis*, vol. 23, no. 3, pp. 264–272, 2010.

[21] B. Kartalovic, D. Okanovic, J. Babic, V. Djordjevic, S. Jankovic, and M. Cirkovic, "Polycyclic aromatic hydrocarbons in smoked ham," *Procedia Food Science*, vol. 5, pp. 144–147, 2015.

[22] J. Li, H. Dong, X. Li, B. Han, C. Zhu, and D. Zhang, "Quantitatively assessing the health risk of exposure to PAHs from intake of smoked meats," *Ecotoxicology and Environmental Safety*, vol. 124, pp. 91–95, 2016.

[23] P. Šimko, "Factors affecting elimination of polycyclic aromatic hydrocarbons from smoked meat foods and liquid smoke flavorings," *Molecular Nutrition and Food Research*, vol. 49, no. 7, pp. 637–647, 2005.

[24] J. Fismes, C. Schwartz, C. Perrin-Ganier, J. L. Morel, A. M. Charissou, and M. J. Jourdain, "Risk of contamination for edible vegetables growing on soils polluted by polycyclic aromatic hydrocarbons," *Polycyclic Aromatic Compounds*, vol. 24, no. 4-5, pp. 827–836, 2004.

[25] R. M. Smith, "Before the injection—modern methods of sample preparation for separation techniques," *Journal of Chromatography A*, vol. 1000, no. 1, pp. 3–27, 2003.

[26] USEPA (US Environmental Protection Agency), *Fedral Register, 59 FR 1788*, U.S. Environmental Protection Agency, Washington, DC, USA, 1994.

[27] U.S. Environmental Protection Agency, *Florisil Cleanup, SW-846: Method 3620c, Revision 3*, U.S. Environmental Protection Agency, Washington, DC, USA, 2007.

[28] U.S. Environmental Protection Agency, *Organic Extraction and Sample Preparation, SW-846: Method 3630c, Revision 3*, U.S. Environmental Protection Agency, Washington, DC, USA, 1996.

[29] M. Anastassiades, S. J. Lehotay, D. Štajnbaher, and F. J. Schenck, "Fast and easy multiresidue method employing acetonitrile extraction/partitioning and "dispersive solid-phase extraction" for the determination of pesticide residues in produce," *Journal of AOAC International*, vol. 86, no. 2, pp. 412–431, 2003.

[30] S. J. Lehotay, K. Maštovská, and S. J. Yun, "Evaluation of two fast and easy methods for pesticide residue analysis in fatty food matrixes," *Journal of AOAC International*, vol. 88, no. 2, pp. 630–638, 2005.

[31] H. R. Norli, A. Christiansen, and E. Deribe, "Application of QuEChERS method for extraction of selected persistent organic pollutants in fish tissue and analysis by gas chromatography mass spectrometry," *Journal of Chromatography A*, vol. 1218, no. 41, pp. 7234–7241, 2011.

[32] D. F. Rawn, J. Judge, and V. Roscoe, "Application of the QuEChERS method for the analysis of pyrethrins and pyrethroids in fish tissues," *Analytical and Bioanalytical Chemistry*, vol. 397, no. 6, pp. 2525–2531, 2010.

[33] Y. Sapozhnikova and S. J. Lehotay, "Multi-class, multi-residue analysis of pesticides, polychlorinated biphenyls, polycyclic aromatic hydrocarbons, polybrominated diphenyl ethers and novel flame retardants in fish using fast, low-pressure gas chromatography–tandem mass spectrometry," *Analytica Chimica Acta*, vol. 758, pp. 80–92, 2013.

[34] N. Omar, J. Bakar, and K. Muhammad, "Determination of organochlorine pesticides in shrimp by gas chromatography–mass spectrometry using a modified QuEChERS approach," *Food Control*, vol. 34, no. 2, pp. 318–322, 2013.

[35] A. Sadowska-Rociek, M. Surma, and E. Cieślik, "Comparison of different modifications on QuEChERS sample preparation method for PAHs determination in black, green, red and white tea," *Environmental Science and Pollution Research*, vol. 21, no. 2, pp. 1326–1338, 2014.

[36] C. Diez, W. A. Traag, P. Zommer, P. Marinero, and J. Atienza, "Comparison of an acetonitrile extraction/partitioning and "dispersive solid-phase extraction" method with classical multi-residue methods for the extraction of herbicide residues in barley samples," *Journal of Chromatography A*, vol. 1131, no. 1-2, pp. 11–23, 2006.

[37] K. Maštovská and S. J. Lehotay, "Evaluation of common organic solvents for gas chromatographic analysis and stability of multiclass pesticide residues," *Journal of Chromatography A*, vol. 1040, no. 2, pp. 259–272, 2004.

[38] F. J. Schenck, P. Callery, P. M. Gannett, J. R. Daft, and S. J. Lehotay, "Comparison of magnesium sulfate and sodium sulfate for removal of water from pesticide extracts of foods," *Journal of AOAC International*, vol. 85, no. 5, pp. 1177–1180, 2002.

[39] A. Müller, E. Björklund, and C. von Holst, "On-line clean-up of pressurized liquid extracts for the determination of polychlorinated biphenyls in feeding stuffs and food matrices using gas chromatography–mass spectrometry," *Journal of Chromatography A*, vol. 925, no. 1, pp. 197–205, 2001.

[40] European Union, "Commission regulation (EC) no. 401/2006 laying down the methods of sampling and analysis for the official control of the levels of mycotoxins in foodstuffs," *Official Journal of the European Union*, vol. L70, pp. 12–34, 2007.

[41] European Union, *Official Journal of the European Union L88*, European Union, Brussels, Belgium, 2007.

[42] R. Carabias-Martínez, E. Rodríguez-Gonzalo, P. Revilla-Ruiz, and J. Hernández-Méndez, "Pressurized liquid extraction in the analysis of food and biological samples," *Journal of Chromatography A*, vol. 1089, no. 1-2, pp. 1–17, 2005.

[43] European Commission, "Setting maximum levels for certain contaminants in foodstuffs," *Official Journal of the European Union*, vol. 364, no. 5, 2006.

[44] I. R. Pizzutti, A. de Kok, M. Hiemstra, C. Wickert, and O. D. Prestes, "Method validation and comparison of acetonitrile and acetone extraction for the analysis of 169 pesticides in soya grain by liquid chromatography–tandem mass spectrometry," *Journal of Chromatography A*, vol. 1216, no. 21, pp. 4539–4552, 2009.

[45] Z. Huang, Y. Li, B. Chen, and S. Yao, "Simultaneous determination of 102 pesticide residues in Chinese teas by gas chromatography–mass spectrometry," *Journal of Chromatography B*, vol. 853, no. 1, pp. 154–162, 2007.

[46] B. E. Richter, B. A. Jones, J. L. Ezzell, N. L. Porter, N. Avdalovic, and C. Pohl, "Accelerated solvent extraction: a technique for sample preparation," *Analytical Chemistry*, vol. 68, no. 6, pp. 1033–1039, 1996.

[47] A. Kouzayha, M. Al Iskandarani, S. Mokh, A. R. Rabaa, H. Budzinski, and F. Jaber, "Optimization of a solid-phase extraction method using centrifugation for the determination of 16 polycyclic aromatic hydrocarbons in water," *Journal of Agricultural and Food Chemistry*, vol. 59, no. 14, pp. 7592–7600, 2011.

[48] U.S. Environmental Protection Agency, *Organic Extraction and Sample Preparation, SW-846: Method 3640a, Revision 1*, U.S. Environmental Protection Agency, Washington, DC, USA, 1994.

[49] European Commission, "Commission regulation no 208/2005: regards polycyclic aromatic hydrocarbons," *Official Journal of the European Union*, vol. 134, pp. 3–5, 2005.

[50] EFSA, "Scientific opinion of the panel on contaminants in the food Chain on a request from the European Commission on Polycyclic Aromatic Hydrocarbons in Food," *EFSA Journal*, vol. 724, pp. 1–114, 2008.

[51] European Commission (EC), *European Commission Regulation: O.J. (L835)*, European Commission, Brussels, Belgium, 2011.

[52] European Commission, *Setting Maximum Levels for Certain Contaminants in Foodstuffs*, Commission Regulation (EU) No. 1881/2006, European Commission, Brussels, Belgium, 2006.

[53] S. Moret, G. Purcaro, and L. S. Conte, "Polycyclic aromatic hydrocarbons in vegetable oils from canned foods," *European Journal of Lipid Science and Technology*, vol. 107, no. 7-8, pp. 488–496, 2005.

19

Characterization and Botanical Differentiation of Monofloral and Multifloral Honeys Produced in Cyprus, Greece, and Egypt using Physicochemical Parameter Analysis and Mineral Content in Conjunction with Supervised Statistical Techniques

Ioannis K. Karabagias,[1] Artemis P. Louppis,[2] Stavros Kontakos,[3] Chryssoula Drouza,[4] and Chara Papastephanou[2]

[1]*Laboratory of Food Chemistry, Department of Chemistry, University of Ioannina, Ioannina 45110, Greece*
[2]*cp Foodlab Ltd., Polifonti 25, 2047 Strovolos, Nicosia, Cyprus*
[3]*Department of Social Administration and Political Science, Democritus University of Thrace, Komotini 69100, Greece*
[4]*Department of Agricultural Sciences, Biotechnology and Food Science, Cyprus University of Technology, 3036 Limassol, Cyprus*

Correspondence should be addressed to Ioannis K. Karabagias; ikaraba@cc.uoi.gr

Academic Editor: Larisa Lvova

Thirty-four honey samples donated by beekeepers and purchased from supermarkets were collected during harvesting years 2010–2014 from Cyprus, Greece, and Egypt. The aims of this study were to characterize honey samples and, if possible, to differentiate honeys according to the honey type on the basis of physicochemical parameter values, mineral content, and their combination using supervised statistical techniques (linear discriminant analysis (LDA)). Physicochemical parameters (colour, pH, free acidity, total dissolved solids, salinity, electrical conductivity, and moisture content) were determined according to official methods, while minerals (Al, As, B, Ba, Be, Ca, Cd, Co, Cr, Cu, Fe, Hg, Mg, Mn, Mo, Ni, P, Pb, Sb, Si, Ti, Tl, V, and Zn) using inductively coupled plasma optical emission spectrometry. The majority of honey samples analyzed met the quality criteria set by the European directive and national decision related to honey. Implementation of multivariate analysis of variance (MANOVA) and LDA on specific physicochemical parameters, minerals, or their combination provided a satisfactory classification of honeys according to floral type. The overall correct classification rate (based on the cross-validation method) was 79.4% using 7 minerals and 91.2% using 8 physicochemical parameters. When the 15 parameters were combined, the classification rate of Egyptian honeys was improved by 25%.

1. Introduction

The ability to determine the origins of honey, the product of *Apis mellifera* honeybees, has become very important for protection of both consumers and beekeepers, as certain types of honey and their place of origin can impact the pricing, which can lead to counterfeits being produced to increase the value of a lesser product.

In that sense, terms such as protected designation of origin (PDO), protected geographical indication (PGI), or traditional specially guaranteed (TSG) [1] have been set to highlight several products' quality produced in different parts of the world, including honey. Some typical examples involve the Spanish PDO "Miel de Granada," the Greek fir tree honey from "Menalon," and the pine-thyme honey (Pefko-thymaromelo) from Crete [2–4]. Such products have the potential to attract consumers, producers, and exporters, and they create a basis for different types of studies, that is, economic studies or studies related to authenticity, using a data set of several physicochemical parameters and chemometric analyses [5, 6].

The usual method to determine the botanical origin of honey is pollen recognition (melissopalynology). However, this approach has certain limitations such as pollen counting procedure and identification, interpretation of results which is difficult and requires trained analysts, considerable time of analysis, or limited geographical origin determination applications [7]. Hence, the combination of pollen analysis with physicochemical and sensory characteristics can overcome the limitations of pollen analysis [8].

On the contrary, the great variability in honey chemical composition is closely related to botanical and geographical origins with regard to soil, specific vegetation, climate conditions, bee species, processing technologies, and honey extraction methods. Thus, these variations, that is, in the mineral content and/or physicochemical parameters of honey, offer a great opportunity for research related to its botanical or geographical discrimination (authentication), using supervised statistical techniques [9, 10].

Besides, in recent years, there has been a rising presence of monofloral honeys in markets, more expensive than multifloral ones, so possible adulteration by honey mixtures must be controlled.

Therefore, the objectives of the present study were to provide information about conventional physicochemical parameters and mineral content of Cypriot, Greek, and Egyptian honeys (which are categorized as multifloral and unifloral honeys) and, if possible, to differentiate these honeys according to the floral type, using a set of specific conventional physicochemical parameters, minerals, or their combination in conjunction with modern and supervised statistical techniques by partial use of melissopalynological data.

To our knowledge, data for the selected physicochemical parameters or minerals have not been reported previously [2-7, 9-13] for these specific honey types, and this constitutes the novelty of the present work.

2. Materials and Methods

2.1. Honey Samples. A total of 34 honey samples were collected during the harvesting periods 2011-2012 from Cyprus, 2010–2012 from Greece, and 2013-2014 from Egypt (Table 1).

Cypriot honeys were supplied by local professional beekeepers from the three different regions: Larnaca, Limassol, and Nicosia (14 multifloral honey samples). According to their declaration, honey samples were multifloral of specific vegetation grown in Cyprus. In particular, honeys from Larnaca consisted of *Acacia*, *Eucalyptus*, *Phoenix*, *Albizia*, rosemary, *Heliotropium europaeum*, *Parkinsonia aculeata*, and *Thymus* spp. Honeys from Limassol consisted of wild rosemary (*Teucrium creticum*), *Rhus coriaria*, *Onopordum acanthium*, *Thymus* spp., *Heliotropium europaeum*, *Eucalyptus*, *Ceratonia*, and *Opuntia ficus-indica*. Finally, honeys from Nicosia consisted of *Rhus coriaria*, *Thymus* spp., wild flowers, *Eucalyptus*, *Teucrium creticum*, *Heliotropium europaeum*, *Parkinsonia aculeata*, and *Phoenix*.

Honey samples from Greece were supplied by professional beekeepers from seven regions: Messinia (1 fir) (*Abies cephalonica*), Lakonia (1 wild thyme) (*Thymus serpyllum*), Rhodes (2 pines and 1 thyme) (*Pinus* and *Thymus* spp., resp.), Arta (2 citrus + erica) (*Citrus* and *Erica* spp.), Ioannina (2 multiflorals), Karditsa (1 fir) (*Abies cephalonica*), Metsovo (1 fir) (*Abies cephalonica*) (1 multifloral), and Thassos (1 pine) (*Pinus* spp.). As in the case of Cypriot honeys, the botanical origin of Greek honey samples was confirmed by the beekeepers.

Finally, Egyptian honeys (8 samples) were purchased from the greater Cairo area. Melissopalynological analysis was carried out according to a previous study [6] for the Egyptian honey samples, since these were purchased from supermarkets. Respective label description was Vitrac pure natural clover honey produced by HERO NFI in Egypt. Results showed that seven of the eight honey samples were classified as clover (*Trifolium alexandrinum*). The percentage of *Trifolium* pollen that was considered as predominant pollen was set at >45%. The percentage of *Trifolium* pollen grains varied between 41 and 91% [14]. All honey samples were stored in dark-coloured glass containers, shipped to the laboratory, and maintained at 4°C until analysis. Each batch of samples was analyzed within the first month of the same harvesting year of collection.

2.2. Reagents and Solutions. All chemicals used in the present study were of analytical grade [10].

2.3. Preparation of Honey Samples for Inductively Coupled Plasma Optical Emission Spectrometry (ICP-OES) Analysis: Instrumentation and Conditions. Preparation of honey samples for ICP-OES analysis and ICP-OES instrumentation and conditions are given in a previous study [10]. Twenty-four minerals were determined including Al, As, B, Ba, Be, Ca, Cd, Co, Cr, Cu, Fe, Hg, Mg, Mn, Mo, Ni, P, Pb, Sb, Si, Ti, Tl, V, and Zn. The emission wavelengths (nm) for all minerals are given in a previous study [10]. The emission wavelength for P was 178.3 nm.

2.4. Physicochemical Parameters. The following physicochemical parameters were determined according to the harmonized methods of the International Honey Commission: pH, moisture content, electrical conductivity, and free acidity [15]. For the determination of honey moisture content, an ATC refractometer (Bellingham + Stanley, UK), covering the range of moisture between 12 and 27%, was used. The interlaboratory variation, determined as the coefficient of variation of independent measurements (RSD_R %), is the quality parameter most often used to compare the precision of a developed analytical method. RSD_R % was calculated as

$$100 \times \left(\frac{R}{X}\right) \times 2.8, \qquad (1)$$

where R is the reproducibility value of independent measurements and X is the average of carrying determinations [15]. The results reported are the mean of three independent determinations ($n = 3$).

2.5. Colour Parameters L^*, a^*, and b^*. The honey surface colour was measured according to the method proposed by

Table 1: Physicochemical parameters of Cypriot, Greek, and Egyptian honeys.

BO	GO	L^*	a^*	b^*	pH	FA (meq/kg)	TDS (mg/L)	Salinity (g/L)	EC (mS/cm)	Moisture (g/100 g)
	Cyprus									
MF	Larnaca	73.72	−4.82	12.80	3.49	14.13	301.00	0.29	0.55	14.10
MF	Larnaca	71.04	−6.27	22.98	3.86	35.01	211.52	0.52	0.37	13.10
MF	Larnaca	71.49	−4.47	15.20	3.91	20.36	235.00	0.42	0.37	16.50
MF	Larnaca	71.69	−5.12	18.32	3.83	22.60	154.70	0.38	0.28	12.50
MF	Limassol	74.63	−4.84	5.20	3.64	12.64	284.00	0.29	0.54	13.45
MF	Limassol	72.16	−4.89	16.77	3.72	26.84	323.00	0.34	0.61	13.70
MF	Limassol	72.59	−4.77	14.76	3.70	21.40	150.50	0.38	0.28	13.40
MF	Limassol	73.48	−5.11	12.66	3.59	19.55	273.00	0.27	0.51	12.50
MF	Limassol	72.50	−3.64	21.49	3.65	29.32	167.80	0.40	0.32	16.10
MF	Limassol	72.92	−5.09	13.91	3.58	47.04	245.00	0.24	0.42	14.05
MF	Nicosia	74.24	−4.95	11.20	3.82	21.48	162.70	0.41	**0.32**	12.10
MF	Nicosia	71.50	−5.01	19.01	3.82	29.02	240.00	0.26	0.46	13.10
MF	Nicosia	75.77	−3.75	10.99	3.87	18.96	112.48	0.48	0.18	12.80
	Mean	72.90^{a}	-4.83^{d}	15.02^{f}	3.73^{i}	24.49^{l}	220.05^{n}	0.36^{q}	0.40^{t}	13.65^{w}
	±SD	1.41	0.65	4.78	0.13	9.18	65.87	0.09	0.13	1.32
	RSD_R %	0.58	14.49	3.73	0.75	11.43	1.27	7.77	3.50	1.03
	Greece									
P	Rhodes	66.47	−4.70	32.68	4.24	22.00	404.00	0.41	0.78	15.50
F	Messinia	69.66	−5.73	26.84	5.05	25.00	346.00	0.54	1.01	17.60
P	Rhodes	66.53	−4.76	32.68	4.39	23.20	425.00	0.46	0.79	15.30
WT	Lakonia	74.17	−4.14	14.76	4.45	30.50	447.00	0.65	0.82	13.60
T	Rhodes	74.19	−4.54	15.25	4.19	14.00	240.00	0.25	0.48	14.45
MF	Ioannina	50.83	8.83	57.68	3.98	40.60	394.86	0.24	0.77	17.20
MF	Metsovo	62.59	−2.31	37.37	4.29	28.00	364.00	0.36	0.73	13.20
P	Thassos	69.56	−5.45	25.94	4.95	32.00	437.00	0.46	1.01	15.05
MF	Ioannina	51.08	8.95	58.12	4.32	41.20	324.00	0.25	0.55	18.90
CE	Arta	71.33	−3.04	13.35	4.81	23.39	294.09	0.87	0.61	17.90
F	Karditsa	72.94	−4.71	18.02	4.68	29.00	678.00	1.25	1.90	17.90
CE	Arta	77.08	−3.13	7.03	3.41	36.40	332.00	0.34	0.64	17.20
F	Metsovo	73.04	−3.85	16.71	4.72	26.50	645.00	0.87	1.25	17.60
	Mean	67.65^{b}	-2.20^{e}	27.42^{g}	4.42^{j}	28.60^{l}	410.07^{o}	0.53^{r}	0.87^{u}	16.26^{x}
	±SD	8.36	5.01	16.18	0.44	7.69	126.38	0.30	0.37	1.83
	RSD_R %	0.62	31.82	2.04	0.63	9.79	0.68	5.28	2.60	1.72
	Egypt									
CL	GCA	76.14	−2.65	7.56	3.38	6.50	112.20	0.11	0.22	16.33
CL	GCA	75.88	−2.85	7.76	3.61	8.00	126.20	0.12	0.25	16.99
CL	GCA	76.05	−3.07	8.34	3.49	6.50	129.40	0.12	0.26	15.96
CL	GCA	76.72	−2.79	6.93	3.46	8.00	132.60	0.13	0.27	16.51
CL	GCA	75.70	−3.06	9.42	3.40	7.00	131.00	0.12	0.26	15.96
CL	GCA	76.69	−3.65	5.88	3.31	7.50	127.20	0.12	0.25	16.84
CL	GCA	76.45	−3.90	6.98	3.27	8.00	117.60	0.11	0.24	16.53
CL	GCA	75.88	−3.14	4.83	3.26	7.00	114.70	0.11	0.23	17.05
	Mean	76.19^{c}	-3.14^{e}	7.21^{h}	3.40^{k}	7.31^{m}	123.86^{p}	0.12^{s}	0.25^{v}	16.52^{x}
	±SD	0.39	0.43	1.42	0.12	0.65	7.87	0.01	0.02	0.43
	RSD_R %	0.55	22.29	7.77	0.82	38.30	2.26	11.67	3.36	1.02

The reported results are the mean of three independent replicates ($n = 3$), with the exception of colour parameters that are the mean of five independent replicates ($n = 5$). Different lowercase letter in each column of the average value represents statistically significant differences for each physicochemical parameter according to the production area ($p < 0.05$). N: the number of honey samples; BO: botanical origin; GO: geographical origin; GCA: greater Cairo area; MF: multifloral; P: pine; F: fir; CE: citrus and erica; WT: wild thyme; T: thyme; CL: clover; FA: free acidity; TDS: total dissolved solids; EC: electrical conductivity; SD: standard deviation; RSD_R %: interlaboratory variation.

Karabagias et al. [6]. Before making measurements, honey solutions were filtered for removal of any coarse particles. The precision of the analytical procedure followed was estimated by calculating RSD_R % values. The results reported are the mean of five independent determinations ($n = 5$).

2.6. Salinity and Total Dissolved Solids (TDS). Salinity and total dissolved solids of a 20% (w/v) honey solution in distilled water were measured at 20°C using a Delta OHM, model HD 3456.2, conductivity meter (Padova, Italy) with 4-ring and 2-ring conductivity/temperature probes. Temperature was measured by 4-wire Pt100 and 2-wire Pt1000 sensors by immersion. The probe was calibrated automatically resorting to the 1413 μS/cm conductivity standard solution (Hanna Instruments, Inc., Woonsocket, USA). Results were expressed as g/L and mg/L, respectively.

The precision of the analytical procedure followed was estimated by calculating RSD_R % values. The results reported are the mean of three independent determinations ($n = 3$).

2.7. Statistical Analysis. All data were subjected to statistical analysis using the SPSS statistics software (v.20.0, SPSS Inc., 2012) by application of multivariate analysis of variance (MANOVA) and linear discriminant analysis (LDA).

Comparison of the means was achieved by MANOVA in order to determine those significant physicochemical parameter or mineral content values that could aid in the geographical differentiation of Cypriot, Greek, and Egyptian honeys. Pillai's trace and Wilks' lambda indices were computed, as inverse quality criteria, to determine a possible significant effect of physicochemical parameter and mineral content values on the geographical origin of honeys. The homogeneity of variability was tested by application of Box's M index, since the number of the collected honey samples from each region was different. Both original and leave-one-out cross-validation methods were used to test the correct classification ability. The cross-validation method provides an additional verification of the developed model of discrimination. In most cases, the correct prediction rates obtained by using the cross-validation method are lower compared to those obtained by the original method but are more reliable [10]. Correlations were obtained by Pearson's correlation coefficient (r) at the confidence level $p < 0.05$.

3. Results and Discussion

3.1. Validity of the Methods Used for the Determination of Physicochemical Parameters and Minerals

3.1.1. Physicochemical Parameters. The methods used for the measurements of pH, moisture content, electrical conductivity, free acidity, total dissolved solids, salinity, and the colour parameters L^*, a^*, and b^* showed different coefficient variation values (RSD_R %) depending on the honey matrix and the nature of analysis (Table 1). However, RSD_R % values of moisture content, electrical conductivity, and free acidity are in conformity to the International Honey Commission guidelines regarding the precision of the analytical method followed for the determination of honey conventional physicochemical parameters [15]. Regarding the determination of physicochemical parameters that are not regulated, that is, colour parameters (especially L^* and b^*), pH, total dissolved solids, and salinity, the method developed showed RSD_R % <10% (Table 1).

3.1.2. Minerals

(1) Calibration Curves and Linearity. The calibration curves were prepared using 8 data points: 50, 100, 150, 200, 250, 300, 350, and 400 μg/kg of the mineral standard solution (ICP multielement standard solution XVI 1.09487.0100 of 100 mL volume and 100 mg/L concentration; Merck, Darmstadt, Germany). The correlation coefficient (R^2) was found to be >0.993 for the regression equation obtained for each mineral that was determined in the present study.

(2) Precision. The repeatability of the sample application and measurement of the peak area of each mineral was evaluated using three replicates of the standard (100 mg/L) on the same day and over three days (three replicates each day). The intra- and interday variations were estimated by relative standard deviation values which were lower than 5% for each analyzed mineral with respect to honey botanical origin (different honey matrixes) (Table 2).

(3) Limit of Detection (LOD) and Limit of Quantification (LOQ). The limit of detection (LOD) and limit of quantification (LOQ) were estimated by spiking a blank sample (ultrapure water obtained from Milli-Q, Millipore, Bedford, MA, USA) three times with the standard mineral solution at low concentrations, and the signal-to-noise ratio was determined. The LOD was defined as 3 : 1 and the LOQ as 10 : 1. The LOD and LOQ values for each mineral are given in Table 2.

3.2. Physicochemical Parameters of Cypriot, Greek, and Egyptian Honeys. A high moisture content in honey (>20 g/100 g) is undesirable because it leads to fermentation during storage. Present moisture content values are within the limits (≤20%) set by the EU [8] for commercial honey samples. Multifloral honeys from Ioannina and Metsovo (Greece) showed variation in moisture content depending on botanical and geographical origins. In addition, moisture content values for fir honeys collected from Messinia, Karditsa, and Metsovo (Greece) were in accordance (≤18.5%) with the Greek decision [16] involving the botanical identification of monofloral honeys. The lower moisture values were recorded for Cypriot honeys, indicative of the climatic conditions and floral type of honeys in this area. Pure clover labeled honey recorded an average value of moisture, slightly higher than that of Greek honeys. However, it should not be forgotten that variations in moisture content, among different honey types and geographical origins, have been reported previously for Turkish [9], Spanish [11, 17], and Greek honeys [6, 11].

Honey samples from Cyprus had lower pH values as compared to Greek or Egyptian honeys since these were multifloral honeys. pH in honey has been reported to range between 3.20 and 4.50 indicating a natural acidity. Present pH values for Greek honeys are in accordance with the previous work in the literature regarding Spanish [17], while Cypriot and Egyptian honeys recorded pH values in agreement with those reported for Malaysian honeys [18].

All honey samples analyzed had FA values below the upper limit (<50 meq/kg) set by the EU [8]. Present FA values are within the range reported previously for Spanish [11, 17], Moroccan [19], and Greek commercial honeys [6, 11]. The higher FA values were recorded for honey samples from Greece. Clover honeys from Egypt recorded pH values in excellent conformity with citrus honeys from Egypt, as shown in the previous work [11].

TABLE 2: Mineral content (mg/kg) of Cypriot, Greek, and Egyptian honeys.

N	BO	GO	Al	B	Ca	Zn	Si	Fe	P	Mn	Mg	Total minerals
		Cyprus										
1	MF	Larnaca	1.69	3.58	23.66	0.88	5.54	1.49	46.25	0.07	9.24	92.40
2	MF	Larnaca	1.63	5.14	93.11	1.10	4.94	0.96	65.77	0.91	18.62	192.18
3	MF	Larnaca	2.19	5.20	91.64	6.94	7.26	3.46	66.10	0.83	23.40	207.02
4	MF	Larnaca	0.49	4.11	54.10	1.61	6.35	1.74	53.95	0.10	20.27	142.72
5	MF	Limassol	1.46	2.12	30.21	1.09	4.94	1.45	40.06	0.08	14.12	95.53
6	MF	Limassol	1.02	3.22	48.34	1.62	5.61	1.52	56.09	0.15	13.05	130.62
7	MF	Limassol	0.53	2.56	49.76	1.93	5.34	1.43	54.26	0.19	11.31	127.31
8	MF	Limassol	1.46	3.35	35.66	0.99	4.69	1.35	51.56	0.27	10.73	110.06
9	MF	Limassol	1.71	4.42	60.88	0.91	6.83	1.77	75.69	0.23	15.59	168.03
10	MF	Limassol	1.60	4.48	58.63	0.93	7.05	1.65	54.16	0.48	12.16	141.14
11	MF	Nicosia	1.48	3.67	47.03	1.42	5.89	1.91	60.69	0.28	19.57	141.94
12	MF	Nicosia	1.77	4.14	69.21	3.77	6.68	3.52	53.62	0.42	13.49	156.62
13	MF	Nicosia	2.13	3.28	143.47	0.86	5.18	1.43	40.97	2.13	19.82	219.27
14	MF	Nicosia	24.43	2.17	27.89	1.55	7.28	3.97	257.19	4.11	80.48	409.07
		Mean	3.11^{a}	3.67^{c}	59.54^{e}	1.83^{h}	5.97^{j}	1.98^{m}	69.74^{o}	0.73^{r}	20.13^{u}	166.70^{x}
		±SD	6.16	0.98	32.13	1.65	0.92	0.94	54.81	1.11	17.88	79.60
		RSD (%)	1.98	0.27	0.54	0.90	0.15	0.47	0.79	1.52	0.89	
		Greece										
15	P	Rhodes	3.21	2.41	25.40	0.97	73.80	3.46	89.29	0.59	53.45	252.58
16	F	Messinia	27.35	2.84	15.22	1.24	8.18	4.16	217.33	4.37	109.53	390.22
17	P	Rhodes	2.93	2.67	23.63	9.30	73.63	3.22	92.67	0.57	53.96	262.58
18	WT	Lakonia	27.37	3.39	24.59	1.01	3.32	2.30	264.31	4.60	88.51	419.40
19	T	Rhodes	1.46	13.29	65.93	1.01	27.64	2.00	55.85	0.26	31.40	198.84
20	MF	Ioannina	9.15	4.72	59.88	3.52	9.58	2.81	138.80	5.54	82.72	316.72
21	MF	Metsovo	1.55	4.75	49.37	1.00	7.21	0.75	62.44	1.13	25.83	154.03
22	P	Thassos	0.40	4.49	37.18	1.48	7.09	0.89	105.75	2.19	26.28	185.75
23	MF	Ioannina	8.30	4.45	56.71	3.45	8.84	2.63	131.12	5.32	80.70	301.52
24	CE	Arta	7.75	1.69	36.07	2.53	96.42	4.11	170.02	2.01	60.20	380.80
25	F	Karditsa	3.28	1.33	30.32	1.20	11.17	1.18	57.25	0.54	20.13	126.40
26	CE	Arta	0.82	4.44	44.11	1.37	11.17	1.19	57.25	0.54	80.13	201.02
		Mean	7.80^{b}	4.21^{c}	39.03^{f}	2.34^{h}	28.17^{k}	2.39^{m}	120.17^{p}	2.31^{s}	59.40^{v}	265.82^{y}
		±SD	9.61	3.10	16.21	2.38	33.04	1.21	67.74	2.07	29.24	96.94
		RSD (%)	1.23	0.74	0.42	1.02	1.17	0.51	0.56	0.90	0.49	
		Egypt										
27	CL	GCA	1.21	0.65	82.23	1.09	9.09	8.45	26.23	0.28	17.49	146.72
28	CL	GCA	1.36	0.52	54.46	0.76	6.58	4.89	15.88	0.22	12.17	96.84
29	CL	GCA	2.07	0.78	77.07	0.55	0.80	2.18	26.33	0.17	18.39	128.34
30	CL	GCA	4.05	0.87	92.63	1.17	11.29	4.50	28.47	0.54	22.04	165.56
31	CL	GCA	0.95	1.40	112.10	0.78	11.54	6.58	39.47	0.55	21.42	194.79
32	CL	GCA	3.35	3.17	59.54	1.23	7.40	4.49	35.29	0.28	17.63	132.38
33	CL	GCA	13.06	2.69	44.79	1.68	5.36	4.82	26.55	0.20	15.24	114.39
34	CL	GCA	1.53	2.18	50.61	1.07	0.81	2.66	30.65	0.19	13.88	103.58
		Mean	3.45^{a}	1.53^{d}	71.68^{g}	1.04^{i}	6.61^{l}	4.82^{n}	28.61^{q}	0.30^{t}	17.28^{w}	135.33^{z}
		±SD	4.03	1.02	23.37	0.35	4.17	2.00	7.01	0.15	3.44	45.56
		RSD (%)	1.17	0.67	0.33	0.34	0.63	0.41	0.25	0.50	0.20	
		LOD	0.41	0.03	0.50	0.02	0.02	0.02	0.10	0.08	0.02	
		LOQ	1.34	0.10	0.17	0.07	0.07	0.07	0.33	0.24	0.07	

The reported results are the mean of three independent replicates ($n = 3$). Different lowercase letter in each column of the average value represents statistically significant differences for each mineral according to the production area ($p < 0.05$). *N*: the number of honey samples; BO: botanical origin; GO: geographical origin; GCA: greater Cairo area; MF: multifloral; P: pine; F: fir; CE: citrus and erica; WT: wild thyme; T: thyme; CL: clover; FA: free acidity; TDS: total dissolved solids; EC: electrical conductivity; SD: standard deviation; RSD: relative standard deviation; LOD: limit of detection; LOQ: limit of quantification.

Electrical conductivity (EC) is a conventional physicochemical parameter related to the botanical origin of honey, and it is used very often in routine honey quality control. According to the EU [8], the maximum limit value of EC for thyme honeys is 0.80 mS/cm, while honeydew honeys often have EC ≥0.80 mS/cm. The Greek decision for monofloral honeys [16] sets the EC values for pine, fir, thyme, and citrus honeys to be ≥0.90, ≥1.0, ≤0.60, and ≤0.45, respectively. Based on the aforementioned directives, pine honey-labeled samples from Rhodes could not be classified as monofloral pine honeys. Mixed citrus honeys from Arta (citrus + erica) recorded EC values slightly greater than 0.60 mS/cm, indicating the correct information regarding the mixed honey botanical origin provided by the beekeepers. Cypriot honeys recorded typical EC values for blossom honeys according to the European directive and Greek decision related to honey origin [8, 16] (Table 1). The variations of EC values among honey samples analyzed are in agreement with previous studies involving Greek, Spanish, Malaysian, Moroccan, and Sardinian honeys of different botanical and geographical origins [6, 11, 17–21].

Total dissolved solids (TDS) is a measure of the combined content of all inorganic and organic substances present in honey in molecular, ionized, or microgranular (colloidal solution) suspended forms. The TDS content of Greek honeys was higher than that of Egyptian and Cypriot honeys. An excellent Pearson's correlation (r) between EC and TDS was obtained, according to the botanical origin of commercial honeys (r = 0.995), indicating that the salts dissolved in honey are the carriers of EC. This is in agreement with the previous work in the literature involving Algerian honeys [22]. To the best of our knowledge, data on TDS content have never been reported before for Cypriot, Greek, or Egyptian honeys.

Salinity is the saltiness or dissolved salt content in an aqueous medium. In general, honeys from Greece recorded higher salinity values as compared to those from Cyprus or Egypt. Additionally, a very strong Pearson's correlation (r) was recorded between EC and salinity (r = 0.927), according to the botanical origin of honeys. Therefore, salinity and TDS content are proposed as supplementary physicochemical markers for the botanical origin differentiation of honey.

Honey colour may vary with respect to several parameters like Maillard reaction products, fructose caramelization, botanical origin, pollen content, and polyphenol or mineral content [18]. Colour parameters (L^* and b^*) varied significantly according to the different botanical origins grouped in the tested geographical zones. In particular, Cypriot and Egyptian honeys were the brightest (higher L^* values) compared to Greek honeys, whereas Greek honeys possessed numerous green and yellow pigments (higher a^* and b^* values, resp.) (Table 2). This is in agreement with the previous work in the literature, regarding Turkish [5], Greek [6], and Spanish [17] honeys. It should also be mentioned that, in a recent study, citrus honeys from Egypt proved to have similar brightness [11] with Egyptian clover honeys of the present study. However, respective comparison of a^* and b^* values indicates substantial differences, originating from the presence of different pigments.

3.3. Mineral Content of Cypriot, Greek, and Egyptian Honeys. The mineral content (mg/kg) of Cypriot, Greek, and Egyptian honeys is given in Table 2. Total mineral content (the sum of individual minerals) followed the order Greece > Cyprus > Egypt. The most dominant minerals were Ca, P, Si, and Mg followed by Zn, Al, B, Fe, and Mn. The higher, in general, mineral content (P, Mg, Si, Fe, and Mn) in honey samples from Greece, as compared to those from Cyprus, is attributed to the presence of honeydew, pine, and fir honey samples (dark-coloured honeys) being collected from these regions. It should also be stressed that the content of Mg, Zn, Al, and Mn of honey samples of the present study was higher compared to that in the previous work involving Egyptian, Spanish, Moroccan, and Greek citrus honeys (light-blossom honeys) [11]. It has been documented previously that dark-coloured honeys possess higher mineral content, as compared to light-blossom honeys [9, 11, 20]. An exception to this was the Ca content of multifloral honeys from Cyprus and clover honeys from Egypt, which was higher compared to that of Greek honeys. This is probably owed to the specific soil conditions in these regions including the vegetation grown. In the recent work [11], the Ca content of Egyptian citrus honeys was lower compared to present results, indicating the strong impact of botanical origin on honey mineral content.

Furthermore, the higher P content (mg/kg) was recorded in wild thyme honey sample no. 4 from Lakonia and mixed citrus honey (citrus + erica) no. 10 from Arta. This indicates the effectiveness of both botanical origin (i.e., the presence of honeydew elements in the wild thyme honey sample) and geographical origin (i.e., certain soil conditions in Lakonia and Arta) to the overall mineral content of honey. This is in very good agreement with the previous work in the literature [20]. Further comparison of present results, regarding P content, with citrus honeys produced in Egypt, Spain, Morocco, and Greece will totally reveal significant differences [11], as in the honey samples analyzed P recorded much higher values (mg/kg) (Table 2).

In addition, small amounts of As (<0.08 mg/kg), Ba (<0.06 mg/kg), Be (<0.06 mg/kg), Cd (<0.05 mg/kg), Co (<0.03 mg/kg), Cr (<0.12 mg/kg), Cu (<0.35 mg/kg), Hg (<0.03 mg/kg), Mo (<0.08 mg/kg), Ni (<0.21 mg/kg), Pb (<0.08 mg/kg), Sb (<0.14 mg/kg), Ti (<0.09 mg/kg), Tl (<0.20 mg/kg), and V (<0.11 mg/kg) were determined in all honey samples analyzed. The level of these minerals is in very good agreement with the previous work in the literature involving Spanish [9, 24], Greek [10, 11], Egyptian [11], Malaysian [18], and Swiss [25] honeys. These small amounts (<1 mg/kg) of toxic heavy metals like Cd and Pb give an additional quality standard to the honey samples analyzed.

Finally, a very strong Pearson's correlation (r = 0.983) was obtained between average values of EC and total mineral content with respect to the specific honey types collected from Cyprus, Greece, and Egypt. This observation shows that the electrical conductivity of honey is owed (among others) to charged compounds such as minerals and highlights at a greater degree the reliability of the two independent methodologies followed in the present study: conventional physicochemical parameter analysis (for the EC determination) and ICP-OES analysis (for the determination of minerals).

3.4. Classification of Cypriot, Greek, and Egyptian Honeys according to Botanical Origin Based on Physicochemical Parameter Values. Pillai's trace = 1.600 (F = 10.659; p value = 0.000) and Wilks' lambda = 0.037 (F = 10.783; p value = 0.000) indices pointed out the existence of a significant effect of the botanical origin (independent variables) on physicochemical parameter values (dependent variables) of the 34 honey samples. Eight of the 9 physicochemical

TABLE 3: Physicochemical parameters and minerals used for the botanical discrimination of Cypriot, Greek, and Egyptian honeys.

Discrimination parameters	Discriminant functions 1	Discriminant functions 2	F	p	Strong physicochemical markers of the provenience of the studied honey types
pH	0.585	0.297	34.979	<0.001	
EC	0.426**	0.284	19.732	<0.001	♦
Moisture	−0.006	0.650	14.725	<0.001	
FA	0.458	−0.226**	21.340	<0.001	♦
TDS	0.535**	0.259	29.097	<0.001	♦
Salinity	0.347**	−0.006	11.272	<0.001	♦
L^*	−0.268**	−0.103	7.059	0.003	♦
b^*	0.318**	0.121	9.942	<0.001	♦
P	0.112**	0.014	7.212	0.003	♦
Fe	−0.092	−0.322	12.282	<0.001	
Ca	−0.086	0.012	4.289	0.023	
Mg	0.147	−0.163	14.381	<0.001	
Mn	0.098**	−0.065	5.913	0.007	♦
Si	0.082	−0.113**	4.788	0.015	♦
B	0.078	0.121**	4.529	0.019	♦

Pooled within-group correlations between discriminating variables and standardized canonical discriminant functions are given. Variables are ordered by the absolute size of correlation within function; **largest absolute correlation between each variable and any discriminant function (♦); FA: free acidity; TDS: total dissolved solids; EC: electrical conductivity.

parameters determined were found to be significant ($p < 0.05$) for the differentiation of honeys (Table 3). Thus, these 8 parameters were subjected to LDA. The results showed that two statistically significant discriminant functions were formed: Wilks' lambda = 0.044, $X^2 = 86.010$, df = 16, and p value = 0.000 for the first function and Wilks' lambda = 0.308, $X^2 = 32.403$, df = 7, and p value = 0.000 for the second. Testing of the uniformity of variability (Box's M index) was not significant at the 95% confidence level (p value = 0.052), showing the existence of uniformity of sample variability for honeys collected from the three regions. The first discriminant function accounted for 72.8% of the total variance, while the second accounted for 26.2%. Both accounted for 99.0% of the total variance, considered an excellent rate.

In Figure 1, it is shown that honeys from Cyprus, Greece, and Egypt are well separated. The first discriminant function differentiates honey samples from Greece, while the second differentiates those from Egypt. The physicochemical parameters that contributed to the first and the second discriminant functions are given in Table 3. The overall correct classification rate was 97.1% for the original method and 91.2% for the cross-validation method, the latter being considered very satisfactory for this method. In particular, the correct classification rate was 100%, 92.3%, and 100% for honeys from Cyprus, Greece, and Egypt using the original method and 100%, 76.9%, and 100% for honeys from Cyprus, Greece, and Egypt, respectively, by using the cross-validation method.

Present results are in agreement with previous works in the literature, in which the use of certain physicochemical parameters (i.e., free acidity, pH, electrical conductivity, ash, hydroxymethylfurfural, diastase activity, and colour attributes) proved to be an effective tool for the geographical and botanical discrimination of several honey types produced in different parts of the world [6, 11, 17, 18, 20].

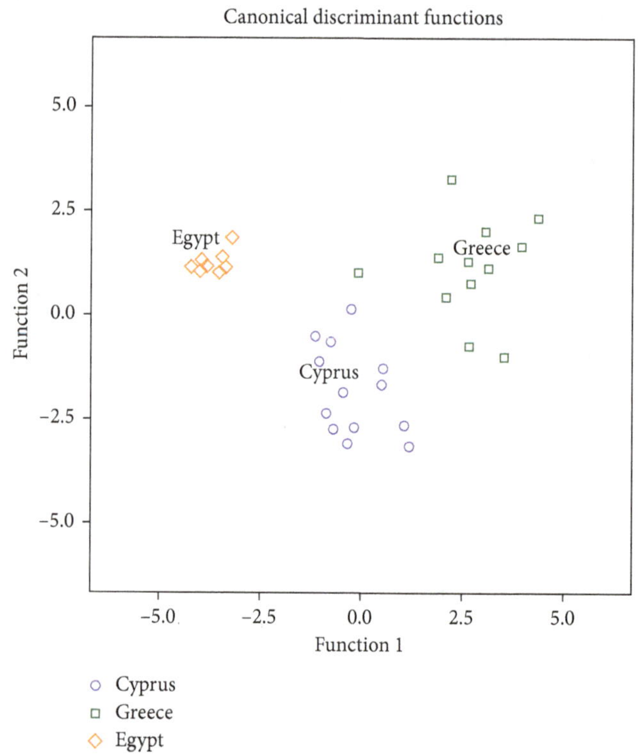

FIGURE 1: Differentiation of Greek, Cypriot, and Egyptian honeys according to botanical origin based on 8 physicochemical parameters.

3.5. Classification of Cypriot, Greek, and Egyptian Honeys according to Botanical Origin Based on Mineral Content Analysis. The next step was to investigate whether minerals could provide a higher classification rate. The 34 honey samples were subjected to MANOVA to determine which minerals were significant for the differentiation of honeys

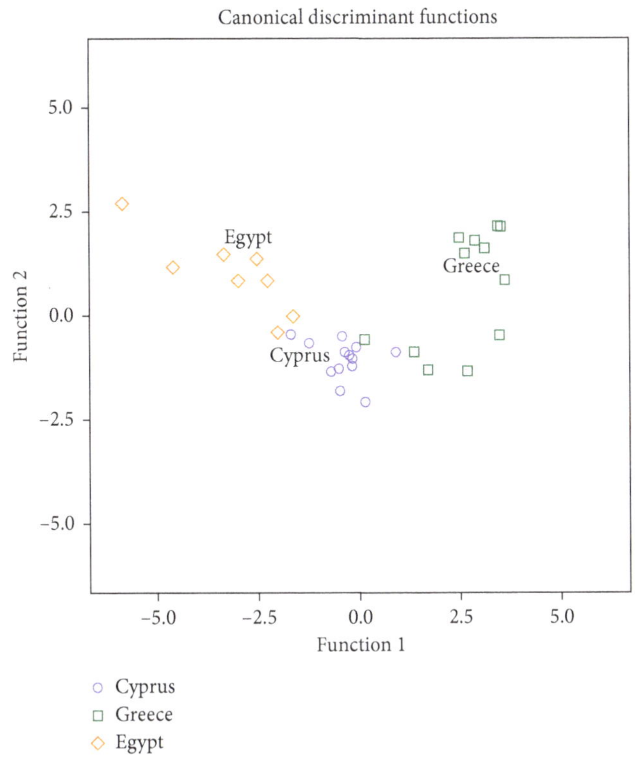

FIGURE 2: Differentiation of Greek, Cypriot, and Egyptian honeys according to botanical origin based on 7 minerals.

from the three different regions. Dependent variables included the 9 minerals, while botanical origin was taken as the independent variable. Pillai's trace = 1.401 (F = 6.235; p value = 0.001) and Wilks' lambda = 0.065 (F = 7.470; p value = 0.001) index values showed the existence of a significant multivariable effect of botanical origin on honey's mineral content. Seven of the 9 minerals (Table 3) were found to be significant ($p < 0.05$) for the differentiation of honeys. Thus, these 7 minerals were subjected to LDA.

Results showed that two statistically significant discriminant functions were formed: Wilks' lambda = 0.083, $X^2 = 69.586$, df = 14, and p value = 0.001 for the first function and Wilks' lambda = 0.518, $X^2 = 18.437$, df = 6, and p value = 0.002 for the second. As in the case of physicochemical parameters, testing of the uniformity of variability (Box's M index) was insignificant at the 95% confidence level showing the existence of uniformity of sample variability for each botanical origin. The first discriminant function accounted for 86.8% of the total variance, while the second accounted for 13.9%. Both accounted for 98.7% of the total variance, which is also considered an excellent rate.

In Figure 2, it is shown that honeys from all the tested regions are clearly separated based on the first and second discriminant functions. The first discriminant function differentiates honeys from Greece and Egypt, whereas the second discriminant function differentiates honeys from Cyprus. The overall correct classification rate was 84.8% using the original method and 79.4% using the cross-validation method, a satisfactory value especially for the second method. More specifically, the original method classified correctly honeys from Cyprus, Greece, and Egypt by 100%, 77%, and 75%, respectively. In the case of the cross-validation method, the respective classification rates were 92.3%, 69.2%, and 75%. The use of certain minerals has the potential to provide accurate classification rates for several honey types produced in different parts of the world for both geographical and botanical origin determination of honey [10, 11, 23–26].

As it can be observed, minerals provided a much lower classification rate as compared to the physicochemical parameters. This finding is in accordance with a recent work in the literature [10, 11] but in contrast to previous works [24, 26], in which the classification rate of Spanish thyme, eucalyptus, orange blossom, rosemary, and heather honeys according to the production area and botanical origin, based on certain minerals and using discriminant analysis, was approx. 100%. Differences in the overall classification rate reported in previous works compared to results of the present study may be attributed to several factors such as botanical origin of honeys, similarities in soil and vegetation conditions in honey production areas, harvesting year, honey processing and extraction methods, the analytical methodology applied, beekeepers' practices/handling, chemometric techniques applied, or any other unpredicted external factor. For example, when Fernández-Torres et al. [26] used certain minerals and principal component analysis to classify Spanish eucalyptus, orange blossom, rosemary, and heather honeys according to the floral type, the classification rate based on the two principal components was 71.59%, best differentiating heather and eucalyptus honeys.

Based on the aforementioned results, the next step was to investigate whether the combination of the significant physicochemical parameters and minerals ($p < 0.05$) could provide a higher classification rate as compared to minerals alone or to improve the classification rate of a specific honey type.

3.6. Classification of Cypriot, Greek, and Egyptian Honeys according to Botanical Origin Based on the Combination of Physicochemical Parameter and Mineral Content Values. Pillai's trace = 1.782 (F = 6.799; p value = 0.000) and Wilks' lambda = 0.007 (F = 8.798; p value = 0.000) index values showed that there was a significant multivariable effect of botanical origin on mineral/physicochemical parameter values. Seven minerals (B, Ca, Si, Fe, P, Mn, and Mg) and 8 physicochemical parameter values (pH, electrical conductivity, moisture, free acidity, total dissolved solids, salinity, L^*, and b^*) served as the significant variables ($p < 0.05$) for the differentiation of honeys. Thus, the combined values of 7 minerals and 8 physicochemical parameters were subjected to LDA.

The results showed that two statistically significant discriminant functions were formed: Wilks' lambda = 0.009, $X^2 = 110.727$, df = 32, and p value = 0.000 for the first function and Wilks' lambda = 0.244, $X^2 = 33.106$, df = 15, and p value = 0.005 for the second.

The first discriminant function accounted for 89.4% of the total variance, while the second accounted for 9.6%. Both accounted for 99% of the total variance. In Figure 3, it is

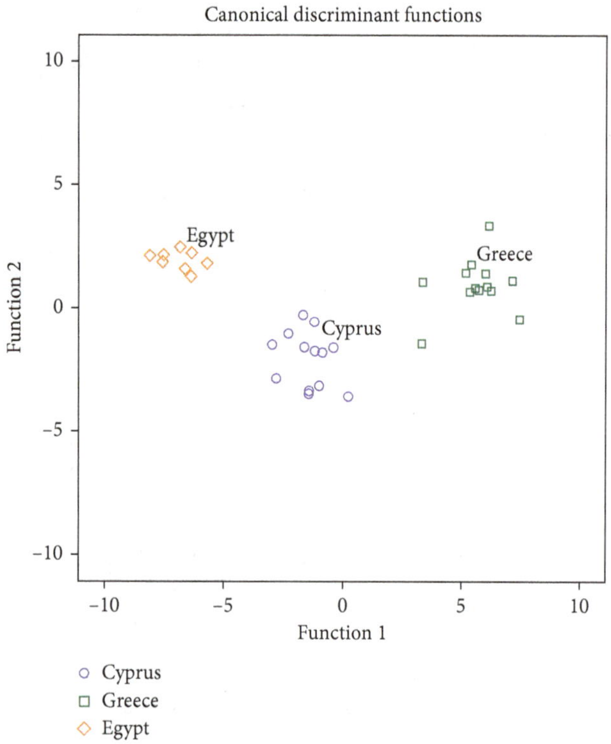

FIGURE 3: Differentiation of Greek, Cypriot, and Egyptian honeys according to botanical origin based on the combination of 8 physicochemical parameters and 7 minerals.

shown that honeys from the investigated regions are clearly separated. The first discriminant function separates honeys from Greece as compared to those from Egypt, while the second discriminant function separates Cypriot honeys. The overall correct classification rate was 100% for the original method for all honey samples and 79.4% for the cross-validation method, a satisfactory value in general for the cross-validation method. The physicochemical parameters and minerals that contributed to the first and the second discriminant functions along with the discriminant function coefficients are given in Table 3. The physicochemical parameters and minerals that were strong markers of the provenience of the studied honeys are those with a largest absolute correlation between each variable and any discriminant function and are indicated with a symbol (Table 3).

As it can be observed, the combination of different physicochemical parameters and minerals did not increase the overall correct classification rate based on the cross-validation method. Therefore, it was not an effective approach in the present study for honey botanical origin differentiation in contrast to the previous work [5]. However, it should be highlighted that the distribution of honey samples in Figure 3 is greatly improved compared to that of minerals (Figure 2). At this point, to avoid any kind of misleading, it should be stressed that the correct classification rate obtained by the original method reflects the distribution of honey samples (Figures 1–3). The correct classification rate was 100% for all the investigated honeys using the original method and 84.6%, 61.6%, and 100% for honeys from Cyprus, Greece, and Egypt, respectively, by using the cross-validation method. The combination of minerals and physicochemical parameters improved only the classification rate of Egyptian honeys by 25% compared to that provided by minerals alone.

4. Conclusion

Botanical origin differentiation of honey produced in a specific country is of great importance for the characterization of "honey uniqueness" which may reflect its price in the global market. Physicochemical parameter analysis, including colour, may accurately provide information about honey origin in cases where melissopalynological analysis cannot be applied. On the contrary, minerals may assist in the nutritional characterization of honey and give additional information about its quality, floral type, and place of origin. The application of physicochemical parameter analysis provided a significantly higher discrimination rate (91.2%) of honeys according to botanical origin, compared to that of minerals (79.4%). However, an overall correct classification rate of ca. 80% using the cross-validation method may be considered satisfactory taking into account the (i) mixed floral honey types investigated, (ii) different harvesting years of honey collection, and (iii) different production countries.

Conflicts of Interest

The authors declare that there are no conflicts of interest regarding the publication of this article.

Acknowledgments

The authors are grateful to local beekeepers from Greece and Cyprus for the donation of honey samples and to Dr. Sofia Karabournioti for her assistance in the melissopalynological analysis of Egyptian honey samples. Professor Michael G. Kontominas is also acknowledged for the access he provided to the Laboratory of Food Chemistry, Department of Chemistry, University of Ioannina, Ioannina 45110, Greece.

References

[1] Regulation (EU) No. 1151/2012 of the European Parliament and of the Council of 21 November 2012 on Quality Schemes for Agricultural Products and Foodstuffs.

[2] A. Thrasyvoulou and J. Manikis, "Some physicochemical and microscopic characteristics of Greek unifloral honeys," *Apidologie*, vol. 26, no. 6, pp. 441–452, 1995.

[3] C. de Alda-Garcilope, A. Gallego-Picó, J. C. Bravo-Yagüe, R. M. Garcinuño-Martínez, and P. Fernández-Hernando, "Characterization of Spanish honeys with protected designation of origin "Miel de Granada" according to their mineral content," *Food Chemistry*, vol. 135, no. 3, pp. 1785–1788, 2012.

[4] *Pefkothymaromelo Kritis, PDO-EL-02142*, May 2016.

[5] H. Z. Senyuva, J. Gilbert, S. Silici et al., "Profiling Turkish honeys to determine authenticity using physical and chemical characteristics," *Journal of Agricultural and Food Chemistry*, vol. 57, no. 9, pp. 3911–3919, 2009.

[6] I. K. Karabagias, A. Badeka, S. Kontakos, S. Karabournioti, and M. G. Kontominas, "Characterisation and classification of Greek pine honeys according to their geographical origin based on volatiles, physicochemical parameters and chemometrics," *Food Chemistry*, vol. 146, pp. 548–557, 2014.

[7] W. Von der Ohe, L. Persano Oddo, M. L. Piana, M. Morlot, and P. Martin, "Harmonized methods of melissopalynology," *Apidologie*, vol. 35, no. S1, pp. S18–S25, 2004.

[8] Council directive 2001/110/EC relating to honey, *Official Journal of the European Communities*, vol. 10, pp. 47–52, 2002.

[9] Z. Can, O. Yildiz, H. Sahin, E. Akyuz Turumtay, S. Silici, and S. Kolayli, "An investigation of Turkish honeys: their physicochemical properties, antioxidant capacities and phenolic profiles," *Food Chemistry*, vol. 180, pp. 133–141, 2015.

[10] I. K. Karabagias, P. A. Louppis, S. Kontakos, C. Papastephanou, and M. G. Kontominas, "Characterization and geographical discrimination of Greek pine and thyme honeys based on their mineral content, using chemometrics," *European Food Research and Technology*, vol. 243, no. 1, pp. 101–113, 2017.

[11] I. K. Karabagias, P. A. Louppis, S. Karabournioti, S. Kontakos, C. Papastephanou, and M. G. Kontominas, "Characterization and geographical discrimination of commercial *Citrus* spp. honeys produced in different Mediterranean countries based on minerals, volatile compounds and physicochemical parameters, using chemometrics," *Food Chemistry*, vol. 217, pp. 445–455, 2017.

[12] A. Thrasyvoulou and S. Bladenopoulou, "Comparative analysis of pine and blossom honeys," in *Proceedings of the 2nd National Beekeeping Congress*, pp. 72–79, Athens, Greece, November 1983.

[13] A. Thrasyvoulou, C. Tananaki, M. Dimou et al., "Quality, identity and safety of Cypriot honey," in *Proceedings of 25th Congress of Hellenic Society of Horticulture Sciences*, PLE-16, pp. 106 -111, Lemesos, Cyprus, November 2011.

[14] S. Karabournioti and I. K. Karabagias, "Pollen characteristics of Egyptian clover honey," *Journal of Botanical Sciences-Research and Reviews*, vol. 6, no. 3, pp. 32–34, 2017.

[15] IHC, "Harmonized methods of the International Honey Commission," in *IHC Responsible for The Methods*, S. Bogdanov, Ed., Swiss Bee Research Centre, Liebefeld, Switzerland, 1997.

[16] Greek Decision 127/2004, "Identification of Greek monofloral pine, fir, chestnut, erica, thyme, citrus, cotton and helianthus honeys," in *Codex Alimentarius*, FEK 239/B/23.2, Greek Ministry of Agricultural and Food Development, Athens, Greece, 2005.

[17] O. Escuredo, M. Fernández-González, and S. M. Carmen, "Differentiation of blossom honey and honeydew honey from Northwest Spain," *Agriculture*, vol. 2, no. 1, pp. 25–37, 2012.

[18] L. S. Chua, L. N. Abdul-Rahaman, M. R. Sarmidi, and R. A. Aziiz, "Multi-elemental composition and physical properties of honey samples from Malaysia," *Food Chemistry*, vol. 135, no. 3, pp. 880–887, 2012.

[19] N. Malika, F. Mohamed, and E. A. Chakib, "Microbiological and physicochemical properties of Moroccan honey," *International Journal of Agriculture and Biology*, vol. 5, pp. 773–776, 2005.

[20] G. L. Petretto, M. Cossu, and M. C. Alamanni, "Phenolic content, antioxidant and physicochemical properties of Sardinian monofloral honeys," *International Journal of Food Science and Technology*, vol. 50, no. 2, pp. 482–491, 2015.

[21] S. A. El Sohaimy, S. H. D. Masry, and M. G. Shehata, "Physicochemical characteristics of honey from different origins," *Annals of Agricultural Sciences*, vol. 60, no. 2, pp. 279–287, 2015.

[22] M. I. Khalil, M. Moniruzzaman, L. Boukraâ et al., "Physicochemical and antioxidant properties of Algerian honey," *Molecules*, vol. 17, no. 9, pp. 11199–11215, 2012.

[23] M. L. González-Miret, A. Terrab, D. Hernanz, M. A. Fernández-Recamales, and F. J. Heredia, "Multivariate correlation between color and mineral composition of honeys and by their botanical origin," *Journal of Agricultural and Food Chemistry*, vol. 53, no. 7, pp. 2574–2580, 2005.

[24] A. Terrab, D. Hernanz, and F. J. Heredia, "Inductively coupled plasma optical emission spectrometric determination of minerals in thyme honeys and their contribution to geographical discrimination," *Journal of Agricultural and Food Chemistry*, vol. 52, no. 11, pp. 3441–3445, 2004.

[25] S. Bogdanov, M. Haldimann, W. Luginbühl, and P. Gallmann, "Minerals in honey: environmental, geographical and botanical aspects," *Journal of Apicultural Research and Bee World*, vol. 46, no. 4, pp. 269–275, 2007.

[26] R. Fernández-Torres, J. L. Pérez-Bernal, M. A. Bello-López, M. Callejón-Mochón, J. C. Jiménez-Sánchez, and A. Guiraúm-Pérez, "Mineral content and botanical origin of Spanish honeys," *Talanta*, vol. 65, no. 3, pp. 686–691, 2005.

A Facile Synthesis of Highly Nitrogen-Doped Carbon Dots for Imaging and Detection in Biological Samples

Qianchun Zhang, Siqi Xie, Yanqun Yang, Yun Wu, Xingyi Wang, Jincheng Wu, Li Zhang, Junyu Chen, and Yuan Wang

School of Biology and Chemistry, Xingyi Normal University for Nationalities, Xingyi 562400, China

Correspondence should be addressed to Qianchun Zhang; qianchunzhang@qq.com and Yuan Wang; 45001490@qq.com

Academic Editor: Subhankar Singha

A facile, green, and high-output hydrothermal synthesis was proposed for the fabrication of highly fluorescent nitrogen-doped carbon quantum dots (N-doped CDs). The nitrogen content in N-doped CDs reached 19.2% and demonstrated strong blue fluorescence emission was obtained with fluorescence quantum yield (QY) of up to 32.9%, which exhibit high fluorescence quantum yield, high photostability, and excellent biocompatibility. The N-doped CDs possess high photostability, low toxicity, and excellent biocompatibility, based on which the N-doped CDs were successfully applied as a fluorescence probe for cell imaging. Moreover, it was then successfully demonstrated for sensitive and selective detection of Fe^{3+} in serum.

1. Introduction

During the past few years, fluorescent CDs have attracted tremendous attention due to their unique optical and electronic properties [1–6]. Compared to conventional semiconductor quantum dots and organic fluorescent dyes, CDs possess several superior features including functionalization, low toxicity, excellent water dispersibility, tunable fluorescence emission, excellent photostability, upconversion, and biocompatibility, thus demonstrating potential application in the fields of bioimaging, in vivo theranostics, drug delivery, light-emitting diodes, photocatalysis solar cells, and heavy metal ion detection [7–11]. In addition, doping CDs with other nonmetallic components, such as N, S, and P, can inject electrons into carbon-based materials and change the electronic transport properties and PL properties [12, 13]. However, in most cases, the QY of the as-synthesized CD was less than 10%, and the QY is a key parameter to evaluate the quality of CDs, which limit the sensitivity and selectivity. So, synthesis of high-fluorescence carbon quantum dots is the direction of development.

The use of N-containing precursors has proved to be an effective route for obtaining N-doped CDs. Chen et al. [14] used 2-azidoimidazole as precursor in a hydrothermal process at 70°C overnight to obtain nitrogen-rich CDs. Lv et al. [15] using ethanediamine and citric acid as precursors obtained N-doped CDs and achieved good results in iron detection. Wang and Zhou. [16] used milk to prepare N-CDs hydrothermally at 180°C for 2 h. In another study, Hsu and Chang [17] found that compounds containing both amino and carboxyl groups are beneficial for synthesizing CDs with high PL quantum yield. Based on the benefits of N-doping in carbon nanostructures, it can be extrapolated that the introduction of N to carbon dots would further enhance their versatile properties. However, most N-doped CDs are unsatisfactory due to harsh synthetic conditions and long reaction times. Thus, a time-saving and eco-friendly synthesis of N-doped CDs is of interest.

Herein, a facile, green, and high-output thermal strategy is proposed for the fabrication of highly fluorescent N-doped CDs. We used L-citrulline as the precursor for a facile and eco-friendly one-step hydrothermal method without the assistance of any chemicals (except pure water) to obtain highly fluorescent N-doped CDs. The as-prepared N-doped CDs exhibit good water solubility, good biocompatibility, and high fluorescence quantum yield (32.9%). Owing to the unique properties of the N-doped CD nanoprobe with good membrane permeability and excellent biocompatibility, it was used for imaging of HeLa cells with high discrimination. Moreover, it was further applicated for detection of Fe^{3+} ions in serum, and the fluorescence intensity exhibited a good linear relationship in the Fe^{3+} concentration range from 0 to 50 μM with a detection limit of about 37 nM.

2. Experimental

2.1. Materials. L-citrulline (98%) and quinine sulphate (98%) were purchased from J&K Scientific Inc. (Beijing, China). 3-(4,5-Dimethylthiazol-2-yl)-2, 5-diphenyltetrazolium bromide (MTT, 98%) was obtained from Sangon Biotechnology Inc. (Shanghai, China). Dimethyl sulphoxide (DMSO) was obtained from Xilong Reagents Company (Guangdong, China). Penicillin-streptomycin, Dulbecco's modified Eagle's medium (DMEM), and foetal bovine serum (FBS) were purchased from Solarbio (Beijing, China). NaCl, KCl, $MgCl_2$, $AlCl_3$, $CaCl_2$, $Cr(NO_3)_3$, $FeCl_2$, $FeCl_3$, $Co(NO_3)_2$, $CuSO_4$, $ZnCl_2$, $Cd(NO_3)_2$, $SrCl_2$, and $Hg(NO_3)_2$ were purchased from Aladdin (Shanghai, China). HeLa cell lines were obtained from Cellcook. Human serum samples were provided by Xing Ying People's Hospital Blood Center (Xingyi, China). Ultrapure water (18.2 MΩ, Millipore Co., USA) was used in all experiments. Other chemical reagents (analytical grade) were purchased from Beijing Chemical Company (Beijing, China).

2.2. Instrumentation and Characterization. The morphologies and sizes of N-doped CDs were characterized by high-resolution transmission electron microscopy (HRTEM, Hitachi-F20) at an accelerating voltage of 200 kV and atomic force microscopy (AMF, Bruker Multimode 8) in the tapping mode. Fourier transform infrared (FT-IR) spectra were obtained on a Nicolet 6700 FT-IR spectrometer using KBr pellets. X-ray photoelectron spectroscopy (XPS) was performed using an ESCALAB 250Xi (Thermo Scientific). X-ray diffraction (XRD) was carried out using a Rigaku diffractometer in the 2θ range 10–80° with step width of 0.02°. UV-Vis absorption spectra were recorded on a DU 800 UV-Vis spectrophotometer. The PL decay curves were obtained on a Leica SP5 FLIM system using a 405 nm laser excitation source. Fluorescence spectroscopy and stability were measured on a PerkinElmer LS 55 with 5/5 nm slit width and equipped with a 1 cm quartz cell. A TGL-20LM-B high-speed refrigerated centrifuge (Hunan Xingke Instrument Co., Ltd., China) was used to purify the N-doped CDs. Cell imaging was carried out using a Leica SP8 confocal laser scanning microscope (Leica, Germany).

2.3. Synthesis of N-Doped CDs. N-doped CDs were synthesized by a facile hydrothermal method. Briefly, 0.50 g·L-citrulline was dissolved in 25 mL ultrapure water and subjected to ultrasonic oscillation for 20 min. The solution was transferred to a Teflon-equipped stainless steel autoclave and reacted at 220°C for 12 h. After the reaction liquid was cooled to room temperature, the reaction liquid was centrifuged at 17,000 rpm for 40 min to separate aggregated particles. The supernatant fluid was removed by filtration with a 0.22 μm filter membrane. The as-prepared N-doped CD solution was stored at 4°C for further use.

2.4. MTT Assay and Intracellular Fluorescence Imaging. Cytotoxicity of the N-doped CDs was investigated with the cancer cell line HeLa by an MTT assay. HeLa cells were seeded in a 96-well plate at a density of 4×10^3 cells per well for 24 h in an incubator (37°C, 5% CO_2). The culture medium was replaced with 100 μL fresh DMEM containing the N-doped CDs at concentrations of 0, 25, 50,100, 250, and 500 μg/mL and incubated in an incubator (37°C, 5% CO_2). After 24 or 48 h, 20 μL MTT (5 mg/mL) solution was added to each cell well, which was incubated for 4 h. Subsequently, the culture medium with MTT was removed and 100 μL DMSO was added, followed by shaking for 10 min. The optical density (OD) of each well at 490 nm was measured on the enzyme-linked immunosorbent detector.

HeLa cells (1×10^5 cells/dish) were seeded in a confocal dish with 100 μL fresh DMEM containing 10% FBS and incubated in an incubator (37°C, 5% CO_2). After 24 h, the N-doped CDs with concentrations of 50, 200, 400, and 600 μg/mL were added to the confocal dish and incubated at 37°C in 5% CO_2 for another 4 h. Subsequently, the adherent cells were carefully washed three times with PBS (0.01 M, pH 7.4). Finally, the laser confocal microscopy imaging of HeLa cells was performed at excitations of 405 and 488 nm. Bright-field images were captured to ensure the locations of fluorescent tag signals.

2.5. Metal Ion Detection by N-Doped CDs. Sources of various metal ions such as Na^+, Mg^{2+}, Al^{3+}, Ca^{2+}, Cr^{3+}, Fe^{2+}, Fe^{3+}, Co^{2+}, Cu^{2+}, Zn^{2+}, Cd^{2+}, Sr^{2+}, K^+, and Hg^{2+} were applied for detection. N-doped CD stock solution (2.0 mg/mL, 250 μL) was mixed with 50 μL of 120 μM solutions of different metal ions to reach the final concentration of 20 μM, respectively. Afterward, the mixtures were recorded under excitation at 360 nm. To evaluate the selectivity of this N-doped CD toward Fe^{3+}, interference assays were performed under identical conditions using above ions and Fe^{3+}, and the N-doped CD stock solution was added to different concentrations of Fe^{3+} solution in a similar manner.

3. Results and Discussion

3.1. Characterization of N-Doped CDs. The size and morphology of N-doped CDs were characterized by HRTEM and AFM. The TEM image (Figure 1(a)) clearly revealed N-doped CDs with spherical morphology, average diameter (inset in Figure 1(a)) of 2.7 nm, and a narrow particle size distribution of 2.3–3.3 nm. HRTEM images (Figure 1(b)) showed that the average lattice spacing of the N-doped CDs was 0.32 nm, in agreement with the (002) diffraction planes of graphite [18, 19]. The AFM 3D image and topography image (Figures 2(a) and 2(b)) indicate that the N-doped CDs had a spherical shape, which is consistent with the TEM results. The average height (Figure 2(c)) of N-doped CDs was 3.16 nm, which is close to the diameter of N-doped CDs measured by TEM (2.68 nm). Hence, in accordance with the previous reports, the N-doped CDs were almost spherical carbon nanoparticles [20, 21]. FT-IR spectroscopy and XPS analyses were performed to study the

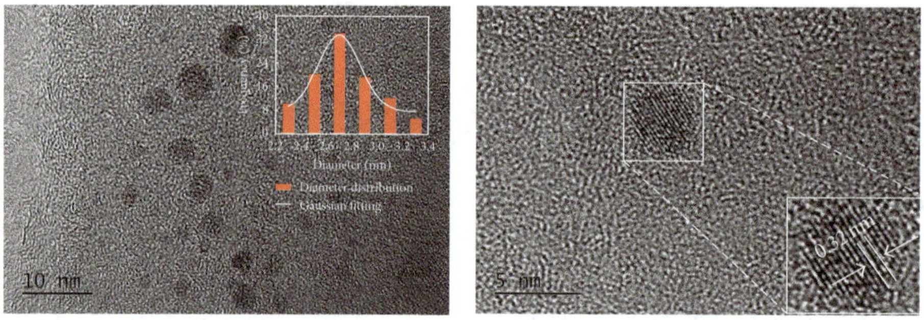

(a) (b)

FIGURE 1: (a) TEM image of N-doped CDs. Inset shows the size distribution of N-doped CDs. (b) HRTEM reveals lattice spacing of N-doped CDs.

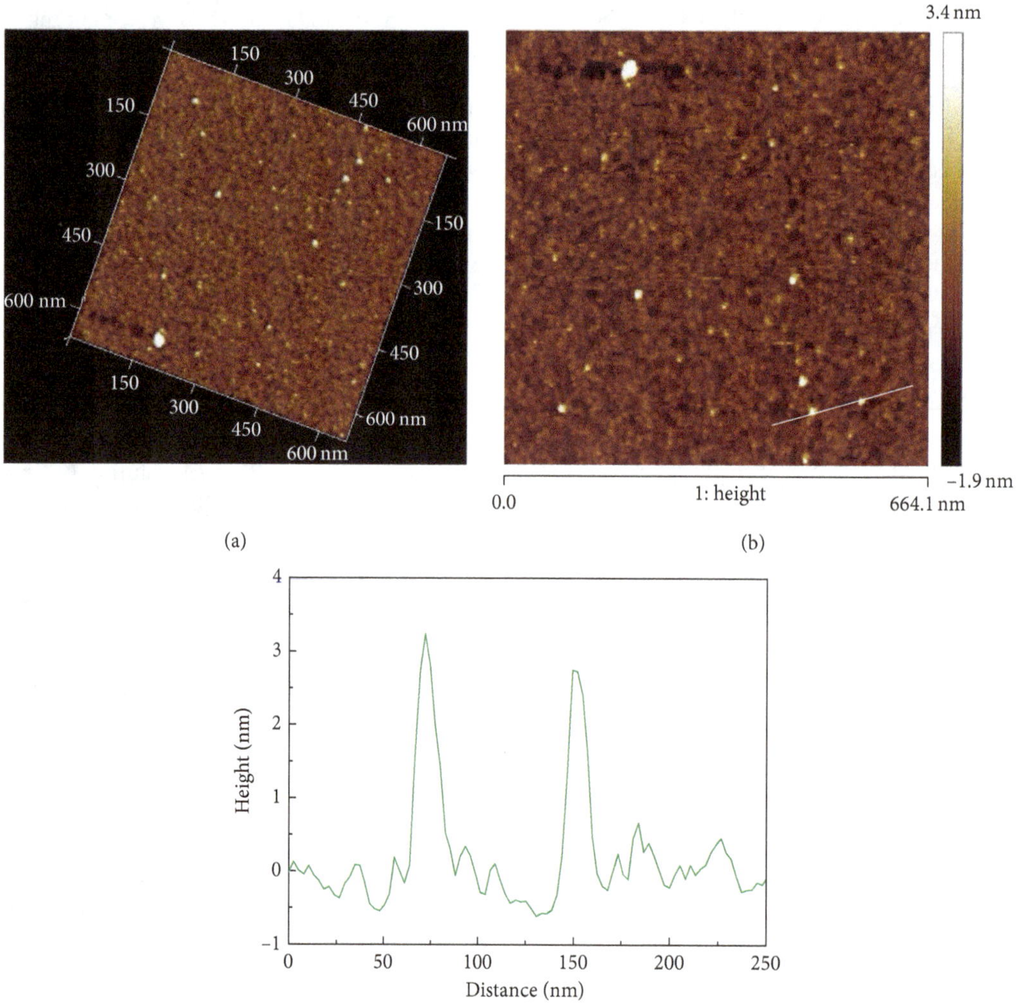

(a) (b) (c)

FIGURE 2: AMF images of N-doped CDs. (a) AMF 3D image, (b) AMF topography image, and (c) height profile along the line in (b).

chemical composition and functional groups of the N-doped CDs. In the FT-IR spectrum in Figure 3(a), the peak at 3433 cm^{-1} is attributed to the stretching vibration of –NH, the peaks at 1625 cm^{-1} indicate the existence of C=C, the peak at 1408 cm^{-1} was assigned to the bending vibration of C–NH (indicating the successful adulteration of nitrogen atoms into the C-dots), and the absorption at 674 cm^{-1} is ascribed to C–H. These FT-IR assignments were further verified by XPS analysis. XPS was used to measure the surface chemical composition and elemental analysis of N-doped CDs. The three main peaks at 284.78, 400.48, and 530.38 eV of the XPS survey spectrum (shown in Figure 3(b)) correspond to C1s, N1s, and O1s, respectively. The N-doped CDs contained 54.5 at.% carbon, 19.2 at. % nitrogen, and 26.3

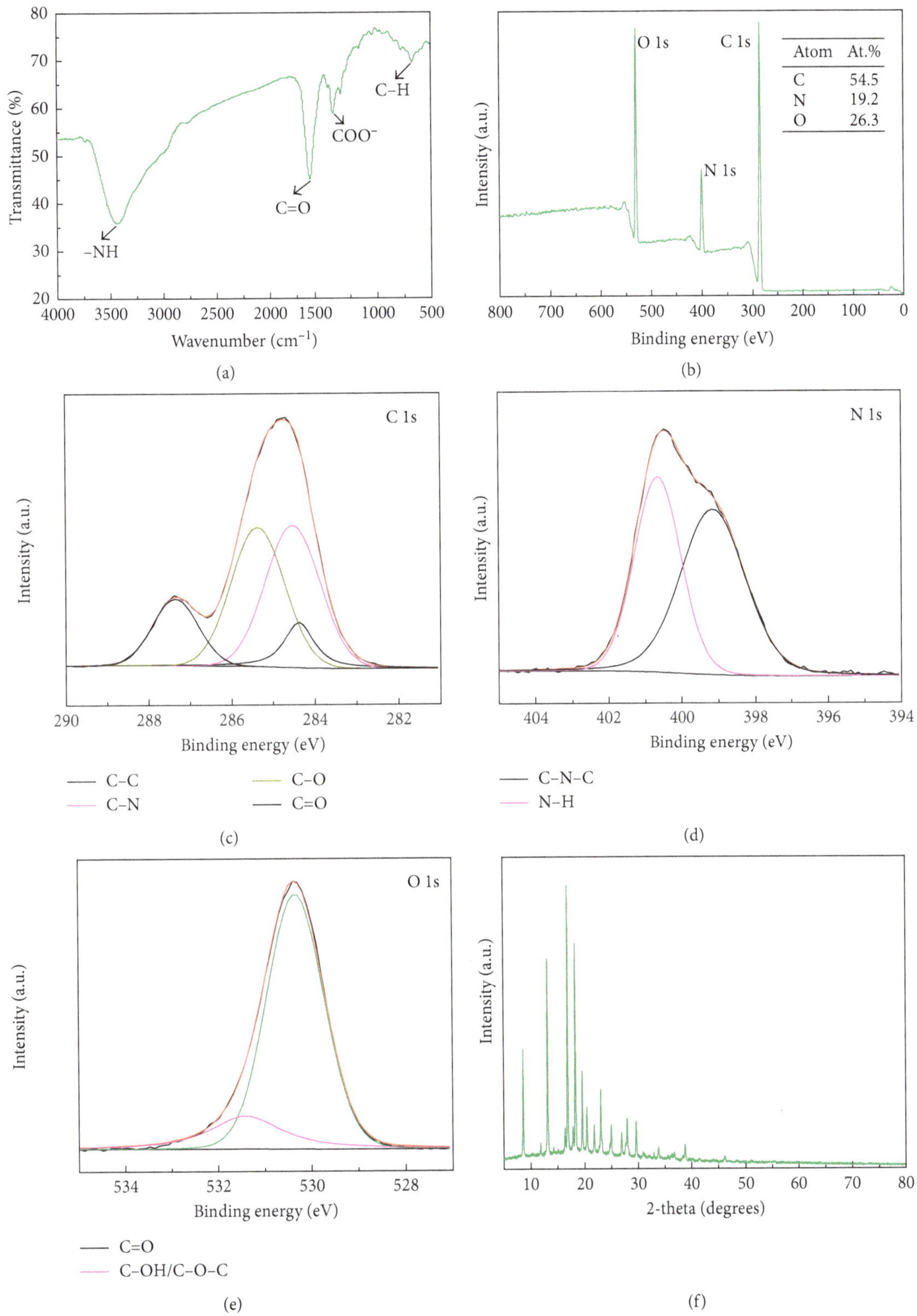

FIGURE 3: (a) FT-IR spectra of N-doped CDs. (b) XPS survey spectrum of N-doped CDs and high-resolution spectra of C1s (c), N1s (d), and O1s (e). (f) XRD pattern of N-doped CDs.

at. % oxygen at the corresponding binding energies given in Figure 3(b). High-resolution XPS spectra of C1s (Figure 3(c)) can be ascribed to four component peaks with binding energies of about 284.38, 284.48, 285.38, and 287.38 eV. Here, the anterior peak located at 284.38 eV reflects the bonding structure of C–C (sp^3) bonds, the peak located at 284.48 eV reflects the bonding structure of the C–N (sp^3) bonds, and the peaks at 285.38 and 287.38 eV are attributed to the C–O (sp^2)

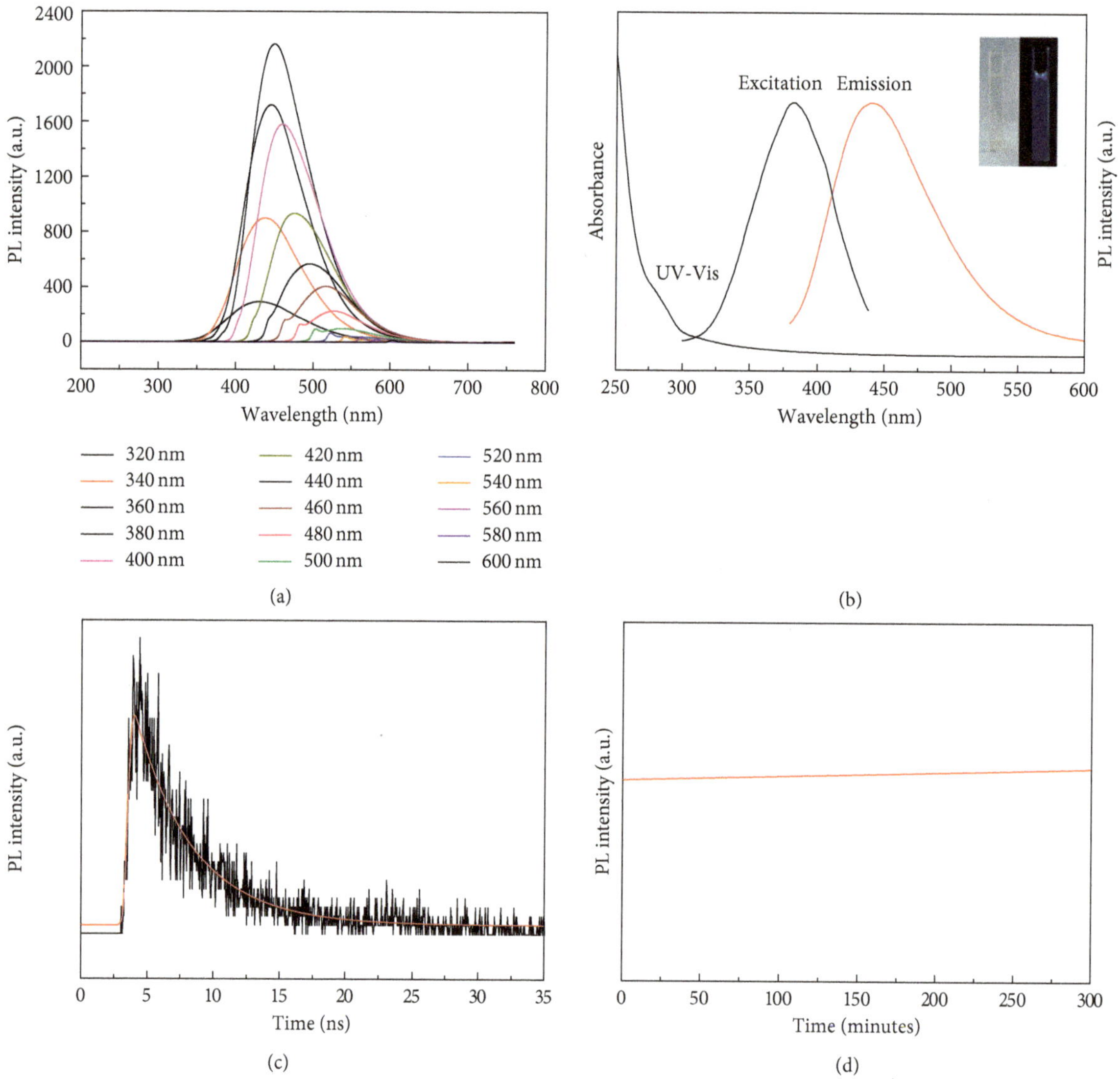

FIGURE 4: (a) UV-Vis absorption spectrum (black line), fluorescence excitation (red line, $\lambda_{em} = 438$ nm), and emission spectra (blue line, $\lambda_{ex} = 377$ nm) of N-doped CDs. The insets show the photograph (left) and fluorescence image of N-doped CD solution under 365 nm UV light (right). (b) Fluorescence emission spectra of the N-doped CDs. (c) The fluorescence decay curve of N-doped CDs. (d) Fluorescence stability of N-doped CDs.

and C=O (sp^2) bonds, respectively [22]. This indicated that the as-prepared N-doped CDs were rich in hydrophilic groups on their surfaces, which was consistent with the corresponding FT-IR spectrum. As shown in Figure 3(d) (partial XPS spectrum of N1s), the N1s peak can also be resolved into two components centred at 399.18 and 400.48 eV; the anterior peak located at 399.18 eV reflects the bonding structure of C–N–C bonds, and the second peak located at 400.48 eV reflects the bonding structure of the N–H bonds [23, 24]. The O1s peak had two components at 530.38 and 531.38 eV for adsorbed oxygen: C=O and C–OH/C–O–C (Figure 3(e)), respectively [21]. Surface functionality analyses via XPS are in agreement with FT-IR results. The above analysis indicated that the N-doped CDs synthesized might have functional groups like -COOH, -OH, and -NH. XRD patterns (Figure 3 (f)) showed many narrow 2θ diffraction peaks at about 8.65°, 11.91°, 13.21°, 16.39°, 16.86°, 17.82°, 18.27°, 19.55°, 20.41°, 21.79°, 23.02°, 24.73°, 24.97°, 26.93°, 27.97°, 29.62°, 33.76°, and 38.75°, suggesting ordered carbon in N-doped CDs.

3.2. Spectral Properties and Cytotoxicity Assay of N-Doped CDs. As displayed in Figure 4(a), the excellent optical properties of the synthesized N-doped CDs were demonstrated by absorption and PL spectroscopy. The N-doped CDs displayed broad UV-Vis absorption, which was attributed to the n-π^* transition in N-doped CDs. The emission wavelength of N-doped C-dots was red-shifted from 430 to 600 nm with excitation wavelength ranging from 320 to 600 nm [25]. In addition, the optimal excitation and emission wavelengths of the N-doped CDs solution were located at 377 and 438 nm (Figure 4(b)). Separately, the N-doped CD aqueous solution emitted strong blue light upon ultraviolet excitation at 365 nm (right inset, Figure 4(b)).

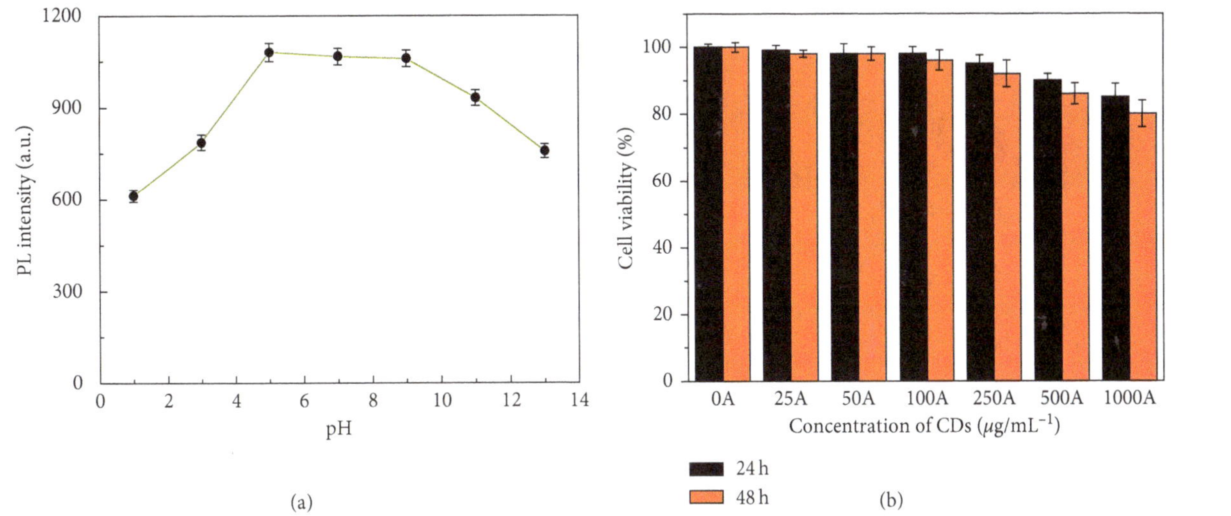

FIGURE 5: (a) PL spectra of N-doped CD aqueous solutions at different pH ($\lambda_{ex} = 365$ nm). (b) Cytotoxicity testing results via an MTT assay.

To further investigate the optical properties of the as-obtained N-doped CDs, the PL excitation spectrum of the N-doped CDs was observed (Figure 4(b)). The spectrum displayed typical excitation wavelength dependence, and the emission wavelength was red-shifted when excited with longer wavelengths. This behaviour of the N-doped CDs has been suggested to be a result of different sizes or the existence of different emissive sites on the surfaces [26]. This excitation-dependent emission property of carbon dots has also been found in the previous reports [27–29]. We further investigated the fluorescence stability of N-doped CDs. Time-correlated single-photon counting (TCSPC) was applied to measure the fluorescence lifetime of the N-doped CDs. As presented in Figure 4(c), the decay lifetime of the N-doped CDs was measured with the previous reports [30], and the calculated average fluorescence lifetime for the N-doped CDs was 4.45 ± 0.05 ns. Moreover, the photostability (Figure 4(d)) of the N-doped CDs synthesized was tested upon continuous excitation at 360 nm for 5 h; the fluorescence remained intact without any photobleaching, which corroborates to reasonably good photostability of N-doped CDs.

To investigate the applicability of N-doped CDs as a fluorescence biomarker in a practical biological environment, the fluorescence stability of N-doped CD aqueous solution was evaluated. As revealed in Figure 5(a), with increase in pH from 3.0 to 5.0, the fluorescence intensity reached a peak and decreased with pH ranging from 5.0 to 13.0. The original solution pH of the N-doped CDs was approximately 9.0. The figure shows that the fluorescence intensity of the N-doped CDs at pH 7.0–9.0 was stronger than that at 9.0 and the fluorescence intensity had not clearly declined at 7.0–9.0. A physiological environment generally has pH of 7.0–8.0, which is beneficial for bioimaging applications. In addition, fluorescence QY of the N-doped CDs was found using quinine sulphate as standard (measured at 350 nm excitation wavelength, QY = 54%). The average QY of N-doped CDs in aqueous solution at room temperature was 32.9%. The high QY is possibly due to the existence of nitrogen-containing functional groups, which are generally excellent auxochromes and greatly enhance photoluminescence.

The biological application of N-doped CDs was also explored. MTT assays were carried out to evaluate the cytotoxicity of the as-prepared N-doped CDs to living cells. As expected, cell viabilities were estimated to be greater than 90% upon addition of N-doped CDs over a wide concentration range (0–500 μg/mL) and after incubation for 48 h (Figure 5(b)). High cell viabilities confirmed the low toxicity, excellent biocompatibility, and great potential of the as-prepared N-doped CDs for imaging in living cells. These also indicate that the as-prepared N-doped CDs can be considered safe for in vitro and in vivo applications.

3.3. *Application of N-Doped CDs*

3.3.1. Imaging of HeLa Cells. Based on these fluorescence properties, experiments were carried out to further demonstrate the availability of the as-prepared N-doped CDs for imaging in cells and plants. Figure 6 shows CLSM images under bright field, 405 nm, 458 nm, and 514 nm excitations of HeLa cells incubated for 4 h at 37°C with 50, 200, 400, and 600 $\mu g \cdot mL^{-1}$ N-doped CDs. As shown in the figure, strong blue and green fluorescence of the HeLa cells can be seen at 405 and 488 nm. More careful observation revealed that the luminescence spots appeared widely in the membrane and cytoplasmic areas of the HeLa cells. In addition, with the increase in the concentration of N-doped CDs, the fluorescence enhancement helped identify tumour cells. According to the previous studies, the cytoplasm-specific property of N-doped CDs should be related to endocytosis [1, 9, 14, 21]. The results indicated that N-doped CDs could be used for in vitro tumour cell labelling via a simple incubation method.

3.3.2. Selectivity and Ratiometric Detection of Fe^{3+} in Serum. As shown in Figure 7(a), under the same conditions, in sharp contrast to Fe^{3+}, other ions including Hg^{2+}, Cr^{3+}, Fe^{2+}, Co^{2+},

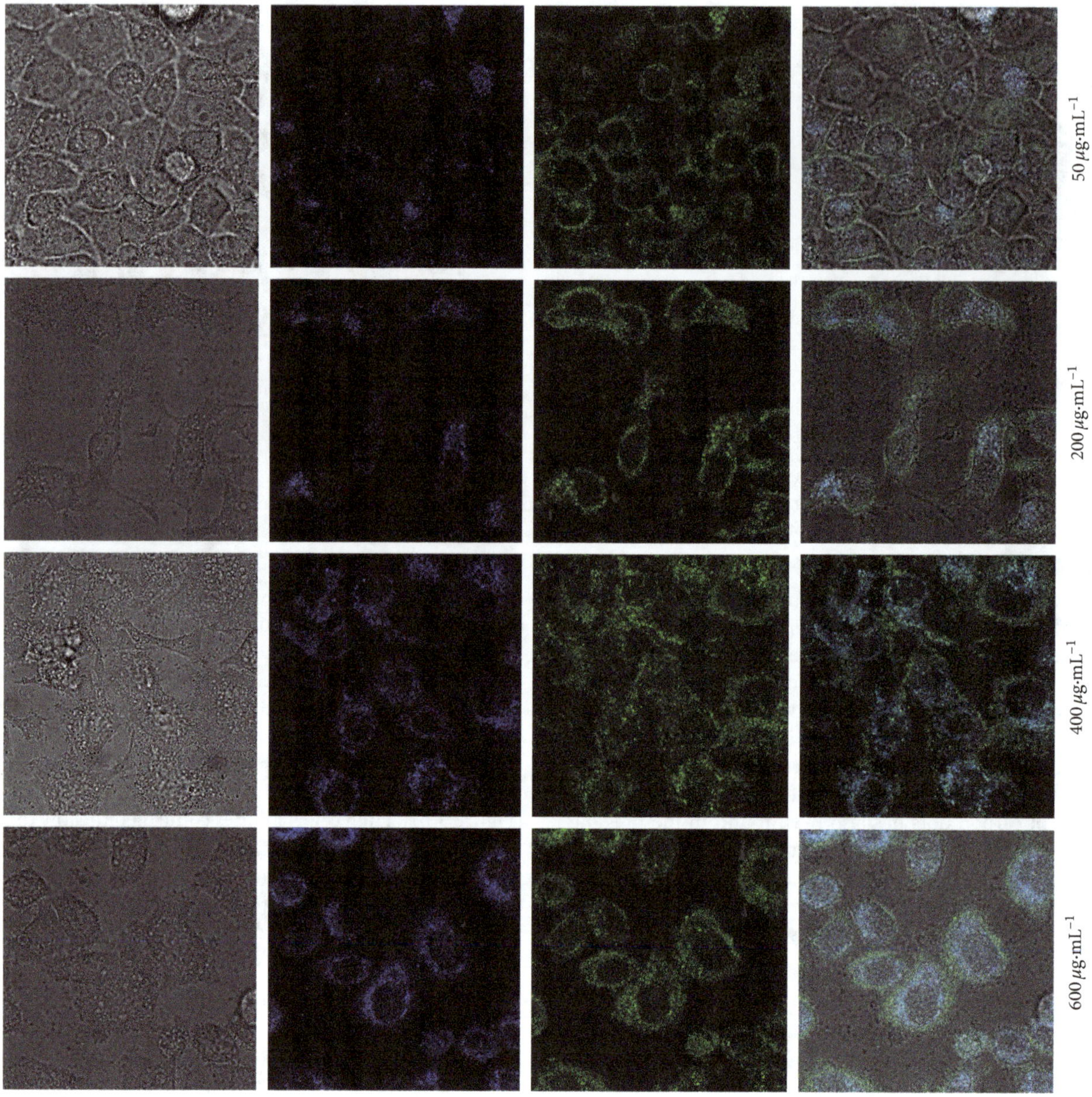

FIGURE 6: Laser scanning confocal microscopy images of HeLa cells with different concentrations of N-doped CDs.

Cd^{2+}, Sr^{2+}, Al^{3+}, Mg^{2+}, Zn^{2+}, Ca^{2+}, K^+, Na^+, and Cu^{2+} showed almost no influence on the spectra of the nanoprobe. The fluorescence could be quenched by Fe^{3+} ion due to the special coordination interaction between Fe^{3+} ion and the hydroxy groups on the surface of CDs, which may contribute to nonradiative electron transfer that involves partial transfer of an electron in the excited state to the d orbital of Fe^{3+}. To further investigate the FL quenching mechanism of N-doped CDs, different concentrations of Fe^{3+} solution were added to N-doped CD stock solution. As shown in Figure 7(b), the FL intensity of N-doped CDs gradually decreased with increasing Fe^{3+} concentration and the inset shows the relationship of the relative fluorescence intensity F/F_0 with Fe^{3+} concentration. Dynamic fluorescence quenching is described by the Stern–Volmer equation [31]. The inset in Figure 6(b) shows a linear relationship ($R^2 = 0.998$) in the range of Fe^{3+} concentration from 0 to 50 μM. The detection limit was 37 nM (signal-to-noise ratio of 3). The results demonstrated that N-doped CDs show promise as a sensitive and selective probe for the detection of Fe^{3+}. In order to demonstrate the analytical performance of the proposed N-doped CDs in complicated biological samples, the capability of the N-doped CDs was evaluated by quantitative detection of Fe^{3+} in human serum, and it was spiked with different concentrations of Fe^{3+} and measured by the proposed method. Table 1 shows that the recoveries were 95.2–112% with the relative standard deviation (RSD, $n = 5$) less than 5.8%, which indicates that the proposed method was sensitive and accurate.

4. Conclusions

We demonstrated a facile and green synthesis method to prepare novel fluorescent N-doped CDs by hydrothermal reaction using amino acid L-citrulline as precursor.

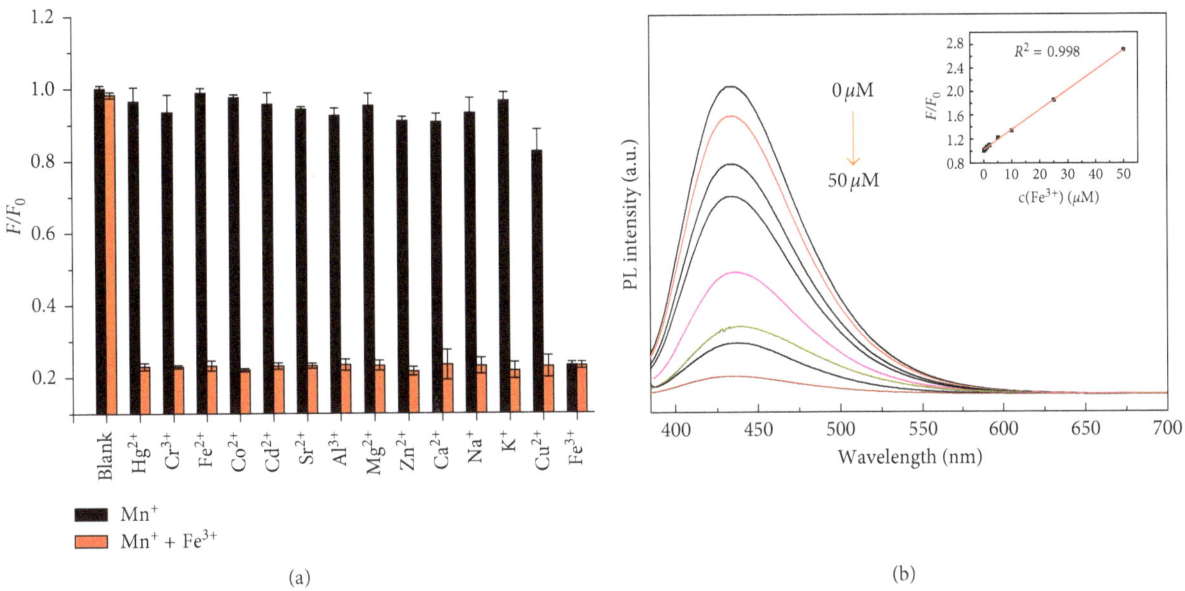

Figure 7: Ion detection of N-doped CDs. (a) Fluorescence recovery efficiency F/F_0 response of the N-doped CDs toward Fe^{3+} and other interference. (b) PL spectra of N-doped CDs with different Fe^{3+} concentrations (0 to 50 μM). Inset: the dependence of F/F_0 on concentration (0 to 50 μM).

Table 1: Determination of iron in human serum using N-doped CDs.

Samples	Added (μmol/L)	Found (μmol/L)	Recovery (%)	RSD% ($n = 5$)
	0.250	0.281	112	5.8
Serum	2.50	2.61	104	4.6
	25.0	23.8	95.2	3.9

The N-doped CD aqueous solution emitted strong blue light under UV irradiation with a fluorescence quantum yield of 32.9%, and the emission wavelength was red-shifted under excitation with longer wavelengths. The fluorescent N-doped CDs acted as novel fluorescence probes that facilitated simultaneous imaging of HeLa cells and sensitivity detection of Fe^{3+} ions. The N-doped CDs showed outstanding overall performance such as outstanding optical properties, good chemical and photochemical stability, inertness to interference of metal ions and biomolecular species, and excellent biocompatibility, which make N-doped CDs a desirable alternative probe for biological imaging, detection, and many other applications.

Conflicts of Interest

The authors declare that there are no conflicts of interest regarding the publication of this paper.

Acknowledgments

This work was supported by the National Natural Science Foundation of China (no. 21505115), the Science and Technology Foundation of the Guizhou Province, Science and Technology Department (nos. LH20147406 and LH20167034), the Natural Science Foundation of the Guizhou Province, Education Department (no. KY(2017)368), the Social Development Foundation of the Science and Technology Bureau of Qianxinan Prefecture (no. 2016-1-29), and the Ph.D. Programs Foundation of Xingyi Normal University for Nationalities (no. 16XYBS20).

References

[1] Y. W. Zheng, D. Yang, X. Wu et al., "A facile approach for the synthesis of highly luminescent carbon dots using vitamin-based small organic molecules with benzene ring structure as precursors," *RSC Advances*, vol. 5, no. 110, pp. 90245–90254, 2015.

[2] V. Georgakilas, J. A. Perman, J. Tucek, and R. Zboril, "Broad family of carbon nanoallotropes: classification, chemistry, and applications of fullerenes, carbon dots, nanotubes, graphene, nanodiamonds, and combined superstructures," *Chemical Reviews*, vol. 115, no. 11, pp. 4744–4822, 2015.

[3] M. J. Meziani, X. L. Dong, L. Zhu et al., "Visible-light-activated bactericidal functions of carbon "quantum" dots," *ACS Applied Materials and Interfaces*, vol. 8, no. 17, pp. 10761–10766, 2016.

[4] S. Mandani, D. Dey, B. Sharma, and T. K Sarma, "Natural occurrence of fluorescent carbon dots in honey," *Carbon*, vol. 119, no. 9, pp. 569–572, 2017.

[5] S. Li, Z. Guo, Y. Zhang, W. Xue, and Z. H. Liu, "Blood compatibility evaluations of fluorescent carbon dots," *ACS Applied Materials and Interfaces*, vol. 7, no. 34, pp. 19153–19162, 2015.

[6] H. Ding, S. B. Yu, J. S. Wei, and H. M. Xiong, "Full-color light-emitting carbon dots with a surface-state-controlled luminescence mechanism," *ACS Nano*, vol. 10, no. 1, pp. 484–491, 2016.

[7] H. Li, W. Q. Kong, J. Liu et al., "Fluorescent N-doped carbon dots for both cellular imaging and highly-sensitive catechol detection," *Carbon*, vol. 91, no. 11, pp. 66–475, 2015.

[8] L. Liu, L. Chen, J. G. Liang, L. Z. Liu, and H. Y. Han, "A novel ratiometric probe based on nitrogen-doped carbon dots and rhodamine b isothiocyanate for detection of $Fe3^{+}$ in aqueous solution," *Journal of Analytical Methods in Chemistry*, vol. 2016, Article ID 4939582, 7 pages, 2016.

[9] S. Y. Lim, W. Shen, and Z. Q. Gao, "Carbon quantum dots and their applications," *Chemical Society Reviews*, vol. 44, no. 1, pp. 362–381, 2015.

[10] Y. B. Song, S. J. Zhu, S. Y. Xiang et al., "Investigation into the fluorescence quenching behaviors and applications of carbon dots," *Nanoscale*, vol. 6, no. 9, pp. 4676–4682, 2014.

[11] J. J. Zhao, M. J. Huang, L. L. Zhang et al., "A unique approach to develop carbon dot-based nanohybrid near-infrared ratiometric fluorescent sensor for the detection of mercury ions," *Analytical Chemistry*, vol. 89, no. 15, pp. 8044–8049, 2017.

[12] K. Jiang, S. Sun, L. Zhang, Y. H. Wang, C. Z. Cai, and H. W. Lin, "Bright-yellow-emissive N-doped carbon dots: preparation, cellular imaging, and bifunctional sensing," *ACS Applied Materials and Interfaces*, vol. 7, no. 41, pp. 23231–23238, 2015.

[13] X. J. Gong, W. J. Lu, Y. Liu et al., "Low temperature synthesis of phosphorous and nitrogen co-doped yellow fluorescent carbon dots for sensors and bioimaging," *Journal of Materials Chemistry B*, vol. 3, no. 33, pp. 6813–6819, 2015.

[14] X. X. Chen, Q. Q. Jin, L. Z. Wu, C. H. Tung, and X. J. Tang, "Synthesis and unique photoluminescence properties of nitrogen-rich quantum dots and their applications," *Angewandte Chemie International Edition*, vol. 53, no. 51, pp. 14447–14450, 2014.

[15] P.F. Lv, Y.X. Yao, H. M Zhou et al., "Synthesis of novel nitrogen-doped carbon dots for highly selective detection of iron ion," *Nanotechnology*, vol. 28, no. 12, pp. 165502–165511, 2017.

[16] L. Wang and H. S. Zhou, "Green synthesis of luminescent nitrogen-doped carbon dots from milk and its imaging application," *Analytical Chemistry*, vol. 86, no. 15, pp. 8902–8905, 2014.

[17] P. C. Hsu and H. T. Chang, "Synthesis of high-quality carbon nanodots from hydrophilic compounds: role of functional groups," *Chemical Communications*, vol. 48, no. 33, pp. 3984–3986, 2012.

[18] Z. H. Kang, E. B. Wang, B. D. Mao et al., "Controllable fabrication of carbon nanotube and nanobelt with a polyoxometalate-assisted mild hydrothermal process," *Journal of the American Chemical Society*, vol. 127, no. 18, pp. 6534-6535, 2005.

[19] J. Lu, J. X. Yang, J. Z. Wang, A. L. Lim, S. Wang, and K. P. Loh, "One-pot synthesis of fluorescent carbon nanoribbons, nanoparticles, and graphene by the exfoliation of graphite in ionic liquids," *ACS Nano*, vol. 3, no. 8, pp. 2367–2375, 2009.

[20] N. Li, A. Than, C. C. Sun et al., "Monitoring dynamic cellular redox homeostasis using fluorescence-switchable graphene quantum dots," *ACS Nano*, vol. 10, no. 12, pp. 11475–11482, 2016.

[21] L. Yang, W. H. Jiang, L. P. Qiu et al., "One pot synthesis of high quantum yield of polyethylene glycol anchored carbon dots and functionalized with nuclear localization signal peptide for cell nucleus imaging," *Nanoscale*, vol. 7, no. 14, pp. 6104–6113, 2015.

[22] X. Q. Niu, G. S. Liu, L. Y. Li, Z. Fu, H. Xu, and F. L. Cui, "Green and economical synthesis of nitrogen-doped carbon dots from vegetables for sensing and imaging applications," *RSC Advances*, vol. 5, no. 115, pp. 95223–95229, 2015.

[23] J. Ju and W. Chen, "Synthesis of highly fluorescent nitrogen-doped graphene quantum dots for sensitive, label-free detection of Fe (III) in aqueous media," *Biosensors and Bioelectronics*, vol. 58, no. 8, pp. 219–225, 2014.

[24] X. W. Dong, Y. J. Su, H. J. Geng et al., "Fast one-step synthesis of N-doped carbon dots by pyrolyzing ethanolamine," *Journal of Materials Chemistry C*, vol. 2, no. 36, pp. 7477–7481, 2014.

[25] H. J. Zhang, Y. L. Chen, M. J. Liang et al., "Solid-phase synthesis of highly fluorescent nitrogen-doped carbon dots for sensitive and selective probing ferric ions in living cells," *Analytical Chemistry*, vol. 86, no. 19, pp. 9846–9852, 2014.

[26] Y. Q. Dong, H. C. Pang, H. B. Yang et al., "Carbon-based dots co-doped with nitrogen and sulfur for high quantum yield and excitation-independent emission," *Angewandte Chemie International Edition*, vol. 52, no. 30, pp. 7800–7804, 2013.

[27] L. Y. Zhang, D. H. Wang, H. W. Huang et al., "Preparation of gold–carbon dots and ratiometric fluorescence cellular imaging," *ACS Applied Materials and Interfaces*, vol. 8, no. 10, pp. 6646–6655, 2016.

[28] A. M. Alam, B. Y. Park, Z. K. Ghouri, M. Park, and H. Y. Kim, "Synthesis of carbon quantum dots from cabbage with down- and up-conversion photoluminescence properties: excellent imaging agent for biomedical applications," *Green Chemistry*, vol. 17, no. 7, pp. 3791–3797, 2015.

[29] J. Hou, J. Yan, Q. Zhao, Y. Li, H. Ding, and L. Ding, "A novel one-pot route for large-scale preparation of highly photoluminescent carbon quantum dots powders," *Nanoscale*, vol. 5, no. 20, pp. 9558–9561, 2013.

[30] J. C. Shen, Q. Li, Y. Zhang, X. J. She, C. F. Wang, and S. Chen, "Nitrogen-doped carbon dots derived from polyamindoamine dendrimer," *RSC Advances*, vol. 6, no. 64, pp. 59702–59707, 2016.

[31] S. Mohapatra, S. Sahu, N. Sinha, and S. K. Bhutia, "Synthesis of carbon dot based photoluminescent probe for selective and ultrasensitive detection of $Hg2^{+}$ in water and living cells," *Analyst*, vol. 140, no. 4, pp. 1221–1228, 2015.

21

Calorimetric Study of *Helix aspersa* Maxima Hemocyanin Isoforms

Svetla Todinova,[1] **Yuliana Raynova**,[2] **and Krassimira Idakieva**[2]

[1]*Institute of Biophysics and Biomedical Engineering, Bulgarian Academy of Sciences, Bl. 21, Acad. G. Bonchev Str., Sofia 1113, Bulgaria*
[2]*Institute of Organic Chemistry with Centre of Phytochemistry, Bulgarian Academy of Sciences, Bl. 9, Acad. G. Bonchev Str., Sofia 1113, Bulgaria*

Correspondence should be addressed to Svetla Todinova; todinova@abv.bg

Academic Editor: Ricardo Jorgensen Cassella

The thermal unfolding of hemocyanin isoforms, β-HaH and α_{D+N}-HaH, isolated from the hemolymph of garden snails *Helix aspersa* maxima, was studied by means of differential scanning calorimetry (DSC). One transition, with an apparent transition temperature (T_m) at 79.88°C, was detected in the thermogram of β-HaH in 20 mM HEPES buffer, containing 0.1 M NaCl, 5 mM $CaCl_2$, and 5 mM $MgCl_2$, pH 7.0, at scan rate of 1.0°C min^{-1}. By means of successive annealing procedure, two individual transitions were identified in the thermogram of α_{D+N}-HaH. Denaturation of both hemocyanins was found to be an irreversible process. The scan-rate dependence of the calorimetric profiles indicated that the thermal unfolding of investigated hemocyanins was kinetically controlled. The thermal denaturation of the isoforms β-HaH and α_{D+N}-HaH was described by the two-state irreversible model, and parameters of the Arrhenius equation were calculated.

1. Introduction

Many natural proteins act as a functional basis of biologic drugs [1]. One of the first steps, required to outline the design of the therapeutic properties of protein drugs, is their thermodynamic characterization [2]. The attainment of desired thermodynamic features of proteins is a relatively new area in biotechnology. Thus, protein stabilization becomes a main goal in medicine and scientific research. The mechanism of folding and unfolding gives knowledge on the function and the possibilities of protein stabilization. Differential scanning calorimetry (DSC) is the most useful technique for analyzing the protein thermal stability. This method measures the heat capacity change, associated with protein thermal denaturation, as a function of temperature. The integral of the excess heat capacity function is the enthalpy in this process. Thermodynamic parameters, such as temperature of unfolding (T_m) and change in enthalpy, give huge information about the structure and functional parameters of the proteins, as well as about the nature of their interactions with other ligands or drug [3–5].

Hemocyanins (Hcs) are oligomeric copper-containing glycoproteins that function as oxygen carriers in the hemolymph of several molluscs and arthropods [6, 7]. Besides their important biological function, a variety of medical applications of molluscan Hcs emerged. Recently, we have established that the Hcs of marine mollusk *Rapana thomasiana* (RtH) and β-Hc isoform of *H. pomatia* (β-HpH) are suitable as potential bioadjuvants for subunit vaccines and that these Hcs could be used as natural adjuvants or protein carriers [8, 9]. The anticancer properties of both molluscan Hcs have been demonstrated on a murine model of colorectal cancer *in vivo* [10]. A number of DSC studies have shown that Hcs possess considerable thermal stability. Melting temperature (T_m) in the range 83–90°C was observed for the RtH [11], β-HpH [12], and the Hc of marine gastropod *Concholepas concholepas* (CCH) [13]. Calorimetric studies demonstrate also the high thermostability of Hcs from arthropods, for example, Hcs of tarantula *Eurypelma californicum* (T_m 90°C) [14] and of horseshoe crab *Limulus polyphemus* (T_m 92°C) [15]. It is considered that the interactions between subunits and the high degree of oligomerization stabilize the quaternary structure of the Hc molecules [15, 16].

In certain species of mollusc, an expression of functionally distinct Hc isoforms has been observed [17]. The physiological relevance of these isoforms is still not clear, but its expression may be connected to the development of species [18]. Heterogeneity at a molecular level has been described first for the Hc from *Helix* species [19]. Three types of Hc molecules or isoforms have been identified in the hemolymph isolated from the Roman snail *Helix pomatia*: β-Hc, α_D-Hc (dissociating α-Hc), and α_N-Hc (non-dissociating α-Hc) [6, 20]. The isoform β-HpH consists of only one type of polypeptide chain (β subunits) compared with two types (α and α' subunits) in each of both α-Hcs [20]. Due to the subunit homogeneity, structural and immunological investigations have been performed primarily on β-HpH. Gielens et al. have identified three types of Hc components in the hemolymph of garden snails *Helix aspersa* [21]. Three isoforms (β-HlH, α_N-HlH, and α_D-HlH) have been reported also for the garden snail *Helix lucorum* [22]. Recently, the Hc of snails *Helix aspersa* maxima (HaH) was isolated and characterized [23, 24]. On the basis of the o-diphenoloxidase activity of HaH, a model of biosensor for quantitative determination of phenols in aqueous solutions was developed [23]. Moreover, this Hc can be produced in large quantities for biotechnology and medical use from snails, bred in special farms under controlled conditions; therefore, the knowledge of its thermal stability is essential.

In the present study, for the first time, the thermal stability of the individual *H. aspersa* maxima Hc isoforms has been characterized by means of DSC. The data are compared with those obtained for other molluscan Hcs in order to acquire further insight into the structural stability of the oxygen-transport proteins from invertebrates.

2. Materials and Methods

2.1. Hemocyanin Preparation. Native HaH was isolated in compliance with the procedure described in [23]. Briefly, the Hc was obtained from the hemolymph, collected from snails *Helix aspersa* maxima, by ultracentrifugation at 180000 ×g (ultracentrifuge Beckman LM-80, rotor Ti 45), for 4 hours, at 4°C. The pellets were resuspended in 50 mM phosphate buffer, pH 7.2, and HaH was further purified by gel filtration chromatography on a Sepharose 4B column (90 × 2.4 cm). Further, HaH was separated into its isoforms (β-HaH and α_{D+N}-HaH) by ion exchange chromatography on a DEAE-Sepharose CL-6B column (32 × 1.2 cm) and equilibrated and eluted with buffer 50 mM Tris-HCl, pH 8.0, using a linear gradient 0.1–0.45 M NaCl. Specific absorption coefficient a_{278} nm = 1.413 ml·mg^{-1}·cm^{-1} for HaH [21] was used for determination of the protein concentration.

2.2. Differential Scanning Calorimetry. Calorimetric measurements were performed using high-sensitivity differential adiabatic scanning microcalorimeter DASM-4 (Biopribor, Pushchino, Russia), with a sensitivity >0.017 mJ·K^{-1}and a noise level < ±0.05 μW. Four different scan rates (0.2, 0.5, 1.0, and 1.5°C min^{-1}) were used. The reversibility of the thermal transitions was checked by examining the reproducibility of the calorimetric trace in a subsequent heating of the sample immediately after cooling following the first scan. In all cases, the thermal denaturation was found to be irreversible; therefore, the thermogram corresponding to the reheating run was used as an instrumental baseline. The transitions were corrected with regard to the difference in heat capacity between the initial and final state by using a linear chemical baseline. The molar excess heat capacity curves, obtained by normalizing with the protein concentrations and the known volume of the calorimetric cell, were smoothed and plotted using the Windows-based software package (Origin). The temperature at the maximum of the excess heat capacity curve was taken as the transition temperature (T_m). DSC measurements were carried out in buffer 20 mM HEPES, containing 100 mM NaCl, 5 mM $CaCl_2$, and 5 mM $MgCl_2$, pH 7.9 (20°C). The temperature coefficient of HEPES buffer is −0.014 ΔpK°C^{-1}.

2.3. Data Analysis. The DSC data were analyzed by means of the nonlinear least-squares method using the Origin 8 software package (Origin Lab Corp.). In all fitting procedures, the correlation coefficient *r*, used as a criterion for the accuracy of fitting, was not less than 0.99. In the calculation of molar quantities, the molecular mass used for the protein was 9000 kDa.

DSC transitions were analyzed in terms of the two-state kinetic model:

$$N \xrightarrow{k} D, \quad (1)$$

which is a limiting case of the Lumry–Eyring model [25]. This model considered only two significantly populated macroscopic states: the initial or native state (*N*), and the final or denatured state (*D*), the transition between which is determined by a strongly temperature-dependent first-order rate constant (*k*). The Arrhenius equation represents the temperature dependence of *k*:

$$k = \exp\left[\frac{E_a}{R}\left(\frac{1}{T^*} - \frac{1}{T}\right)\right], \quad (2)$$

where E_a is the activation energy of the denaturation process, *R* is the gas constant, and T^* is the temperature at which *k* is equal to 1 min^{-1}.

In this case, the excess heat capacity C_p^{ex} is given by the following equation [26]:

$$C_p^{ex} = \frac{1}{v}\Delta H \exp\left\{\frac{E_a}{R}\left(\frac{1}{T^*} - \frac{1}{T}\right)\right\} \times \exp\left\{-\frac{1}{v}\int_{T_0}^{T} \exp\left[\frac{E_a}{R}\left(\frac{1}{T^*} - \frac{1}{T}\right)\mathrm{d}T\right]\right\}, \quad (3)$$

where $v = \mathrm{d}T/\mathrm{d}t$ (K min^{-1}) is a scan rate value and ΔH is the enthalpy difference between the denatured and native states.

3. Results and Discussion

3.1. DSC of Purified HaH Isoforms. The three Hc isoforms (β-HaH, α_D-HaH, and α_N-HaH), identified in the hemolymph

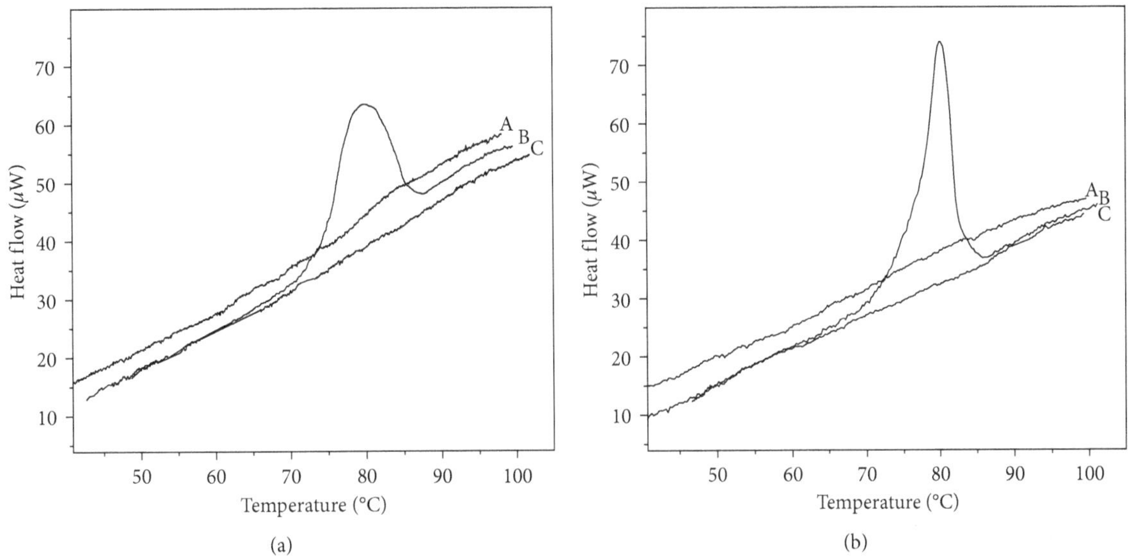

FIGURE 1: Original calorimetric recording of the apparent heat capacity of isoforms β-HaH (a) and α_{D+N}-HaH (b) as a function of the temperature (line B); reheating scan (line C); and baseline obtained with buffer in both cells of the calorimeter (line A). The experimental curves shown refer to β-HaH (2.88 mg·ml^{-1}) and α_{D+N}-HaH (3 mg·ml^{-1}) in buffer 20 mM HEPES, containing 100 mM NaCl, 5 mM $CaCl_2$, and 5 mM $MgCl_2$ (pH 7.06 at 80°C), at scan rate of 1.0°C min^{-1}.

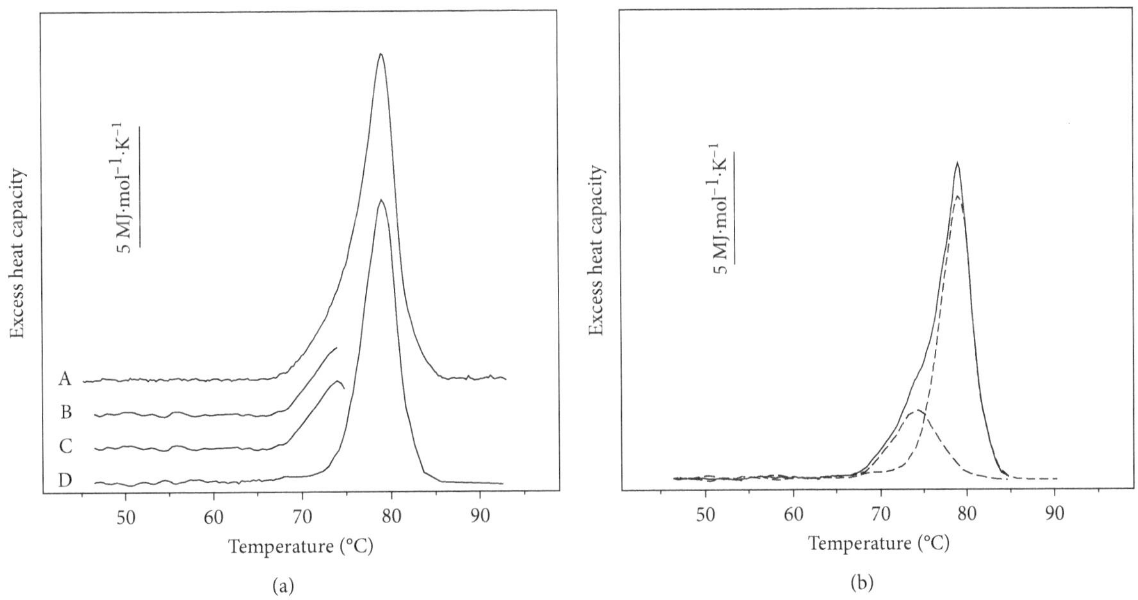

FIGURE 2: Example of the successive annealing process. (a) Complete calorimetric scan of α_{D+N}-HaH in buffer 20 mM HEPES (pH 7.06 at 80°C), at scan rate of 0.5°C min^{-1} (line A); the first scan of a new sample with the same composition, stopped at 74.5°C (line B); result of subtracting the following (second scan) DSC line from the first one (line C); and the second calorimetric scan stopped at 90°C (line D). (b) Result of the deconvolution of the experimental contour (solid line), which provides two individual components (dash lines).

of snails *H. aspersa* maxima, were isolated by ion exchange chromatography. The isoform β-HaH (pI 5.2) eluted as a single symmetric peak, followed by a peak containing both electrophoretically similar α-isoforms (pI 4.6), which eluted together as α_{D+N}-HaH, as previously described by Raynova and coworkers [23]. At neutral pH and in the presence of Ca^{2+} and Mg^{2+}, these Hc molecules occur as a didecamers of ~450 kDa subunits.

The thermogram of the isoform β-HaH in buffer 20 mM HEPES, pH 7.0, containing 100 mM NaCl, 5 mM $CaCl_2$, and 5 mM $MgCl_2$, at a heating rate of 1.0°C min^{-1}, is shown in Figure 1(a) (line B). Heat absorption was observed between 70 and 87°C, with an apparent transition temperature (T_m) at 79.88°C. Value of 145.0 MJ mol^{-1} was calculated for the calorimetric enthalpy (ΔH_{cal}) by an integration of the heat capacity curve. Judging from the absence of any transitions upon rescanning the samples, the thermal denaturation of β-HaH was irreversible (Figure 1(a), line C).

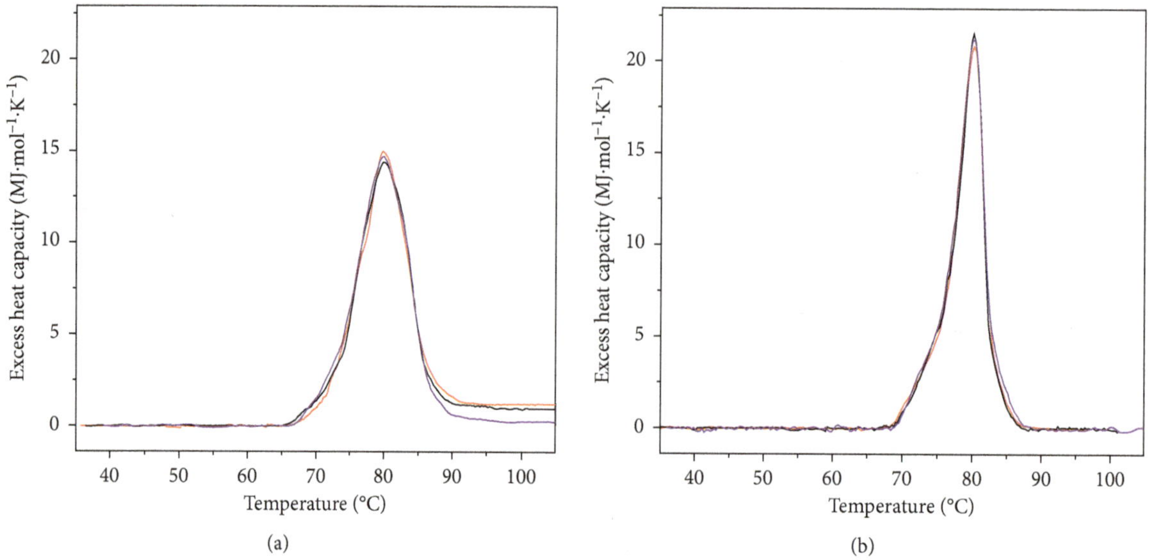

FIGURE 3: (a) C_p transition curve of β-HaH in buffer 20 mM HEPES (pH 7.06 at 80°C), recorded at scan rate of 1°C min^{-1} and different protein concentrations: 1.01 mg·ml^{-1} (blue line), 1.91 mg·ml^{-1} (red line), and 2.88 mg·ml^{-1} (black line). (b) C_p transition curve of α_{D+N}-HaH in buffer 20 mM HEPES (pH 7.06 at 80°C), recorded at scan rate of 1°C min^{-1} and different protein concentrations: 2.0 mg·ml^{-1} (blue line), 3.0 mg·ml^{-1} (black line), and 4.52 mg·ml^{-1} (red line).

Analysis of the heating profile of isoform α_{D+N}-HaH, run in the same experimental conditions, revealed asymmetry of the shape, due to two overlapping transitions. In the thermogram of α_{D+N}-HaH, one main transition with an apparent T_m of 80°C was observed (Figure 1(b), line B). Minor transition at ca. 76°C can also be discerned in the thermogram. The thermal denaturation of the Hc isoforms α_{D+N}-HaH also was found to be irreversible (Figure 1(b), line C). Irreversibility of the thermal denaturation has been observed in DSC measurements of all studied to date Hcs [11–14, 27] and limits the use of a standard equilibrium thermodynamic analysis.

In view of the complex oligomeric Hc structure, the presence of more than one structural domains undergoing thermal unfolding more or less independently of each other is not unexpected. An analysis of DSC contours of the isoform α_{D+N}-HaH was attempted with a successive annealing procedure, which was shown to be useful for the experimental deconvolution of completely complicated or partially irreversible thermal transitions [28, 29]. Successive annealing procedure was applied in the analysis of the thermograms of the other studied Hcs [11–13]. Figure 2 shows an example of this procedure, which provides two irreversible transitions. At first, α_{D+N}-HaH was heated from 35°C up to 90°C to determine the number and T_m of expected transitions (Figure 2(a), line A). After that, a new sample of the same composition was heated slightly above the shoulder, which appeared at ~74°C (Figure 2(a), line B) and after cooling to 35°C was reheated up to 90°C (Figure 2(a), line D). To determine the ascending part of the first transition, subtracting of the second scan from the first one was done (Figure 2(a), line C). The descending part of the curve (line C) was obtained by extrapolation (Figure 2(b), dash lines).

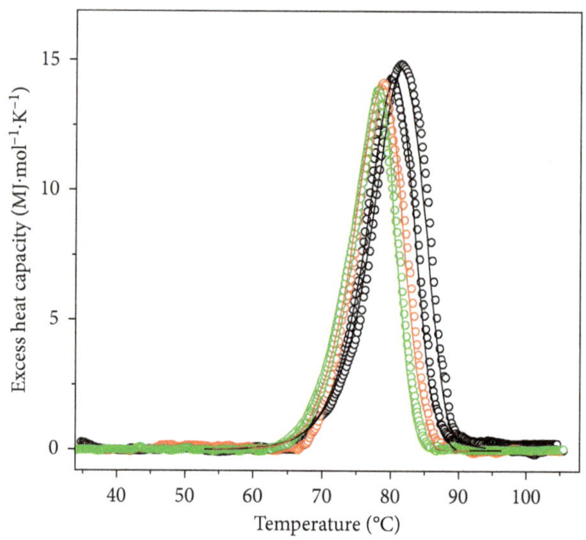

FIGURE 4: Dependence of the C_p transition curves of β-HaH in buffer 20 mM HEPES (pH 7.06 at 80°C), with a scan rate: 0.24°C min^{-1}, green symbols; 0.5°C min^{-1}, red symbols; 1.0°C min^{-1}, black symbols; 1.5°C min^{-1}, blue symbols. In all cases the protein concentration was 2.88 mg·ml^{-1}. The solid lines represent the theoretical fitting curves based on (3).

3.2. Concentration Dependence of the Temperature of Denaturation. As is mentioned above, the quaternary structure of the Hc molecule represents oligomer of twenty subunits (didecamer). Dissociation of the Hc molecule to individual subunits may take place during the denaturation process. According to the equilibrium thermodynamics, any change in the oligomerization state of the protein during the denaturation process should produce a concentration dependence of T_m. In order to determine whether this was occurring, DSC traces were collected at various protein concentrations, at a constant scan rate of 1°C min^{-1}. DSC experiments showed that the T_m and calorimetric enthalpy

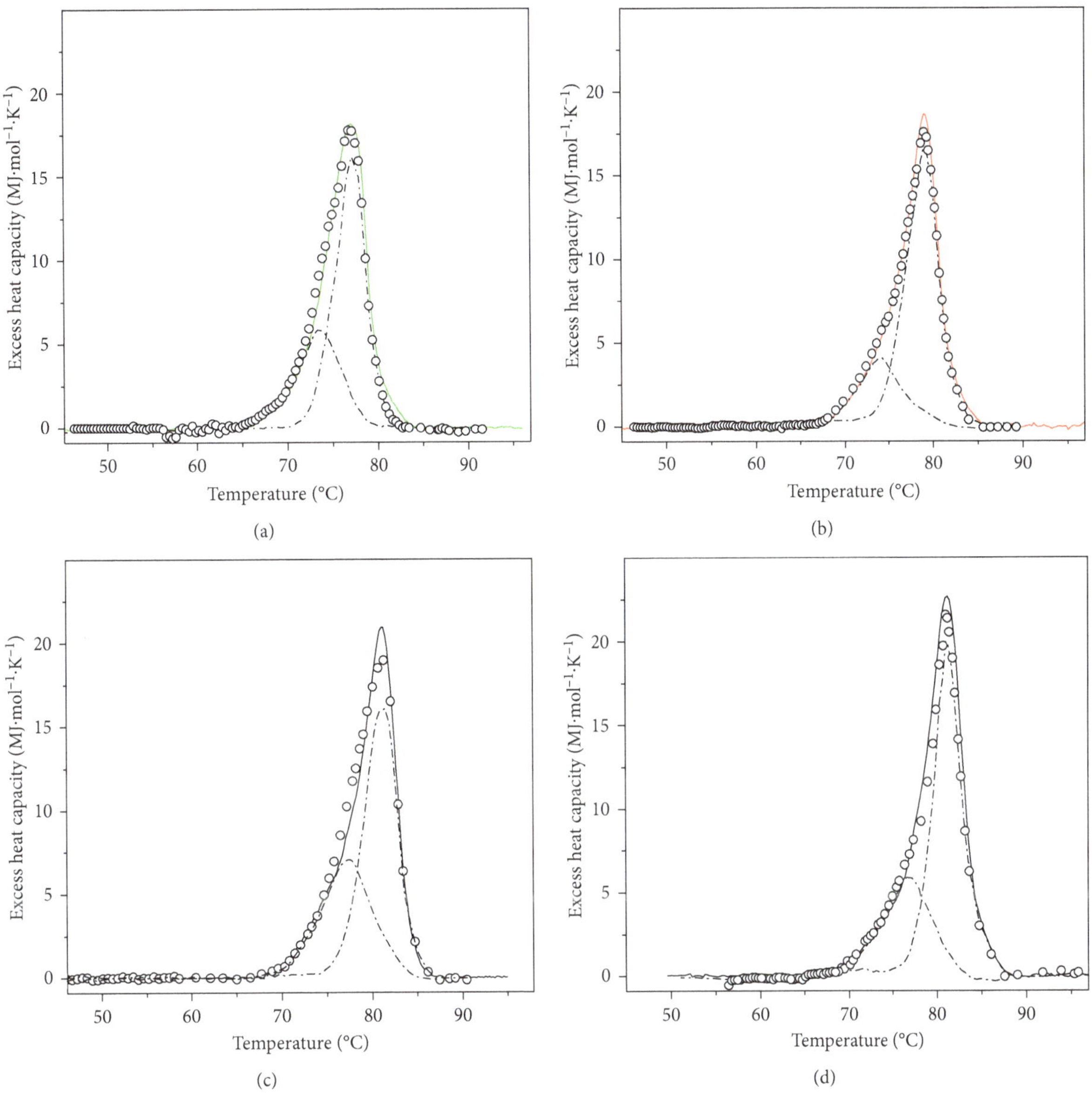

FIGURE 5: Dependence of the C_p transition curves of α_{D+N}-HaH isoforms in buffer 20 mM HEPES (pH 7.06 at 80°C), with a scan rate: 0.2°C min^{-1} (a); 0.5°C min^{-1} (b); 1.0°C min^{-1} (c); 1.5°C min^{-1} (d). Continuous lines depict the experimental data; dash dot lines show the experimental individual components, result of the annealing process, and symbols represent the result of the sum of the corresponded individual components. In all cases, the protein concentration was 3 mg·ml^{-1}.

(ΔH_{cal}) values for the thermal unfolding of isoforms β-HaH and α_{D+N}-HaH were independent of the protein concentration (Figures 3(a) and 3(b)); therefore, denatured proteins remained in the same oligomerization state as the native ones. Consistent with our previous studies, the heat absorption observed in the DSC curves (T_m~80°C) is connected to the melting of the compact Hc quaternary structure without simultaneous dissociation into subunits [11–13]. This indicates that the interactions between the subunits, constituting the Hc molecules, are strong enough to prevent dissociation before the rate-determining step of the process of thermal unfolding.

3.3. Scan Rate Dependence. In the case of irreversible thermal denaturation of proteins, the calorimetric profiles are scan rate dependent, and the process is under kinetic control. Figure 4 shows the excess heat capacity function versus temperature profiles of the isoform β-HaH at four different scan rates (0.24, 0.5, 1.0, and 1.5°C min^{-1}). The transition temperature of the irreversible thermal denaturation of β-HaH is scan rate dependent; the maximum of the DSC profiles is shifted toward lower temperatures with a reduction in scan rate (Figure 4).

The transition temperature of α_{D+N}-HaH isoforms is also scan rate dependent, indicating that the process of thermal unfolding is kinetically controlled (Figures 5(a)–5(d)).

As it is evident from Figures 4 and 5, the DSC traces for both HaH isoforms were strongly dependent on the scan rate, which prompted us to analyze this nonequilibrium process based on the simplest model of irreversible thermal denaturation of proteins (1). The excess heat capacity functions were analyzed by fitting the data to the two-state

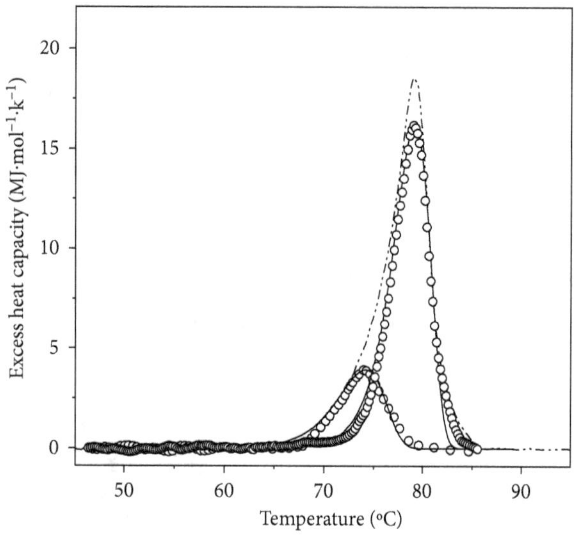

FIGURE 6: C_p transition curve of of α_{D+N}-HaH isoforms in buffer 20 mM HEPES (pH 7.06 at 80°C) recorded at scan rate 0.5°C min^{-1}. The solid lines represent the theoretical fitting curves based on (3), lines with symbols show the curves obtained from the deconvolution procedure, and the dash dot line represent the experimental C_p transition curve.

TABLE 1: Arrhenius equation parameters, estimated for the two-state irreversible model of the thermal denaturation of isoform β-HaH, at neutral pH.

Parameter	Temperature scan rate (°C min^{-1})			
	0.24	0.5	1.0	1.5
T_m (°C)	77.90	78.80	79.88	81.40
ΔH_{cal} (MJ·mol^{-1})	118.6	127.4	145.0	154.0
E_a (kJ·mol^{-1})	307 ± 7	316 ± 4	302 ± 6	306 ± 8
T^*(°C)	84.6 ± 0.5	84.8 ± 0.3	85.0 ± 0.4	84.9 ± 0.3
r	0.992	0.993	0.996	0.994

TABLE 2: Arrhenius equation parameters, estimated for the two-state irreversible model of the thermal denaturation of isoform α_{D+N}-HaH, at neutral pH.

Parameter	Temperature scan rate (°C min^{-1})			
	0.2	0.5	1.0	1.5
First transition				
T_m (°C)	73.4	74.0	76.4	77.0
ΔH_{cal} (MJ·mol^{-1})	31.10	34.07	39.10	41.27
E_a (kJ·mol^{-1})	424 ± 9	434 ± 6	418 ± 5	423 ± 7
T^* (°C)	80.8 ± 0.5	79.0 ± 0.3	80.0 ± 0.4	79.0 ± 0.3
r	0.989	0.985	0.988	0.987
Second transition*				
T_m (°C)	77.1	79.02	80.07	81.20
ΔH_{cal} (MJ·mol^{-1})	69.9	79.0	79.8	81.4
E_a (kJ·mol^{-1})	586 ± 12	564 ± 8	596 ± 10	601 ± 14
T^* (°C)	83.8	83.9	83.7	83.6
r	0.995	0.996	0.993	0.994

*Main transition.

TABLE 3: Parameters for the thermal denaturation of HaH isoforms obtained by DSC at scan rate of 1°C min^{-1}. Comparison with data obtained for Hcs of *R. thomasiana* [11], *H. pomatia* [12], *C. concholepas* [13], and *H. rubra* [27].

Hemocyanin	T_m^* (°C)	ΔHcal (MJ·mol^{-1})	E_a (kJ·mol^{-1})
H. aspersa β-Hc (β-HaH)	79.88	145 ± 1.0	302.0 ± 6
H. aspersa α_{D+N}-Hc (α_{D+N}-HaH)	80.07	79.8 ± 1.0	596.0 ± 10
R. thomasiana Hc (RtH)	82.4	195 ± 1.0	597.0 ± 20
H. pomatia β-Hc (β-HpH)	84.0	190 ± 1.0	521.0 ± 7
C. concholepas Hc (CCH)	78.0	120 ± 1.0	323.0 ± 2
H. rubra Hc	78.9	—	535.0 ± 65

*Main transition.

irreversible model according to (3) (Figures 4 and 6). Applying this analysis on the observed transitions of the two HaH isoforms, the kinetic parameters presented in Tables 1 and 2 were obtained.

The thermal denaturation of Hcs from different species has been described by the two-state irreversible model, and the parameters of the Arrhenius equation have been calculated, allowing comparison of transitions during Hc denaturation [11–13, 26, 27]. The activation energy value of 302.0 ± 6 (kJ·mol^{-1}) and 596.0 ± 10 (kJ·mol^{-1}) was determined for β-HaH and α_{D+N}-HaH, respectively, which is in good agreement with values reported for Hcs in related species (Table 3). The structural organization of these Hcs is similar (didecamer), so it is difficult to propose a structural basis for the difference in their thermal stability. From the other hand, a difference in the carbohydrate content of Hcs has been observed [30]. Hence, a correlation between the thermal stability of Hcs and their carbohydrate content could be expected [31]. Actually, the most thermostable β-Hc of *H. pomatia* contains 7% (w/w) carbohydrates [32], while carbohydrate content of 2.6% (w/w) has been determined for RtH [30]. Data about the carbohydrate content of the Hcs of gastropods *C. concholepas* and *H. aspersa* maxima, however, are still not available in the literature.

The present study confirms that the two-state irreversible model, built especially on experimental examinations on small globular proteins will hold true for larger oligomeric proteins as well [33]. Nevertheless, as Lyubarev and Kurganov noted in [34], that experimental data are satisfactorily described by the two-state model, the real mechanism of protein denaturation can be more complex. Therefore, it should be stressed that the obtained values for the activation energy of the denaturation process are apparent and characterize the initial rate-limiting step of the process.

4. Conclusions

The results of the present study on the isoforms of the Hc, isolated from snails *Helix aspersa* maxima, allow classifying these Hcs as thermostable proteins (T_m~80°C). The high degree of oligomerization of hemocyanin molecules, in general, is probably one of the reasons for their increased thermal stability. The data obtained will facilitate the further investigation of therapeutic properties and applications of these dioxygen-binding proteins.

Conflicts of Interest

The authors declare that there are no conflicts of interest regarding the publication of this paper.

References

[1] T. Morrow and L. H. Felcone, "Defining the difference: what makes biologics unique," *Biotechnology Healthcare*, vol. 1, no. 4, pp. 24–26, 2004.

[2] A. Buchanan, F. Ferraro, S. Rust et al., "Improved drug-like properties of therapeutic proteins by directed evolution," *Protein Engineering, Design and Selection*, vol. 25, pp. 631–638, 2012.

[3] S. Krumova, S. Todinova, M. Tileva et al., "Thermal stability and binding energetics of thymidylate synthase ThyX," *International Journal of Biological Macromolecules*, vol. 91, pp. 560–567, 2016.

[4] K. Huus, S. Havelund, H. B. Olsen, M. van de Weert, and S. Frokjaer, "Chemical and thermal stability of insulin: effects of zinc and ligand binding to the insulin zinc-hexamer," *Pharmaceutical Research*, vol. 23, pp. 2611–2620, 2006.

[5] S. Todinova, M. Guncheva, and D. Yancheva, "Thermal and conformational stability of insulin in the presence of imidazolium-based ionic liquids," *Journal of Thermal Analysis and Calorimetry*, vol. 123, pp. 2591–2598, 2016.

[6] G. Préaux and C. Gielens, "Hemocyanins," in *Copper Proteins and Copper Enzymes*, R. Lontie, Ed., vol. 2pp. 159–205, CRC Press, Boca Raton, FL, USA, 1984.

[7] K. E. van Holde and K. I. Miller, "Hemocyanins," *Advance in Protein Chemistry*, vol. 47, pp. 1–81, 1995.

[8] A. Tchorbanov, K. Idakieva, N. Mihaylova, and L. Doumanova, "Modulation of the immune response using *Rapana thomasiana* hemocyanin," *International Immunopharmacology*, vol. 8, pp. 1033–1038, 2008.

[9] V. Gesheva, K. Idakieva, N. Kerekov et al., "Marine gastropod hemocyanins as adjuvants of non-conjugated bacterial and viral proteins," *Fish and Shellfish Immunology*, vol. 30, no. 1, pp. 135–142, 2011.

[10] V. Gesheva, S. Chausheva, N. Mihaylova et al., "Anti-cancer properties of gastropodan hemocyanins in murine model of colon carcinoma," *BMC Immunology*, vol. 15, no. 1, pp. 34–44, 2014.

[11] K. Idakieva, K. Parvanova, and S. Todinova, "Differential scanning calorimetry of the irreversible denaturation of Rapana thomasiana (marine snail, Gastropod) hemocyanin," *Biochimica et Biophysica Acta*, vol. 1748, no. 1, pp. 50–56, 2005.

[12] K. Idakieva, C. Gielens, N. I. Siddiqui et al., "Irreversible thermal denaturation of β-hemocyanin of *Helix pomatia* and its substructures studied by differential scanning Calorimetry," *Zeitschrift für Naturforschung A*, vol. 62, no. 9, pp. 499–506, 2007.

[13] K. Idakieva, P. Nikolov, I. Chakarska, V. L. Shnyrov, and N. Genov, "Spectroscopic properties and conformational stability of *Concholepas concholepas* hemocyanin," *Journal of Fluorescence*, vol. 18, no. 3-4, pp. 715–725, 2008.

[14] R. Sterner, T. Vogl, H. J. Hinz et al., "Extreme thermostability of tarantula hemocyanin," *FEBS Letters*, vol. 364, no. 1, pp. 9–12, 1995.

[15] K. Idakieva, Y. Raynova, F. Meersman, and C. Gielens, "Phenoloxidase activity and thermostability of *Cancer pagurus* and *Limulus polyphemus* hemocyanin," *Comparative Biochemistry and Physiology*, vol. 164, no. 3, pp. 201–209, 2013.

[16] K. Idakieva, F. Meersman, and C. Gielens, "Reversible heat inactivation of copper sites precedes thermal unfolding of molluscan (*Rapanathomasiana*) hemocyanin," *Biochimica et Biophysica Acta*, vol. 1824, no. 5, pp. 731–738, 2012.

[17] B. Lieb and J. Markl, "Evolution of molluscan hemocyanins as deduced from DNASequencing," *Micron*, vol. 35, no. 1-2, pp. 117–119, 2004.

[18] K. Streit, D. Jackson, B. M. Degnan, and B. Lieb, "Developmental expression of two *Haliotis asinina* hemocyanin isoforms," *Differentiation*, vol. 73, no. 7, pp. 341–349, 2005.

[19] K. Heirwegh, H. Borginon, and R. Lontie, "Separation and absorption spectra of α- and β-hemocyanin of *Helix pomatia*," *Biochimica et Biophysica Acta*, vol. 48, no. 3, pp. 517–526, 1961.

[20] R. Lontie, "Components, functional units, and active sites of *Helix pomatia* hemocyanin," *Life Chemistry Reports*, no. 1, pp. 109–120, 1983.

[21] C. Gielens, J. De Sadeleer, G. Preaux, and R. Lontie, "Identification, separation and cheracterization of the hemocyanin components of *Helix aspersa*," *Comparative Biochemistry and Physiology Part B: Comparative Biochemistry*, vol. 88, pp. 181–186, 1987.

[22] L. Velkova, I. Dimitrov, H. Schwarz et al., "Structure of hemocyanin from garden snail *Helix lucorum*," *Comparative Biochemistry and Physiology Part B: Biochemistry and Molecular Biology*, vol. 157, no. 1, pp. 16–26, 2010.

[23] Y. Raynova, L. Doumanova, and K. Idakieva, "Phenoloxidase activity of *Helix aspersa* maxima (garden snail, Gastropod) hemocyanin," *Protein Journal*, vol. 32, no. 8, pp. 609–618, 2013.

[24] Y. Raynova, S. Todinova, D. Yordanov, and K. Idakieva, "SDS-induced phenoloxidase activity of *Helix aspersa* maxima hemocyanin," *Bulgarian Chemical Communications*, vol. 46, pp. 111–116, 2014.

[25] R. Lumry and H. Eyring, "Conformational changes in proteins," *Journal of Physical Chemistry*, vol. 58, no. 2, pp. 110–120, 1954.

[26] B. I. Kurganov, A. E. Lyubarev, J. M. Sanchez-Ruiz, and V. L. Shnyrov, "Analysis of differential scanning calorimetry data for proteins, criteria of validity of one-step mechanism of irreversible protein denaturation," *Biophysical Chemistry*, vol. 69, no. 2-3, pp. 125–135, 1997.

[27] G. Marshall, P. Valtchev, F. Dehghani, and G. W. Gomes, "Thermal denaturation and protein stability analysis of *Haliotis rubra* hemocyanin," *Journal of Thermal Analysis and Calorimetry*, vol. 123, no. 3, pp. 2499–2505, 2016.

[28] V. Shnyrov, G. Zhadan, and I. Akoev, "Calorimetric measurements of the effect of 330-MHz radiofrequency radiation on human erythrocyte ghosts," *Bioelectromagnetics*, vol. 5, no. 4, pp. 411–418, 1984.

[29] V. Shnyrov and P. Mateo, "Thermal transitions in the purple membrane from *Halobacterium halobium*," *FEBS Letters*, vol. 324, no. 2, pp. 237–240, 1993.

[30] K. Idakieva, S. Stoeva, W. Voelter, and C. Gielens, "Glycosylation of *Rapana thomasiana* Hemocyanin. Comparison with Other Prosobranch (Gastropod) Hemocyanins," *Comparative Biochemistry and Physiology Part B: Biochemistry and Molecular Biology*, vol. 138, no. 3, pp. 221–228, 2004.

[31] C. Wang, M. Eufemi, C. Turano, and A. Giartosio, "Influence of the carbohydrate moiety on the stability of glycoproteins," *Biochemistry*, vol. 35, no. 23, pp. 7299–7307, 1996.

[32] E. J. Wood, M. F. Chaplin, C. Gielens, J. De Sadeleer, G. Preaux, and R. Lontie, "Relative molecular mass of the polypeptide chain of β-haemocyanin of *Helix pomatia* and carbohydrate composition of the functional units," *Comparative Biochemistry and Physiology Part B: Comparative Biochemistry*, vol. 82, no. 1, pp. 179–186, 1985.

[33] A. E. Lyubarev and B. I. Kurganov, "Study of irreversible thermal denaturation of proteins by differential scanning Calorimetry," *Recent Research Developments in Biophysical Chemistry*, vol. 2, pp. 141–165, 2001.

[34] A. E. Lyubarev and B. I. Kurganov, "Analysis of DSC data relating to proteins undergoing irreversible thermal denaturation," *Journal of Thermal Analysis and Calorimetry*, vol. 62, no. 1, pp. 51–62, 2000.

22

Voltammetric Behavior, Identifying and Quantitatively Determining Iron-Based Nanoparticles, and Evaluating their Stability in Simulated Solutions of Gastric Juice

N. M. Dubova,[1] G. B. Slepchenko,[1] I. A. Khlusov,[2,3,4] M. S. Ostapenko,[1] and E. A. Nesterov[5]

[1]*Physical and Analytical Chemistry Department, National Research Tomsk Polytechnic University, Tomsk, Russia*
[2]*Experimental Physics Department, National Research Tomsk Polytechnic University, Tomsk, Russia*
[3]*Laboratory of Immunology and Cell Biotechnologies, Immanuel Kant Baltic Federal University, Kaliningrad, Russia*
[4]*Morphology and General Pathology Department, Siberian State Medical University, Tomsk, Russia*
[5]*Laboratory No. 31 of the Nuclear Reactor, National Research Tomsk Polytechnic University, Tomsk, Russia*

Correspondence should be addressed to G. B. Slepchenko; slepchenkogb@mail.ru

Academic Editor: Bengi Uslu

The electrochemical behavior of Fe_3O_4 nanoparticles, iron nanoparticles coated with carbon, and diazonium salts by voltammetry methods using a carbon paste electrode (CPE) has been studied. There has been developed a voltammetric method for identifying and quantifying solid-phase iron-based nanoparticles in the background electrolyte of 0.02 mole/dm^3 Trilon B (pH 3.5) with the use of the CPE. Investigations of nanoparticles' stability with various coatings in a simulated solution of gastric juice have been carried out. Nanoparticles stability has been evaluated on the basis of determining Fe(III) ions in a simulated solution after contacting with nanoparticles within different periods of time using the method of inversion voltammetry. It has been shown that nanoparticles coated with carbon and salts of arendiazonium are the most resistant to aggressive media.

1. Introduction

The unique physical and chemical properties of metal nanoparticles (including iron-based nanoparticles) make it possible to use them in various sectors of the economy because of a wide range of potential use in biomedical, optical, and electronic spheres [1, 2]. In biology and medicine, nanomaterials are used, in particular, to develop facilities for directed delivery of drugs to the pathological focus of the body, as a magnetically controlled sorbent of toxins. The effect or transport of nanoparticles is often realized in an electrochemically aggressive biological environment, for example, the digestive tract (saliva and gastric and intestinal acid), which contributes to the degradation of nanoparticles with excreting ions that possess their own additional toxicity for the organism causing multiple organ pathology [3]. In connection with expanding the technological use of nanoparticles, there is growing threat of environmental pollution, in which their intake with water and food, alongside with air, is one of the main ways of getting nanoobjects into the human body and animals.

To reduce the harmful effects of nanoparticles on the human body, studies are underway to apply various protective coatings to their surface, for example, with carbon materials and organic substances [4–8]. Therefore, the problems of identifying and quantifying both the nanoparticles themselves and the products of their chemical interaction with those media in which they have been placed are acute.

In scientific literature are presented the main methods used to evaluate the physical and chemical characteristics of the functional properties of nanomaterials. Such investigations were carried out, for example, by the authors of [9–14]. To identify and determine qualitatively solid phases and various substances, highly sensitive, fairly inexpensive

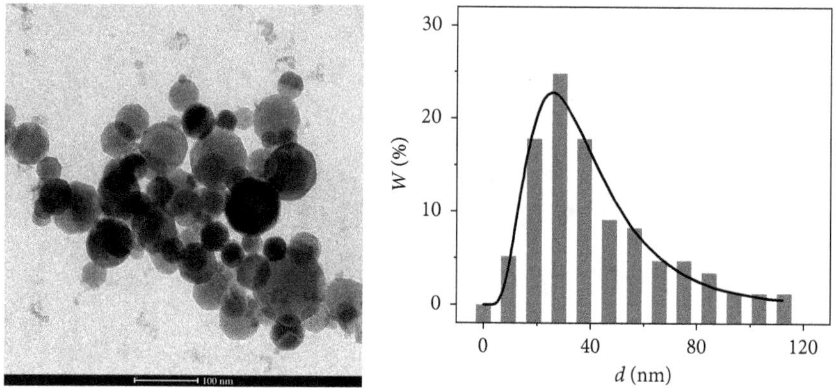

(a) (b)

FIGURE 1: Bright field TEM image (a) and size distribution (b) of Fe_3O_4 nanoparticles.

voltammetric methods employing an electrode with carbon paste are used [15–19]. It should be noted that, at present, there are limited data on the determination of iron-based nanoparticles using the voltammetric method with carbon paste electrode (CPE) in various sites, including biological fluids [20]. Therefore, an urgent problem is the development of methodologies for the quantitative determination and identification of iron-based nanoparticles in biological fluids using the voltammetric method.

The *work objective* is studying the electrochemical behavior of iron-based nanoparticles with various coatings using voltammetry methods, developing methods for their identification and quantitative determination, and studying stability of iron nanoparticles with various coatings in simulated solutions of gastric juice.

2. Experimental

In this work, there has been used a voltammetric complex for analytical measurements. The analytical voltammetric complex STA TU 4215-001-20694097-98 (Russia, Tomsk, ITM LLC) presents a compact device consisting of an electronic unit, a measuring unit with three electrochemical cells. The STA complex is completely computerized. The processing of the obtained data has been carried out using the established programs "STA." As the indicator electrode, a carbon paste electrode (CPE), there has been used a gold-graphite electrode. The CPE is a graphite rod with the diameter of 1.5–2.0 mm pressed into a fluoroplastic case with the diameter of 5-6 mm, so that the length of the protruding part of the fluoroplastic case is 1-2 mm. For producing the carbon paste, a powder of spectral graphite of C-4 grade with a particle size of 100–200 microns has been mixed with silicone oil (1.0 g of powder and 0.5 ml of oil) and introduced into the protruding part of the fluoroplastic case. In the gold-graphite electrode, a graphite rod with the working surface diameter of 3–5 mm, impregnated with a mixture of polyethylene and paraffin in the ratio of 1 : 1, has been used. Before depositing gold films in the "in situ" mode, the graphite surface has been ground with grinding skins and a filter, and then electrochemically degreased by conducting a cathodic (at −1.0 V) and anodic (at +1.2 V) polarization with the interval of 1–2 s within 60 s.

A saturated silver chloride electrode has been used as the reference electrode.

For studies, there have been used nanopowders: sample 1, iron oxide Fe_3O_4 (average particle diameter 25 nm); sample 2, iron in a carbon shell (average particle diameter 5–10 nm) obtained by gas-phase synthesis at the Institute of Metal Physics, Ural Branch of the Russian Academy of Sciences (Yekaterinburg); and sample 3, iron coated with a layer of carbon and salts of arendiazonium tosylates obtained by a team of employees of the Department of Biotechnology and the Engineering School of New Production Technologies [21].

3. Results and Discussion

Figure 1 shows the TEM image and the size distribution of Fe_3O_4 nanoparticles.

Transmission electron microscopy (TEM) has been carried out on the transmission microscope FEI Tecnai 20 G2 TWIN at the Laboratory of Electron Microscopy of NSTU in Novosibirsk.

Transparent electron microscopy (TEM, Tecnai 20 G2 TWIN, FEI Company, USA) has been used to study the morphology and size distribution of MNP.

According to TEM, Fe_3O_4 nanoparticles have a polyhedral shape with rounded edges (Figure 1(a)). The analysis of the TEM images shows a lognormal distribution in size with the average value of 25 (6–110) nm (Figure 1(b)).

Electrochemical transformations of Fe_3O_4 nanoparticles, carbon-coated nanoparticles, and iron nanoparticles coated with a layer of carbon and diazonium salts from the surface of the CPE under its cathodic and anodic polarization have been studied. All the studies have been carried out in the background electrolyte with 0.02 mol/dm^3 Trilon B solution (pH 3.5).

The cathodic and anodic volt-ampere curves of electric transformations of Fe_3O_4 nanoparticles (sample 1) have been recorded in the direct current mode (1st derivative) with the potential change at the rate of 80–90 mV/s in the range from +1.0 to −1.2 V cathodic scanning, from −1.2 to +1.0 V (anodic scanning). The reproduced analytical signal of Fe_3O_4 nanoparticles has been observed only on the anodic

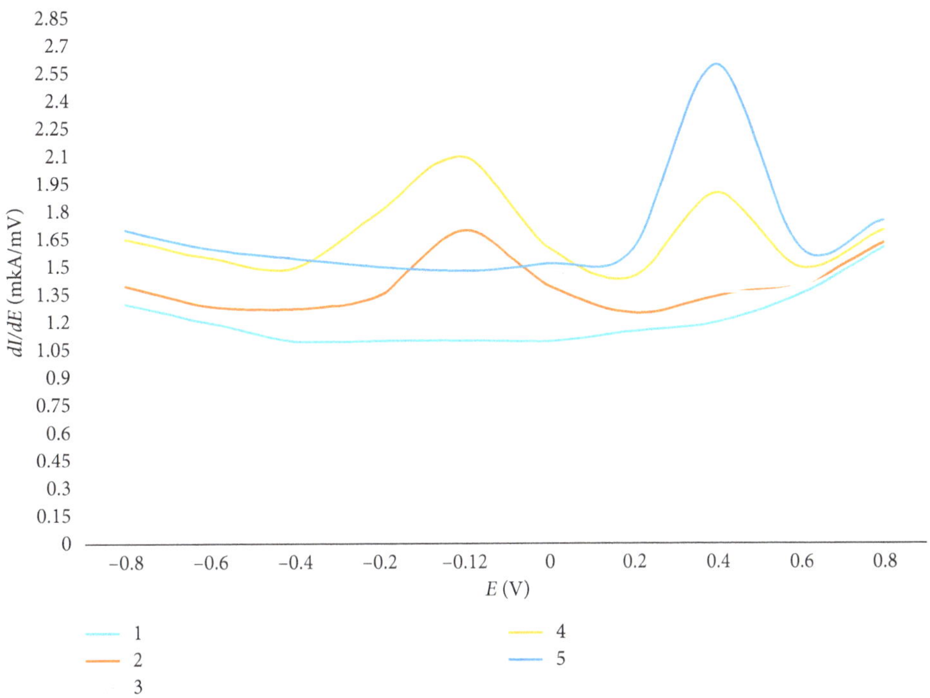

FIGURE 2: Anodic votammograms of nanoparticles from the CPE. 1, background 0.02 M Trilon B (pH 3.5); 2, Fe_3O_4 nanoparticles; 3, iron nanoparticles with carbon coating; 4, iron nanoparticles with carbon and diazonium salts coating; 5, diazonium salts.

voltammogram at the potential $E = -0.12$ V (Figure 2, curve 2). Since the anodic peak on the voltammogram occurs only after the cathodic polarization of the CPE, it is most likely associated with the electric oxidation of iron(II) to iron(III) from the nanoparticles to the CPE surface.

The anodic voltammogram of iron-coated carbon nanoparticles (sample 2) has also shown, as one would expect, only one anodic signal at the potential, $E_p = -0.12$ V (Figure 2, curve 3), coinciding with the value of the anodic peak potential of Fe_3O_4 nanoparticles (sample 1). The value of the anodic peak potential has not practically changed with increasing the amount of nanoparticles in the CPE. Therefore, as an analytical signal for identifying and quantitative determining solid-phase Fe_3O_4 nanoparticles (sample 1) and carbon-coated iron nanoparticles (sample 2), we selected the anodic peak at the potential $-(0.12 \pm 0.01)$ V, which has been recorded in the anodic mode after cathodic polarization of the electrode.

To study the impact of the background electrolyte on the shape of the current-voltage curves of Fe_3O_4 nanoparticles (sample 1) and carbon-coated iron nanoparticles (sample 2), the CPE has been kept with different contents of nanoparticles in the background electrolyte without applying a potential to the electrode for different periods of time (1 to 30 minutes).

Keeping the CPE with nanoparticles of samples 1 and 2 in the background electrolyte within the identified time has not led to changing voltammetric curves, there has been observed a permanent value of the anodic peak potential. The value of the maximum anodic current with increasing the CPE keeping time with Fe_3O_4 nanoparticles in the back has not practically decreased. An insignificant decrease of the anodic current has been observed only after keeping the CPE with Fe_3O_4 nanoparticles in the background electrolyte within 20 minutes and longer but has not exceeded the error of the signal measurement. This fact proves the absence of chemical dissolution (degradation) of nanoparticles from the CPE surface into the background electrolyte and confirms the fact that the analytical signal is directly associated with nanoparticles electric transformations from the CPE. Besides, keeping samples 1 and 2 in the background electrolyte has not led to appearance of additional signals and the anodic peak potential shift. This proves the absence of forming some new electrically active compounds on the CPE surface with Fe_3O_4 nanoparticles and iron nanoparticles with carbon coating.

On the anodic current-voltage curve of iron nanoparticles coated with a layer of carbon and diazonium salts (sample 3), besides a peak (at potential $-(0.12 \pm 0.01)$), there has been another additional anodic signal at the potential $E = (0.45 \pm 0.05)$ (Figure 2, curve 4) that is obviously related to electric transformations of diazonium salts. This fact is confirmed by the voltammogram shown in Figure 2, curve 5. There is presented a voltammogram of diazonium salts from the CPE. The signal value has increased with increasing the concentration of diazonium salts in the CPE. Thus, the analytical signal at $E_p = -(0.12 \pm 0.01)$ V of Fe_3O_4 nanoparticles (sample 1) with the carbon coating from the CPE surface (sample 2) on the anodic voltammogram and the analytical signal at the potential $E = (0.45 \pm 0.05)$ V of iron nanoparticles coated with a layer of carbon and diazonium salts (sample 3) can be used for their identification.

The value of anodic peaks at $E_p = -(0.12 \pm 0.01)$ V in samples 1 and 2 increased linearly with increasing the amount (weight fraction %) of nanoparticles in the CPE (Figure 3)

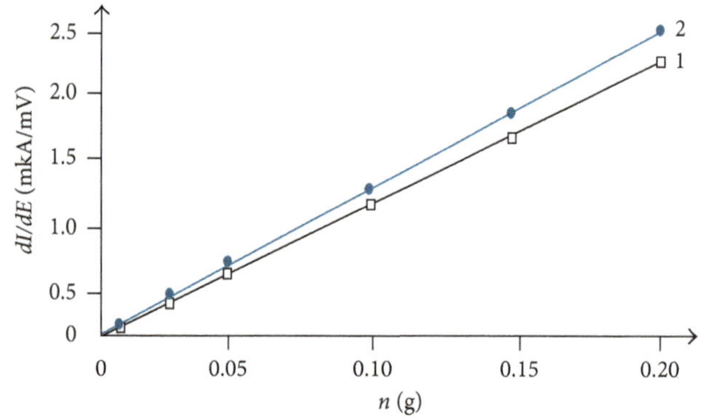

FIGURE 3: The anodic peak value dependence at $E_p = -(0.12 \pm 0.01)$ V on the quantitative amount of nanoparticles in the CPE: 1, Fe_3O_4 nanoparticles; 2, iron nanoparticles in the carbon shell.

that allows using them for the quantitative determination of the nanoparticles studied.

We estimated the quantitative number of Fe_3O_4 nanoparticles and iron nanoparticles in the carbon shell after contact with a physiological solution (0.5 M NaCl) for 20 minutes using the solid-phase voltammetry method in accordance with the calibration diagram constructed for the mass fraction of the Fe_3O_4 nanoparticle (Figure 3, curve 1) and the mass fraction of iron nanoparticles in the carbon shell (Figure 3, curve 2) in the carbon paste in the range (0.3–15)%. An estimation of the amount of nanoparticles was carried out after separation from a physiological solution by filtration, and then they were placed in CPE and analytic signals were recorded. An estimation of the correctness of the results of the voltammetric determination of nanoparticles in physiological solution ($P = 0.95$; $n = 3$) by the "introduced and found" method was obtained, and satisfactory convergence was obtained with an error of determination no more than 15%. These studies show the possibility of using the method of voltammetry with the CPE not only for identifying iron-based nanoparticles but also for their quantitative determination.

As it was mentioned above, in addition to the development of methods for identifying and quantifying solid-phase iron-based nanoparticles, it is of great interest to study stability of nanoparticles with various coatings (carbon and diazonium salts) in the solution of gastric juice. It has been natural to assume that when applying nanoparticles to the simulated solution of gastric juice, partial chemical dissolution will occur. We have proposed a method for studying stability (degradation) of iron nanoparticles after different periods of time of nanoparticles contacting with the model gastric juice by means of determining dissolved iron(III) in it by the method of stripping voltammetry (SV).

When studying various nanoparticles' degradation, the samples have been located in the simulated solution of gastric juice (0.1 M of HCl) and kept within different periods of time (from 1 to 22 hours). The analysis of the simulated solution of gastric juice after contacting with nanoparticles for the content of iron(III) ions has been carried out after filtering (their separation from the simulated solution of gastric juice) by the SV method with the use of the method of additives in the following conditions: the direct current mode of voltammograms (with registration of the 1st derivative); the polarizing voltage for electric accumulation $E_e = -1.0$ V; the speed of linear change of the potential 80 mV/s; and the background electrolyte 0.01 mol/dm^3 Trilon B. In these conditions, the potential of the analytical signal on the gold-graphite electrode (the anode peak potential of the trilonate complex of iron oxidation) is $E_p = -0.15$ V. Figure 4 shows the differential VA-curve of the simulated solution of gastric juice (curve 1) against the background Trilon B after contacting with nanoparticles.

The anodic peak potential obtained from the studied analyzed solutions (that are in contact with nanoparticles) coincides with the anodic peak potential of iron(III) ions and increases with addition of the standard solution of iron(III) (Figure 4, curve 2) that somehow proves the existence in the studied solution of iron(III), that is, chemical dissolution of nanoparticles.

In Figure 5, curve 1, there is presented the dependence of the anodic current of iron ions value on the time of keeping nanoparticles without coating in the solution 0.1 M of HCl.

The studies carried out have shown that when Fe_3O_4 nanoparticles are uncoated (sample 1), the chemical dissolution of nanoparticles and the appearance of iron(III) ions in the solution of 0.1 M HCl occur almost immediately. The number of iron(III) ions transferred to the solution increases by approximately 10 times with increasing the keeping time to 120 minutes (Figure 5, curve 1).

For iron-containing nanoparticles with carbon coating (sample 2), the appearance of iron(III) ions in the test solution of gastric juice was observed only after keeping them in 0.1 M HCl solution within 1 hour (Figure 5, curve 2) which indicates greater stability of such nanoparticles in the simulated solution in comparison with iron nanoparticles without carbon coating.

When iron nanoparticles coated with a layer of carbon and diazonium salts (sample 3) and were placed in the simulated solution of gastric juice, there was observed increasing chemical stability of nanoparticles in the time intervals studied. Iron(III) ions appeared in the simulated solution after 2 hours, and their concentration in the test solution decreased by approximately 3–5 times in comparison with the previous case (Figure 5, curve 3).

Studying stability of nanoparticles with different coatings has shown that they are most resistant to aggressive media (e.g., gastric juice), and consequently, nanoparticles coated with carbon and arendiazonium salts are the least harmful to the human body.

Thus, for evaluating stability of nanoparticles with various coatings, there is applicable method of inversion voltammetry that allows in a wide range of the determined contents of iron(III) ions evaluating it with a sufficient error.

The voltammetry method of solid phases is applicable for identifying and quantitative determining solid-phase iron-based nanoparticles.

It is simple, and it does not require a large number of reagents and labor intensity and can be used in any chemical laboratory, especially at present, when there are

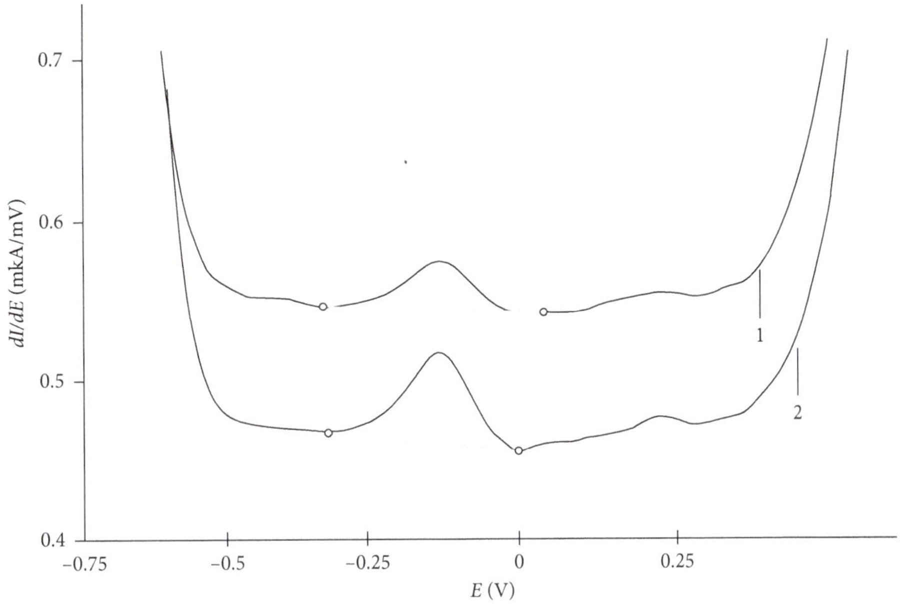

FIGURE 4: Differential voltammograms of the simulated solution of gastric juice (1) and the simulated solution with addition of iron(III) $C = 0.1\ \text{mg/dm}^3$ (2). Background is −0.02 M Trilon B gold-graphite electrode.

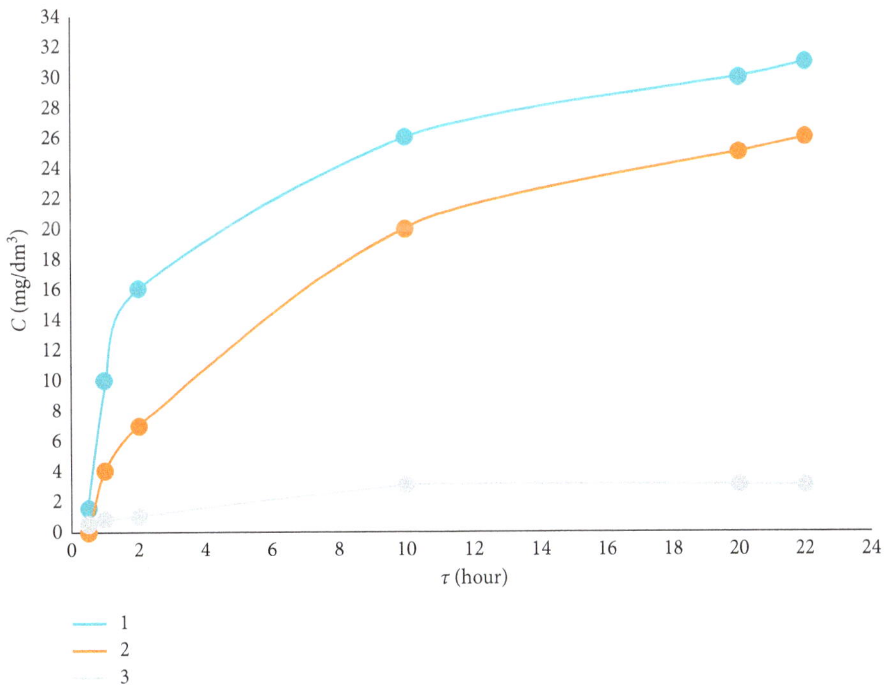

FIGURE 5: Fe(III) concentration in the simulated solution of gastric juice after keeping nanoparticles within different periods of time in 0.1 M HCl: 1, Fe_3O_4 nanoparticles; 2, iron nanoparticles with the carbon coating; and 3, iron nanoparticles coated with carbon and diazonium salts. Background: 0.02 M Trilon B gold-graphite electrode.

produced analyzers with controlling and data processing. This method can be used in pharmaceutical research and in technological control of producing nanomaterials.

4. Conclusion

There has been studied the electrochemical behavior of Fe_3O_4 nanoparticles, iron-based nanoparticles with different coatings, the voltammetry method for solid phases, the voltammetry method for identifying and quantitative determining solid-phase iron-based nanoparticles, and the method for their determining in various objects including model biological objects.

By the SV method, there has been studied stability of iron-based nanoparticles in the simulated solutions of gastric juice.

Conflicts of Interest

The authors declare that they have no conflicts of interest for this research.

Acknowledgments

This work was financed by the State Task (Science) 1.1343.2014 and was supported by Tomsk Polytechnic University CE Program VIU-Laboratory No. 31 ЯР-221/2018. The authors express gratitude to Professor A. Ye Yermakov, Cand. Phys.-Math. M. A. Uymin (Institute of Metal Physics of the RAS UD, Yekaterinburg), and Cand. Chem. T. A. Fedushchak (Institute of Oil Chemistry of the RAS SD, Tomsk), for the presented nanodimensional samples of magnetite and iron in the carbon shell.

References

[1] R. Taylor, S. Coulombe, T. Otanicar et al., "Small particles, big impacts: a review of the diverse applications of nanofluids," *Journal of Applied Physics*, vol. 113, no. 1, article 011301, 2013.

[2] Y. L. Hewakuruppu, L. A. Dombrovsky, C. Chen et al., "Plasmonic "pump–probe" method to study semi-transparent nanofluids," *Applied Optics*, vol. 52, no. 24, pp. 6041–6050, 2013.

[3] A. E. Chekanova, T. Sorkina, V. N. Nikiforov et al., "New environmental nontoxic agents for preparation of core-shell magnetic nanoparticles," *Mendeleev Communication*, vol. 19, no. 2, pp. 72–74, 2009.

[4] H. Shagholani and S. M. Ghoreishi, "Synthesis of nano-composition of poly acrylic acid/chitosan coated-magnetite nanoparticles to investigation of interaction with BSA and IGG proteins," *International Journal of Nanomaterials, Nanotechnology and Nanomedicine*, vol. 3, no. 1, pp. 27–33, 2017.

[5] B. Thiesen and A. Jordan, "Clinical applications of magnetic nano particles for hyperthermia," *International Journal of Hyperthermia*, vol. 24, no. 6, pp. 467–474, 2008.

[6] F. K. H. Van Landeghem, K. Maier-Hauff, A. Jordan et al., "Post-mortem studies in glioblastoma patients treated with thermotherapy using magnetic nanoparticles," *Biomaterials*, vol. 30, no. 1, pp. 52–57, 2009.

[7] T. Ji, X. Xu, X. Wang, Q. Zhou, and G. Chen, "Monodisperse water-stable Sio_2-coated fluoride upconversion nanoparticles with tunable shell thickness," *International Journal of Nanomaterials, Nanotechnology and Nanomedicine*, vol. 3, no. 1, pp. 15–18, 2017.

[8] C.-Y. Wu, C.-H. Lin, and Y.-C. Chen, "Using glucose-bound Fe_3O_4 magnetic nanoparticles as photothermal agents for targeted hyperthermia of cancer cells," *Journal of Nanomedicine and Nanotechnology*, vol. 6, no. 1, p. 264, 2015.

[9] B. Kalska-Szostko, U. Klekotka, K. Piekut, and D. Satula, "Stability of Fe_3O_4 nanoparticles in various simulated solutions," *Colloids and Surfaces A: Physicochemical and Engineering Aspects*, vol. 450, no. 1, pp. 15–24, 2014.

[10] D. V. Shtansky, "Transmission electron microscopy of high resolution in nanotechnology research," *Russian Journal of General Chemistry*, vol. 46, no. 5, pp. 81–88, 2002.

[11] V. A. Chebanov and S. I. Mironchenko, "Methods and evaluation nanoparticle characterization in pharmaceutical," *Success of Modern Science*, vol. 6, pp. 111-112, 2014.

[12] L. D. Belyakova, O. G. Larionov, A. A. Revina, and L. N. Kolomiets, *Modern Problems of Physical Chemistry of Nanomaterials: Digest of Articles*, 2008.

[13] C. Yao, Y. Xin-wei, L. Jian-Ping, Z. Guo-Peng, and P. Yuan-Fang, "Methods for determining the content of SiO_2 nanoparticles in composite coatings," *Diandu yu Tushi: Electroplat*, vol. 24, no. 4, pp. 27–29, 2005.

[14] A. M. Shipilin, *Mossbauer Spectroscopy of Iron Oxide Nanoparticles*, Author's abstract of dis. doctor of physical and mathematical sciences, Yaroslavl, Russia, 2000.

[15] V. A. Tsurin, A. E. Ermakov, M. A. Uimin et al., "Synthesis, structure, and magnetic properties of iron and nickel nanoparticles encapsulated into carbon," *Physics of the Solid State*, vol. 56, no. 2, pp. 287–300, 2014.

[16] A. A. Firooz, B. H. Nia, J. Beheshtian, and M. Ghalkhani, "Voltammetric sensor based on Fe-doped ZnO and TiO_2 nanostructures-modified carbon-paste electrode for determination of levodopa," *Journal of Electronic Materials*, vol. 46, no. 10, pp. 5657–5663, 2017.

[17] M. Dingab, Y. Zhouab, X. Liangab et al., "An electrochemical sensor based on graphene/poly (brilliant cresyl blue) nanocomposite for determination of epinephrine," *Journal of Electroanalytical Chemistry*, vol. 763, pp. 25–31, 2016.

[18] Y. Nikodimos and M. Amare, "Electrochemical determination of metronidazole in tablet samples using carbon paste electrode," *Journal of Analytical Methods in Chemistry*, vol. 2016, Article ID 3612943, 7 pages, 2016.

[19] A. C. O. Silva, L. C. F. de Oliveira, A. V. Delfino, M. R. Meneghetti, and F. C. de Abreu, "Electrochemical study of carbon nanotubes/nanohybrids for determination of metal species Cu^{2+} and Pb^{2+} in water samples," *Journal of Analytical Methods in Chemistry*, vol. 2016, Article ID 9802738, 12 pages, 2016.

[20] N. M. Dubova, G. B. Slepchenko, L. Ho Shi, and I. O. Maximchuk, "Investigation of iron-based nanoparticles action by solid-phase voltammetry," *Procedia Chemistry*, vol. 10, pp. 378–382, 2014.

[21] V. D. Filimonov, E. A. Krasnokytskaya, O. K. Poleshchuk, and Y. A. Lesina, "Theoretical analysis of reactions of electrophilic iodination and chlorination of benzene and polycyclic arenes in density functional theory approximation," *Russian Journal of Organic Chemistry*, vol. 44, pp. 681–687, 2008.

23

Characterization of Urban Particulate Matter by Diffusive Gradients in Thin Film Technique

Michaela Dufka and Bohumil Dočekal

Institute of Analytical Chemistry of the Czech Academy of Sciences, v.v.i., Veveří 97, 602 00 Brno, Czech Republic

Correspondence should be addressed to Bohumil Dočekal; docekal@iach.cz

Academic Editor: Fábio Rodrigo Piovezan Rocha

A diffusive gradient in thin films (DGT) technique was employed in characterization of the particulate matter related to the urban area suffering from heavy traffic. Kinetics of mobilization metals fluxes from the metal-contaminated particulate matter was investigated. To monitor responses of the particulate matter sample, DGT probes of various thickness of diffusion layer were deployed in aqueous model suspensions of the particulate matter for different time periods. Particulate matter samples and exposed DGT resin gels were decomposed in a mixture of nitric and hydrochloric acid in a microwave pressurized PTFE-lined system. Total content of some traffic-related elements (Cd, Co, Cu, Mo, Ni, Pb, Pd, Pt, Rh, Sb, and V) was determined by inductively coupled plasma mass spectrometry. DGT measurements revealed that two metals pools associated with particles could be recognized, which can be characterized as high soluble fraction and almost insoluble fraction. DGT-measured metal fluxes from the labile pool showed significant difference in mobilization and resupply fluxes of individual selected elements, which might reflect the origin of selected metals and their speciation in particulate matter. The DGT technique can be applied as a useful tool for characterization of metals mobilization from the particulate matter.

1. Introduction

The particulate matter (PM) has received growing attention since many epidemiological studies demonstrated an association between exposure to particles and adverse human health effects [1, 2]. Particulate pollution encompasses emissions from both natural and man-made sources. The main source of PM in urban areas is road transport in addition to the burning of fossil fuels in power stations and factories. The particulate matter emitted from on-road motor vehicles includes complex mixtures of metals from tires, brakes, parts wear, engine emissions, and dust from road surfaces. Fine particulate material containing platinum group elements (PGE) is also introduced into the environment due to utilization of three-way catalytic converters [3].

The PM comprises a mixture of numerous components, including various organic substances, elemental carbon, and inorganic substances. The metallic fraction is associated with several adverse health effects. The content of transition metals in PM is significantly affected by vehicle volume and speed, type of engine and its operation conditions, road type, rush time periods, neighboring environment and meteorological conditions, etc. [4, 5]. As urban traffic is continuously increasing, traffic-generated atmospheric pollutant loads will impose an even greater impact on human and ecosystem health. Epidemiological studies pointed out a number of adverse health effects associated with atmospheric particles originating from transportation [6–8].

Conventional analytical metal measurements in atmospheric particles usually entail the determination of total metal concentrations. Nevertheless, it is often necessary to quantify specific metallic forms, species, since bioavailability, solubility, geochemical transport, and metal cycles largely depend on physical-chemical speciation [9]. Consequently, toxicity of transition metals associated with atmospheric PM depends on solubility [10], speciation, and kinetics of metal release from the solid phase. The use of water as a leaching agent was recommended [9] because these conditions are more similar to environmental conditions of extraction by rain water or to conditions in the lung environment, when PM is inhaled.

In recent decades, batchwise equilibrium-based single [11] or sequential extraction schemes [9] have been consolidated as analytical tools for fractionation analysis to assess the ecotoxicological significance of metal ions in solid environmental samples. However, they cannot simply provide information about the kinetics of the leaching process. Moreover, the ratio of the solid phase mass to volume of leaching agent, particularly water, can considerably influence the yield of extraction. Taking into account that naturally occurring processes take predominantly place under dynamic conditions, recent trends have been focused on the development of alternative methods aimed at mimicking environmental events more correctly than their classical batchwise static operational and sequential extraction counterparts [12].

A comprehensive survey of these developments in assessment of bioaccessible trace metal fractions in airborne particulate matter is recently reported in a review [13]. Dynamic extractions procedures provide a continuous flow of the leaching agent through the sample; thereby, soluble species are dissolved and subsequently detected offline or online providing information about the kinetics of the leaching process. The risk of sample losses as well as contamination is reduced since troublesome and time-consuming sample handling steps of batchwise procedures such as sample shaking, filtration, or centrifugation can be avoided. In literature, dynamic extraction procedures based on the concepts of stirring chambers, rotating coiled columns, packed microcartridges, and flow-through microdialysis are used mostly for the analysis of samples like fly ash, rock, soil, and sediments [12, 14]. Application of these procedures to fractionation studies (e.g., [15]) is limited because airborne particulate matter, unlike soil, sediments, and fly ash, is only available with very limited sample amounts collected on filters, typically at submilligram to milligram level [11]. To overcome problems associated with this special kind of sample, various approaches were introduced [16–18]. Instead of whole sample only aliquots of the sample in the form of either circular parts or punches of the sampling filter substrate with homogeneously distributed particulate matter were used, which enables loading of sample into the extraction unit.

Another group of methods for determination of the water-soluble fraction of metals in atmospheric aerosols can be based on the continuous sampling of atmospheric aerosols into deionized water [10, 19]. By using the aerosol growth technology, saturated water vapor condenses on the PM forming droplets, which can be collected in various impactors. The sampled particles are leached in interaction with condensed water in time, so that the soluble fraction of elements can be subsequently determined after separation of insoluble part of particles by suitable high-sensitive detection technique.

A new approach, the diffusive gradients in thin film (DGT) technique [20], has been developed to assess the bioavailability of metals in environmental systems such as sediments and soils [21]. DGT technique can measure the fluxes of labile species in water, soil, and sediment compartments in situ under undisturbed conditions and without pretreatment. In the uptake process, metals diffuse from pore water through the diffusion layer to the binding phase, typically Chelex 100 resin gel [20, 22], which is selective to transition metals. As in an organism uptake, the DGT uptake reduces metal concentration in the aqueous phase in the vicinity of the DGT probe, so that it induces resupply of metals associated with the solid phase. Thus, DGT can measure the metal fraction that is accessible from the complete metal pool in the solid phase system [21, 23].

Interpretation of DGT-induced fluxes indicates the degree of depletion of the solution metal concentration in the vicinity of the DGT probe (as determined by the thickness of the diffusive layer), the kinetics of desorption of metals, and the extent of the pool(s) of labile metal species in the particulate phase [24, 25].

The scope of this study was to investigate the kinetics of metals resupplied from an urban particulate matter into the aqueous phase by employing the DGT technique. This approach represents a useful tool in estimation of the water-soluble, mobile fraction of metals in particulate matter. Until now, to the best of our knowledge, no results have been reported concerning utilization of the DGT technique for characterization of atmospheric particulate matter in direct contact of the probe with the moisturized PM, providing information on the metal contamination and on the kinetics of metal release.

2. Materials and Methods

2.1. Particulate Matter Samples. Within a period of years 2006–2016, the total suspended particulate matter (TSPM) was sampled in annual service intervals from filters of an air-conditioning unit in the building of the Institute of Analytical Chemistry of the CAS, v.v.i., in Brno, which is situated at the heavy traffic area on the Veveri street, Brno, Czech Republic, GPS location: 49°12′27.958″N, 16°35′27.845″E. In this study, the sample collected within the period of March 2008 to February 2009 was characterized in more detail. Concentration of the PM 2.5 fraction was found [26] within the range of 20–32 $\mu g{\cdot}m^{-3}$ for this sampling site.

2.2. Preparation of DGT Probes. The polyacrylamide hydrogels used in this study were prepared according to the procedure developed by Zhang and Davison [20]. Gel solution contained 15% by volume of acrylamide (Sigma-Aldrich, Germany) and 0.3% by volume of patented agarose cross-linker (DGT Research, Ltd., Lancaster, United Kingdom). *N,N,N′,N′*-Tetramethylethylenediamine (TEMED; Sigma-Aldrich) was used as catalyst. A freshly prepared solution of ammonium peroxydisulfate (10%, Sigma-Aldrich) was used as an initiator for polymerization. For preparation of polyacrylamide diffusive gels (APA), 10 μL TEMED, and 28 μL ammonium peroxydisulfate solutions were added to 4 mL gel solution. This solution was immediately cast between two glass plates separated alternatively by a plastic spacer of 0.3, 0.5, 1.0, 1.5, and 2.0 mm thickness, respectively, and allowed to set in an oven at 42–46°C for 60 min. The gel was then removed from glass plates and hydrated in ultrapure water at least for 24 h before use. Ultrapure water was exchanged three times during hydration period. Finally, the gels exhibited a stable thickness of 0.45, 0.80, 1.6, 2.35, and 3.2 mm (tolerance ± 0.01 mm), respectively, according to spacers used. Gel discs of 25 mm in diameter were cut from hydrated gel strips using a round plastic knife and then stored in 0.01 $mol \cdot L^{-1}$ sodium nitrate solutions at room temperature.

The resin gel consisted of 0.5 g of Chelex 100 ion-exchange resin (Na form, 100–200 wet mesh; Bio-Rad Laboratories, USA) in 2.5 mL of the gel solution. After one-day swelling of the resin in the gel solution, 3.75 μL TEMED and 15 μL ammonium peroxydisulfate solutions were added. Contrary to preparation of the diffusive gel, a half amount of the initiator was added to the mixture to prolong the setting time and consequently to allow settling of the resin beads on one side of the gel. Discs of resin gels were cut by the same way as diffusive gels and then stored in ultrapure water at 4°C. They reached a thickness of 0.44 ± 0.01 mm.

The DGT sampling probes were assembled by inserting the resin gel inside the piston-type DGT probe (R-SDU, DGT Research Ltd., Lancaster, UK) with resin beads oriented towards the sampler exposition window. Afterwards, the resin gel was covered with the diffusive gel and a wet protective membrane filter (0.45 μm in pore size, 25 mm in diameter, and 0.13 mm in thickness (Supor®-450, Pall Corporation, Michigan, USA)), both components representing the overall diffusive layer thickness (Δg). Finally, the DGT probes were carefully closed. Blanks of resin gels were treated in the same way including probe preparation.

2.3. Deployment Experiments. The sample of PM taken from the textile filter of the air conditioner was gently stirred in order to prepare sufficient amount of homogeneous representative model sample for comparative experiments. A mixture of 20 g PM and 25 g water was equilibrated at laboratory temperature of 23°C for 24 h. The set of three DGT probes packed with polyacrylamide diffusive and Chelex 100 resin gels were gently immersed into the surface of the PM suspensions, making sure that there were no air bubbles between the PM slurry and the exposition window of the DGT probe. The probes with diffusive gel thicknesses of 1.6 mm were deployed at constant room temperature of 23°C for various time periods (short term up to 24 hours, long term up to 60 days). DGT devices were then retrieved from the PM suspension and rinsed with ultrapure water to remove all the particles adhered on the filter membrane. Analogously, the probes with diffusive gels of different thicknesses (0.45, 0.8, 1.6, 2.35, and 3.2 mm, resp.), that is, DGT probes with specific decreasing demand, were also used in a series of experiments in duration of 8 h of deployment to measure responses of PM.

2.4. Apparatus and Sample Analysis. In the analysis of the PM sample, pore water solution was collected from experimental suspensions by centrifugation in plastic tubes for 30 min. The supernatant was filtered through a 0.45 μm membrane filter Supor-450 to remove colloidal components, then stabilized by addition of nitric acid, and stored at 4°C before analysis.

Prior ICP MS analysis, resin gels (0.24 g) and samples of PM (0.1 g) were digested in a mixture of concentrated subboil grade acids consisting of 2 mL·HNO_3 and 0.5 mL·HCl. The samples were treated in precleaned PTFE (110 mL) reaction vessels by using focused microwave (MW) energy in one-stand closed high-pressure autoclave unit (UniClever, Plazmatronika, Wroclaw, Poland). In the first predigestion step, the sample aliquot was left to react at laboratory temperature for 5 min. Subsequently, it was heated in three steps, each in duration of 5 min, applying 70, 90, and 100 percent of microwave power (150 W) controlling working pressure within limits of 3.5/3.2, 4.0/3.5, and 4.5/4.2 MPa, respectively. After cooling down period of 10 min, the digests were diluted with ultrapure water to the final mass of 10 g. To achieve complete decomposition of PM samples inclusive silicates, 0.5 mL HF (Suprapur grade, Merck, Darmstadt, Germany) was also added to the mixture of subboil grade nitric and hydrochloric acids. Blank samples of acids ($n = 4$) and resin gels ($n = 6$) were processed analogously.

Ultrapure water, prepared by Ultra Clear system (SB Barsbüttel, Germany), was exclusively used throughout the work. The subboil grade HNO_3 and HCl were obtained by purifying concentrated analytical reagent grade acids (Penta, Prague, Czech Republic) in the subboil quartz distillation system model MSBQ 2 (Maasen, Eningen, Germany).

Inductively coupled plasma mass spectrometry was applied as a multielement, high-sensitive technique in the determination of metals in sample solutions. An Agilent 7700 Series ICP MS with MicroMist Nebulizer was employed under operating conditions summarized in Table 1. The Octopole Reaction System (ORS) of the 7700 ICP MS was operated in helium collision mode (He mode) to exclude potential isobaric interferences according to the US EPA Method 6020A, validated by multilaboratory studies [27]. Possible spectral overlaps were checked by matrix interference and recovery tests [28]. For these tests, a series of synthetic-matrix samples and standards were prepared from the certified single-element standard stock solution Astasol® (Analytika, Prague, Czech Republic) to obtain background equivalent concentrations of elements, especially for noble metals. The analysis of real PM sample solutions revealed that the matrix elements were present in solutions below the critical interfering concentrations to cause spectral overlaps when using the He mode of the collision cell. Potential overlaps on $^{103}Rh^{+}$ by $^{87}Sr^{16}O^{+}$, on $^{105}Pd^{+}$ by $^{89}Y^{16}O^{+}$, and $^{179}Hf^{16}O^{+}$ on $^{195}Pt^{+}$ were not found. When applicable, flame and electrothermal atomic absorption spectrometry was also applied to verify the results. In these comparative measurements, the bias of results was below 10% of relative standard deviation of values found by ICP MS.

The multielement calibration standards for multipoint (level) calibration were prepared by mixing of certified Astasol standards. For fast screening of the digested PM, the semiquantitative analysis mode of ICP MS was applied. In

TABLE 1: Operating conditions applied in the analysis of samples by the Agilent 7700 ICP MS instrument.

Parameter	Value
Plasma mode	Normal
RF power	1600 W
Sampling depth	7 mm
Carrier gas flow	0.6 L/min
Dilution gas flow	0.55 L/min
Spray chamber temperature	2°C
Extraction lens 1	0 V
Extraction lens 2	200 V
He cell gas flow rate	4.8 mL/min
Internal standards (online addition)	^{45}Sc, ^{209}Bi
Monitored isotopes	
In semiquantitative mode	^{7}Li, ^{9}Be, ^{23}Na, ^{24}Mg, ^{27}Al,^{39}K, ^{45}Sc,^{47}Ti, ^{51}V, ^{52}Cr, ^{55}Mn, ^{56}Fe, ^{59}Co, ^{60}Ni, ^{63}Cu, ^{66}Zn, ^{75}As, ^{78}Se, ^{87,88}Sr, ^{89}Y, ^{95}Mo, ^{107}Ag, ^{111}Cd, ^{115}In, ^{118}Sn, ^{121}Sb, ^{137}Ba, ^{140}Ce, ^{179}Hf, ^{205}Tl, ^{206}Pb
In conventional analytical mode for selected traffic-related elements	^{51}V, ^{59}Co, ^{60}Ni, ^{63}Cu, ^{95}Mo, ^{103}Rh, ^{105}Pd, ^{111}Cd, ^{121}Sb, ^{195}Pt, ^{206}Pb

this instance, a one-point calibration was used, in which a single multielement calibration standard (1 $\mu g{\cdot}L^{-1}$ Li, Mg, Co, Ce, Y, Tl + 10 $\mu g{\cdot}L^{-1}$ As) was introduced before sample analysis to update the response factors.

The level of overall blanks related to analysis of PM samples and to analysis of DGT resin disks was negligible. Except platinum group metals, the blank level was 4 to 5 orders of magnitude lower than that of sample solutions, so that it could be omitted in evaluation of results. Based on 3-sigma criterion of blank fluctuation, limits of detection of selected traffic-related elements of 23 $ng{\cdot}L^{-1}$ Cd, 3 $ng{\cdot}L^{-1}$ Co, 55 $ng{\cdot}L^{-1}$ Cu, 90 $ng{\cdot}L^{-1}$ Mo, 74 $ng{\cdot}L^{-1}$ Ni, 14 $ng{\cdot}L^{-1}$ Pb, 11 $ng{\cdot}L^{-1}$ Pd, 0.14 $ng{\cdot}L^{-1}$ Pt, 0.05 $ng{\cdot}L^{-1}$ Rh, 26 $ng{\cdot}L^{-1}$ Sb, and 7 $ng{\cdot}L^{-1}$ V in aqueous media (supernatant, PM digest) and 0.13 ng Cd, 0.07 ng Co, 1.1 ng Cu, 2 ng Mo, 8 ng Ni, 0.82 ng Pb, 0.56 ng Pd, 0.002 ng Pt, 0.001 ng Rh, 14 ng Sb, and 0.22 ng V in DGT resin gel disks were achieved, respectively.

3. Results and Discussion

3.1. Characterization of the PM Sample. Particulate air pollution constitutes a complex mixture of particles, present in the atmosphere as solids or liquids that vary in mass, number, size, shape, surface area, chemical composition as well as reactivity, acidity, solubility, and origin [2].

Content of the inorganic fraction in the PM sample was assessed by charring approximately 1 g aliquots of the dry PM in ambient air atmosphere in three steps at 200°C for 1 h, 500°C for 3 h, and 900°C for 5 h. In this instance, residual ash represented 57.9 ± 0.5% ($n = 3$) of the original sample mass. The organic fraction of the PM contained probably some parts of plants, pollen species, common organic compounds as saccharides, organic acids, PAH, and high -molecular-weight alkanes [29].

Preliminary examination of the PM sample was performed by using a semiquantitative mode of ICP MS in which content of up to 80 elements could be estimated. The main components of the PM sample were as follows: Al (7.5%), Fe (4.2%), K (1.4%), Mg (0,75%), Na (0.59%), Zn (0.40%), Ti (0.25%), Ba (0.08%), and Mn (0.06%). The metallic elements Ca, Mg, Al, Si, Fe, and Mn are typical markers for soil and resuspended dust sources in the atmosphere [30]. Crustal and other elements were obviously the major contributors to composition of the PM. The results of conventional ICP MS determination of the total content of selected minor, traffic-related elements (Cd, Co, Cu, Mo, Ni, Pb, Pd, Pt, Rh, Sb, and V) are summarized in Table 2. Platinum group elements were found at similar concentration ratio as in German urban areas [3].

In natural media, metal contaminants undergo reactions with water, ligands dissolved in it, and with surface sites of the solid material. The metal partitioning is usually characterized by distribution coefficient, which is the ratio of adsorbed metal concentration ($mg{\cdot}kg^{-1}$) to the dissolved metal concentration ($mg{\cdot}L^{-1}$) at equilibrium [31]. In this work, the mobilization release of metals into water under static conditions is described by the reciprocal value, that is, by elution distribution coefficient K_d, which reflects the metal distribution between the aqueous phase and the PM during leaching of samples with water. The K_d values were evaluated (Table 2) from the pore water metal concentration (mixture of 20 g PM and 25 g water in equilibrium, phase ratio $r \sim 1.25$) and the content of metal in the PM (solid phase). The values differ among individual traffic-related metals, decreasing exactly in the following order: Cd > Sb > Cu > Pd > Mo > Co~Ni > Pb~V~Rh > Pt. These differences might reflect the origin and the speciation of individual selected metals.

3.2. DGT Experiments. The basic theory of DGT technique assumes that transport in both the diffusion layer of the thickness Δg (the thickness of polyacrylamide gel and filter membrane) and in the pore water of the slurry of PM was solely driven by molecular diffusion and that all labile metal

Table 2: Characteristics of traffic-related metals in the PM listed according to source categories.

Origin/element	Total content ($\mu g \cdot g^{-1}$) ($n = 3$)	Soluble fraction (%) ($n = 5$)	K_d ($g \cdot mL^{-1}$) ($n = 10$) Mean ± SD	DGT-measured mobile metal fraction (%) ($n = 3$)	Change of the DGT-measured resupply rate (%) ($n = 5$)
Disk brake					
Cu	625 ± 12	9.9 ± 0.5	0.079 ± 0.004	13.4 ± 2.4	14.3 ± 0.2
Sb	105 ± 16	32.8 ± 1.2	0.263 ± 0.010	1.8 ± 0.2	4.4 ± 1.7
Car body					
Cd	10.1 ± 0.2	38.3 ± 2.5	0.306 ± 0.021	9.0 ± 1.8	18.7 ± 5.8
Co	30.6 ± 0.8	4.4 ± 0.2	0.035 ± 0.002	31.0 ± 3.3	56.9 ± 18.2
Ni	146 ± 4	4.3 ± 0.2	0.034 ± 0.002	22.6 ± 2,6	27.5 ± 2.8
Pb	411 ± 8	2.3 ± 0.2	0.019 ± 0.002	5.7 ± 1.2	10.3 ± 0.1
Power fuel					
Mo	53.9 ± 0.4	7.0 ± 0.3	0.056 ± 0.002	5.8 ± 0.5	12.1 ± 2.7
V	30.4 ± 0.7	2.3 ± 0.1	0.018 ± 0.001	17.4 ± 2.1	23.6 ± 7.2
Catalytic converter[a]					
Pd	0.186 ± 0.004	8.8 ± 0.1	0.0700 ± 0.0004	26.0 ± 8.2	7.7 ± 0.9
Pt	0.016 ± 0.002	0.22 ± 0.02	0.0017 ± 0.0001	<20	n.d.[b]
Rh	0.034 ± 0.001	2.3 ± 0.1	0.0184 ± 0.0006	53.9 ± 3.8	19.2 ± 3.6

Note. Solubility is represented by the water-soluble fraction ($n = 5$) of the total metal content after 1-day extraction ($r = 1.25\ mL \cdot g^{-1}$), K_d is the average ($n = 10$) distribution coefficient for aqueous extraction, the mobile metal fraction is the percentage of the C_{DGT}-measured concentration from the overall metal concentration in the aqueous phase for the lowest demand of the probe ($\Delta g = 3.33$ mm, 8 h, $n = 3$) applied in the DGT experiment, and change of DGT-measured resupply rate is the percentage decrease of the metal flux per day within one-day short-term experiment ($\Delta g = 1.73$ mm, $n = 5$). n: number of replicates in evaluation of data from series of experiments; [a]limits of detection (LOD) of 11 $ng \cdot L^{-1}$ Pd, 0.14 $ng \cdot L^{-1}$ Pt, and 0.05 $ng \cdot L^{-1}$ Rh for determination of these metals in aqueous media (supernatant, digest) and 0.56 ng·Pd, 0.002 ng·Pt, and 0.001 ng·Rh in DGT resin gel disks were achieved, respectively; [b]n.d.: not determined.

species in the pore water had a single self-diffusion coefficient D that is related to free metal ions. If the binding of a metal to the resin is strong and fast in relation to the transport rate of metal species to the resin layer by diffusion, the average flux F, related to the unit area of the resin during the deployment, is given by the following equation [20, 22]:

$$F = \frac{M}{(At)} = \frac{Dc}{\Delta g}, \tag{1}$$

where M is the mass of the metal trapped (metal uptake) in the resin (mol), A is the area of the DGT exposition window (cm^{-2}) and t is the deployment time (s), D is the diffusion coefficient ($cm^2 \cdot s^{-1}$), Δg is the thickness of diffusion layer (cm), and c ($mol \cdot L^{-1}$) is the metal concentration in the pore water solution at the sampling window of the probe. This equation can be applied in evaluation and interpretation of DGT-measured data, in assessment of kinetic phenomena of metal release from the PM.

The plots of DGT-measured uptake and flux of lead versus deployment time (Figures 1(a)–1(c)) are presented as examples of results of DGT experiments. Similar results were obtained also for other monitored metals listed in Table 2. The DGT uptake sharply decreases during a short deployment period of time. After few hours of deployment, the resupply rate from the solid phase to the solution is evidently too slow to sustain metal concentration in the aqueous phase, that is, to follow the demand of the DGT probe. Consequently, the metal flux into the DGT probe asymptotically approaches zero (Figure 1(c)). Based on the overall concentration of a metal in the aqueous phase, uptake yields below and close to only one percent of the nominal value were recorded by the DGT probes. This observation documents that diffusive processes from the PM slurry are predominantly responsible for the resupply of metals into the solution adjacent the DGT probe [32]. This phenomenon was also reflected by the effect of the demand of the DGT probe, controlled by the diffusive gel thickness Δg (Figure 1(d)), even for a very thick diffusive layer thickness of 3.33 mm. Initial DGT-measured resupply fluxes of Cd, Co, Cu, Mo, Ni, Pb, Pd, Pt, Rh, Sb, and V reached values of 3.7, 8, 160, 2.7, 27, 4, 1.1×10^{-2}, 1×10^{-5}, 1.1×10^{-3}, 8.4, and 1.9 $nmol \cdot cm^{-2} \cdot day^{-1}$, respectively.

Generally, two metal pools associated with particles could be recognized, that is, a labile pool characterized by weak interactions (high solubility fraction) and an immobile, inert pool (insoluble fraction) incorporated strongly in the solid phase of the PM. Similar dissolution behavior was also observed for some elements in natural and anthropogenic particulate matter samples [15].

The general consensus is that ionic, water-soluble metal forms are the most bioavailable species, having the potential to access cells and organs of biota. Solubility and mobility (mobile metal fraction) of individual traffic-related metals are presented in Table 2. As documented by these data and

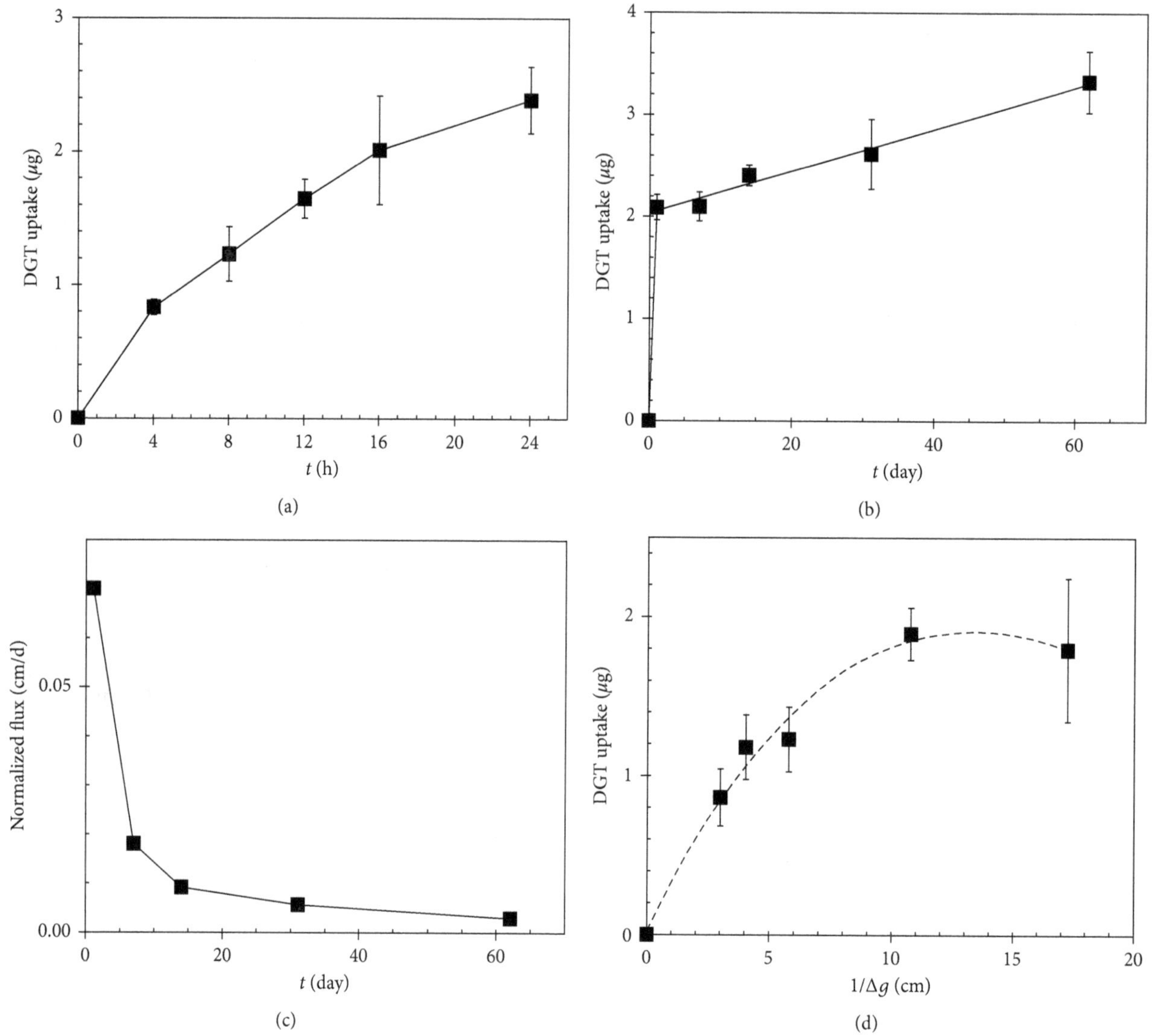

FIGURE 1: Mass of Pb found in the resin gel of the DGT probe immersed in the PM suspension ($n = 3$) for short (a) and long (b) deployment time periods ($\Delta g = 1.73$ mm). The metal flux into the DGT probe (c) is normalized for the metal concentration in the pore water solution. The effect of the diffusive layer thickness (8 h), representing the response of PM to the DGT probe demand, onto Pb uptake is also depicted (d).

results of DGT experiments (Figure 1), the soluble fraction of the metal pool is released within a very short time period, that is, within preparation of aqueous PM suspension and its equilibration period of one day. It is interesting to note that overall solubility of metals differs among elements of interest. It exceeded 30% of the total metal content of Cd and Sb, whereas solubility of other elements did not reach 10%. The residual fraction of the metal pool was almost insoluble. The kinetics of the subsequent release of metals was very slow even within the period of several months.

Particulate matter is a complex, heterogeneous mixture. It encompasses many different chemical components and physical characteristics. Many of them have been cited as potential contributors to toxicity [33]. Each component might have multiple sources. Therefore, identification and quantification of specific components or source-related mixtures represent one of the most challenging areas of environmental health research. The main source of PM in urban areas is the road transport [2]. Consequently, the metals of interest are divided into the source categories. It is evident from the comparison of the data in Table 2 that low content and minor soluble fraction do not represent low mobility, as documented for elements Co, Pd, and Rh. In this instance, mobility of elements can be expressed by the DGT-measured mobile fraction of the leachate, presented in Table 2, within the first stage of deployment experiments, for example, by the slope of a tangent in Figure 1(d). This mobile fraction reaches approximately 31, 26, and 54% of the total soluble fraction of Co, Pd, and Rh in aqueous leachate, respectively.

Resupply fluxes can be estimated from the short-term DGT experiments, from relation of calculated metal flux into the DGT probe and duration of deployment period. Calculated metal fluxes should be normalized [20] by the metal pore water concentration, as shown in Figure 2, in order to

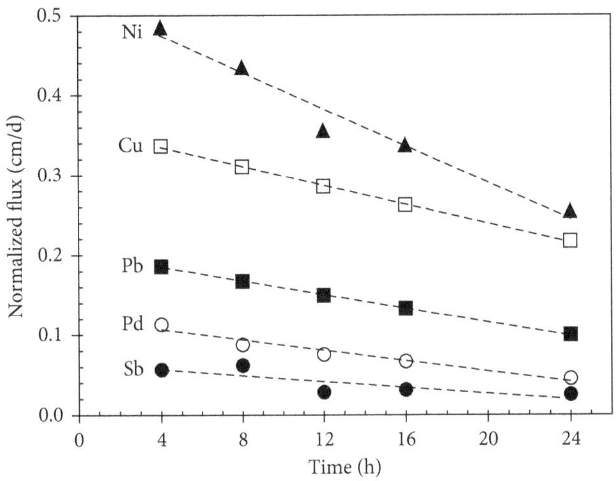

Figure 2: Comparison of normalized metal fluxes of Cu, Ni, Pb, Pd, and Sb into DGT probes during short-term deployment experiments ($\Delta g = 1.73$ mm; $n = 3$).

make comparison of these relations possible. Quasilinear relationship was found between the flux and the deployment time for most of the metals investigated. This decrease of the normalized flux (Table 2), expressed by the slope of the linear function, reflects depletion of a pool capacity of the mobile metal fraction with duration of deployment. The capability of resupplying metal is low when the slope is high. The slope of these lines decreased in the following order: Co, Ni, V, Rh, Cd, Cu, Mo, Pb, Pd, and Sb. In addition, the level of the normalized flux in Figure 2 represents the mobile fraction of the metal pool. In this graph, it is reflected by the virtual intercept of the line with ordinate axis. These plots can provide very interesting characteristics of PM with respect to behavior of individual elements in interaction with aqueous phase. Significant differences can be found among the elements investigated. For example, when comparing results presented in Figure 2 with the data in Table 2, they show for Sb total content of 104 $\mu g \cdot g^{-1}$ Sb, relatively high solubility (32.8%), very low mobility of soluble species (1.8%), and relatively well-balanced resupply kinetics (buffering capacity) both latter assessed by DGT experiments. On the other hand, results for Pb exhibit higher total content of 411 $\mu g \cdot g^{-1} \cdot$Pb, low solubility (2.3%), higher mobility (5.7%), and slower resupply kinetics.

Similar results were also obtained in series of experiments for other PM samples taken within a period of years 2006–2016 on the same sampling site. They are not presented in this communication, because no significant trends or changes in these data have been observed.

4. Conclusion

The results of aqueous leaching and simultaneous DGT experiments showed that DGT technique can be successfully applied for characterization of total suspended particulate matter taken in urban area, for characterization of transition metal soluble fractions and their dissolution kinetics. DGT probes deployed in the moisturized PM disturb metal concentration close to the sampling window of the probe, so that the DGT-measured data can monitor for the response of the PM, that is, for the kinetics of metals release from the solid phase. The study of mobilization of important metals related to traffic, to various car components and fuel, showed considerable differences in metal fractionation, that is, in their content and solubility, in mobility of soluble species. DGT resupply experiments revealed that most metals in the particulate matter from Brno city were present in two fractions, in a very soluble mobile one and in an almost insoluble immobile one. This observation is in contrast to the typical behavior of soils, which exhibit more balanced resupply fluxes, that is, higher buffering metal pools for deployment time periods of several days [34, 35].

In this work, homogenized PM, representing annual accumulation on large textile fiber filters in air-conditioning system, were taken for pilot model DGT experiments on particulate slurries. Nevertheless, based on these results, the DGT technique can be successfully applied in future in a modified arrangement for measurements with routinely used filters, for example, nitrate cellulose disks of the relevant diameter, for monitoring purposes, for assessment of soluble, mobile fractions of toxic elements in aerosols of interest.

Conflicts of Interest

The authors declare that there are no conflicts of interest regarding the publication of this paper.

Acknowledgments

This work was performed with the Institutional Support RVO 68081715 and the Project no. 503/10/2002 of the Czech Science Foundation.

References

[1] C. A. Pope, R. T. Burnett, M. J. Thu et al., "Lung cancer, cardiopulmonary mortality, and long-term exposure to fine particulate air pollution," *Journal of the American Medical Association*, vol. 287, no. 9, pp. 1132–1141, 2002.

[2] F. J. Kelly and J. C. Fussell, "Size, source and chemical composition as determinants of toxicity attributable to ambient particulate matter," *Atmospheric Environment*, vol. 60, pp. 504–526, 2012.

[3] F. Zereini, H. Alsenz, C. L. S. Wiseman et al., "Platinum group elements (Pt, Pd, Rh) in airborne particulate matter in rural vs. urban areas of Germany: concentrations and spatial patterns of distribution," *Science of the Total Environment*, vol. 416, pp. 261–268, 2012.

[4] E. Apeagyei, M. S. Bank, and J. D. Spengler, "Distribution of heavy metals in road dust along an urban-rural gradient in Massachusetts," *Atmospheric Environment*, vol. 45, no. 13, pp. 2310–2323, 2011.

[5] B. A. Lesniewska, B. Godlewska-Zylkiewicz, B. Bocca, S. Caimi, S. Caroli, and A. Hulanicki, "Platinum, palladium and rhodium content in road dust, tunnel dust and common grass in Biaystok area (Poland): a pilot study," *Science of the Total Environment*, vol. 321, pp. 93–104, 2004.

[6] D. W. Dockery, C. A. Pope, X. P. Xu et al., "An association between air pollution and mortality in six U.S. cities," *New*

England Journal of Medicine, vol. 329, no. 24, pp. 1753–1759, 1993.

[7] J. Schwartz, D. W. Dockery, and L. M. Neas, "Is daily mortality associated specifically with fine particles?," *Journal of the Air & Waste Management Association*, vol. 46, no. 10, pp. 927–939, 1996.

[8] J. M. Samet, F. Dominici, F. C. Curriero, I. Coursac, and S. L. Zeger, "Fine particulate air pollution and mortality in 20 US cities, 1987–1994," *New England Journal of Medicine*, vol. 343, no. 24, pp. 1742–1749, 2000.

[9] A. J. Fernandez, M. Ternero, F. J. Barragan, and J. C. Jimenez, "An approach to characterization of sources of urban airborne particles through heavy metal speciation," *Chemosphere-Global Change Science*, vol. 2, no. 2, pp. 123–136, 2000.

[10] M. Vojtesek, P. Mikuska, and Z. Vecera, "Analysis of water-soluble fraction of metals in atmospheric aerosols using aerosol counterflow two-jets unit and chemiluminescent detection," *International Journal of Environmental Analytical Chemistry*, vol. 92, no. 4, pp. 432–449, 2012.

[11] M. D. Santos, D. Gómez, L. Dawidowski, E. Gautier, and P. Smichowski, "Determination of water-soluble and insoluble compounds in size classified airborne particulate matter," *Microchemical Journal*, vol. 91, no. 1, pp. 133–139, 2009.

[12] M. Miro, E. H. Hansen, R. Chomchoei, and W. Frenzel, "Dynamic flow-through approaches for metal fractionation in environmentally relevant solid samples," *Trends in Analytical Chemistry*, vol. 24, no. 8, pp. 759–771, 2005.

[13] A. Mukhtar and A. Limbeck, "Recent developments in assessment of bio-accessible trace metal fractions in airborne particulate matter: a review," *Analytica Chimica Acta*, vol. 774, pp. 11–25, 2013.

[14] M. Miró and W. Frenzel, "The potential of microdialysis as an automatic sample-processing technique for environmental research," *Trends in Analytical Chemistry*, vol. 24, pp. 324–333, 2005.

[15] K. V. Desboeufs, A. Sofikitis, R. Losno, J. L. Colin, and P. Ausset, "Dissolution and solubility of trace metals from natural and anthropogenic aerosol particulate matter," *Chemosphere*, vol. 58, no. 2, pp. 195–203, 2005.

[16] J. J. Morales-Riffo and P. Richter, "Rapid determination of inorganic elements in airborne particulate matter by using acidified subcritical-water extraction and inductively-coupled plasma–optical-emission spectrometry," *Analytical and Bioanalytical Chemistry*, vol. 380, no. 1, pp. 129–134, 2004.

[17] A. Mukhtar and A. Limbeck, "On-line determination of water-soluble zinc in airborne particulate matter using a dynamic extraction procedure coupled to flame atomic absorption spectrometry," *Journal of Analytical Atomic Spectrometry*, vol. 25, no. 7, pp. 1056–1062, 2010.

[18] A. Limbeck, C. Wagner, B. Lendl, and A. Mukhtar, "Determination of water soluble trace metals in airborne particulate matter using a dynamic extraction procedure with on-line inductively coupled plasma optical emission spectrometric detection," *Analytica Chimica Acta*, vol. 750, pp. 111–119, 2012.

[19] C. B. Kidwell and J. M. Ondov, "Development and evaluation of a prototype system for collecting sub-hourly ambient aerosol for chemical analysis," *Aerosol Science and Technology*, vol. 35, no. 1, pp. 596–601, 2001.

[20] H. Zhang and W. Davison, "Performance-characteristics of diffusion gradients in thin-films for the in-situ measurement of trace-metals in aqueous-solution," *Analytical Chemistry*, vol. 67, no. 19, pp. 3391–3400, 1995.

[21] R. Bade, S. Oh, and W. S. Shin, "Diffusive gradients in thin films (DGT) for the prediction of bioavailability of heavy metals in contaminated soils to earthworm (Eisenia foetida) and oral bioavailable concentrations," *Science of the Total Environment*, vol. 416, pp. 127–136, 2012.

[22] W. Davison and H. Zhang, "In-situ speciation measurements of trace components in natural-waters using thin-film gels," *Nature*, vol. 367, no. 6463, pp. 546–548, 1994.

[23] H. Zhang, F. J. Zhao, B. Sun, W. Davison, and S. P. McGrath, "A new method to measure effective soil solution concentration predicts copper availability to plants," *Environmental Science and Technology*, vol. 35, no. 12, pp. 2602–2607, 2001.

[24] H. Zhang, W. Davison, B. Knight, and S. McGrath, "In situ measurements of solution concentrations and fluxes of trace metals in soils using DGT," *Environmental Science and Technology*, vol. 32, no. 5, pp. 704–710, 1998.

[25] M. P. Harper, W. Davison, and W. Tych, "DIFS—a modelling and simulation tool for DGT induced trace metal remobilisation in sediments and soils," *Environmental Modelling and Software*, vol. 15, no. 1, pp. 55–66, 2000.

[26] K. Krumal, P. Mikuska, M. Vojtesek, and Z. Vecera, "Seasonal variations of monosaccharide anhydrides in PM1 and PM2.5 aerosol in urban areas," *Atmospheric Environment*, vol. 44, no. 39, pp. 5148–5155, 2010.

[27] S. Wilbur and C. Jones, *Application Note-Simple, Reliable Analysis of High Matrix Samples According to US EPA Method 6020A Using the Agilent 7700x/7800 ICP-MS*, Agilent Technologies, Inc., Santa Clara, CA, USA, 2015.

[28] N. Sugiyama, *Application Handbook*, Agilent Technologies, Inc., Santa Clara, CA, USA, 2013.

[29] Y. Jia, S. Bhat, and M. P. Fraser, "Characterization of saccharides and other organic compounds in fine particles and the use of saccharides to track primary biologically derived carbon sources," *Atmospheric Environment*, vol. 44, no. 5, pp. 724–732, 2010.

[30] K. Cheung, N. Daher, W. Kam et al., "Spatial and temporal variation of chemical composition and mass closure of ambient coarse particulate matter (PM10–2.5) in the Los Angeles area," *Atmospheric Environment*, vol. 45, no. 16, pp. 2651–2662, 2011.

[31] J. D. Allison and T. L. Allison, *Partition Coefficients for Metals in Surface Water, Soil and Waste*, US EPA Washington, DC, USA, 2005, http://www.epa.gov/athens/publications/reports/Ambrose600R05074PartitionCoefficients.pdf.

[32] H. Zhang, W. Davison, S. Miller, and W. Tych, "In situ high-resolution measurements of fluxes of Ni, Cu, Fe, and Mn and concentrations of Zn and Cd in porewaters by DGT," *Geochimica et Cosmochimica Acta*, vol. 59, no. 20, pp. 4181–4192, 1995.

[33] T. M. C. M. de Kok, H. A. L. Driece, J. G. F. Hogervorst, and J. J. Briedé, "Toxicological assessment of ambient and traffic-related particulate matter: a review of recent studies," *Mutation Research*, vol. 613, no. 2-3, pp. 103–122, 2006.

[34] V. Rezacova-Smetkova, B. Docekal, and H. Docekalova, "Application of diffusive gradients in thin film technique to soil characterization," *Chemické Listy*, vol. 99, no. 8, pp. 594–599, 2005.

[35] P. S. Hooda, H. Zhang, W. Davison, and A. C. Edwards, "Measuring bioavailable trace metals by diffusive gradients in thin films (DGT): soil moisture effects on its performance in soils," *European Journal of Soil Science*, vol. 50, no. 2, pp. 285–294, 1999.

24

Development and Validation of an LC-MS/MS Method and Comparison with a GC-MS Method to Measure Phenytoin in Human Brain Dialysate, Blood, and Saliva

Raphael Hösli,[1,2] Stefan König,[3] and Stefan F. Mühlebach[1]

[1]*Clinical Pharmacy and Epidemiology, Hospital Pharmacy, University of Basel, Spitalstrasse 26, CH-4031 Basel, Switzerland*
[2]*Spitalzentrum Biel, Apotheke, Vogelsang 84, CH-2501 Biel-Bienne, Switzerland*
[3]*Division of Forensic Medicine, University of Bern, Bühlstrasse 20, CH-3012 Bern, Switzerland*

Correspondence should be addressed to Stefan F. Mühlebach; stefan.muehlebach@unibas.ch

Academic Editor: Federica Bianchi

Phenytoin (PHT) is one of the most often used critical dose drugs, where insufficient or excessive dosing can have severe consequences such as seizures or toxicity. Thus, the monitoring and precise measuring of PHT concentrations in patients is crucial. This study develops and validates an LC-MS/MS method for the measurement of phenytoin concentrations in different body compartments (i.e., human brain dialysate, blood, and saliva) and compares it with a formerly developed GC-MS method that measures PHT in the same biological matrices. The two methods are evaluated and compared based on their analytical performance, appropriateness to analyze human biological samples, including corresponding extraction and cleanup procedures, and their validation according to ISO 17025/FDA Guidance for Industry. The LC-MS/MS method showed a higher performance compared with the GC-MS method. The LC-MS/MS was more sensitive, needed a smaller sample volume (25 μL) and less chemicals, was less time consuming (cleaning up, sample preparation, and analysis), and resulted in a better LOD (<1 ng/mL)/LOQ (10 ng/mL). The calibration curve of the LC-MS/MS method (10–2000 ng/mL) showed linearity over a larger range with correlation coefficients $r^2 > 0.995$ for all tested matrices (blood, saliva, and dialysate). For larger sample numbers as in pharmacokinetic/pharmacodynamic studies and for bedside as well as routine analyses, the LC-MS/MS method offers significant advantages over the GC-MS method.

1. Introduction

Sensitive and specific quantification methods are of critical importance when monitoring individualized drug therapy in patients or investigating drug concentration in forensic toxicology [1]. Critical dose drugs but also newly developed and designed complex drugs require analytical methods to check for effective drug delivery to target tissues and to minimize toxicity in sensitive organs or cells. When such drugs have to be used in patients with varying pharmacokinetics (PK) (e.g., ICU patients), an appropriate therapeutic drug monitoring (TDM), which allows, for example, to correlate the drug concentration in easy accessible plasma samples with those in the tissue of action, becomes even more relevant for a safe and efficient drug treatment [2].

Phenytoin (PHT) belongs to the most widely prescribed drugs to prevent and control most types of seizure disorders and to treat epilepsy [3]. It is one of the most often used critical dose drugs where insufficient or excessive dosing can have severe consequences such as seizures or toxicity. Thus, the monitoring and precise measuring of PHT concentrations in patients is crucial [4, 5]. As an example, in forensic toxicology, epilepsy patients under PHT treatment who have been involved in an accident have to be analyzed in order to verify whether the PHT concentration was adequate or possibly the reason for the accident [6]. However, there are several characteristics of PHT including a relatively low therapeutic index, difficult pharmacokinetics (PK) and pharmacodynamics (PD), saturable oxidative biotransformation, and the nonlinear clearance, which complicate a therapeutic drug monitoring (TDM) aimed at preventing intoxication of patients or treatment failures [7].

Thus, researchers and practitioners are interested in specific, sensitive, robust, and cost-effective methods to identify PHT concentrations in patients. Thereby, several compartments to measure the PHT concentration could be addressed such as blood, saliva, and CNS fluid (microdialysate). The correlation of PHT in different body compartments is not yet completely understood and has only recently been addressed by researchers who have compared the measurement of PHT in these different compartments with a GC-MS method [8]. While the GC-MS has long been the standard method in forensic testing, LC-MS/MS methods have become more common, as they generally offer some advantages over GC-MS [9]. Recently, researchers have developed an LC-MS/MS method to measure PHT in one specific body compartment (i.e., blood plasma or serum) [10]. Missing, however, is a thorough comparison of the performance of these two analytical methods in the detection and analysis of PHT in different body compartments (i.e., blood, saliva, and samples from brain tissue microdialysis).

The aim of the present study was to develop and validate an LC-MS/MS method for the measurement of PHT concentrations in different body compartments such as blood and saliva, as well as samples from brain tissue microdialysis often used in neurology and neurosurgery, where antiepileptic therapy is often mandatory [11, 12], and to compare its efficiency with a formerly developed GC-MS method [8]. The fact that this established GC-MS method measured PHT in the same biological matrices (i.e., blood, saliva, and human brain dialysate) enables a reliable comparison with regard to the performance of GC-MS versus LC-MS/MS in measuring PHT in different body compartments. The two methods are evaluated and compared based on their analytical performance, appropriateness to analyze human biological samples, including corresponding extraction and cleanup procedures, and their validation according to ISO 17025/FDA Guidance for Industry [13]. Finally, the suitability of the two analytical methods for PK/PD studies, bedside measurement, and forensic use is discussed. In addition, the LC-MS/MS method developed in the current study is compared with an established LC-MS/MS method which measured PHT in blood plasma samples [10].

2. Materials and Method

2.1. Chemicals and Samples Used for the Development of the LC-MS/MS Method and Its Validation. PHT reference substance was purchased from Desitin Pharma GmbH (Liestal, Switzerland) and from the European Pharmacopoeia (PHT Ph. Eur. Standard, EDQM, Strasbourg, France). The IS for LC-MS/MS was PHT-D_{10} (PHT D-10, $C_{15}H_2D_{10}N_2O_2$, MW = 262.33) in methanol (MeOH) (100 μg/mL) from Cerilliant (Round Rock, TX).

Calcium chloride, perchloric acid, citric acid monohydrate, potassium chloride, magnesium chloride hexahydrate, sodium chloride, sodium hydroxide, and the solvents (methanol, acetic acid 100%, and acetone) were of analytical grade and purchased from Merck (Darmstadt, Germany).

Artificial cerebrospinal fluid (aCSF; dialysate solution) was prepared according to M Dialysis AB (Stockholm, Sweden) [14]. Blood CPDA-1 (anticoagulant citrate phosphate dextrose adenine solution; to simplify only named blood in the following) was obtained from the Blood Donor Center (Bern, Switzerland). Saliva was obtained from one of the investigators. 20–60 μL PHT-containing samples from patients collected from a 2 μL/min flow rate brain microdialysis and 2 mL of CPDA containing PHT patient blood samples were provided by the Department of Neurosurgery (Kantonsspital Aarau AG, Switzerland and Centre Hospitalier Universitaire Vaudois, Switzerland). All biological samples (blood and dialysates) were frozen and stored at −24°C. Before sample analysis, the samples were thawed at room temperature for 30 minutes and homogenized by shaking with a vortex for one minute.

2.2. Internal Standards, Calibrator Standard System Suitability Testing, and Sample Preparation. The internal standard (IS) stock solution was prepared by adding 100 μL of the PHT-D_{10} (100 μg/mL) to 9900 μL of MeOH. 5 mL of this solution was added to 95 mL of 1 M perchloric acid aqueous solution to get the final concentration of 50.0 ng/mL, which is used as IS working solution. The PHT reference stock solution (1.00 mg/mL) was used to obtain eight calibration (Cal) solutions with concentrations of 2000, 1000, 500, 250, 100, 50, 20, and 10 ng/mL PHT. 20 μL of these Cal solutions were added to 980 μL of the biomatrices to get the Cal working solutions. For quality control (QC), solutions with 1600, 400, 30, and 10 ng/mL PHT were prepared out of PHT reference stock solution (1.00 mg/mL).

The IS working solution of 75 μL was added either to an aliquot of 25 μL Cal working solution, QC solutions, or 25 μL sample from patients containing PHT. The sample preparation for the LC-MS/MS consisted of pipetting 75 μL of IS working solution to 25 μL sample into a deep well plate (0.6 mL, Chemie Brunschwig AG, Basel, Switzerland) covered by a sealing mat (Silicone, Chemie Brunschwig AG, Basel, Switzerland). The well plates were rigorously shaken for 5 minutes and then centrifuged (4.500 U/min; Mikro 22R, Hettich Instruments, Andreas Hettich AG, Bäch, Switzerland) for 30 minutes at about 8°C (Figure 1). The processed samples were ready for the LC-MS/MS analysis.

2.3. LC-MS/MS Settings. The prepared samples were placed into the autosampler (Dionex WPS-3000TSL Olten, Switzerland) which was set at 8°C. With a100 μL syringe from the autosampler, 10 μL of the prepared samples was injected into a 130 μL loop. The solvent rack (Dionex SRD-3600, Olten, Switzerland) carried the mobile phase A (H_2O + HCOOH (100 + 0.1, v + v)) and phase B (MeCN + HCOOH (100 + 0.1, v + v)). These mobile phases were delivered by three pumps (binary pump 1 (flow 0.350 mL/min) and isocratic pump 2 (flow 0.200 to 1.000 mL/min) (Dionex pump HPG-3200A, Olten, Switzerland), and binary pump 3 (Dionex pump ISO-3100A, Olten, Switzerland)) connected to a triple stage quadrupole mass spectrometer with linear ion-trap capability (3200 QTrap, Analyst Software Version 1.5.1, Applied

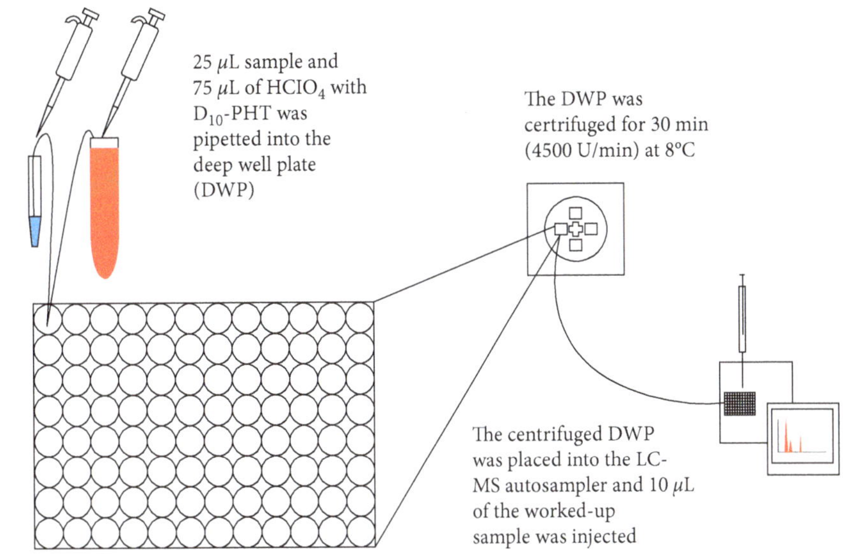

Figure 1: Sample preparation for the LC-MS/MS analyses for blood, saliva, and aCSF samples.

Table 1: Settings of the HPLC program.

Time (minutes)	Pump 1 (main column (MC))			Pumps 2 and 3 (trapping column (TC))			
	%B	Flow (μL/min)	Comments	%B	Flow (μL/min)	Flow pump 5 ($H_2O + 0.1\%$ HCOOH) (μL/min)	Switching valve
0	35		Start MS and pumps	50	300	800	TC → waste, MC → MS (loading)
0.5	35		Start gradient	50	300	800	TC → MC → MS (eluting)
0.6	↓	500					
1	97.5			50	20	20	
2	97.5						TC → waste, MC → MS
2.5	35			50	300	800	Reequilibration

Biosystems/MDS Sciex, Toronto, Canada) (Table 1). For the mass spectrometric detection, SRM scan mode (selective reaction monitoring) was used. SRM transitions and mass spectrometric conditions were as follows: transition: 253.1 → 182.2 (PHT) and 253.1 → 192.2 (PHT-D_{10}); orifice (V): 36; collision energy (eV): 41 (PHT) and 51 (PHT-D_{10}); and dwell time (msec): 100. Electrospray ionization was performed in positive ion mode for the analyte and the IS. The following instrument parameters for ionization were used: ion source voltage: 5000 volt, curtain gas: 25, gas 1: 40 and gas 2: 60; and the CAD gas was set to 5 (arbitrary units for the gas settings). As trapping column, a Phenomenex Gemini Polar column (2.0 × 10 mm, 5 μm; Brechbühler AG, Schlieren, Switzerland) temperated to room temperature was used. The main column Phenomenex Synergy Polar RP column (2.0 × 50 mm; Brechbühler AG, Schlieren, Switzerland) was placed into the column oven (Cluzeau Info Labo CrocoCil) set on 50°C with a column thermostat (Dionex TCC-3100, Olten, Switzerland) including switching valve (Figure 2). This system was operated by Analyst Software (Version 1.5.1, AB Sciex, Toronto, Canada).

2.4. Validation of the LC-MS/MS Method according to ISO 17025/FDA Guidance for Industry. The validation was carried out according to ISO 17025/FDA Guidance for Industry including selectivity, sensitivity, accuracy, recovery of PHT, reproducibility and suitability of the calibration curves, stability of PHT, and matrix effects. The selectivity and sensitivity (absence of PHT) were verified by analyzing blank samples without PHT (extraction and matrix effects). For the accuracy, QCs and Cal samples were analyzed. The recovery of PHT was analyzed by measuring QCs at different levels. The reproducibility and suitability of the calibration curves was measured by a complete series of Cal 1 to Cal 8 (LC-MS/MS) analyses. The limit of detection (LOD) and the limit of quantification (LOQ) were analyzed using Cal 1 (LC-MS/MS 10 ng/mL PHT). The LOD was checked as a signal-to-noise (S/N) ratio of more than 4 : 1. The LOQ was considered as 5 times the response to a blank sample. The stability tests consisted of the freeze-thaw stability of PHT, which was determined after 3 freeze-thaw cycles. The short-term stability was analyzed by keeping the samples thawed at ambient temperature for at least 6 hours, frozen for at least 12 hours at −25°C ± 5°C, and again thawed, worked-up, and analyzed. Postpreparative stability was evaluated to determine whether an analytical run can be reinjected in the case of instrument failure and, furthermore, whether the preparation of a large number of samples could be done at once. Therefore, one of the validation runs was analyzed a second time after 7 days. The described criteria for Cal curves, QC, accuracy, and precision had to be met.

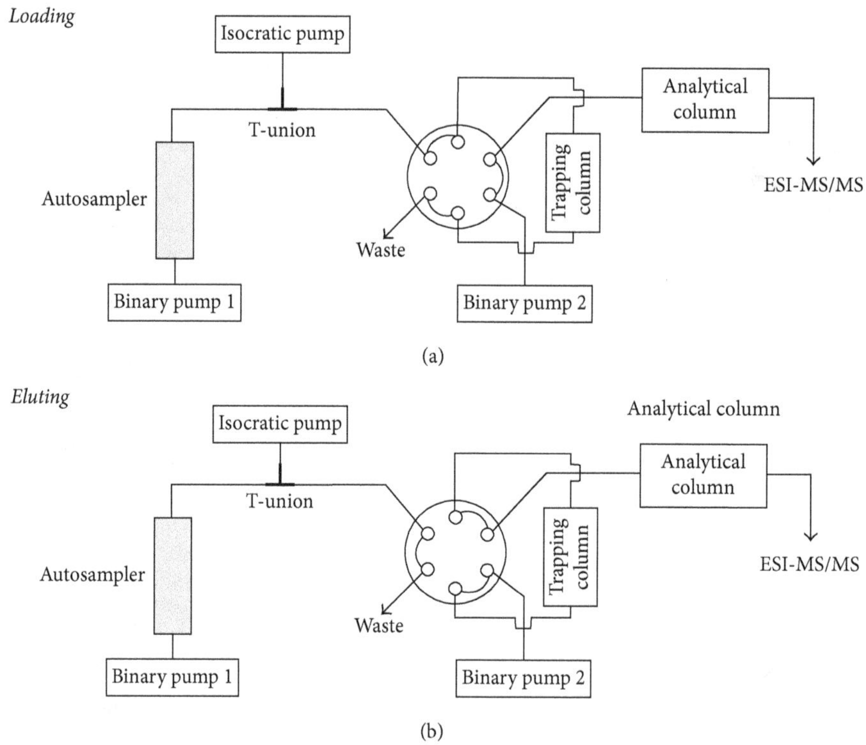

FIGURE 2: LC-MS/MS settings.

Matrix effects were analyzed by comparing the calibration curves generated with the three matrices aCSF, blood, and saliva. PHT microdialysis and blood samples from patients were analyzed to demonstrate the suitability of the method for biological samples from patients.

2.5. Comparison of the LC-MS/MS and the GC-MS Method. The LC-MS/MS method was evaluated and compared with the GC-MS method published by Hösli et al. [8] with regard to its analytical performance, appropriateness to analyze human biological samples, including corresponding extraction and cleanup procedures, and its validation according to ISO 17025/FDA Guidance for Industry.

The statistical data were calculated with Microsoft Excel and IBM SPSS Statistics 22. To compare the different matrices, a one-way ANOVA was calculated. The corresponding values were checked for significance by *t*-tests.

3. Results

3.1. Validation of the LC-MS/MS Method. The retention time (RT) for PHT and for PHT-D_{10} (IS) was about 2.8 min (Figure 3). The selectivity and sensitivity were checked; all blank samples were negative. The recovery of PHT after precipitation with $HClO_4$ was 89.5% for QC1 (10 ng/mL) and 97.1% for QC3 (1600 ng/mL) compared to the amount found in unprepared samples (=100%). The LOD calculated as *S/N* ratio of 4 : 1 for this method in aCSF, saliva, and blood was set at <1 ng/mL. The LOQ calculated as 5 times the response/blank was 10 ng/mL PHT. For the accuracy, the Cal 1 to Cal 8 were assessed. The calibrator values showed min–max deviations of 1–8% for Cal 2 (20 ng/mL) to Cal 8 (2000 ng/mL) with 3% for Cal 1 (10 ng/mL). The calibration curves for all three matrices were linear. The regression coefficients (r^2) of the three different matrices were $r^2_{\text{blood}} = 0.996$ ($n = 3$), $r^2_{\text{dialysate}} = 0.997$ ($n = 6$), and $r^2_{\text{saliva}} = 0.995$ ($n = 3$). Reinjection after 7 days showed no difference in accuracy. The sample volume needed was 25 μL. The sample preparation time was about 2 min per sample (6 hours for 182 samples). The run time for one LC-MS/MS analyses was 7 min.

3.2. Comparison of the LC-MS/MS with the GC-MS Method. After validation of the LC-MS/MS method, it was compared with the referred GC-MS method [8]. Table 2 shows the comparative results of the two methods for their analytical performance, appropriateness to analyze human biological samples, including corresponding extraction and cleanup procedures, and its validation according to ISO 17025/FDA Guidance for Industry (Table 2).

The selectivity and the sensitivity were met by both methods, and the recovery showed no differences (Table 2). But the accuracy differed between the two methods. The GC-MS method showed a higher variation at Cal 1 (20%) than the LC-MS/MS method (Cal 1 = 3%). As expected, the biggest difference in terms of analytical performance between the two methods was observed by the LOQ (GC-MS = 50 ng/mL; LC-MS/MS = 10 ng/mL) and LOD (GC-MS = 15 ng/mL; LC-MS/MS = <1 ng/mL) (Table 2).

Both methods showed linear regression coefficients (r^2) higher than 0.995 in all three different matrices for the PHT calibration curve. The calibration range of the LC-MS/MS

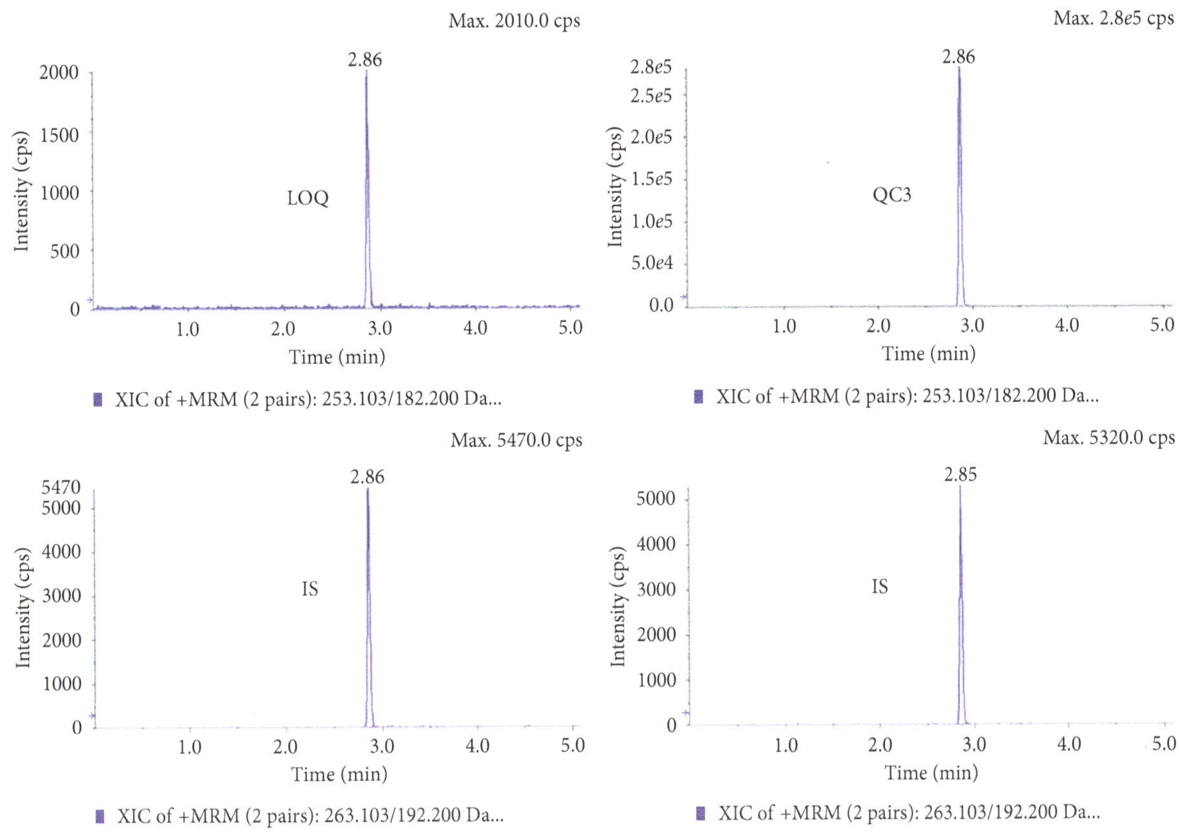

FIGURE 3: Chromatogram of phenytoin (illustrated for LOQ (10 ng/mL) and QC3 (400 ng/mL)) with PHT-D_{10} as IS (50 ng/mL).

TABLE 2: Comparison of the GC-MS [8] versus LC-MS/MS method.

Criterion	GC-MS	LC-MS/MS
Retention time	PHT 15.12 min, IS MPPH 16.15 min	PHT and PHT-D_{10} 2.8 min
Selectivity/sensitivity (absence of PHT)	Good peak differentiation and quantification of PHT. All blank samples were negative (no presence of PHT)	All blank samples were negative (no presence of PHT)
Recovery	94.1% for QC2 (100 ng/mL) 94.3% for QC5 (1000 ng/mL)	89.5% for QC1 (10 ng/mL) 97.1% for QC3 (1600 ng/mL)
LOD (calculated as S/N ratio of 4 : 1)	15 ng/mL	<1 ng/mL
LOQ (calculated as 5 times the response/blank)	50 ng/mL	10 ng/mL
Accuracy	The calibrator values showed min–max percent deviations of 1–20% for Cal 1 (50 ng/mL) to Cal 6 (1200 ng/mL)	The calibrator values showed min–max percent deviations of 1–8% for Cal 1 (10 ng/mL) to Cal 8 (2000 ng/mL)
Regression coefficient, r^2	$r^2_{\text{blood}} = 0.998$ ($n = 2$) $r^2_{\text{dialysate}} = 0.999$ ($n = 8$) $r^2_{\text{saliva}} = 0.999$ ($n = 2$)	$r^2_{\text{blood}} = 0.996$ ($n = 3$) $r^2_{\text{dialysate}} = 0.997$ ($n = 6$) $r^2_{\text{saliva}} = 0.995$ ($n = 3$)
Calibration range	50–1200 ng/mL	10–2000 ng/mL
Run time per analysis	30 min	7 min
Injection volume of the sample	2.0 μL	10 μL
Sample preparation time	5 h for 25 samples	6 h for 182 samples
Stability of the processed samples	Dried extracts were stable for ≥4 weeks (min/max deviation 4%). No effect by reinjection and storage (33 h) on the autosampler	Reinjection after 7 days showed no difference in accuracy
Sample volume needed	50 μL	25 μL

(from 10 ng/mL to 2000 ng/mL) is twice as large as of the GC-MS (50 ng/mL to 1200 ng/mL). The stability of the samples after extraction and cleaning up was demonstrated for both methods. The sample preparation procedure is demonstrated in Figure 1 for LC-MS/MS and Figure 4 for GC-MS.

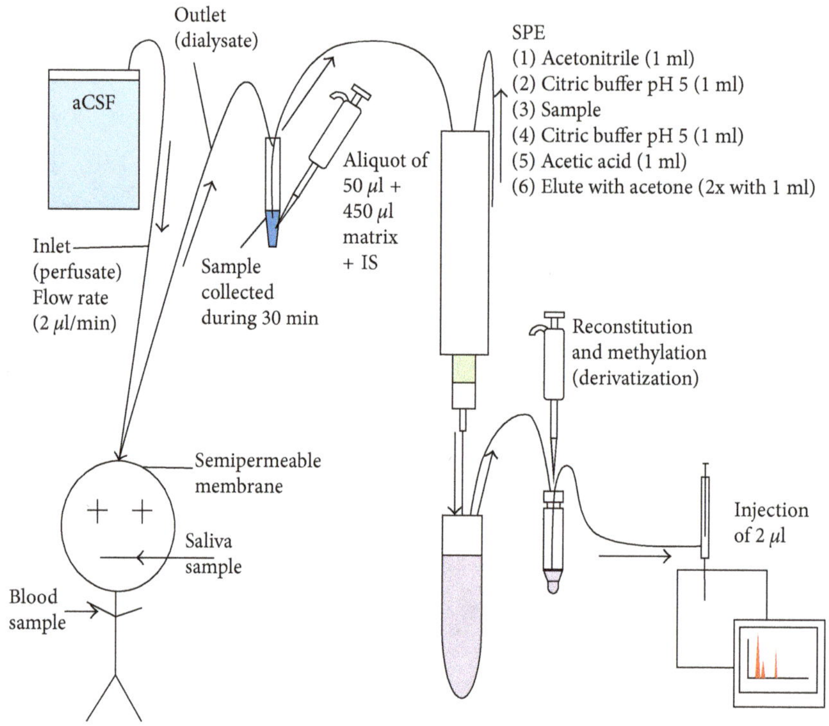

Figure 4: Sample preparation for the GC-MS analyses [8].

3.3. Comparison of the LC-MS/MS Method with a Formerly Established LC-MS/MS Method. Recently, a LC-MS/MS method has been developed which measures PHT in blood plasma or serum [10]. For the measurement of PHT in blood, the newly validated LC-MS/MS method can hence also be compared with this recently published study. The two methods show some similarities such as an identical IS (100 μg/mL PHT-d10), similar sample volumes needed (25 μL versus 20 μL [10]), and a comparable retention time (2.8 min versus approximately 2.1 min [10]). Both methods showed linear regression coefficients (r^2) higher than 0.99 in the blood matrix. The accuracy was similar as both studies showed deviations of <10%. With regard to the calibration range and the calibration solution, the two LC-MS/MS methods differ. While the LC-MS/MS method developed in this study showed a calibration range from 10 ng/mL to 2000 ng/mL, the calibration curve of the published LC-MS/MS method [10] ranged from 100 ng/mL to 4000 ng/mL. The calibration solution in the current study was the respective biological matrix (e.g., blood). In the published study [10], phosphate-buffered saline was used as the calibration solution.

4. Discussion

In this study, a LC-MS/MS method to measure PHT in different biological samples was successfully validated and compared with a similarly validated GC-MS method [8]. Overall, the LC-MS/MS method showed to be a more specific analytical method with a higher general performance (Table 2). The LC-MS/MS method needed less sample volume, less chemicals, and less analytical time and therefore resulted in less costs for the sample preparation.

Concerning the LOD, there was a huge difference between the two methods. The LOD of the LC-MS/MS method was 15 times better than the one of the GC-MS methods: the LOD of the LC-MS/MS method was <1 ng/mL compared to 15 ng/mL for the GC-MS method (increments by a factor of ten). Similarly, the difference in LOQ was 5 times lower in LC-MS/MS (10 ng/ml) compared to GC-MS (50 ng/mL). The LOQ for the LC-MS/MS could be set even lower than 10 ng/ml PHT (Cal 1). The FDA guidelines which claim a minimal reproducibility at the LOQ level of 20% were well below (deviation to target PHT amount: <8% in aCSF ($n = 6$), <4% in blood ($n = 3$), <9% in saliva ($n = 3$); accuracy: aCSF 103%, blood 101%, and saliva 106%). The LOQ of the GC-MS method and hence the lowest concentration level (Cal 1 at 50 ng/ml) of the calibration curve showed a deviation value of 19%. The LC-MS/MS method, in contrast, showed a value of only 3% deviation at the lowest Cal (10 ng/mL). This difference is of high importance, as samples with even lower concentrations could be reliably analyzed.

The calibration range (from 10 ng/mL to 2000 ng/mL) of the LC-MS/MS method was twice as large as of the GC-MS method (50 ng/mL to 1200 ng/mL). This indicates that the LC-MS/MS method is more powerful and effective over a larger range of concentration, since the linearity is given over a larger area (10 ng/mL–2000 ng/mL) compared to the GC-MS method (50 ng/mL–1200 ng/mL).

As IS, two different substances were used. MPPH as a structurally related compound was used for the GC-MS method. As IS for LC-MS/MS, deuterated PHT (PHT-D_{10}) was used, which is the same molecule as PHT and differs only by the molecular mass (+1). All the physicochemical processes upon cleanup and analysis are identical or highly similar for PHT and PHT-D_{10}. MMPH, however, could be chemically affected in a different way than PHT, which could lead to a systematic bias in a given situation [15].

Regarding the sample preparation procedure, the LC-MS/MS (Figure 1) showed an important advantage compared to the GC-MS method as it only needs 3 steps of sample preparation compared to 11 steps necessary for the GC-MS method including a solid-phase extraction (SPE) and derivatization with a more critical chemical trimethylsulfonium (TMSH) (Figure 4). This resulted in significant shortening of the overall analysis: Preparation of the samples before injection for GC-MS is about ten times more time consuming than for the LC-MS/MS. For the GC-MS method, researchers needed 5 hours to prepare 25 samples (5 samples/h), whereas for the LC-MS/MS method 182 samples were prepared in 6 hours (30.3 samples/h), which corresponds to 6 times the amount of prepared samples per hour compared to the GC-MS method.

From the exposure side, the volumes are much larger and the exposure to the chemicals are more prolonged with the GC-MS method compared to the sample cleanup for the LC-MS/MS method. Especially, the derivatization agent TMSH is critical to handle because of toxicity. The risk of serious and even irreversible effects through inhalation, skin contact, or eye exposure is well known. TMSH is also considered to be teratogenic. Therefore, the potential health risk for the laboratory staff handling the samples can be reduced by the LC-MS/MS method and the elimination of a safety critical agent.

The amount of biological samples needed for the GC-MS method (50 μL) was twice as much as for the LC-MS/MS (25 μL). The sample volume is a critical point for PK/PD studies, where, for example, by continuing dialysis from brain in neurosurgical patients only small volumes of samples per time point/period are available. For 50 μL dialysate about 25 minutes collecting time is necessary at the usual flow rate of ~2 μL per minute [12, 16]. Therefore, not a requested specific time point, but a rather large time segment is represented which can influence the requested results. The reduced sample volume needed (25 μL) for the LC-MS/MS analyses reduces the dialyses time needed per sample to about 15 minutes. The smaller the dialysis time, the more precise correlations of the respective tissue concentration with plasma/blood samples can be made.

Furthermore, LC-MS/MS also has the shorter run time. The time needed for 100 GC-MS analyses would be approximately 50 hours. The LC-MS/MS method, in contrast, needs only 11 hours and 40 minutes for 100 analyses. This is a time saving of more than 38 hours. While this may not be highly relevant for forensic purposes, for bedside and routine analyses (real-time) and PK/PD studies with larger numbers of samples, this factor is relevant. Also, when the time between taking a sample and the result needed is short, as it is in TDM to adjust subsequent dosing for PHT treatment, this time saving is crucial.

The costs for one way materials per sample was about 50% lower for the LC-MS/MS compared to the GC-MS method. Especially because no SPE device was needed. Also, the reduced work load for the laboratory technician must be considered as an imported cost factor.

Finally, the appropriateness of the method also depends on the biological matrix. Both methods can generally be used to measure PHT in blood and saliva, as the sample volume is less limiting. As mentioned before, however, for dialysates, the most difficult aspect is to get enough sample volume. Therefore, the LC-MS/MS method needing only half of the sample volume compared to the GC-MS method is more suited for microdialysate measurements. With respect to the LOD/LOQ, the LC-MS/MS method is also better suited for PK/PD studies, as it allows to include patients with low PHT dosages.

In addition, the newly established LC-MS/MS method was compared with a recently published LC-MS/MS method [10]. While this study measured PHT only in one body compartment (i.e., blood plasma or serum), the current LC-MS/MS method was developed and validated for the measurement of PHT in different body compartments (i.e., blood, saliva, and samples from brain tissue microdialysis often used in neurology and neurosurgery). The calibration range of the published LC-MS/MS method [10] (from 100 ng/mL to 4000 ng/mL) is appropriate for the measurement of PHT in blood plasma. As the PHT concentrations in brain tissue dialysates are much smaller than in blood plasma, the LC-MS/MS method of the current study was more appropriate for such samples, showing a lower calibration range from 10 ng/mL to 2000 ng/mL. Finally, as the aim of this study was to measure PHT in different biological matrices, a general substitute solution for blood plasma such as phosphate-buffered saline [10] could not be used. Instead, the fluid of the respective body compartment was used as calibration solution (e.g., artificial cerebrospinal fluid (aCSF) for the measurement of PHT in the brain tissue dialysates). This also eliminates a potential analytical bias due to matrix effects.

5. Conclusion

In this study, a LC-MS/MS method to measure PHT in different biological samples (i.e., human brain dialysate, blood, and saliva) was developed and validated under circumstances that ensured a high comparability with an established GC-MS method [8]. Overall, the study concludes that LC-MS/MS is not only better performing in human PHT concentration measuring or comparable drug PK/PD studies but is the only one to be used for bedside analysis. The time-consuming sample preparation and the long run time of the GC-MS method delay the result, which is critical in TDM. The higher sensitivity, the smaller needed sample volume, the better LOD/LOQ, the less time-consuming cleaningup and sample preparation procedure, and the shorter run time make the LC-MS/MS method the preferred analytical procedure.

Conflicts of Interest

The authors declare that they have no conflicts of interest.

References

[1] S. Deeb, D. A. McKeown, H. J. Torrance, F. M. Wylie, B. K. Logan, and K. S. Scott, "Simultaneous analysis of 22 antiepileptic drugs in postmortem blood, serum and plasma

using LC–MS-MS with a focus on their role in forensic cases," *Journal of Analytical Toxicology*, vol. 38, no. 8, pp. 485–494, 2014.

[2] S. L. von Winckelmann, I. Spriet, and L. Willems, "Therapeutic drug monitoring of phenytoin in critically ill patients," *Pharmacotherapy*, vol. 28, no. 11, pp. 1391–1400, 2008.

[3] E. H. Grover, Y. Nazzal, and L. J. Hirsch, "Treatment of convulsive status epilepticus," *Current Treatment Options in Neurology*, vol. 18, no. 3, 2016.

[4] M. F. Wu and W. H. Lim, "Phenytoin: a guide to therapeutic drug monitoring," *Proceedings of Singapore Healthcare*, vol. 22, no. 3, pp. 198–202, 2013.

[5] A. Tobler, R. Hösli, S. Mühlebach, and A. Huber, "Free phenytoin assessment in patients: measured versus calculated blood serum levels," *International Journal of Clinical Pharmacy*, vol. 38, no. 2, pp. 303–309, 2016.

[6] L. Nilsson, B. Y. Farahmand, P. G. Persson, I. Thiblin, and T. Tomson, "Risk factors for sudden unexpected death in epilepsy: a case-control study," *The Lancet*, vol. 353, no. 9156, pp. 888–893, 1999.

[7] E. Martin, T. N. Tozer, L. B. Sheiner, and S. Riegelman, "The clinical pharmacokinetics of phenytoin," *Journal of Pharmacokinetics and Biopharmaceutics*, vol. 5, no. 6, pp. 579–596, 1977.

[8] R. Hösli, A. Tobler, S. König, and S. Mühlebach, "A quantitative phenytoin GC-MS method and its validation for samples from human ex situ brain microdialysis, blood and saliva using solid-phase extraction," *Journal of Analytical Toxicology*, vol. 37, no. 2, pp. 102–109, 2013.

[9] E. R. Perez, J. A. Knapp, C. K. Horn, S. L. Stillman, J. E. Evans, and D. P. Arfsten, "Comparison of LC–MS-MS and GC–MS analysis of benzodiazepine compounds included in the drug demand reduction urinalysis program," *Journal of Analytical Toxicology*, vol. 40, no. 3, pp. 201–207, 2016.

[10] J. Peat, C. Frazee, and U. Garg, "Quantification of free phenytoin by liquid chromatography tandem mass spectrometry (LC/MS/MS)," *Methods in Molecular Biology*, vol. 1383, pp. 241–246, 2016.

[11] E. P. Thelin, L. H. Keri, P. J. Carpenter, and A. H. Hutchinson, "Microdialysis monitoring in clinical traumatic brain injury and its role in neuroprotective drug development," *AAPS Journal*, vol. 19, no. 2, 2017.

[12] Y. Yamamoto, M. Danhof, and C. M. Elizabeth de Lange, "Microdialysis: the key to physiologically based model prediction of human CNS target site concentrations," *AAPS Journal*, vol. 19, no. 4, 2017.

[13] FDA, *Guidance for Industry, Bioanalytical Method Validation*, U.S Department of Health and Human Services, Food and Drug Administration, Center for Drug Evaluation and Research (CDER), Center for Veterinary Medicine (CVM), Rockville, MD, 2001.

[14] Perfusion Fluid, 2017, http://www.mdialysis.com/iscus/iscus-international/products/catheter-accessories/perfusion-fluid.

[15] J. Wieling, "LC-MS-MS experiences with internal standards," *Chromatographia Supplement*, vol. 55, pp. 107–113, 2002.

[16] H. K. Kimelberg, "Water homeostasis in the brain: basic concepts," *Neuroscience*, vol. 129, no. 4, pp. 851–860, 2004.

25

Identification of the Related Substances in Ampicillin Capsule by Rapid Resolution Liquid Chromatography Coupled with Electrospray Ionization Tandem Mass Spectrometry

Lei Zhang,[1] Xian Long Cheng,[1,2] Yang Liu,[1] Miao Liang,[1] Honghuan Dong,[1] Beiran Lv,[1] Wenning Yang,[1] Zhiqiang Luo,[1] and Mingmin Tang[1]

[1] *School of Chinese Materia Medica, Beijing University of Chinese Medicine, No. 6 Zhonghuan South Road, Wangjing, Chaoyang District, Beijing 100102, China*
[2] *Institute for the Control of Traditional Chinese Medicine and Ethnic Medicine, National Institutes for Food and Drug Control, State Food and Drug Administration, 2 Tiantan Xili, Beijing 100050, China*

Correspondence should be addressed to Yang Liu; liuyang@bucm.edu.cn

Academic Editor: Josep Esteve-Romero

Rapid Resolution Liquid Chromatography coupled with Electrospray Ionization Tandem Mass Spectrometry (RRLC-ESI-MS^n) was used to separate and identify related substances in ampicillin capsule. The fragmentation behaviors of related substances were used to identify their chemical structures. Finally, a total of 13 related substances in ampicillin capsule were identified, including four identified components for the first time and three groups of isomers on the basis of the exact mass, fragmentation behaviors, retention time, and chemical structures in the literature. This study avoided time-consuming and complex chemosynthesis of related substances of ampicillin and the results could be useful for the quality control of ampicillin capsule to guarantee its safety in clinic. In the meantime, it provided a good example for the rapid identification of chemical structures of related substances of drugs.

1. Introduction

Ampicillin is an important semisynthetic β-lactam antibiotic and it is still widely used nowadays because of its good efficacy in urinary tract infections, respiratory infections, and other diseases caused by germs and bacteria. In recent years, the requirement of quality control for related substances in chemicals became stricter no matter in structure confirmation or content limitation. Ampicillin was especially degradable in presence of aqueous solution or humid storage environment, which would lead to the formation of a variety of degradation products [1]. These related substances (the related substances previously reported were shown in Table 1) would have a great influence on the quality of the products and clinical medication safety.

Although there has been much research on the related substances of ampicillin [2, 3], it is not completely explicit so far. To ensure the clinical safety and meet the new requirement of related substances in chemicals [4], it is still necessary to conduct further studies to develop a rapid and efficient method to describe in more detail the related substances of ampicillin capsule.

Many analytical methods including high-performance liquid chromatography (HPLC) [1, 5], high-performance capillary electrophoresis (HPCE) [6], high-performance liquid chromatography-atmospheric pressure chemical ionization mass (HPLC-APCI-MS) [7], and high-performance liquid chromatography-electrospray mass spectrometry (HPLC-ESI-MS) [8, 9] had been utilized for the analysis of ampicillin. Among these methods, LC-ESI-MS had been shown to be a powerful technique for the analysis of ampicillin and its related substances due to its excellent ability in separation and identification.

In this paper, a simple, rapid, and sensitive Rapid Resolution Liquid Chromatography coupled with Electrospray Ionization Tandem Mass Spectrometry (RRLC-ESI-MS^n) method was established for the identification of the related substances in ampicillin capsule. The result suggested that this technique

Table 1: The structures of the known related substances of ampicillin.

Number	Name of related substances	Chemical structure
1	6-Aminopenicillanic acid (6-APA)	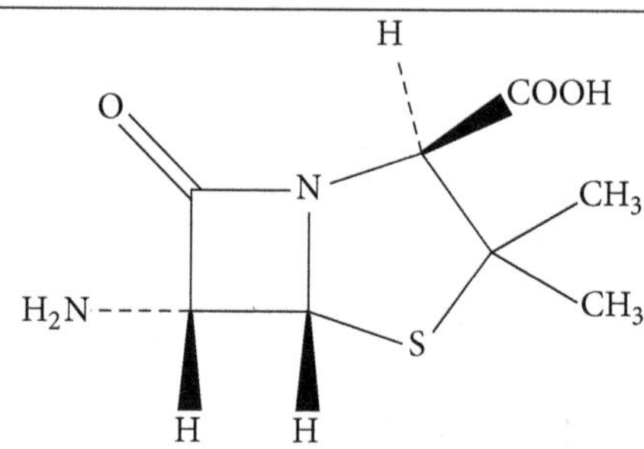
2	L-Ampicillin	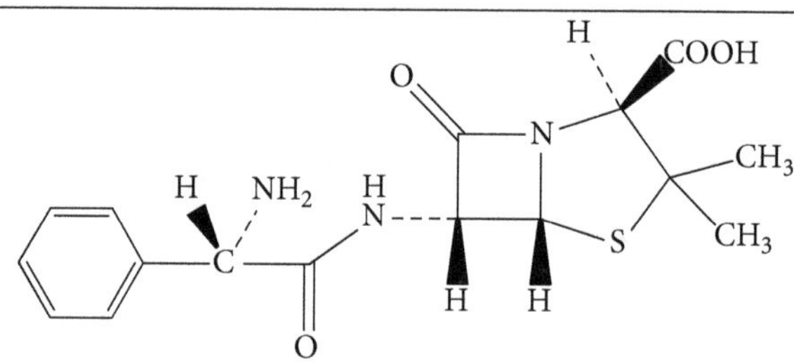
3	Diketopiperazines of ampicillin	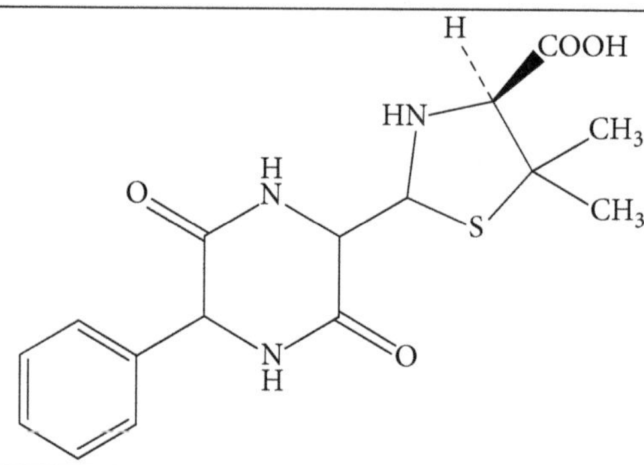
4	Ampicilloic acid	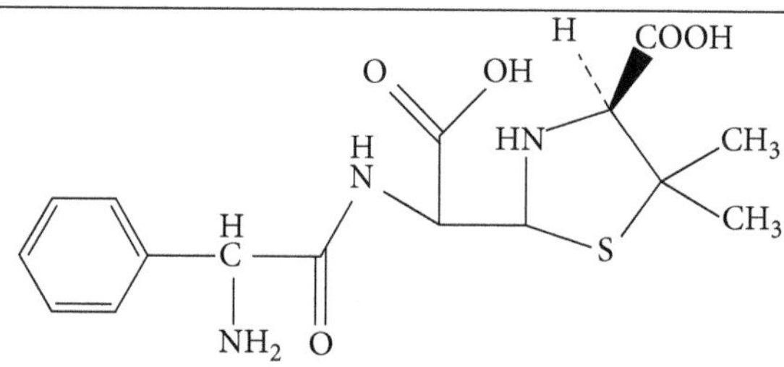
5	Ampilloic acid	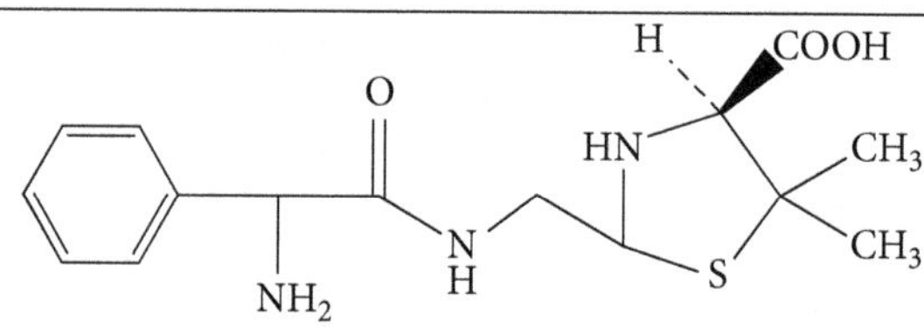
6	Ampicillinyl-D-phenylglycine	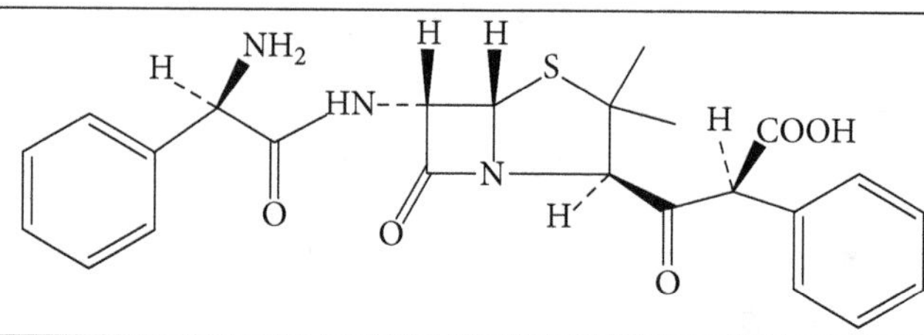

Table 1: Continued.

Number	Name of related substances	Chemical structure
7	(3R,6R)3,6-Diphenylpiperazine-2,5-dione	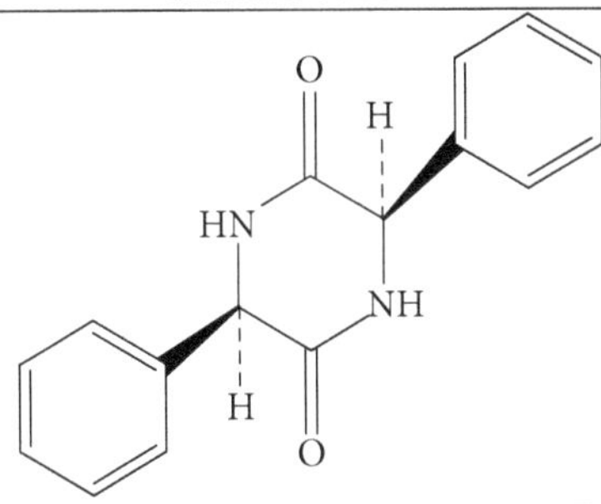
8	3-Phenylpyrazin-2-ol	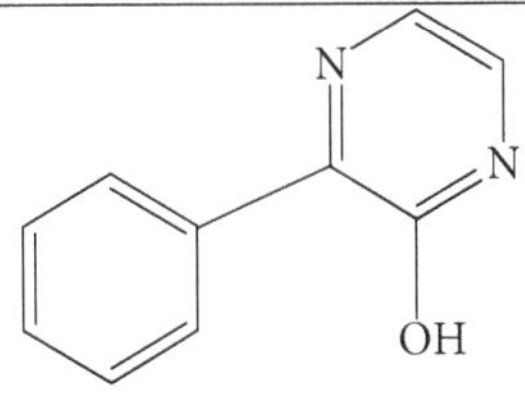
9	D-Phenylglycylampicillin	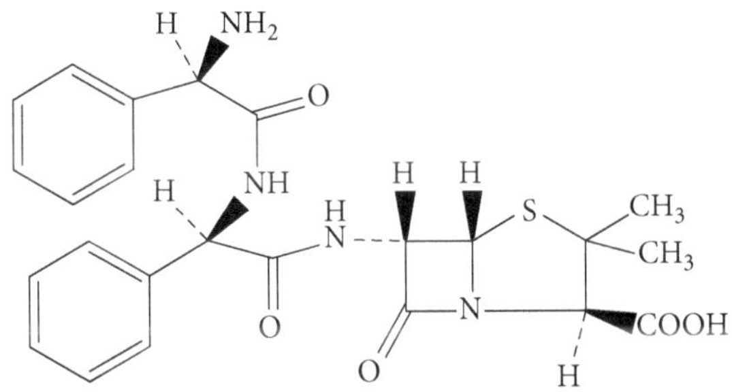
10	N-Pivaloyl-6-APA	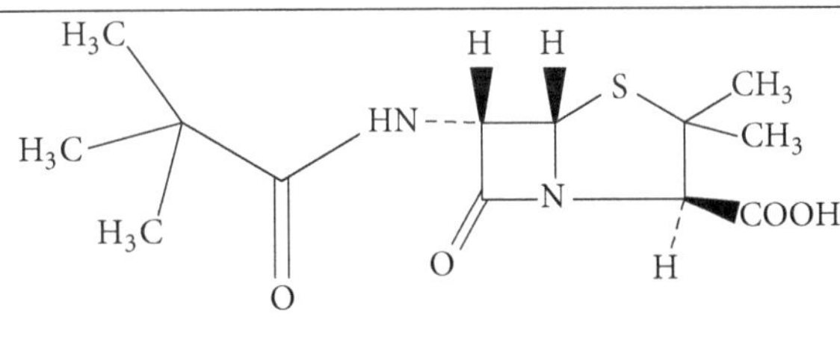
11	N-Pivaloylphenylglycine	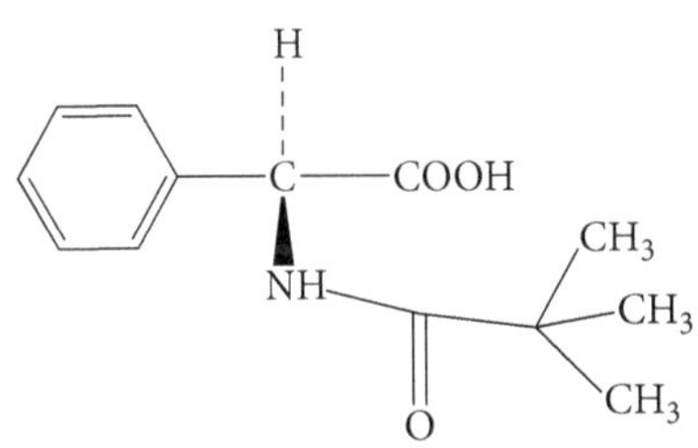
12	D-Phenylglycine	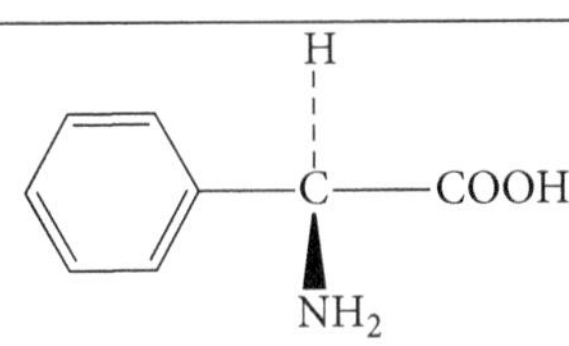

Table 1: Continued.

Number	Name of related substances	Chemical structure
13	Open-cycle dimer	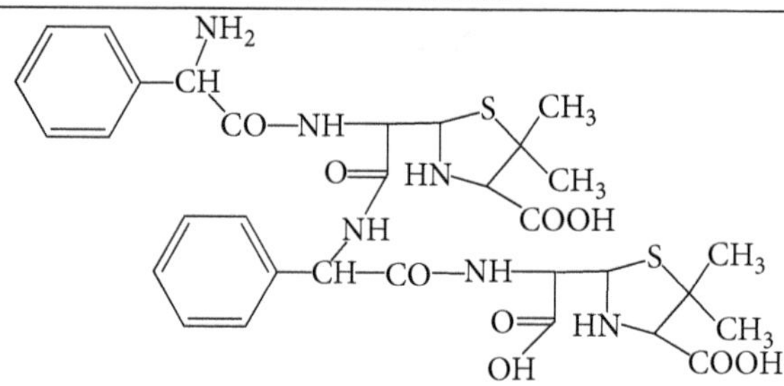
14	Closed-cycle dimer	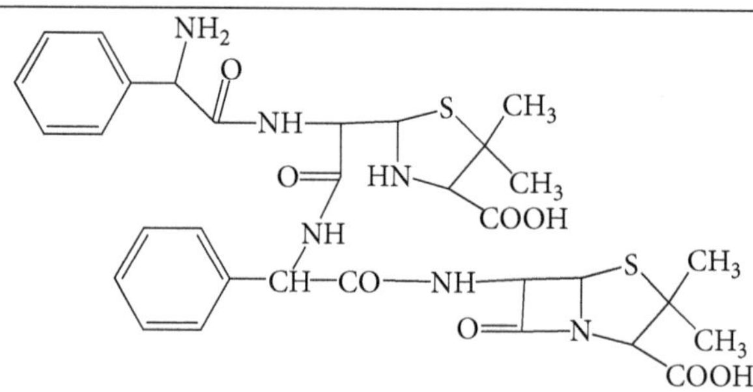
15	Open-cycle trimer	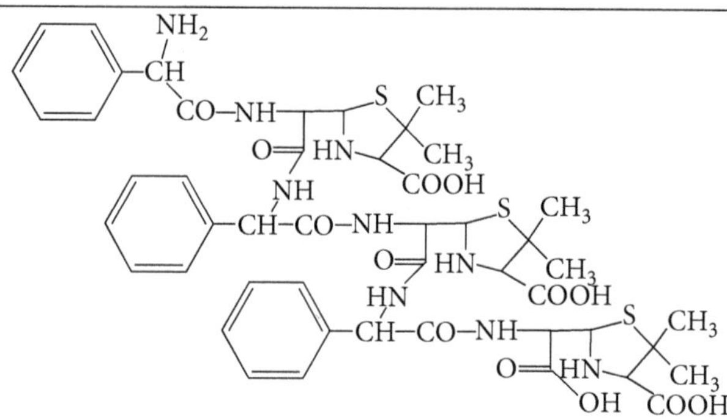
16	Closed-cycle trimer	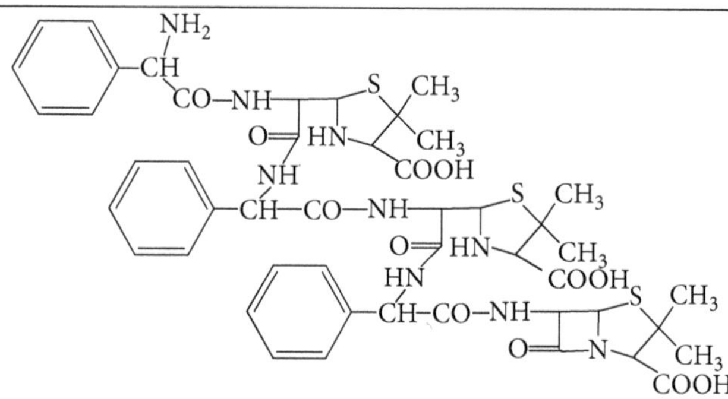
17	Open-cycle tetramer (n = 2)	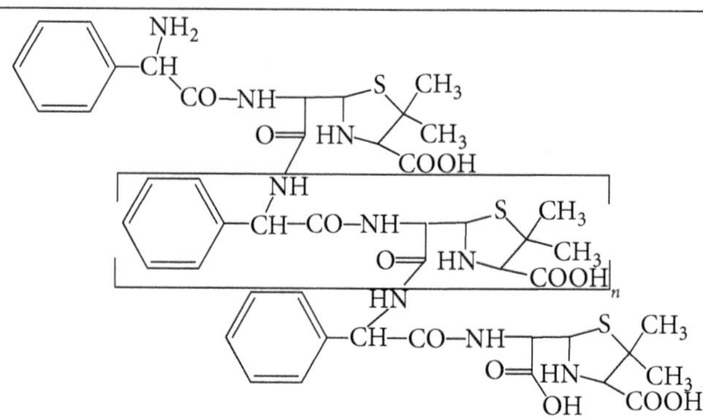

Table 1: Continued.

Number	Name of related substances	Chemical structure
18	Closed-cycle tetramer ($n = 2$)	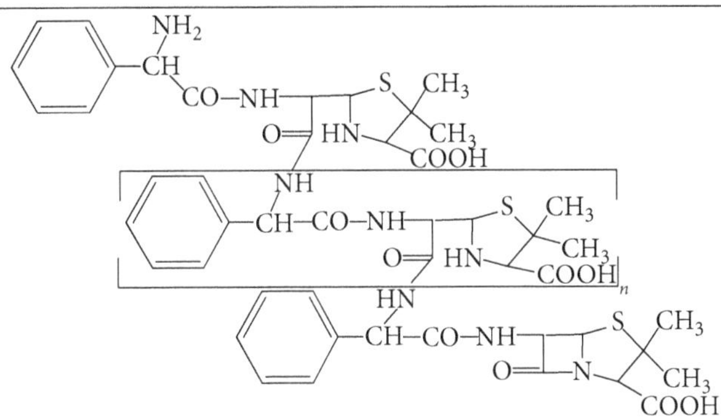

could facilitate rapid and accurate identification of related substances in ampicillin capsule.

2. Experimental

2.1. Chemicals and Materials. Methanol (HPLC grade) was purchased from Fisher Scientific (Pittsburgh, PA, USA). Formic acid (HPLC grade) was obtained from Acros Organics (Geel, Belgium). Deionized water was further purified with a Milli-Q water system (Bedford, Massachusetts, USA). Ampicillin capsule was purchased from DAVA Pharmaceuticals. Inc. (Huntsville, AL, USA).

The chromatographic separation was performed with an Agilent 1200 series Rapid Resolution Liquid Chromatography system (Agilent Technologies, USA), equipped with a binary pump, a microvacuum degasser, a high-performance autosampler, a column compartment, a diode array detector, and a MS detector. The samples were separated on a 1.8 μm Agilent Zorbax XDB-C_{18} column (50 mm × 4.6 mm) at a flow rate of 0.4 mL·min^{-1}. The mobile phases consisted of 0.1% formic acid solution (A) and methanol (B). The optimized RRLC elution conditions were as follows: 0–2 min, 10% B; 2–10 min, 10–20% B; 10–20 min, 20–50% B; 20–25 min, 50% B; 25-25.1 min, 50–10% B; 25.1–30 min, 10% B. DAD spectra were acquired over a scan range of 190–400 nm. The sample volume injected was 1 μL. Agilent 6320 mass spectrometer with an Agilent ChemStation to control and process the data was performed with the ESI source in positive ion mode. The vaporizer temperature was maintained at 300°C. The temperature of the drying gas was set at 350°C. The flow rate of the drying gas and the pressure of the nebulizing gas were set at 12 L·min^{-1} and 35 psi, respectively. The capillary voltage was kept at 3.5×10^3 V. The mass spectrometer scanned from a mass-to-charge ratio (m/z) 100–900.

2.2. Preparation of Sample. The contents of ampicillin capsule (equivalent to 10 mg Ampicillin) were dissolved in 10 mL methanol and then filtered through a 0.22 μm syringe filter. And an aliquot (1 μL) of the filtrate was subjected to RRLC-ESI-MS^n for analysis.

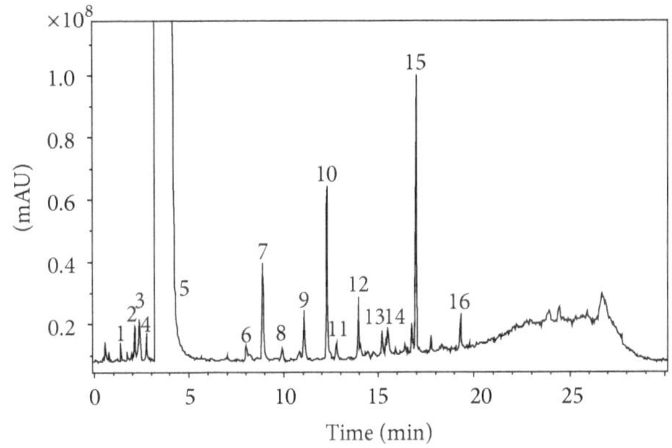

Figure 1: The total ion chromatography of ampicillin capsule (DAVA). The peaks were numbered according to their retention time.

3. Results and Discussion

3.1. Investigation of the Fragmentation Patterns of Ampicillin. It was necessary to study the characterization of the mass spectra of the parent drug to identify the molecular structure of the related substances in ampicillin capsule. Identifications were based on the fact that the related substances of ampicillin usually contain structural fragments and analogous cleavage characteristic of the parent drug. Structural information and fragmentation mechanisms had been deduced from ions in the mass and collision spectra. This knowledge was useful in the analysis and identification of related substances in ampicillin capsule. We utilized knowledge of characteristic fragment ions of ampicillin and its related substances to identify their structures. Figures 1 and 2 showed the detailed total ion chromatography (TIC) and mass spectrum of ampicillin and its related substances, respectively.

Ampicillin yielded an abundant ion in the ESI mass spectrum at m/z 350.1. The ESI mass spectrum of this ion (m/z 350.1) was shown in Figure 2. The fragment ions at m/z 106.2 and 160.0 were reported to arise from the benzylamine group and the thiazolidine ring. The fragment ion at m/z 192.0 was

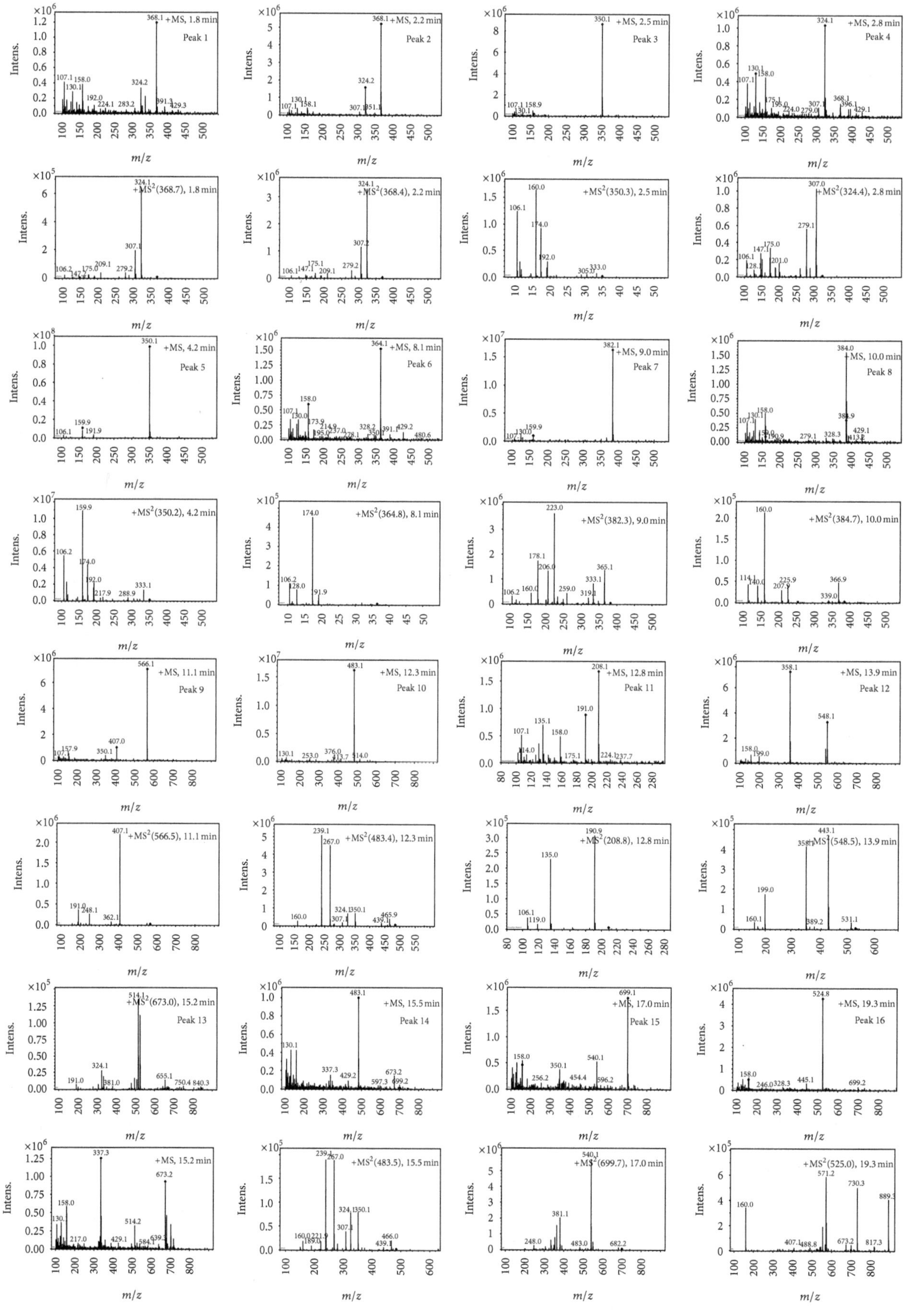

FIGURE 2: The mass spectrum of 16 chemicals in ampicillin capsule.

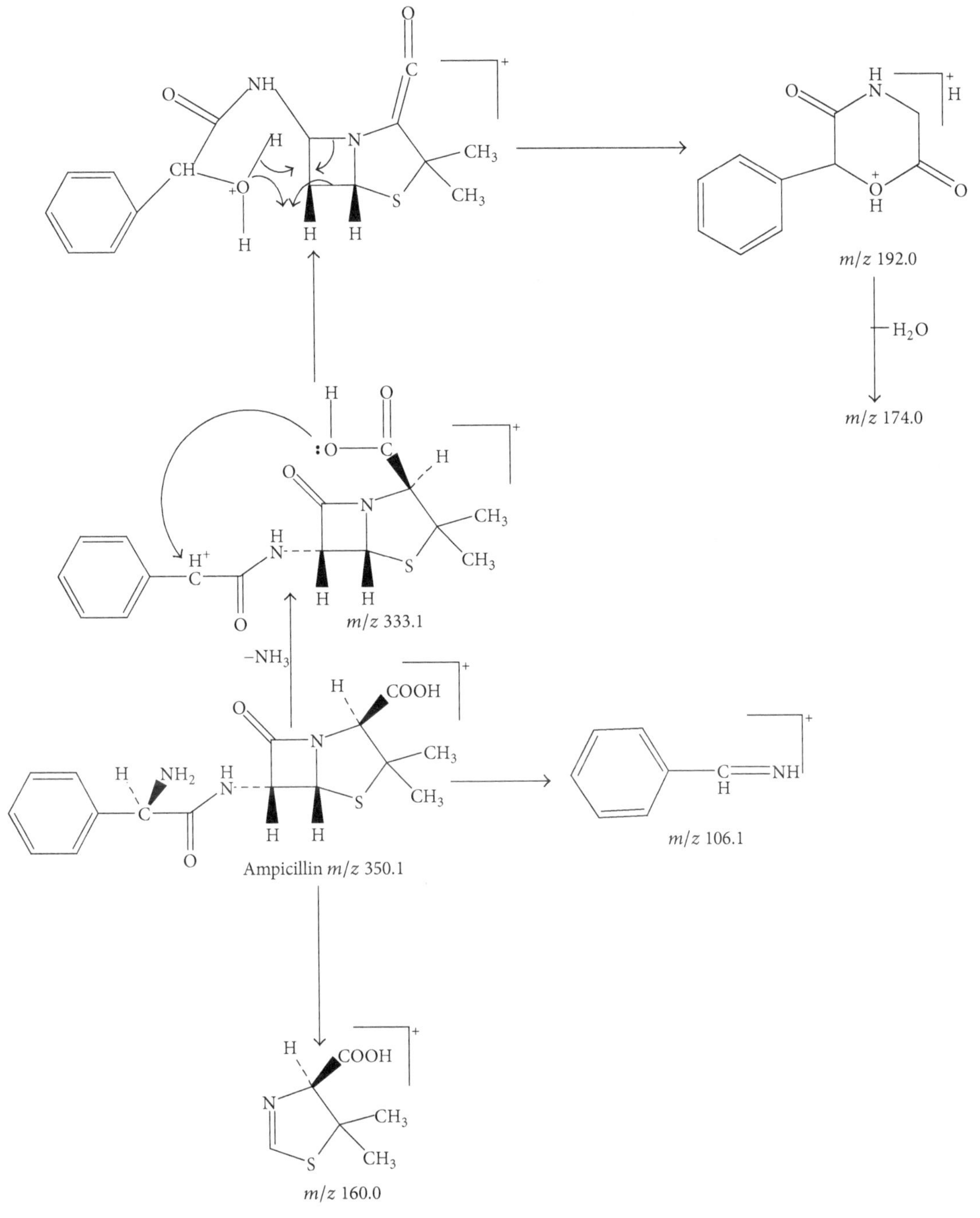

FIGURE 3: Proposed fragmentation pathways and characteristic ions of protonated ampicillin (m/z 350).

proposed to arise as a result of losing a $-NH_2$ group at the benzylamine side chain followed by an oxygen rearrangement and cleavage of the β-lactam ring. The fragment ion at m/z 174.0 could be attributed to the loss of H_2O from the fragment at m/z 192.0, but it might arise from other pathways. The proposed fragmentation pathways of ampicillin were shown in Figure 3.

3.2. Identification of the Known Related Substances in Ampicillin Capsule. This part of the investigation focused on the characterization of the ESI-MS properties of the parent drug and its known related substances. Table 2 showed the chromatographic and mass spectral characteristics of the detected related substances in ampicillin capsule.

Peak 1 and Peak 2 showed the same MS data. All of them produced protonated quasimolecular ion at m/z 368.1 $[M + H]^+$, major ions at m/z 324.1, 307.1, 279.2, and 175.1 in ESI^+ mode. Based on diagnostic ions (m/z 324.1, 307.1, and 175.1) and comparison with the published literature of known related substances of ampicillin [10], Peak 1 and Peak 2 were identified as isomers of ampicilloic acid. (5S, 6R) or (5R, 6R) ampicilloic acids were the two groups of ampicilloic

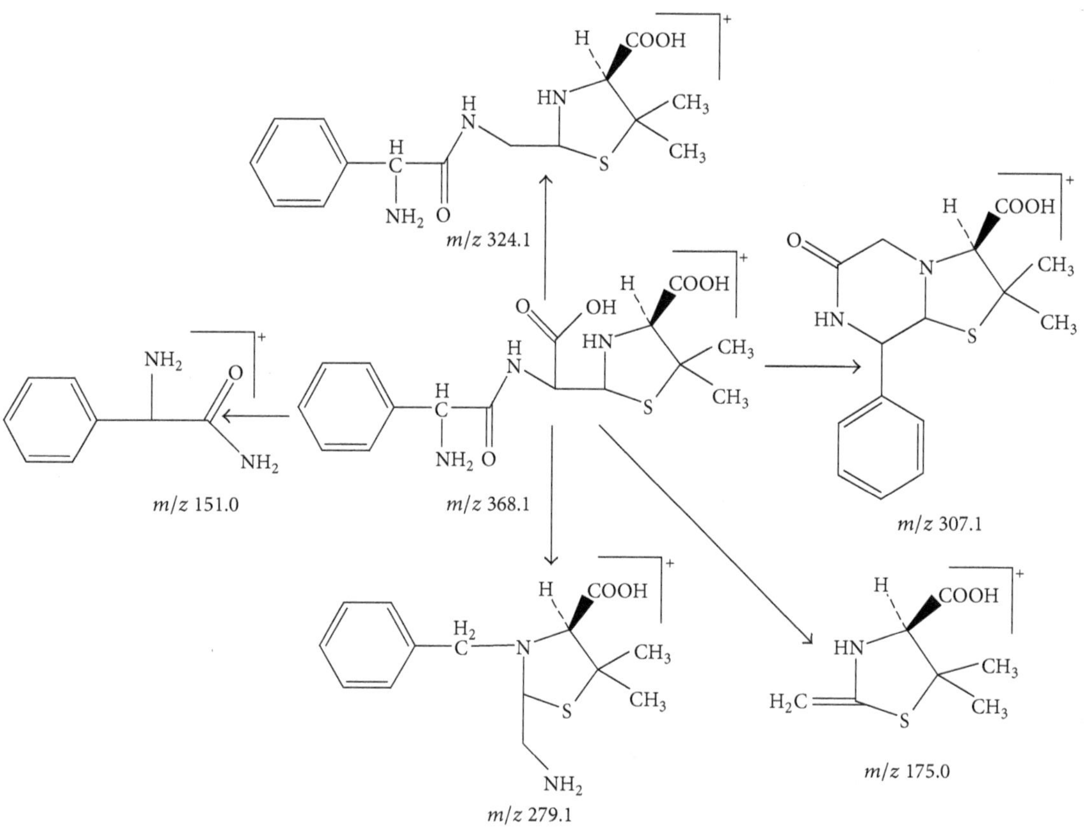

FIGURE 4: Proposed fragmentation pathway for the fragmentation ions of ampicilloic acid (m/z 368).

TABLE 2: Results of identification of the known related substances in ampicillin capsule.

Peak number	RT (min)	MS (m/z)	MS^2 (m/z)	Identification
1	1.8	368.1 $[M + H]^+$	324.1; 307.1; 279.2	(5S, 6R) ampicilloic acid
2	2.2	368.1 $[M + H]^+$	324.1; 307.2; 279.2; 175.1	(5R, 6R) ampicilloic acid
3	2.5	350.1 $[M + H]^+$	333.0; 192.0; 174.0; 160.0; 106.1	L-Ampicillin
4	2.8	324.1 $[M + H]^+$	307.0; 279.1; 201.0; 175.0; 147.1; 128.1; 106.1	(5S) or (5R) ampilloic acids
5	4.2	350.1 $[M + H]^+$	333.1; 192.0; 174.0; 159.9; 106.2	Ampicillin
6	8.1	364.1 $[M + K^+H]^+$	191.9; 174.0; 128.0; 106.2	(5S) or (5R) ampilloic acids
7	9.0	382.1 $[M + MeOH]^+$	331.1; 223.0; 206.0; 160.0; 106.2	Diketopiperazines of ampicillin
10	12.3	483.1 $[M + H]^+$	439.1; 350.0; 267.0; 239.1	D-Phenylglycylampicillin
15	17.0	699.1 $[M + H]^+$	540.1; 381.1; 248.0	Closed-cycle dimer
16	19.3	524.8 $[1/2M + H]^+$	889.3; 730.3; 571.2; 160.0	Closed-cycle trimer

acid isomers which were reported to be the metabolites and degradation products of ampicillin [1]. According to the retention behavior in reversed-phase chromatography of Peak 1 and Peak 2 and the related literature [1], Peak 1 and Peak 2 were tentatively identified as (5S, 6R) ampicilloic acid and (5R, 6R) ampicilloic acid, respectively. Figure 4 showed the proposed MS fragmentation pathway for the fragmentation ions of ampicilloic acid.

Peak 3 produced a protonated molecular ion at m/z 350.1 $[M + H]^+$, fragment ions at m/z 333.0 $[M - NH_3]^+$, 192.0, 174.0, 160.0, and 106.1 in ESI^+ mode. Peak 5 was clearly identified as ampicillin based on comparison of its retention time and mass spectrometric data with reference standards [8]. Peak 3 showed the same fragment ions, fragmentation pattern, and characteristic ions as Peak 5. Therefore, we could conclude that Peak 3 was an isomer of Peak 5. Considering that Peak 3 had a much shorter retention time than Peak 5, and with the comparison of related substances reported in the literature [1], Peak 3 was tentatively identified as L-ampicillin. Figure 5 showed the proposed MS fragmentation pathway for the fragmentation ions of L-ampicillin.

Peak 4 gave a protonated molecular ion $[M + H]^+$ with an m/z value of 324.1, major fragment ions at m/z 307.0, 279.1, 201.0, 175.0, 147.1, 128.1, and 106.1 in ESI^+ mode. Peak 4 was tentatively identified as (5R) or (5S) ampilloic acid based on its characteristic ions (m/z 324.1, 307.0, 279.1, 128.1, and

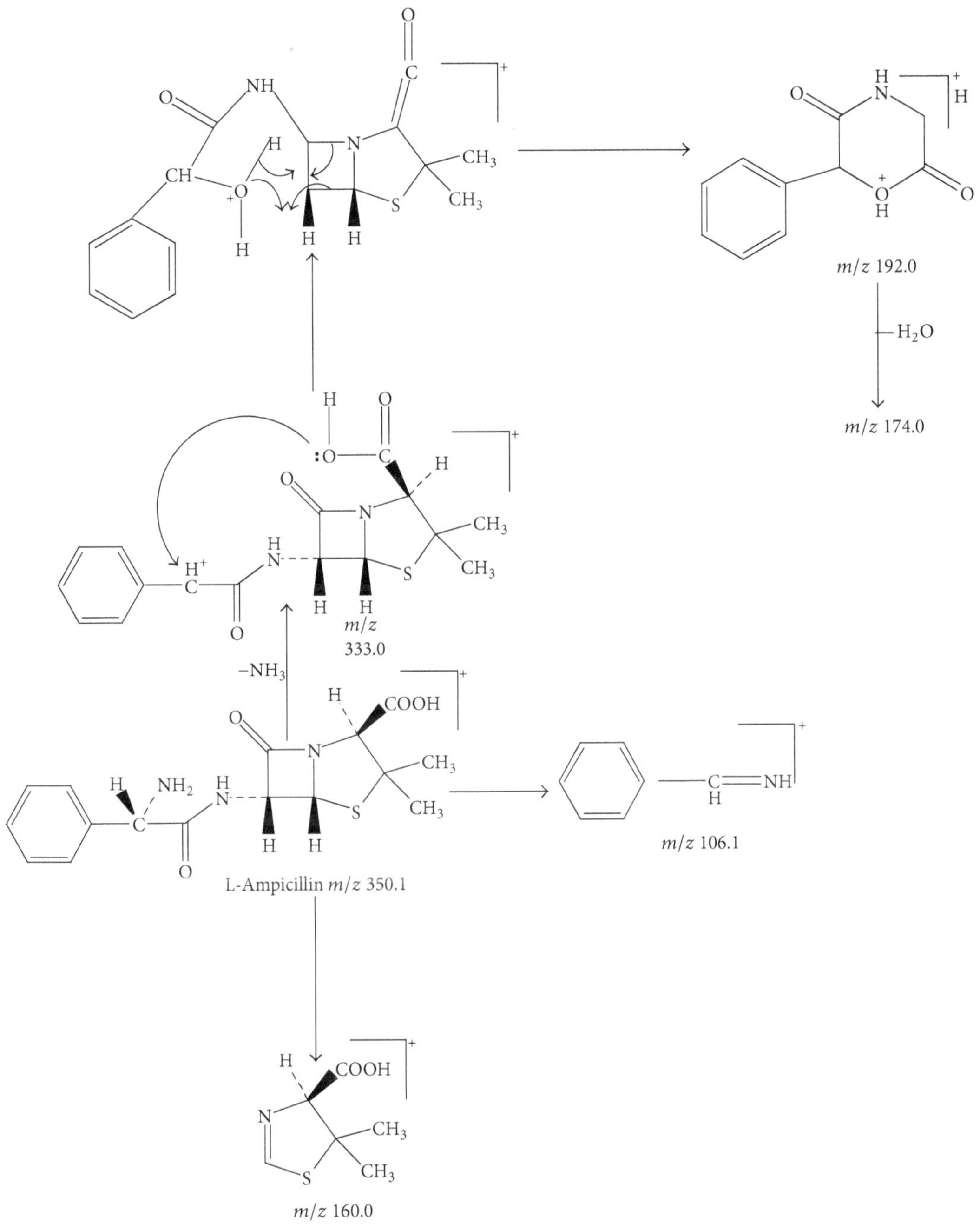

FIGURE 5: Proposed fragmentation pathways and characteristic ions of protonated L-ampicillin (m/z 350).

106.1) and comparison with the published literature of known related substances of ampicillin [10]. Peak 6 produced major fragment ions at m/z 364.1 $[M + K + H]^+$, 191.9, 174.0, 128.0, and 106.2. They all had similar MS fragmentation patterns (m/z 191.9, 174.0, 128.0, and 106.2). By comparison with the published literature [1], Peak 6 was tentatively identified as (5R) or (5S) ampilloic acid. (5S) or (5R) ampilloic acids were the isomers of ampilloic acids which were reported to be metabolites or degradation products of ampicillin [1]. However, the exact structure of these two components could not be determined due to the limited information. Figure 6 showed the proposed MS fragmentation pathway for the fragmentation ions of ampilloic acids.

Peak 7 had a molecular weight of 350 ($[M + MeOH+ H]^+$, m/z 382.1) and five major fragment ions were observed at m/z 331.1, 223.0 $[191 + MeOH]^+$, 206.0 $[174 + MeOH]^+$, 160.0, and 106.2. As it was reported [8], the two characteristic fragment ions m/z 160.0 and 106.2 were the representative fragment ions of ampicillin. By comparison with the published literature [7], this component was tentatively identified

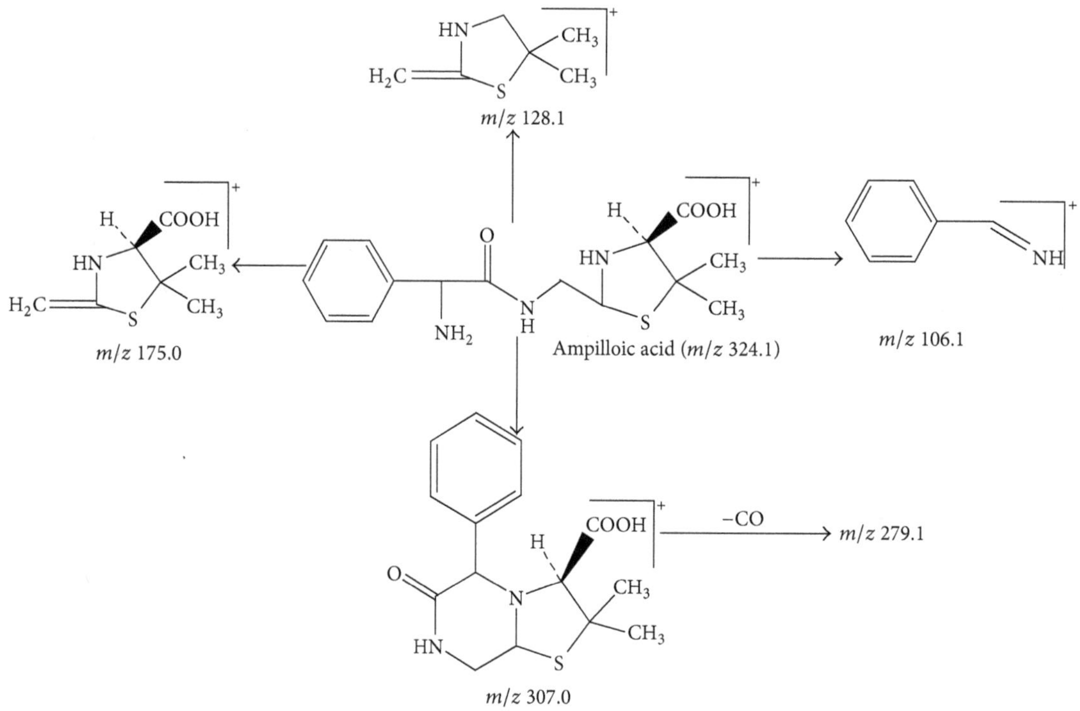

FIGURE 6: Proposed fragmentation pathway for the fragmentation ions of ampilloic acids.

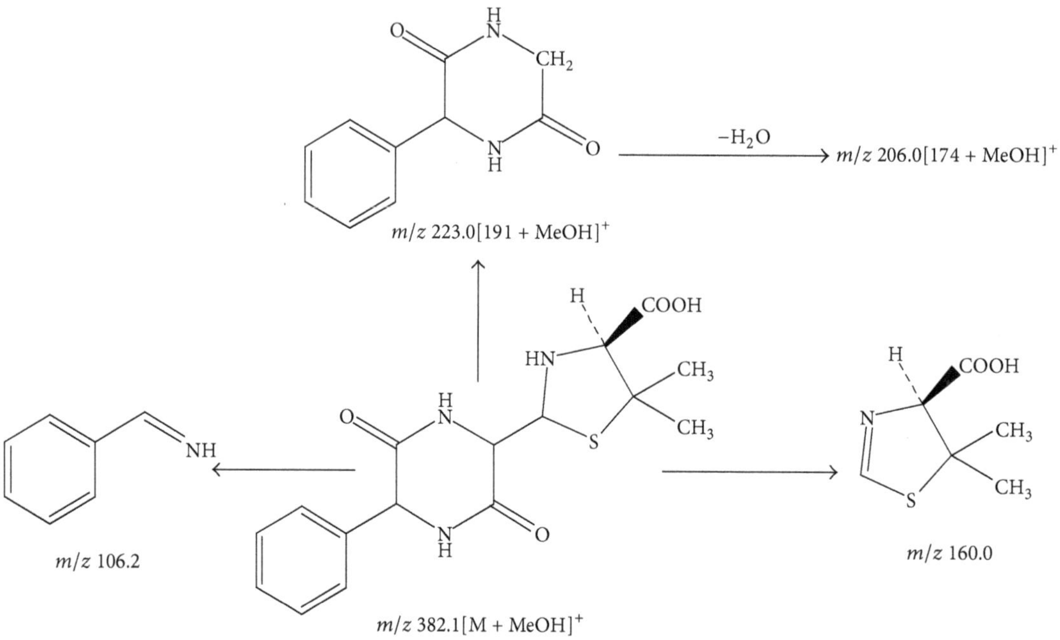

FIGURE 7: Proposed fragmentation pathway for the fragmentation ion of diketopiperazines of ampicillin.

as diketopiperazines of ampicillin. Figure 7 showed the proposed MS fragmentation pathway for the fragmentation ion of diketopiperazines of ampicillin.

Peak 10 produced a protonated molecular ion at m/z 483.1 $[M + H]^+$, the major fragment ions at m/z 439.1, 350.0, 267.0, and 239.1 in ESI^+ mode. Fragment ion at m/z 439.1 could be attributed to loss of one –COOH from the ion (m/z 483.1). Based on the mass spectra, Peak 10 was identified as D-phenylglycylampicillin [1, 11]. Figure 8 showed the proposed MS fragmentation pathway for the fragmentation ion of D-phenylglycylampicillin.

Peak 15 had a molecular weight of m/z 699.1 $[M + H]^+$ and three major fragment ions m/z 540.1, 381.1, and 248.0 in ESI^+ mode. Upon collision-induced dissociation (CID),

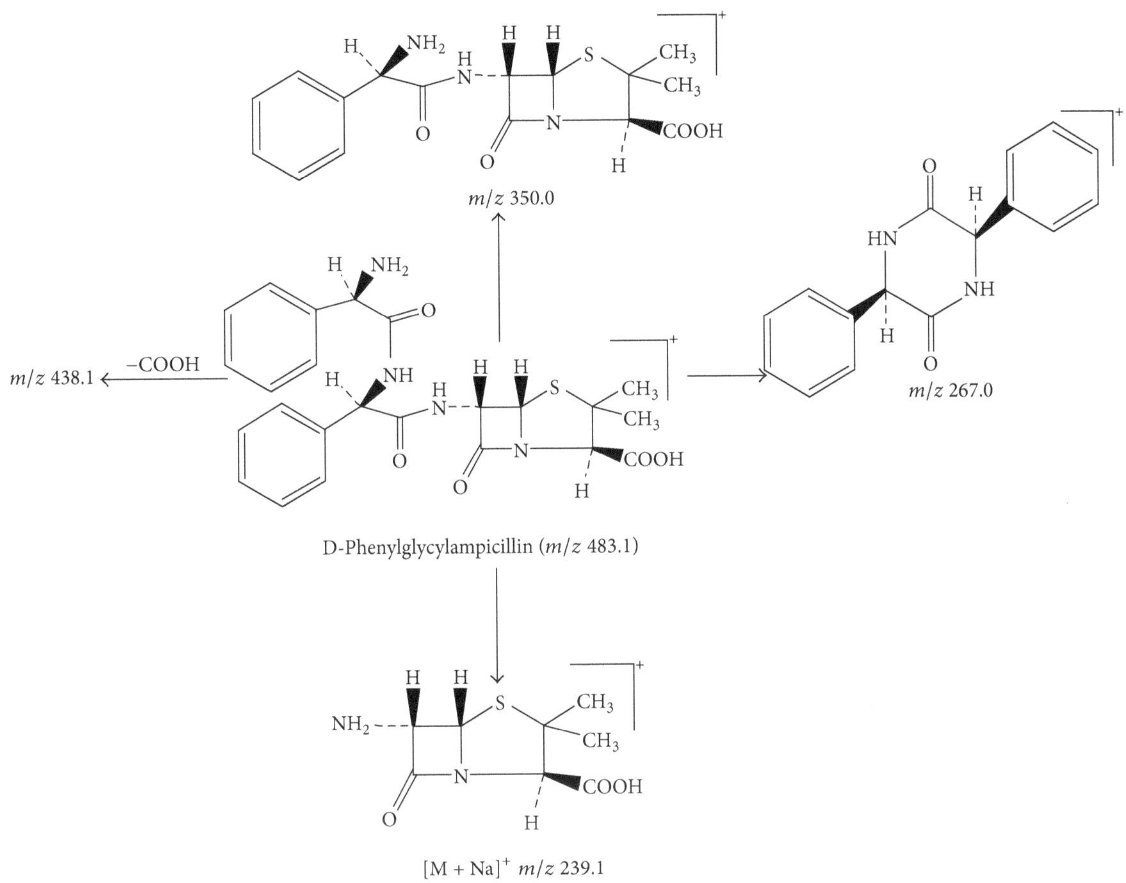

FIGURE 8: Proposed fragmentation pathway for the fragmentation ion of D-phenylglycylampicillin.

TABLE 3: Results of identification of the unknown related substances in ampicillin capsule.

Peak number	RT (min)	MS (m/z)	MS^2 (m/z)	Identification
9	11.1	566.1 $[M + H]^+$	407.1; 248.1; 191.0	Ampicilloic acid and 6-APA oligomer
12	13.9	548.1 $[M + H]^+$	443.1; 358.1; 199.0	6-APA ampicillin amide
13	15.2	673.2 $[M + H]^+$	655.1; 514.1; 324.1; 191.0	Ampilloic acids and ampicilloic acids oligomer
14	15.5	483.1 $[M + H]^+$	439.1; 350.1; 239.1; 160.0	Isomer of D-phenylglycyl ampicillin

the ion at m/z 699.1 eliminated one molecule of thiazolidine ring to produce m/z 540.1. The m/z 540.1 ion could further lose one molecule of thiazolidine ring successively to give significant m/z 381.1 fragment ion. Thus, Peak 15 was identified as closed-cycle dimer based on the published literature [1]. Closed-cycle dimer was the main cause of allergy, so that we must control the amount of this related substance in ampicillin capsule.

Peak 16 produced major fragment molecular ions at m/z 524.8 $[M + H]^+$, 889.3, 730.3, 571.2, and 160.0 in ESI$^+$ mode. Upon CID, the ion at m/z 1048 $[M]^+$ eliminated one molecule of thiazolidine ring to produce m/z 889.3. The m/z 889.3 ion could lose one molecule of thiazolidine ring successively to give significant m/z 730.3 ion. The m/z 730.3 ion could further lose one molecule of thiazolidine ring successively to give significant m/z 571.2. The fragment ion m/z 160.0 is characteristic thiazolidine ring of ampicillin. Thus, Peak 16 was identified as closed-cycle trimer based on the published literature [1]. Closed-cycle trimer was also the main cause of allergy, so that we must control the amount of this related substance in ampicillin capsule.

3.3. Identification of the Unknown Related Substances in Ampicillin Capsule. This part of the investigation was to identify the chemical structures of unknown related substances which were not yet reported in ampicillin capsule based on the mass fragment characterization and cleavage pathways of ampicillin and its known related substances. By means of the RRLC-ESI-MS^n experiments, in this part, chemical structures of four related substances were tentatively identified in ampicillin capsule for the first time. Table 3 showed the chromatographic and mass spectral characteristics of the above unknown related substances detected by RRLC-ESI-MS^n in ampicillin capsule.

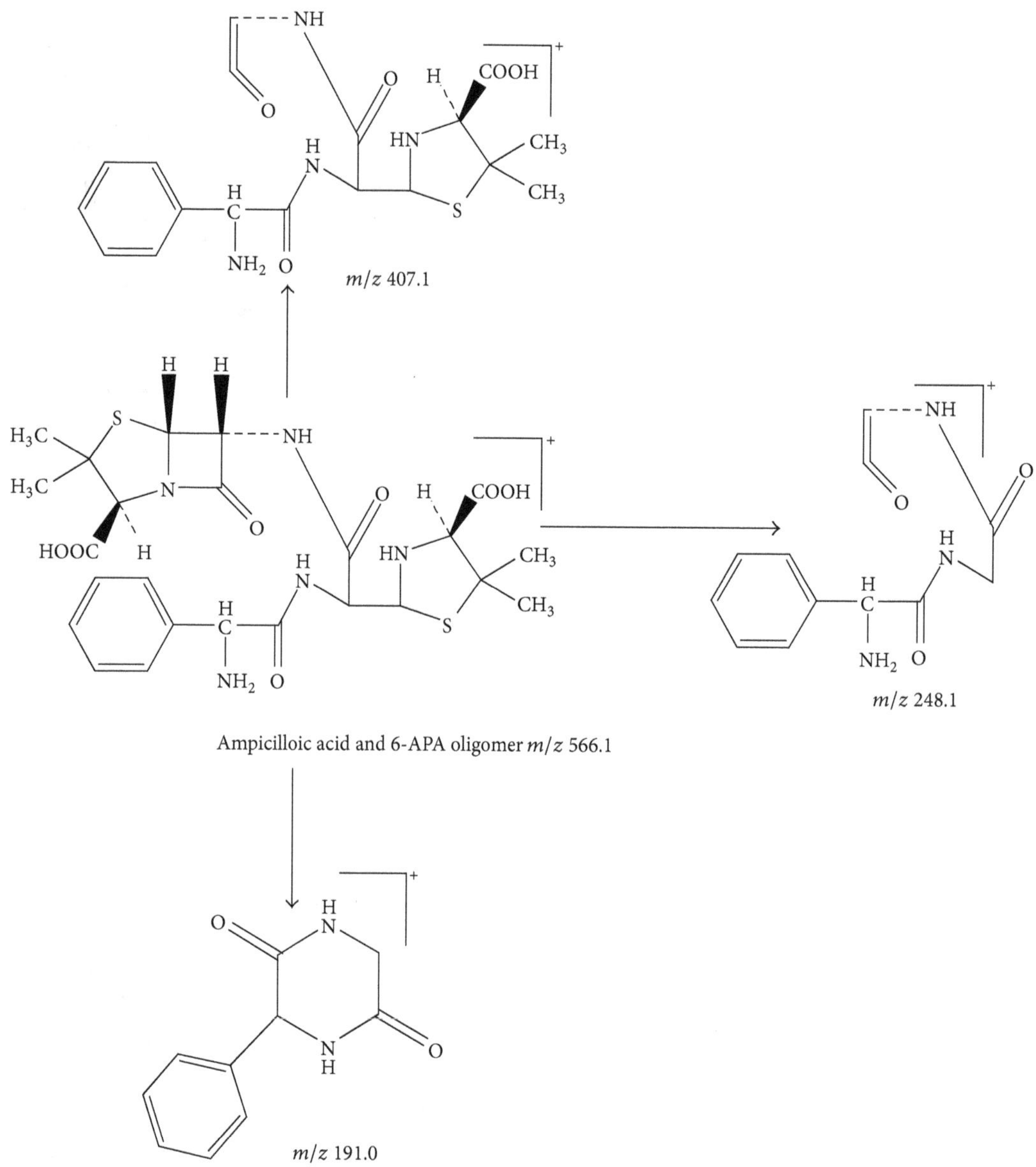

FIGURE 9: Proposed fragmentation pathway for the fragmentation ion of ampicilloic acid and 6-APA oligomer.

Peak 9 had a molecular weight of m/z 566.1 $[M + H]^+$ and three major fragment ions m/z 407.1, 248.1, and 191.0 in ESI^+ mode. The fragment ions at m/z 407.1 and 248.1 eliminated one molecule of thiazolidine ring successively from ion at m/z 566.1 $[M + 1]^+$. The fragment ion at m/z 191.0 probably should be a characteristic fragment ion of ampicillin piperazine-2,5-dione [12]. Thus, Peak 9 was identified tentatively as ampicilloic acid and 6-aminopenicillanic acid (6-APA) oligomer. Figure 9 showed the proposed MS fragmentation pathway for the fragmentation ion of ampicilloic acid and 6-APA oligomer.

Peak 12 produced a protonated molecular ion at m/z 548.1 $[M + H]^+$ and three major fragment ions at 443.1, 358.1, and 199.0. Based on fragment ions, Peak 12 was tentatively identified as 6-APA ampicillin amide. Figure 10 showed the proposed MS fragmentation pathway for the fragmentation ion of 6-APA ampicillin amide.

Peak 13 had a molecular weight of 673.2 $[M + H]^+$ and four major fragment ions m/z 655.1, 514.1, 324.1, and 191.0. The fragment ions at m/z 655.1 and 514.1 were attributed to the loss of one water molecule (18 Da) and one molecule of thiazolidine ring from ion at m/z 673.2. The fragment ion at m/z 324.1 was molecular weight of ampilloic acids. The fragment ion at m/z 191.0 was the fragment ion of ampicilloic acids. Peak 13 was identified tentatively as ampilloic acids and ampicilloic acids oligomer. Figure 11 showed the proposed MS fragmentation pathway for the fragmentation ion of ampilloic acids and ampicilloic acids oligomer.

Peak 14 produced a protonated molecular ion at m/z 483.1 $[M + H]^+$, which was identified as the other isomer of

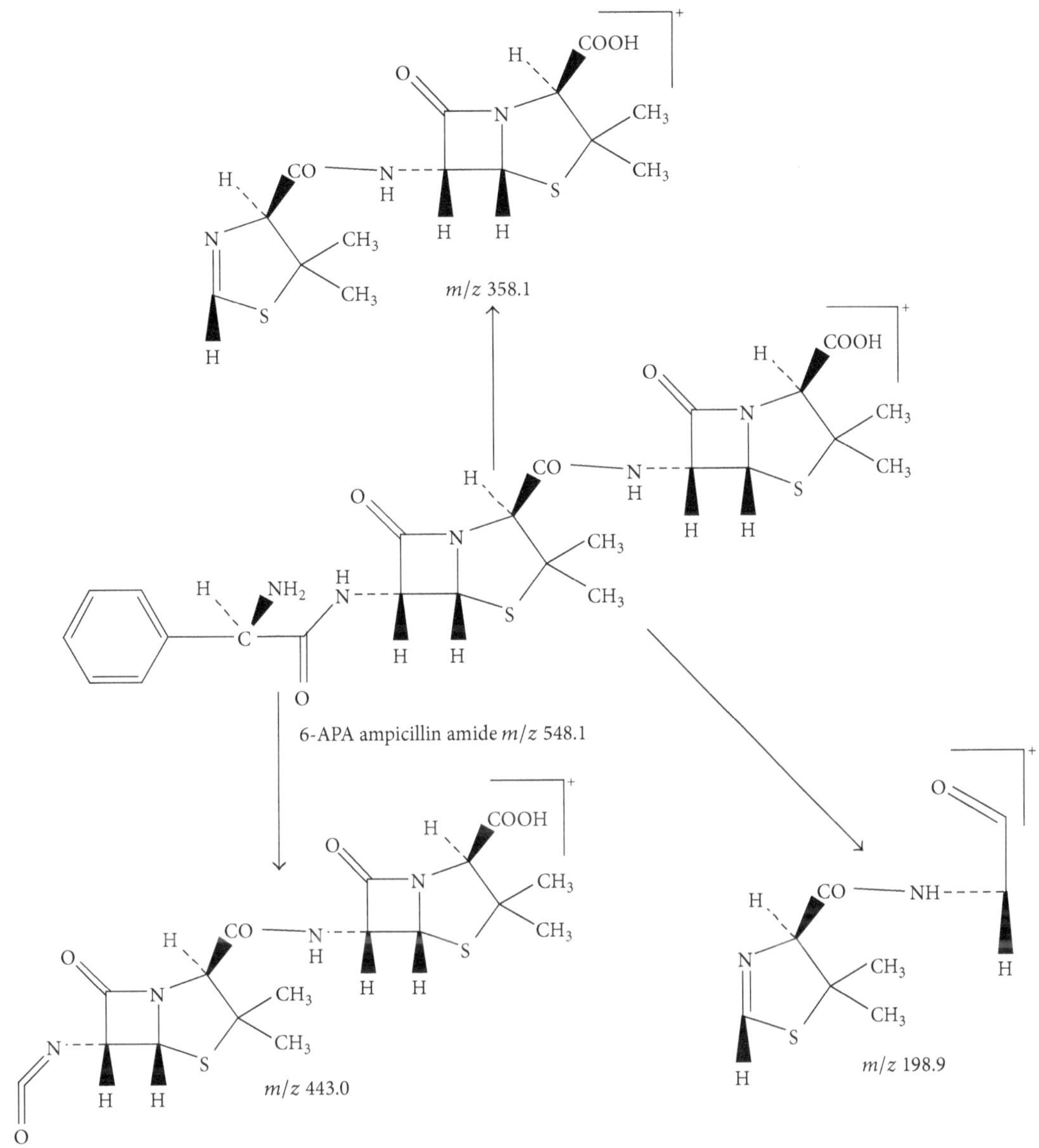

FIGURE 10: Proposed fragmentation pathway for the fragmentation ion of 6-APA ampicillin amide.

D-phenylglycylampicillin because Peak 14 and Peak 10 both showed the same diagnostic ions at m/z 439.1, 350.1, 239.1, and 160.0.

4. Conclusion

The RRLC-ESI-MS^n technique was successfully established to rapidly determine and identify the structures of the related substances in ampicillin capsule. RRLC is efficient in separating chemical compounds in a mixture, and MS provides abundant information for structural elucidation of the compounds when tandem mass spectrometry is applied [13]. Although ampilloic acids, ampicilloic acid, and closed-cycle dimer had been investigated previously by LC-MS method, MS information and characteristic diagnostic ions of a number of components in ampicillin capsule were described simultaneously for the first time. None of the previously reported methods have led to so much chemical information on the related substances in ampicillin capsule. The results of this study had identified 13 out of 15 related substances in ampicillin capsule. Unfortunately, three groups of isomers (Peak 1 and Peak 2, Peak 4 and Peak 6, and Peak 10 and Peak 14) and condensation of amino and carboxyl groups (Peak 9, Peak 12, and Peak 13) could not be identified fully by current RRLC-ESI-MS^n information. Peak 8 and Peak 11 had not yet been identified based on current mass spectra information. In summary, this investigation had provided an example of the rapid identification of related substances in ampicillin capsule. The meaningful information for the related substances in ampicillin capsule could lead to the development of the understanding of the quality and safety of the drug.

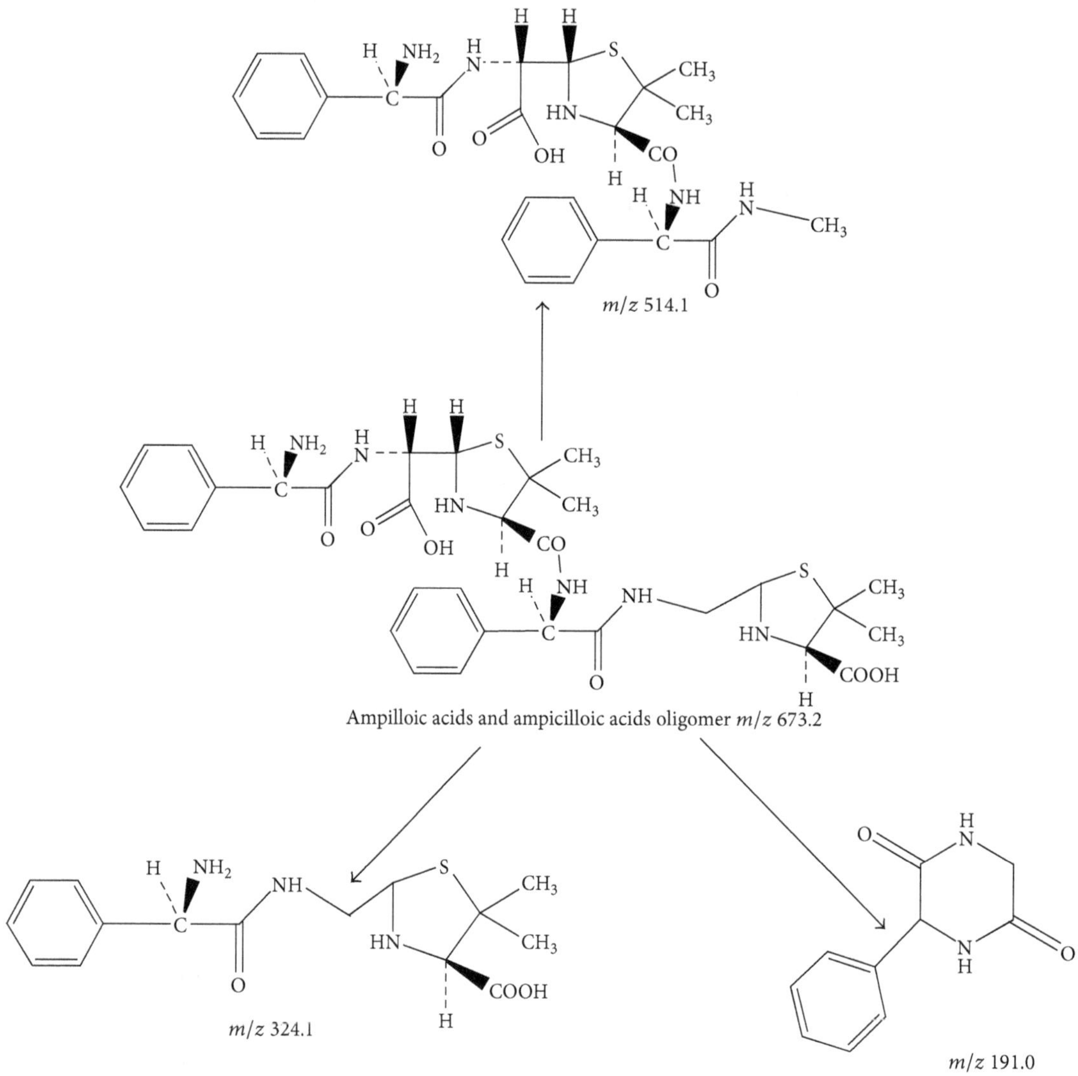

FIGURE 11: Proposed fragmentation pathway for the fragmentation ion of ampilloic acids and ampicilloic acids oligomer.

Conflict of Interests

The authors declare that there is no conflict of interests.

Authors' Contribution

Lei Zhang and Xian Long Cheng contributed equally in this work.

References

[1] Y. Zhu, E. Roets, Z. Ni, M. L. Moreno, E. Porqueras, and J. Hoogmartens, "Evaluation of liquid chromatography methods for the separation of ampicillin and its related substances," *Journal of Pharmaceutical and Biomedical Analysis*, vol. 14, no. 5, pp. 631–639, 1996.

[2] M. Shamsipur, Z. Talebpour, H. R. Bijanzadeh, and S. Tabatabaei, "Monitoring of ampicillin and its related substances by NMR," *Journal of Pharmaceutical and Biomedical Analysis*, vol. 30, no. 4, pp. 1075–1085, 2002.

[3] O. Shakoor and R. B. Taylor, "Analysis of ampicillin, cloxacillin and their related substances in capsules, syrups and suspensions by high-performance liquid chromatography," *Analyst*, vol. 121, no. 10, pp. 1473–1477, 1996.

[4] EMA, "Guideline on setting specifications for related impurities in antibiotics," http://www.ema.europa.eu/docs/en_GB/document_library/Scientific_guideline/2012/07/WC500129997.pdf.

[5] C. Larsen and H. Bundgaard, "Polymerization of penicillins. V. Separation, identification and quantitative determination of antigenic polymerization products in ampicillin sodium preparations by high performance liquid chromatography," *Journal of Chromatography*, vol. 147, pp. 143–150, 1978.

[6] C. Q. Niu and S. Q. Zhu, "Separation and determination of ampicillin polymers by high performance capillary electrophoresis," *Acta Pharmaceutica Sinica*, vol. 32, no. 3, pp. 207–209, 1997.

[7] S. Horimoto, T. Mayumi, K. Aoe, N. Nishimura, and T. Sato, "Analysis of β-lactam antibiotics by high performance liquid chromatography-atmospheric pressure chemical ionization mass spectrometry using bromoform," *Journal of Pharmaceutical and Biomedical Analysis*, vol. 30, no. 4, pp. 1093–1102, 2002.

[8] R. F. Straub and R. D. Voyksner, "Determination of penicillin G, ampicillin, amoxicillin, cloxacillin and cephapirin by high-performance liquid chromatography—electrospray mass spectrometry," *Journal of Chromatography*, vol. 647, no. 1, pp. 167–181, 1993.

[9] E. Verdon, R. Fuselier, D. Hurtaud-Pessel, P. Couëdor, N. Cadieu, and M. Laurentie, "Stability of penicillin antibiotic residues in meat during storage ampicillin," *Journal of Chromatography A*, vol. 882, no. 1-2, pp. 135–143, 2000.

[10] S. Suwanrumpha and R. B. Freas, "Identification of metabolites of ampicillin using liquid chromatography/thermospray mass spectrometry and fast atom bombardment tandem mass spectrometry," *Biomedical and Environmental Mass Spectrometry*, vol. 18, no. 11, pp. 983–994, 1989.

[11] B. P. Commission and G. Britain, *British Pharmacopoeia 2011*, Stationery Office, 2010.

[12] S. Suwanrumpha, D. A. Flory, R. B. Freas, and M. L. Vestal, "Tandem mass spectrometric studies of the fragmentation of penicillins and their metabolites," *Biomedical and Environmental Mass Spectrometry*, vol. 16, no. 1-12, pp. 381–386, 1988.

[13] M. Ye, J. Han, H. Chen, J. Zheng, and D. Guo, "Analysis of phenolic compounds in rhubarbs using liquid chromatography coupled with electrospray ionization mass spectrometry," *Journal of the American Society for Mass Spectrometry*, vol. 18, no. 1, pp. 82–91, 2007.

26

Multivariate Calibration and Model Integrity for Wood Chemistry Using Fourier Transform Infrared Spectroscopy

Chengfeng Zhou,[1,2] Wei Jiang,[1,3] Qingzheng Cheng,[1] and Brian K. Via[1,2]

[1] *Forest Products Development Center, School of Forestry and Wildlife Sciences, Auburn University, 520 Devall Drive, Auburn, AL 36849, USA*
[2] *Center for Bioenergy and Bioproducts, Biosystems Engineering, Auburn University, 520 Devall Drive, Auburn, AL 36849, USA*
[3] *College of Textiles, Qingdao University, 308 Ningxia Road, Qingdao 266071, China*

Correspondence should be addressed to Brian K. Via; bkv0003@auburn.edu

Academic Editor: Miren Lopez de Alda

This research addressed a rapid method to monitor hardwood chemical composition by applying Fourier transform infrared (FT-IR) spectroscopy, with particular interest in model performance for interpretation and prediction. Partial least squares (PLS) and principal components regression (PCR) were chosen as the primary models for comparison. Standard laboratory chemistry methods were employed on a mixed genus/species hardwood sample set to collect the original data. PLS was found to provide better predictive capability while PCR exhibited a more precise estimate of loading peaks and suggests that PCR is better for model interpretation of key underlying functional groups. Specifically, when PCR was utilized, an error in peak loading of $\pm 15\ cm^{-1}$ from the true mean was quantified. Application of the first derivative appeared to assist in improving both PCR and PLS loading precision. Research results identified the wavenumbers important in the prediction of extractives, lignin, cellulose, and hemicellulose and further demonstrated the utility in FT-IR for rapid monitoring of wood chemistry.

1. Introduction

As the most abundant fibrous material, wood is utilized in numerous areas including textile [1], paper making [2], building construction [3], composites [4], and bioenergy [5]. The chemical composition of wood including cellulose, hemicellulose, and lignin plays an important role when evaluating the utility of a feedstock for various product streams. Rapid assessment of biomass would open up opportunities for categorizing the raw material to an appropriate end use, allowing for better process control, or assist in the selection of better silvicultural or genetic management strategies for improved product performance and forest health [6]. Recently, near-infrared (NIR) spectroscopy was found to be a good tool for fast and quantitative analyses of chemistry components in plants [7–11]. NIR has been primarily successful due to its rapidity, precision, and low cost.

These same chemometric methods commonly used for NIR have been found to also be useful for Fourier transform infrared (FT-IR) spectroscopy [12]. Some researchers have studied the ability to predict secondary biomass properties with FT-IR [8, 13, 14]. Compared to NIR, FT-IR has some advantages as follows. First, not only is it advantageous for quantitative analysis, but one can also determine key functional groups associated with a particular trait of interest. Also, FT-IR spectroscopy is much more common for standard laboratory analysis of polymeric materials than NIR and is used in research institutes, laboratories, universities, and companies.

Principal components regression (PCR) and partial least squares (PLS) are the most common chemometric techniques used to construct prediction models and the loadings of the model can be further used for interpretation on contributing functional groups [15]. During woody tissue analysis, coefficients/loadings within the models are often used to interpret the relationship between wood chemistry functional groups and key traits. It was found in our previous study with NIR that PLS performed better for prediction while PCR was better for model interpretation and wavenumber selection [16]. Given that PLS optimizes both X and Y matrix for optimal prediction, it is hypothesized that PCR may also be the best for interpretation with FT-IR. Such work has not been done for the midinfrared region and is the subject of this research.

The objective of this paper was to investigate whether FT-IR coupled with chemometrics yields good predictive equations of wood chemical composition and whether the PLS modeling method introduces additional error to the loading plots.

We assumed that the loadings/coefficients in PLS modeling will decrease in precision and consequently increase in variance as a result of optimization for prediction. To test this alternative hypothesis (H_a), it was necessary to obtain the residuals between the locations of the local peak loading (PLS and PCR) and the best representative band assignment was obtained from the literature. The mathematical expression was listed as follows:

$$W - \mathrm{BA}_L = R, \tag{1}$$

where W represents the wavenumber obtained through PLS or PCR analysis, BA_L is the best representative band assignment obtained from the literature, and R represents the residual between W and BA_L. Then, the variance of the residuals will be further analyzed under the following hypothesis constructs:

$$\begin{aligned} H_0&: \sigma^2_{\text{PLS-}R} = \sigma^2_{\text{PCR-}R}, \\ H_a&: \sigma^2_{\text{PLS-}R} > \sigma^2_{\text{PCR-}R}, \end{aligned} \tag{2}$$

where σ^2 represents the variance of R obtained from PLS or PCR models. Differences for variance between model loadings will be tested by the F-test or t-test of R [17].

2. Experimental Procedures

2.1. Materials and Sample Preparation. Acetone (ACS Grade), sulfuric acid (98%, w/w), acetic acid (100%, w/w), sodium chlorite (High Purity Grade), and sodium hydroxide (Solution Reagent Grade, 50% w/w) were purchased from the VWR Company (Atlanta, GA, USA).

All of the hardwood samples were composed of four different genera, including 4 *Eucalyptus* samples, 9 cotton wood samples, 12 aspen samples, and 12 poplar samples. These wood samples were dried in the air for two weeks and then ground to 40 mesh and 80 mesh using Wiley mill. After that, the 80-mesh samples were used for FT-IR spectra collection and the 40-mesh samples were used for wet chemistry analysis.

2.2. Wood Chemistry. The extractives and lignin content were measured with National Renewable Energy Laboratory (NREL) standards [18, 19]. The cellulose and hemicellulose content were measured by traditional wet chemistry methods. The brief schematic diagram is shown in Figure 1, and the detailed processes were presented as follows.

2.2.1. Extractives. One hundred fifty mL of acetone was used to extract 5 g of sample for 6 h to get acetone based extractives solution. It was then dried to remove acetone and extractives for gravimetric analysis, and the extractive-free sample was reserved for the next step.

2.2.2. Lignin. A 72% (w/w) sulfuric acid treatment at 30°C for 2 h was used to prehydrolyze the extractive-free sample. The solution was then diluted to 4% sulfuric acid with distilled water, sealed in a bottle, and placed in an autoclave for 1 h at 121°C. After that, 10 mL supernatant was taken out from the bottle. Five mL supernatant was diluted to 20 mL to measure the acid-soluble lignin using ultraviolet and visible spectrophotometer (UV-Vis), and then the residual was filtered and oven dried to measure lignin content by gravimetric methods.

2.2.3. Cellulose and Hemicellulose. In order to obtain the percentage of cellulose and hemicellulose, we have to calculate the holocellulose content first. We used a delignification procedure to determine the holocellulose content. First, 2 g of extractive-free samples was placed equally in two conical flasks (500 mL) with 320 mL of distilled water in each flask. Second, the flasks were placed in a water bath (75°C) and then 1 mL of acetic acid and 20 mL of 15% (w/w) sodium chlorite were added to each flask on a 1 h cycle for 4 h. After 4 h, the residues were filtered with filter paper and then oven dried for 3 h to test the holocellulose content. Then, 1.5 g of oven dried holocellulose was placed in a 250 mL conical flask. One hundred mL of 17.5% sodium hydroxide was stirred in the flask and the air was replaced with nitrogen and then the flask was immediately sealed with aluminum foil. The flask was then placed in a water bath at 20°C and stirred occasionally until the reaction was complete. The solution was then filtered through a preweighed filter paper and washed with 500 mL of distilled water. The sample was then oven dried at 105°C for 12 h and weighed. The residue was determined as cellulose and the hemicellulose content was considered to be the difference in holocellulose and cellulose.

2.3. FT-IR Acquisition. Oven dried samples, used for FT-IR spectra collection, were placed in a dessicator and allowed to cool to room temperature to avoid spectra fluctuation caused by rapidly changing temperatures [20]. The 80-mesh oven dried samples were placed on the diamond plate of the FT-IR machine (PerkinElmer spectrum ATR 400 FT-IR/FT-NIR spectrometer, Waltham, MA, USA) and given a pressure of 70 ± 2 psi for spectra collection. The spectra covered the range of 4,000–650 cm^{-1} at a spectral resolution of 4 cm^{-1}. Each spectrum was collected from an average of 4 scans and no zero filling.

2.4. Chemometric Analysis. Chemometric techniques, PCR and PLS modules in Spectrum Quant + software, were used for model construction. Models were adopted using

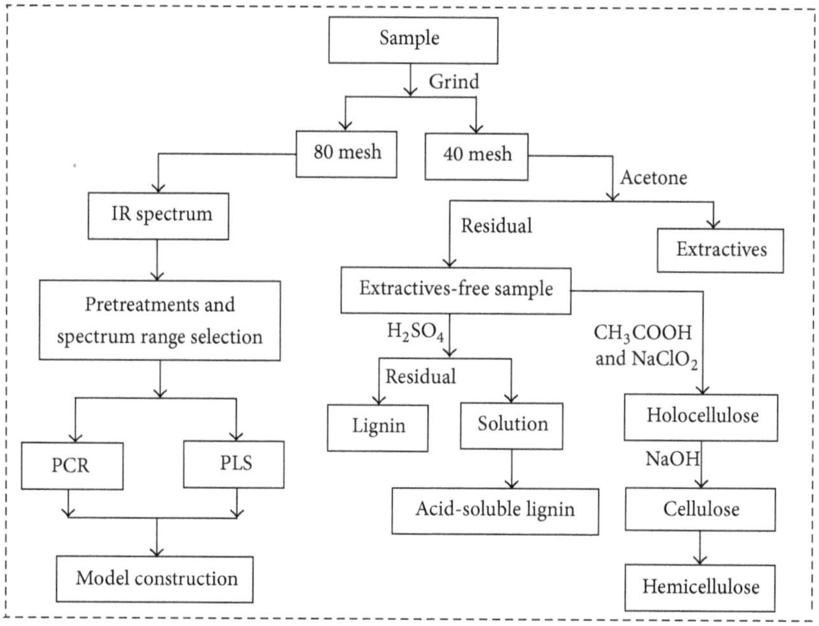

FIGURE 1: Schematic diagram for the wet chemical and FT-IR analysis process.

unprocessed spectra (raw) and first derivative (FD). Thirty-one samples were used to construct models and 6 samples were used for validation. Cross validation on all 37 samples was also run to ensure similar results and confirm validation results. While the populations for calibration and validation were randomly selected, the distribution of the data was checked to ensure a similar mean and range between the two populations. The coefficient of determination (r^2), root mean square error of calibration (RMSEC), and root mean square error of prediction (RMSEP) [17] were used to estimate the predictive performance of the models in this work. And the residual predictive deviation (RPD) was also measured to decide whether models were good at prediction, screening, or interpretation [21].

The principal component (PC) related to the chemical constituent during multivariate modeling was utilized for PCR coefficient/loading plots. The coefficients (y-axis) were connected via a smooth line in Origin software (Northampton, MA, USA) and then plotted against the wavenumbers (x-axis). For PLS, the regression coefficients plot was generated to represent the relationship between all of the absorbance and the specific chemical constituent. The wavenumbers were then chosen and compared to the literature.

3. Results and Discussion

3.1. Assessment of Loading Plots. The coefficients/loadings within the models are effective ways to interpret the relationship between wood chemistry functional groups and corresponding wood chemistry percent. For extractives, lignin, and cellulose loading plots, application of the first-derivative pretreatment was made prior to PLS and PCR execution. This was found to improve loading plots as evidenced by more parallel patterns (Figures 2, 3, and 4). The positive effect of the first-derivative pretreatment within the PLS coefficient plots was probably ascribed to the removal of the baseline shift present in the raw spectra. However, the positive effect of the first-derivative pretreatment was not apparent during the prediction of hemicellulose (Figure 5). This suggests that, during model development, application of the first derivative can generally improve the precision of the loadings, but not always.

It was also apparent that there was a shift in the location of the loading between native and first-derivative based data sets. For example, the wavenumber at 1640 cm^{-1}, 1508 cm^{-1}, and 1269 cm^{-1} shifted to 1649 cm^{-1}, 1514 cm^{-1}, and 1282 cm^{-1} for the extractive calibration models when the first-derivative pretreatment was used. This loading shift at 15 to 35 cm^{-1} could also be seen for lignin (1211 to 1227 cm^{-1}), cellulose (1187 to 1207 cm^{-1}), and hemicellulose (1633 to 1652 cm^{-1}) (Figures 3, 4, and 5). These shifts will be studied statistically in Table 1.

3.2. Wavenumber Assignments. Important wavenumbers for the prediction of extractives, lignin, cellulose, and hemicellulose were identified through the loading plots in Figures 2, 3, 4, and 5, and the functional groups important for a given wavenumber were further confirmed by the literature.

3.2.1. Extractives. The sensitivity of the bands at 1730 cm^{-1} was attributable to the C=O stretching vibrations produced by the ester carbonyl. These peaks appear when the lipophilic fraction of extractive is studied; they may come from fat, wax compounds or in esterified resin acids [22–25]. The peak near 1600 cm^{-1} can be assigned to either the C=C stretching or

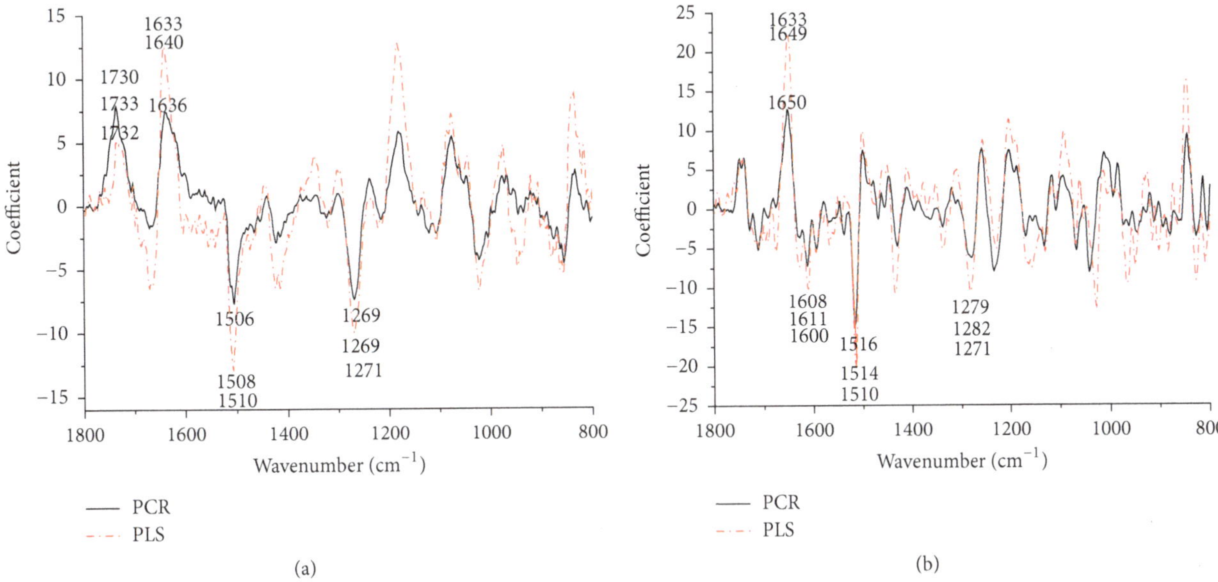

FIGURE 2: Coefficients by wavenumber for PCR and PLS for extractives prediction (a) when raw spectra were processed and (b) when a first-derivative pretreatment was processed. PC numbers 9, 2, 3, and 1 were chosen for PCR-raw, PLS-raw, PCR-derivative, and PLS derivative, respectively ($\alpha = 0.05$).

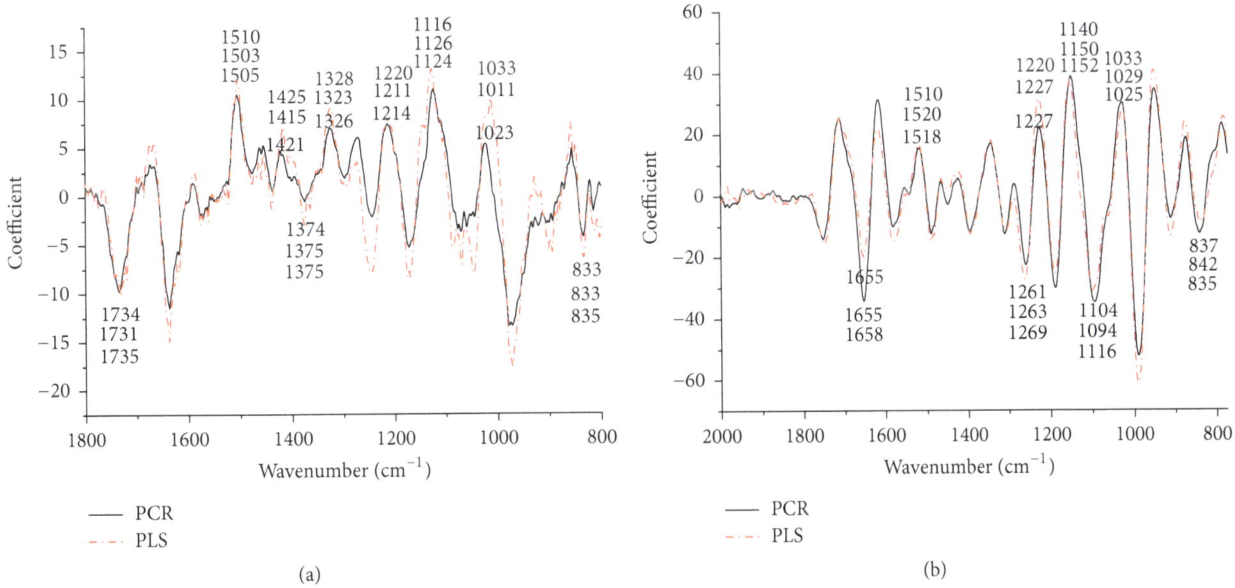

FIGURE 3: Coefficients by wavenumber for PCR and PLS for lignin prediction (a) when raw spectra were processed and (b) when a first-derivative pretreatment was processed. PC numbers 6, 2, 2, and 1 were chosen for PCR-raw, PLS-raw, PCR-derivative, and PLS derivative, respectively ($\alpha = 0.05$).

an aromatic ring deformation mode [26]. The strong peak at 1510 cm^{-1} is assigned to the deformation vibration within benzene rings [22, 27]. This peak is characteristic of aromatic compounds in wood and wood extractives [28, 29]. The weak olefinic double bond stretching is exhibited at 1633 cm^{-1} [26]. The strong band at 1271 cm^{-1} was due to carbon single bonded oxygen but is more likely to be an interaction band between carbon single bonded oxygen stretch and in-plane carbon single bonded hydroxyl bending in carboxylic acids, which is usually masked by a methylene scissoring band due to the methylene group attached to the carbonyl [22, 24].

3.2.2. Lignin. Coefficients by wavenumber for PCR and PLS for lignin prediction are displayed in Figure 3. C=O stretching and aromatic skeletal vibration were important loadings at 1735, 1658, 1328, 1510, and 1425 cm^{-1}, respectively

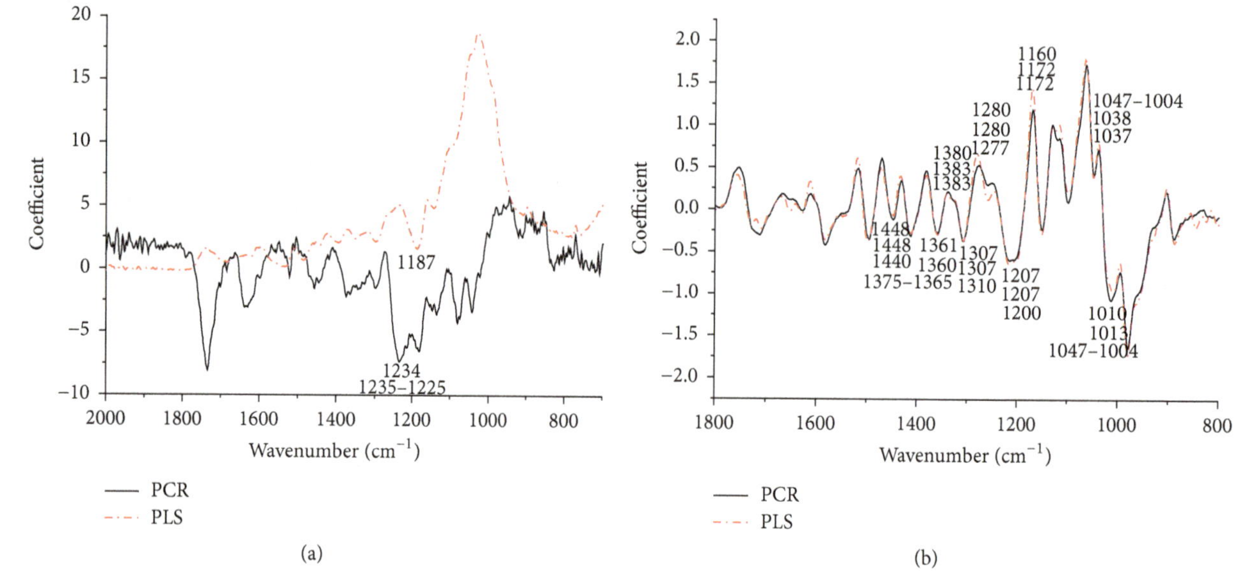

FIGURE 4: Coefficients by wavenumber for PCR and PLS for cellulose prediction (a) when raw spectra were processed and (b) when a first-derivative pretreatment was processed. PC numbers 8, 3, 1, and 2 were chosen for PCR-raw, PLS-raw, PCR-derivative, and PLS derivative, respectively ($\alpha = 0.05$).

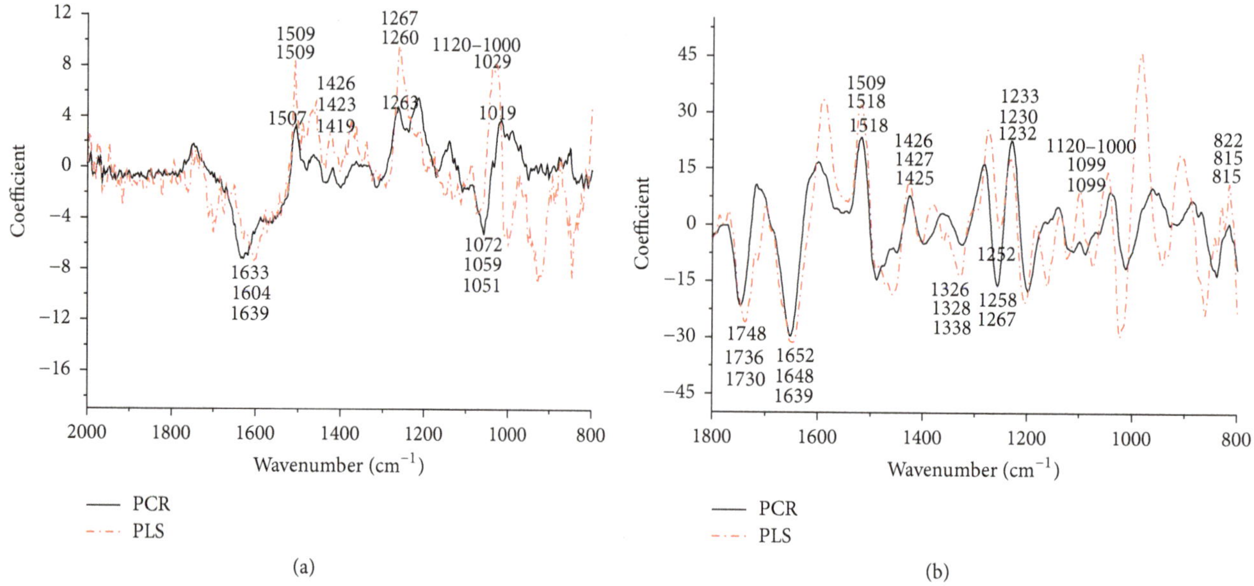

FIGURE 5: Coefficients by wavenumber for PCR and PLS for hemicellulose prediction (a) when raw spectra were processed and (b) when a first-derivative pretreatment was processed. PC numbers 3, 3, 1, and 1 were chosen for PCR-raw, PLS-raw, PCR-derivative, and PLS derivative, respectively ($\alpha = 0.05$).

[30–33]. Phenolic OH and aliphatic C-H in methyl groups were important based on the loading at 1375 cm^{-1}. The G ring showing with carbonyl stretching was a key functional group that was important based on the loading at 1269 cm^{-1}. The C-C, C-O, and C=O stretch were important due to the variation in the loading at 1220 cm^{-1}. The C-H in-plane deformation of the G ring plus secondary alcohols and C=O stretch were evident based on the loading at 1140 cm^{-1}. Aromatic C-H deformation in the S ring was based on the loading at 1116 cm^{-1} [30, 34]. The aromatic C-H in-plane deformation plus C-O deformation in primary alcohols plus the C=O stretch was based on the loading at 1033 cm^{-1}. C-H out-of-plane deformation in positions 2 and 6 of S rings was based on the loading at 835 cm^{-1} [30–32].

3.2.3. Cellulose. As shown in Figure 4, the wavenumbers demonstrate that the O-H in-plane bending (1440 and 1220 cm^{-1}), C-H bending (1380, 1375–1365, and 1280 cm^{-1}),

TABLE 1: Hypothesis testing of (2) through (a) t-test and (b) F-test.

(a) t-test

	PCR	PLS
Mean R	0.06	−0.49
Variance	55.5	92.6
Standard deviation	7.45	9.62
95% CI	0.06 ± 2.1	−0.49 ± 2.8
Observations	49	49
Degrees of freedom	48	48
F value	1.67	
P for t-test	0.0796*	

(b) F-test

	PCR	PLS
Mean R	0.06	−0.49
Variance	55.5	92.6
Standard deviation	7.45	9.62
95% CI	0.06 ± 2.1	−0.49 ± 2.8
Observations	49	49
Degrees of freedom	48	48
F value	0.60	
P $(F < f)$ one tail	0.0398*	
F critical one tail	0.62	

* This means that the t-test and F-test were significant with 95% confidence.

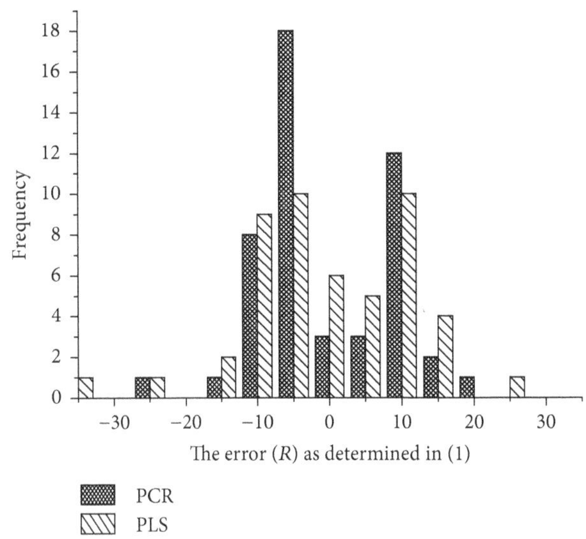

FIGURE 6: Frequency of R from PLS and PCR loadings of wood chemistry models.

and CH_2 wagging at 1310 cm^{-1} were important in the prediction of cellulose. The C-O-C asymmetric stretching, assigned to cellulose, appears at 1160 cm^{-1} and the C-O stretch in cellulose occurred at 1047–1004 cm^{-1} [35, 36].

3.2.4. Hemicellulose. The loadings corresponding to FT-IR spectra of the hemicelluloses are shown in Figure 5(a). The peak multiplicity between 1120 and 1000 cm^{-1} is a typical characteristic of carbohydrates. The signal at 1051 and 1008 cm^{-1} corresponding to the glycosidic (C-O-C) stretching cannot be clearly distinguished due to the multiplicity of the peaks in that region. Signals at 1453, 1426, and 1338 cm^{-1} are attributed to -CH_2 symmetric bending, CH and OH bending, and -CH wagging, respectively [37]. The signal at 1639 cm^{-1} is due to the absorbed water, but it can also reveal the presence of conjugated carbonyl groups that either are present in the polyphenolic structure of lignin or exist in uronic acids or result from carbohydrate oxidation and acetylated residues [38]. Hardwood xylan is heavily O-acetylated; therefore, the absorption of carbonyl groups is mainly contributed by xylan in hemicellulose. The characteristic peak at 1509 cm^{-1} which indicates the aromatic skeletal vibration of lignin and, correspondingly, the band at 1267 cm^{-1} associated with guaiacyl nuclei and the band at 1233 cm^{-1} related to syringyl nuclei of plane at positions 2, 5, and 6 in G units could not be distinguished. The peak at 822 cm^{-1} was attributed to C-H out of plane at positions 2 and 6 of S units and all positions of H units were not detected [39].

3.3. Interpretation of Significant Coefficients and Loadings Error Assessment. As we discussed in Section 3.2, C=O and C=C were key functional groups that were based on the loadings at 1730 and 1600 cm^{-1} for the prediction of extractives. For lignin prediction, the important loadings of OH and CH appeared at 1375 cm^{-1}, and C-C, C-O, and C=O loadings were assigned at 1220 cm^{-1}. For cellulose prediction, the C-H and CH_2 bond was important based on loadings at 1380, 1280, and 1310 cm^{-1}. Hemicellulose yielded -CH_2, CH, and OH bending and -CH wagging bond corresponding to the loadings at 1453, 1426, and 1338 cm^{-1}.

From this study, it is apparent that error in band assignments exists and can be quantified as BA_L. The distribution of error (R) for PCR exhibited a distribution closer to normality than PLS. Furthermore, PCR demonstrated a higher frequency of lower error in the loading estimates which supports the hypothesis that PCR provides better precision in wavenumber assignment (Figure 6). Finally, both PLS and PCR ($\alpha = 0.05$) overlapped with zero when a confidence interval test was performed (Table 1), which is indicative of lack in bias for either PLS or PCR.

Hypothesis testing was used to test the precision of peak loading location. The statistical results of the F-test indicated that the variance of R for PLS was greater than that of PCR. This means that the alternative hypothesis (H_a) was correct and that PCR is a better tool for assignment or interpretation of wavelengths and corresponding functional groups through multivariate modeling. Thus, PLS proved to be a better multivariate tool for prediction while PCR was better for interpretation. This was supported by another study in which two-dimensional correlation analysis and waterfall plots were used for detecting positional fluctuations of spectral changes. It was shown that 2D correlation analysis of the spectra was clear in defining the very characteristic cluster pattern both for the band position shifts and for partitioning out

TABLE 2: Calibration and predictive results of FT-IR based multivariate models.

Algorithm	Chemical components	Wavenumber ranges	Raw spectra			First derivative		
			R^2	RMSEP	RPD	R^2	RMSEP	RPD
PLS	Extractives	1750–1250 cm^{-1}	73.51	1.19	1.19	86.66	0.34	4.18
	Lignin	1800–800 cm^{-1}	87.76	1.05	2.30	90.06	0.50	4.83
	Cellulose	1500–1000 cm^{-1}	56.74	1.38	1.53	85.62	0.80	1.72
	Hemicellulose	1750–800 cm^{-1}	79.31	2.25	1.73	92.90	1.90	2.04
PCR	Extractives	1750–1250 cm^{-1}	57.93	0.87	1.69	71.04	0.59	2.41
	Lignin	1800–800 cm^{-1}	73.70	1.51	1.60	87.36	0.82	2.94
	Cellulose	1500–1000 cm^{-1}	52.61	1.59	1.32	48.15	0.94	1.34
	Hemicellulose	1750–800 cm^{-1}	30.09	4.14	0.94	52.69	3.34	1.16

two overlapped bands. Their study also found principal components analysis (PCA) to be very sensitive to peak shifts and helped to justify PCR for shift identification [40]. Practically speaking, this study found that the error (±2 standard deviations in peak loadings, Table 1) for PCR loadings was approximately ±15 cm^{-1}. This suggests that when loadings from PCR plots are used for interpretation, one can expect this estimate to be ±15 cm^{-1} of the true value.

In this study, transforming the spectra with the first derivative was found to improve loading plot precision. It is perhaps even possible to use PLS with competitive performance in loading precision although this study found PCR to still perform better.

3.4. Predictive Diagnostics. The predictive results for the chemical composition in wood were shown in Table 2. It should be noticed that the assignment of the peaks is a very important reference for modeling optimization. During the model construction, coupled with the peak assignment in the previous section, the optimal wavenumber ranges of extractives, lignin, cellulose, and hemicellulose for building the FT-IR model were found to be 1750–1250 cm^{-1}, 1800–800 cm^{-1}, 1500–1000 cm^{-1}, and 1750–800 cm^{-1}, separately. It was found that PLS always outperformed PCR in predictive diagnostics with the first-derivative pretreatment often improving calibration statistics. Models with a higher r^2, a lower RMSEP, and a higher RPD were deemed to generally be the best performers. RPD has particularly been utilized as a way of classifying whether a model was used for screening, prediction, or actual measurement [7]. The best predictive results (RPD value) for extractives, lignin, cellulose, and hemicellulose were 4.18, 4.83, 1.72, and 2.04, which were performed by PLS and first-derivative pretreatment. The RPD values demonstrate that the FT-IR prediction models for extractives and lignin were good enough to conduct quantitative analysis, while cellulose and hemicellulose models were more appropriate for screening or for measurement of population statistics during process control [41]. The predictive ability of calibration models on new samples is presented in Figure 7. Several chemical components could be predicted from a single measurement in a rapid and precise way.

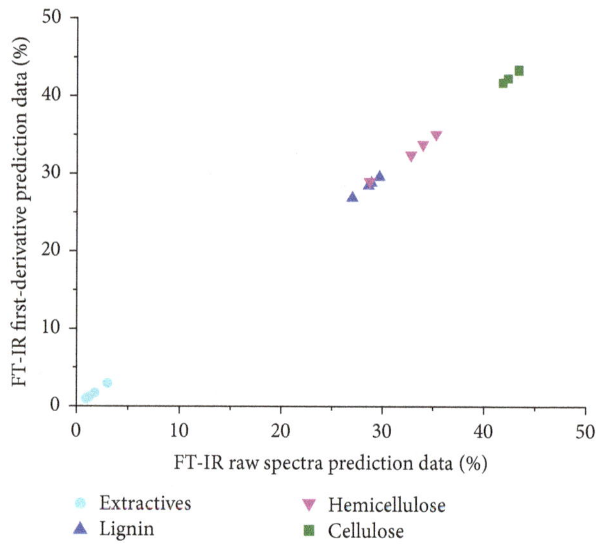

FIGURE 7: Chemical content (%, w/w) predicted by multivariate models with FT-IR (PLS) for validation samples.

4. Conclusion

PLS and PCR were paired with FT-IR spectroscopy to determine which methods were better for interpretation and prediction of wood chemistry. Important wavenumbers to extractives, lignin, cellulose, and hemicellulose were identified by analyzing the loading plots and this was compared to the literature. It was found that PCR performed better than PLS for interpretation when FT-IR spectroscopy was used. Research also found that the chemical composition content in wood could be predicted by FT-IR models with extractives and lignin models exhibiting an RPD > 4, which suggests that these models are adequate for quantitative analysis of individual samples.

Conflict of Interests

The authors declare that there is no conflict of interests regarding the publication of this paper.

Acknowledgments

This work was most supported by the Agriculture and Food Research Initiative (AFRI) CAP: "Southeast Partnership for Integrated Biomass Supply Systems," the Department of Energy grant titled "High Tonnage Forest Biomass Production Systems from Southern Pine Energy Plantations." Other smaller contributions included the Hatch in which these calibration equations will be utilized to partition out feedstocks of high lignin for adhesive production and energy applications. Regions Bank is also acknowledged for their financial support.

References

[1] C. M. Mak, C. W. M. Yuen, S. K. A. Ku, and C. W. Kan, "Changes in surface morphology of Tencel fabric during the fibrillation process," *Journal of the Textile Institute*, vol. 97, no. 3, pp. 241–245, 2006.

[2] I. A. Weinstock, R. H. Atalla, R. S. Reiner et al., "A new environmentally benign technology for transforming wood pulp into paper, engineering polyoxometalates as catalysts for multiple processes," *Journal of Molecular Catalysis A: Chemical*, vol. 116, no. 1-2, pp. 59–84, 1997.

[3] B. R. Ellingwood, D. V. Rosowsky, Y. Li, and J. H. Kim, "Fragility assessment of light-frame wood construction subjected to wind and earthquake hazards," *Journal of Structural Engineering*, vol. 130, no. 12, pp. 1921–1930, 2004.

[4] J. Bodig and B. A. Jayne, *Mechanics of Wood and Wood Composites*, Van Nostrand Reinhold, 1981.

[5] M. Giampietro, S. Ulgiati, and D. Pimentel, "Feasibility of large-scale biofuel production," *BioScience*, vol. 47, no. 9, pp. 587–600, 1997.

[6] L. Eckhardt, M. A. Sword Sayer, and D. Imm, "State of pine decline in the southeastern United States," *Southern Journal of Applied Forestry*, vol. 34, no. 3, pp. 138–141, 2010.

[7] W. Jiang, G. Han, Y. Zhang, and M. Wang, "Fast compositional analysis of ramie using near-infrared spectroscopy," *Carbohydrate Polymers*, vol. 81, no. 4, pp. 937–941, 2010.

[8] T. Schultz and D. Burns, "Rapid secondary analysis of lignocellulose: comparison of near infrared (NIR) and fourier transform infrared (FTIR)," *Tappi Journal*, vol. 73, no. 5, pp. 209–212, 1990.

[9] A. Vergnoux, M. Guiliano, Y. Le Dréau, J. Kister, N. Dupuy, and P. Doumenq, "Monitoring of the evolution of an industrial compost and prediction of some compost properties by NIR spectroscopy," *Science of the Total Environment*, vol. 407, no. 7, pp. 2390–2403, 2009.

[10] N. Wang, X. Zhang, Z. Yu, G. Li, and B. Zhou, "Quantitative analysis of adulterations in oat flour by FT-NIR spectroscopy, incomplete unbalanced randomized block design, and partial least squares," *Journal of Analytical Methods in Chemistry*, vol. 2014, Article ID 393596, 5 pages, 2014.

[11] Y. Meng, S. Wang, R. Cai, B. Jiang, and W. Zhao, "Discrimination and content analysis of fritillaria using near infrared spectroscopy," *Journal of Analytical Methods in Chemistry*, vol. 2015, Article ID 752162, 8 pages, 2015.

[12] D. A. Burns and E. W. Ciurczak, *Handbook of Near-Infrared Analysis*, CRC Press, 2007.

[13] B. K. Via, O. Fasina, and H. Pan, "Assessment of pine biomass density through mid-infrared spectroscopy and multivariate modeling," *BioResources*, vol. 6, no. 1, pp. 807–822, 2011.

[14] T. Trafela, M. Strlič, J. Kolar et al., "Nondestructive analysis and dating of historical paper based on IR spectroscopy and chemometric data evaluation," *Analytical Chemistry*, vol. 79, no. 16, pp. 6319–6323, 2007.

[15] X.-F. Xu, L.-X. Nie, L.-L. Pan et al., "Quantitative analysis of *Panax ginseng* by FT-NIR spectroscopy," *Journal of Analytical Methods in Chemistry*, vol. 2014, Article ID 741571, 6 pages, 2014.

[16] B. K. Via, C. Zhou, G. Acquah, W. Jiang, and L. Eckhardt, "Near infrared spectroscopy calibration for wood chemistry: which chemometric technique is best for prediction and interpretation?" *Sensors*, vol. 14, no. 8, pp. 13532–13547, 2014.

[17] J. Neter, M. H. Kutner, C. J. Nachtsheim et al., *Applied Linear Statistical Models, Volume 4*, Irwin, Chicago, Ill, USA, 1996.

[18] A. Sluiter, R. Ruiz, C. Scarlata, J. Sluiter, and D. Templeton, "Determination of extractives in biomass: laboratory analytical procedure (LAP)," Tech. Rep. NREL/TP-510-42619, National Renewable Energy Laboratory, Golden, Colo, USA, 2005.

[19] A. Sluiter, B. Hames, R. Ruiz et al., "Determination of structural carbohydrates and lignin in biomass: laboratory analytical procedure," Tech. Rep. NREL/TP-510-42618, National Renewable Energy Laboratory, Golden, Colo, USA, 2008.

[20] L. G. Thygesen and S.-P. Lundqvist, "NIR measurement of moisture content in wood under unstable temperature conditions. Part 2. Handling temperature fluctuations," *Journal of Near Infrared Spectroscopy*, vol. 8, no. 3, pp. 191–199, 2000.

[21] W. Jiang, G. Han, B. K. Via, M. Tu, W. Liu, and O. Fasina, "Rapid assessment of coniferous biomass lignin–carbohydrates with near-infrared spectroscopy," *Wood Science and Technology*, vol. 48, no. 1, pp. 109–122, 2014.

[22] D. Williams, I. Fleming, and E. Pretsch, *Spectroscopic Methods in Organic Chemistry*, McGraw-Hill, 1989.

[23] A. Holmgren, B. Bergström, R. Gref, and A. Ericsson, "Detection of pinosylvins in solid wood of scots pine using fourier transform raman and infrared spectroscopy," *Journal of Wood Chemistry and Technology*, vol. 19, no. 1-2, pp. 139–150, 1999.

[24] M. Nuopponen, T. Vuorinen, S. Jämsä, and P. Viitaniemi, "The effects of a heat treatment on the behaviour of extractives in softwood studied by FTIR spectroscopic methods," *Wood Science and Technology*, vol. 37, no. 2, pp. 109–115, 2003.

[25] R. Sun and J. Tomkinson, "Characterization of hemicelluloses isolated with tetraacetylethylenediamine activated peroxide from ultrasound irradiated and alkali pre-treated wheat straw," *European Polymer Journal*, vol. 39, no. 4, pp. 751–759, 2003.

[26] E.-M. A. Ajuong and M. C. Breese, "Fourier transform infrared characterization of Pai wood (*Afzelia africana* Smith) extractives," *Holz als Roh—und Werkstoff*, vol. 56, no. 2, pp. 139–142, 1998.

[27] D. Lin-Vien, N. B. Colthup, W. G. Fateley, and J. G. Grasselli, *The Handbook of Infrared and Raman Characteristic Frequencies of Organic Molecules*, Elsevier, 1991.

[28] B. H. Nielsen, J. Rodrigues, H. Pereira, and et al, "Rapid determination of the lignin content in Sitka spruce (Picea sitchensis (Bong.) Carr.) wood by Fourier transform infrared spectrometry," *Holzforschung*, vol. 53, no. 6, pp. 597–602, 1999.

[29] M. Schwanninger, J. C. Rodrigues, H. Pereira, and B. Hinterstoisser, "Effects of short-time vibratory ball milling on the shape of FT-IR spectra of wood and cellulose," *Vibrational Spectroscopy*, vol. 36, no. 1, pp. 23–40, 2004.

[30] S. Kubo and J. F. Kadla, "Hydrogen bonding in lignin: a fourier transform infrared model compound study," *Biomacromolecules*, vol. 6, no. 5, pp. 2815–2821, 2005.

[31] O. Faix and O. Beinhoff, "FTIR spectra of milled wood lignins and lignin polymer models (DHP's) with enhanced resolution obtained by deconvolution," *Journal of Wood Chemistry and Technology*, vol. 8, no. 4, pp. 505–522, 1988.

[32] O. Faix, "Classification of lignins from different botanical origins by FT-IR spectroscopy," *Holzforschung*, vol. 45, no. 1, pp. 21–28, 1991.

[33] A. T. Martínez, G. Almendros, F. J. González-Vila, and R. Fründ, "Solid-state spectroscopic analysis of lignins from several Austral hardwoods," *Solid State Nuclear Magnetic Resonance*, vol. 15, no. 1, pp. 41–48, 1999.

[34] C. G. Boeriu, D. Bravo, R. J. A. Gosselink, and J. E. G. Van Dam, "Characterisation of structure-dependent functional properties of lignin with infrared spectroscopy," *Industrial Crops and Products*, vol. 20, no. 2, pp. 205–218, 2004.

[35] C. Y. Liang and R. H. Marchessault, "Infrared spectra of crystalline polysaccharides. II. Native celluloses in the region from 640 to 1700 cm.$^{-1}$," *Journal of Polymer Science*, vol. 39, no. 135, pp. 269–278, 1959.

[36] D. Fengel and M. Ludwig, "Möglichkeiten und grenzen der FTIR-spektroskopie bei der charakterisierung von cellulose. I: vergleich von verschiedenen cellulosefasern und bakterien-cellulose," *Das Papier*, vol. 45, no. 2, pp. 45–51, 1991.

[37] R. C. Sun and J. Tomkinson, "Characterization of hemicelluloses obtained by classical and ultrasonically assisted extractions from wheat straw," *Carbohydrate Polymers*, vol. 50, no. 3, pp. 263–271, 2002.

[38] P. Reyes, R. T. Mendonça, J. Rodríguez, P. Fardim, and B. Vega, "Characterization of the hemicellulosic fraction obtained after pre-hydrolysis of pinus radiata wood chips with hot-water at different initial PH," *Journal of the Chilean Chemical Society*, vol. 58, no. 1, pp. 1614–1618, 2013.

[39] M.-F. Li, S.-N. Sun, F. Xu, and R.-C. Sun, "Ultrasound-enhanced extraction of lignin from bamboo (*Neosinocalamus affinis*): characterization of the ethanol-soluble fractions," *Ultrasonics Sonochemistry*, vol. 19, no. 2, pp. 243–249, 2012.

[40] S. R. Ryu, I. Noda, C.-H. Lee, P. H. Lee, H. Hwang, and Y. M. Jung, "Two-dimensional correlation analysis and waterfall plots for detecting positional fluctuations of spectral changes," *Applied Spectroscopy*, vol. 65, no. 4, pp. 359–368, 2011.

[41] B. Via, "Characterization and evaluation of wood strand composite load capacity with near infrared spectroscopy," *Materials and Structures*, vol. 46, no. 11, pp. 1801–1810, 2013.

27

Determination of Sunset Yellow and Tartrazine in Food Samples by Combining Ionic Liquid-Based Aqueous Two-Phase System with High Performance Liquid Chromatography

Ou Sha,[1,2,3] Xiashi Zhu,[2] Yanli Feng,[3] and Weixing Ma[3]

[1] *Analysis and Test Centre of Jiangsu Marine Resources Development Research Institute, Lianyungang 222001, China*
[2] *College of Chemistry & Chemical Engineering, Yangzhou University, Yangzhou 225002, China*
[3] *School of Chemistry and Chemical Engineering, Huaihai Institute of Technology, Lianyungang 222005, China*

Correspondence should be addressed to Ou Sha; 7993259@163.com

Academic Editor: Verónica Pino

We proposed a simple and effective method, by coupling ionic liquid-based aqueous two-phase systems (IL-ATPSs) with high performance liquid chromatography (HPLC), for the analysis of determining tartrazine and sunset yellow in food samples. Under the optimized conditions, IL-ATPSs generated an extraction efficiency of 99% for both analytes, which could then be directly analyzed by HPLC without further treatment. Calibration plots were linear in the range of 0.01–50.0 μg/mL for both Ta and SY. The limits of detection were 5.2 ng/mL for Ta and 6.9 ng/mL for SY. This method proves successful for the separation/analysis of tartrazine and sunset yellow in soft drink sample, candy sample, and instant powder drink and leads to consistent results as obtained from the Chinese national standard method.

1. Introduction

Tartrazine (Ta) and sunset yellow (SY) are synthetic food colors used most extensively as food additives, to improve the appearance, color, and texture of foods [1]. When added in excess, however, these synthetic food colors can be pathogenic [2]. To ensure food safety, the Chinese government has imposed rigorous standards on the permitted levels for various synthetic food colors [3]. For the same reason, it is important to develop effective methods for analyzing synthetic colors in food.

Various techniques have been applied for analyzing synthetic colors in food samples, including spectrophotometry [4], capillary electrophoresis [5], differential pulse polarography [6], liquid chromatography [7], and others [8], which all require an optimal extraction method to concentrate an analyte from a small amount of food sample containing miscellaneous components [9, 10]. Solid phase extraction (SPE) and liquid-liquid extraction (LLE) have been reported for separating synthetic food colors from different matrices [11, 12]. For SPE, analytes are separated following complicated and time-consuming absorption and desorption steps involving the use of toxic and volatile reagents, such as methanol, acetic acid, and ammonia, that are used in the current Chinese national standard method [13]. In contrast, an aqueous two-phase system (ATPS) has attracted increasing attention in that it minimizes the potential interferences from other components and enables the simultaneous extraction and concentration of analytes [14]. By far, most reported ATPSs are generated by mixing two solutions of polymers, such as dextran T500 and polyethylene glycol (PEG) system [15], or by adding a high-concentration salt solution to an aqueous polymer solution [16]. The partition behavior of analytes between the polymer-rich phase and the aqueous phase can be controlled and optimized with a judicious choice of phase system. However, most of polymer-rich phase is highly viscous and opaque, rendering the follow-up analysis difficult. To address these issues, the development of a simple and environmentally friendly method with minimal use of volatile and toxic solvents used is of great importance.

Ionic liquid (IL) is a green solvent and a potential replacement for traditional volatile solvents [17]. In 2003,

Gutowski and coworkers first generated several ATPSs from different IL and salts and determined the corresponding phase diagrams [18]. The identity of IL is important for the separation/analysis of a specific analyte in IL-based ATPS [19]. An optimal IL-based APTS should possess the following features: minimal emulsion formation, low viscosity, rapid phase separation, high extraction efficiency and low cost, and so forth [20]. In this study, we aim to develop a simple and "green" IL-based ATPS extraction method with high efficiency and couple it with HPLC for the analysis of Ta and SY from food samples. For this purpose, we selected 1-butyl-3-methylimidazolium bromide ([C_4MIM]Br), which is not easy to emulsify and presents a lower viscidity compared with polymer-salt ATPS. Besides, [C_4MIM]Br was diffluent in water and methanol (the HPLC mobile phase). Analytes (Ta + SY) could be extracted into IL phase and analyzed by HPLC with no further sample treatment. We showed the success of this method in the separation/analysis of Ta and SY from soft drink sample, candy sample, and instant powder drink.

2. Experimental

2.1. Reagents and Apparatus. All reagents were of analytical grade or higher in purity and purchased from Sinopharm Chemical Reagent Co., Ltd., Shanghai, China, unless otherwise specified. HPLC grade methanol was purchased from Merck (Germany). The standard stock solutions of the colorants, tartrazine (Ta; C.I. Food Yellow 4; 0.5 mg/mL), and sunset yellow (SY; C.I. Food Yellow 3; 0.5 mg/mL) were obtained from the National Research Center for Certified Reference Materials (Beijing, China) and were both prepared in distilled water to a final concentration of 100 μg/mL. All glass containers were stored in 10% (v/v) nitric acid overnight and rinsed with distilled water before use. ILs (1-alkyl-3-methylimidazolium bromide, [C_nMIM]Br, $n = 4, 6, 8$) were synthesized as described previously [21].

All spectral measurements were carried out using a model UV-2501 spectrophotometer (Shimadzu, Japan). Phase separation of the sample solution was achieved with a centrifuge (Model TDL80-2B, Shanghai Anting Science Instrument Factory, Shanghai, China). Chromatographic analyses were performed on an Agilent 1260 HPLC system which consisted of a 1260 infinity quaternary pump with degasser gradient pump, a 1260 infinity variable wavelength detector, a 1260 infinity manual injector, and an open LAB CDS chemstation workstation (Agilent, USA).

2.2. Preparation of Phase Diagram. Phase diagrams were determined using the cloud-point method [22]. A certain of [C_nMIM]Br was put into a 10.0 mL centrifugal tube. K_2HPO_4 solution or the other tested salt solution was added dropwise to the test tube until a turbidity formed, indicating the formation of a two-phase system. Thereafter, water was added dropwise to the test tube to obtain a clear one-phase system. More salt solution was added again to afford a two-phase system. The composition of this mixture was noted and so on. The bimodal curve was applied to characterize the phase diagram [23]. In the region above the bimodal curve, the system is divided into two phases; in the region below the bimodal curve, the system is of a homogeneous phase.

2.3. IL-Based ATPS Extraction. 0.3 mL of [C_4MIM]Br and 1.0 mL of the mixed standard solution (100 μg/mL SY + 100 μg/mL Ta) or 1.0–3.0 mL of the sample solution were placed in a 10.0 mL centrifugal tube. The mixture was diluted to 4.0 mL with distilled water and then 3.0 g K_2HPO_4 was added. The phase separation quickly occurred in the tube after gentle blending. The tube was centrifuged for 2 min at 3500 rpm to ensure a complete phase separation. 20.0 μL of [C_4MIM]Br phase was withdrawn using a microsyringe and injected into HPLC for quantification.

2.4. Chromatographic Conditions. Chromatographic separation was achieved on a Zorbax ODS column (4.6 mm × 150 mm × 5 μm) associated with a guard column packed with the same bonded phase. The composition of the mobile phase at time zero (the time of injection) was 85% ammonium acetate (20 mM, pH 4.5) and 15% methanol. Over the next 5 min, the percentage of methanol was gradually increased to 40%. Thereafter, the mobile phase was changed to ammonium acetate-methanol (2 : 98; v : v) within 5 min. Finally, the chromatographic system was equilibrated during 5 min before the next injection. A flow rate of 0.8 mL/min was used throughout the 15 min run. Chromatography was performed at 30°C. Dual UV wavelength mode was used, with Ta monitored at 420 nm and SY at 480 nm. The mobile phase was filtered through 0.45 μm micropore filter membrane prior to use.

2.5. The Extraction Parameter. The distribution behaviors of Ta and SY between IL phase and salt phase were characterized by the extraction efficiency (E), partition coefficient (K), and phase ratio (R).

The parameters E, K, and R were defined as follows:

$$E = \frac{C_{\mathrm{IL}} V_{\mathrm{IL}}}{C_{\mathrm{aq}} V_{\mathrm{aq}} + C_{\mathrm{IL}} V_{\mathrm{IL}}} \times 100\%,$$
$$K = \frac{C_{\mathrm{IL}}}{C_{\mathrm{aq}}}, \quad (1)$$
$$R = \frac{V_{\mathrm{IL}}}{V_{\mathrm{aq}}},$$

where C_{IL} and C_{aq} are the concentration of Ta or SY in IL phase and in salt phase, respectively, and V_{IL} and V_{aq} are the volume of IL phase and K_2HPO_4 phase, respectively.

The spectra and absorbance of Ta or SY in [C_4MIM]Br/K_2HPO_4 system were determined as an example to study the distribution behavior and extraction efficiency. During each experiment, the IL phase containing the analyte was collected to measure the UV-visible spectra on the UV-2501 UV-vis spectrophotometer. The absorbance of Ta and SY was measured at 420 nm and 480 nm, respectively. The blank containing the same phase composition but without analyte was used as reference solution.

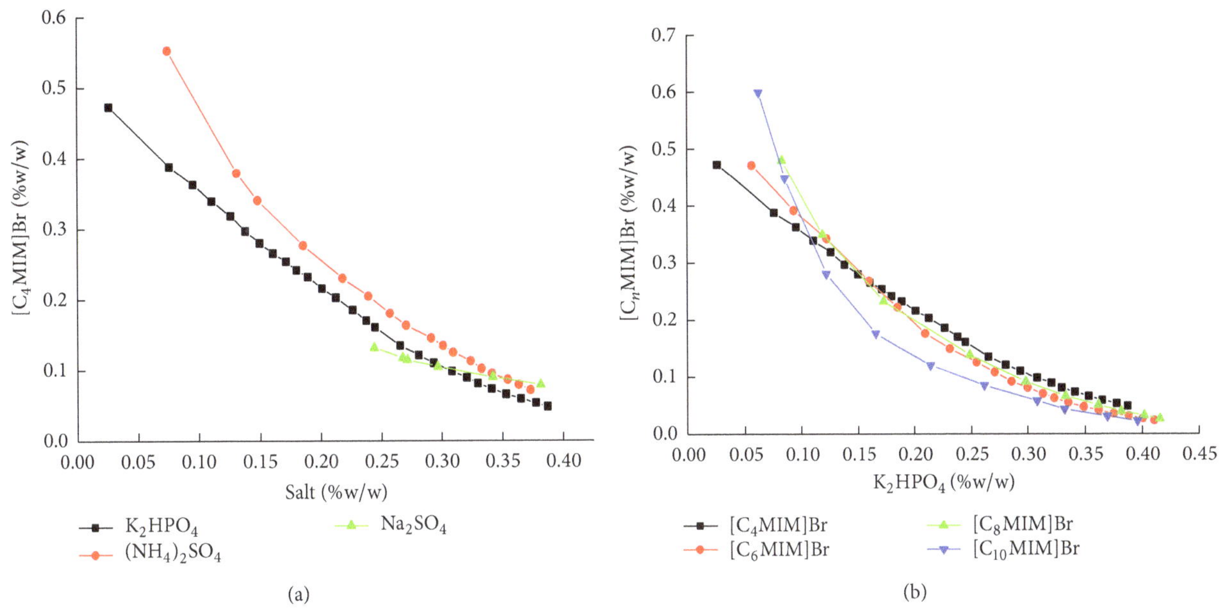

FIGURE 1: Phase diagrams of different salts (a) and $[C_nMIM]Br$ (b) for the ILs/salt/water systems at 25°C.

2.6. Sample Analysis. All samples, including soft drink, candy, and instant powdered drink, were obtained from a local market. Appropriate amounts (0.1–5 g) of the samples were dissolved in 15 mL of water. The carbonated drinks were degassed by ultrasonication for 5 minutes to remove the carbon dioxide. A warming process (50°C, 30 min) was used for the complete dissolution of the sugar-based candy. These solutions were centrifuged and the upper solutions were filtered through 0.45 μm micropore filter membrane. The filtrate was transferred to volumetric flask of 25.0 mL, volume adjusted to 25.0 mL using distilled water, and ready for IL-ATPS extraction.

The Chinese national standard method (GB/T 5009.35-2003) was also used for separation/analysis of Ta and SY in different samples. Briefly, the sample solution was adjusted to pH 6.0 with 200 g/L citric acid solution and heated up to 60°C. Then this solution was stirred with polyamide powder for 5 min and filtered through Buchner filter using double-decked filter papers. After filtering, the polyamide powder was washed with deionized water for 3–5 times, followed with the mixture of methanol-formic acid for 3–5 times. The colorants absorbed by polyamide powder were eluted with 15 mL of eluting solution (30% ammonia solution mixed with ethanol in the volume atio of 3 : 7), adjusted to pH 7.0 by acetic acid, and evaporated to near dryness, following which distilled water was added to the volume of 5.0 mL for HPLC analysis.

3. Results and Discussion

3.1. Phase Diagrams. In this paper the bimodal curve for the aqueous two phase systems of $[C_4MIM]Br$ with a series of salts at 25°C was investigated. Two replicate measurements were performed for each point. Results showed that IL-salt ATPS could be formed by adding appropriate amount of different salts, such as $(NH_4)_2SO_4$, Na_2SO_4, and K_2HPO_4. Other salts, such as K_3PO_4, KH_2PO_4, NaCl, and NaH_2PO_4, cannot separate $[C_4MIM]Br$ solution into two phases. As shown in Figure 1(a), the phase separation capability of salts is K_2HPO_4 > $(NH_4)_2SO_4$ > Na_2SO_4, consistent with their solubility in water at 25°C, suggesting a correlation between the phase separation capability of salts and their solubility in water [24]. Consequently, K_2HPO_4 was chosen in the following studies.

Methylimidazolium bromides with four different carbon-chain length, namely, $[C_4MIM]Br$, $[C_6MIM]Br$, $[C_8MIM]Br$, and $[C_{10}MIM]Br$, were selected to investigate the effect of carbon-chain length on phase diagrams of IL-salt ATPS. The bimodal curves determined at 25°C for the IL/K_2HPO_4 system were shown in Figure 1(b). When the weight percentage of K_2HPO_4 was more than 0.15%, the phase separation capability of IL followed the order of $[C_4MIM]Br \approx [C_6MIM]Br \approx [C_8MIM]Br < [C_{10}MIM]Br$. In view of the lower synthetic cost and lower viscosity of $[C_4MIM]Br$ than that of $[C_{10}MIM]Br$, $[C_4MIM]Br$ was chosen for further study.

3.2. Optimization of Extraction System. The extraction efficiency mainly depends on the structure of analyte and its affinity towards the extractant (i.e., partitioning coefficient), phase ratio, and the number of extractions in the liquid-liquid extraction system. It was also often used to estimate the migration ability of analyte between the two phases and the separation performance of extractant. Higher separation efficiency can be achieved with a greater partitioning coefficient K and a lower phase ratio R [25].

In this paper the effects of ionic liquid, salt, and temperature on extraction efficiency, partition coefficient, and phase ratio were assessed to obtain the optimal parameter.

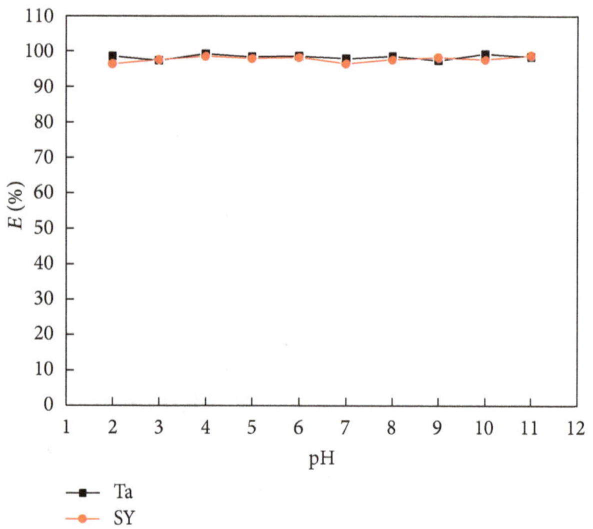

FIGURE 2: The effect of pH on the extraction efficiency (m_{SY} = 100 μg, m_{Ta} = 100 μg, V_{IL} = 0.30 mL).

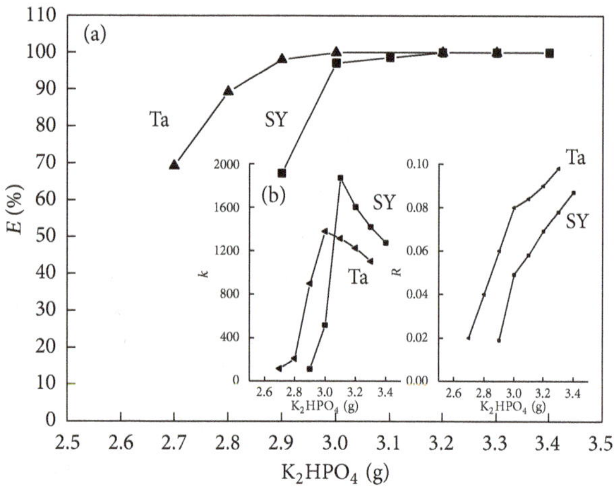

FIGURE 3: Effects of K_2HPO_4 amount on extraction efficiency, partition coefficient, and phase ratio (m_{SY} = 100 μg, m_{Ta} = 100 μg, V_{IL} = 0.30 mL).

3.2.1. The pH Effect. The effect of pH on IL-ATPS extraction ability of [C_4MIM]Br for Ta and SY was studied in the pH range of 2.0–11.0 by the addition of Britton-Robinson buffer solution in the presence of 0.3 mL [C_4MIM]Br and 3.0 g K_2HPO_4 (Figure 2). It was foundthat the extraction efficiency remained relatively constant over the pH range and the pH has little effect on the extraction of Ta and SY.

3.2.2. Effect of K_2HPO_4. With a fixed volume of [C_4MIM]Br at 0.3 mL, [C_4MIM]Br-K_2HPO_4 ATPS was formed when the amount of K_2HPO_4 was over 2.7 g in weight for Ta and 2.9 g in weight for SY, respectively. The influence of K_2HPO_4 concentration on the extraction efficiency, partition coefficient, and phase ratio was studied with 0.3 mL [C_4MIM]Br added and the results obtained were shown in Figure 3. It revealed that the extraction efficiency obtained was above

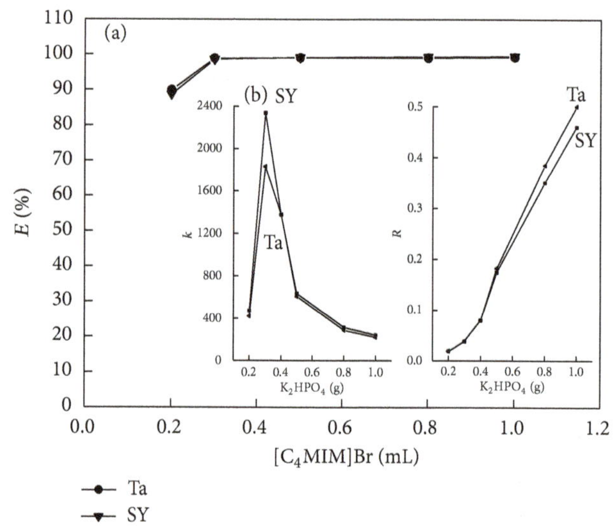

FIGURE 4: Effects of [C_nMIM]Br amount on extraction efficiency, partition coefficient, and phase ratio (m_{SY} = 100 μg, m_{Ta} = 100 μg, $m_{K_2HPO_4}$ = 3.0 g).

95% with K_2HPO_4 of 3.00–3.40 g (Figure 3(a)). The partition coefficient of Ta and SY was increased to a maximum value when the amount of K_2HPO_4 was set at 3.0 g and 3.1 g, respectively, then decreased with the increased amount of K_2HPO_4. The phase ratio of Ta and SY was increased with the amount of K_2HPO_4 (Figure 3(b)). In order to achieve quantitative extraction and higher separation efficiency, 3.0 g of K_2HPO_4 was used in all following experiments for the simultaneous extraction of SY and Ta.

3.2.3. Effect of [C_4MIM]Br Amount. The amount of [C_4MIM]Br used in the preconcentration procedure is a critical factor for obtaining a high extraction performance. Therefore, the extraction system was carefully studied to determine the lowest IL-phase volume necessary for achieving the best extraction. In this system, the ATPS of [C_4MIM]Br-K_2HPO_4 could not be achieved with 3.2 g K_2HPO_4 if the volume of [C_4MIM]Br was less than 0.2 mL.So the effect of the volume of [C_4MIM]Br on extraction efficiency was studied in the range of 0.2–1.0 mL. It was found that the extraction efficiency calculated was more than 95% in the concentration range of 0.3–1.0 mL (Figure 4(a)). The partition coefficient of Ta and SY got a maximum value when 0.3 mL [C_4MIM]Br was added, and then decreased with amount of IL, whereas the phase ratio was increased with the increased amount of [C_4MIM]Br (Figure 4(b)). Considering the extraction efficiency, enrichment factors, and the low consumption of [C_4MIM]Br, 0.3 mL [C_4MIM]Br IL was used to achieve a higher extraction efficiency in the subsequent experiments.

3.2.4. Effect of Temperature. The effect of temperature on the extraction efficiency and phase ratio of Ta and SY by IL-salt APTS was also studied (Figure 5). The results revealed that the extraction efficiency of Ta and SY was near to 100% within

TABLE 1: Analytical figures of merit for Ta and SY using HPLC-UV method.

Analyte	Slope of the calibration graph ±SD ($n = 3$)	Correlation coefficient r	Limit of detection (ng/mL)	Intraday precision RSD (%) ($n = 6$)[b]	Interday precision RSD (%) ($n = 3$)[a]
Ta	751.36 ± 2.14	0.9975	5.2	3.2	5.6
SY	838.51 ± 1.23	0.9967	6.9	2.4	4.8

[a]three independent calibration graphs obtained in three different days.
[b]concentration of analyte was 0.5 μg mL.

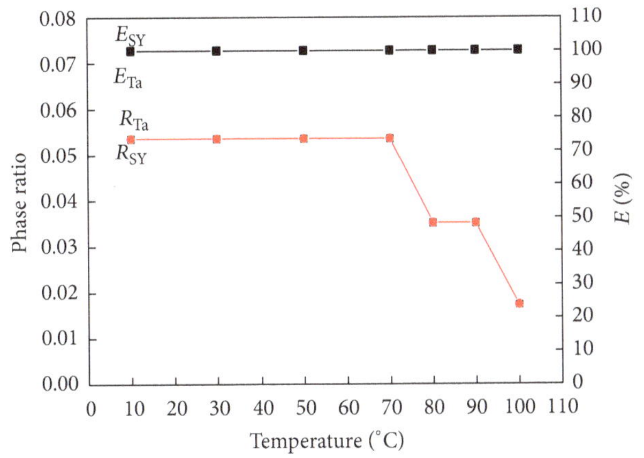

FIGURE 5: The effect of temperature on the extraction efficiency and phase ratio.

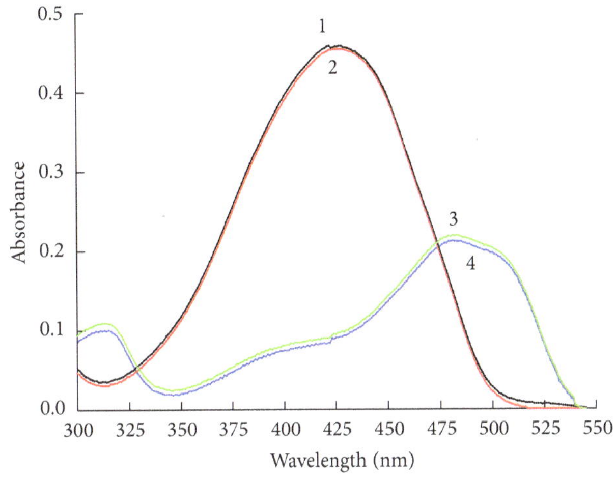

FIGURE 6: The UV-Vis spectra of colorants before and after extraction ($m_{SY} = 100\,\mu g$, $m_{Ta} = 100\,\mu g$, $V_{IL} = 0.30$ mL, $m_{K_2HPO_4} = 3.0$ g).

the entire tested temperature range from 10°C to 70°C. The phase ratio was stable within this range and decreased when temperature was over 70°C. The curve of extraction efficiency and phase ratio of sunset yellow was superposed practically compared with that of tartrazine. The reason was that higher temperatures result in more [C_4MIM]Br redissolved into K_2HPO_4-enriched bottom phase. Accordingly, [C_4MIM]Br concentration in top phase decreased. But the decrease of [C_4MIM]Br in top phase does not affect the extraction efficiencies of Ta and SY. This new extraction system can afford a wide temperature range for extraction of Ta and SY. In the following experiments, the temperature was set at room temperature of 25°C for the extraction of Ta and SY.

3.3. UV-Vis Spectrometric Studies of the Ta and SY in the [C_4MIM]Br-Rich Top Phase. The absorption spectra recorded for Ta (curve 1 and curve 2) and SY (curve 3 and curve 4) before and after [C_4MIM]Br/K_2HPO_4 extraction, as scanned in the wavelength range of 350 nm–700 nm against the reagent blank, respectively, were shown in Figure 6. The spectra of these two colorants remained unaltered when they were extracted into the [C_4MIM]Br-rich top phase and practically superimposed before and after [C_4MIM]Br/K_2HPO_4 ATPS extraction. This observation clearly indicated that no direct chemical (bonding) interactions were involved between colorants and ionic liquid in the IL-ATPS.

3.4. Analytical Performance. Under the optimized conditions described above, analytical parameters of the proposed HPLC-UV-Vis method such as linearity, sensitivity, correlation coefficients, limits of detection, and precisions were evaluated. Calibration graphs were constructed by plotting the peak area versus the concentration of analytes. For the tested Ta and SY, linearity was observed in the concentration range of 0.01–50 μg/mL. The other parameters are summarized in Table 1. The results demonstrate that the method is sensitive for the determination of Ta and SY in all studied samples. The intraday precision of the proposed method was tested with 6 repeated injections of solution containing Ta and SY standards at the concentration level of 0.5 μg/mL. The relative standard deviations (RSD) were ≤3.2%. The obtained RSD values indicate a satisfactory precision of the proposed method. The interday precision, expressed as RSD of the slopes of the calibration graphs obtained in three different days, was ≤5.6%. Due to the fact that the RSD is higher than 5%, the calibration graphs should be registered the same day as the real sample is analyzed.

3.5. Sample Analysis. Under the optimal conditions, the proposed [C_4MIM]Br/K_2HPO_4 APTS-HPLC method has been adopted to determine Ta and SY in candy samples. Typical chromatograms of Ta and SY in the standard solution and candy sample were shown in Figure 7. Curves (a) and (b) were the chromatograms of the mixed standard solution (Ta + SY) before and after IL-salt ATPS extraction, respectively. Curve (c) was chromatogram of candy sample by direct sampling without treatment. Curves (d) and (e)

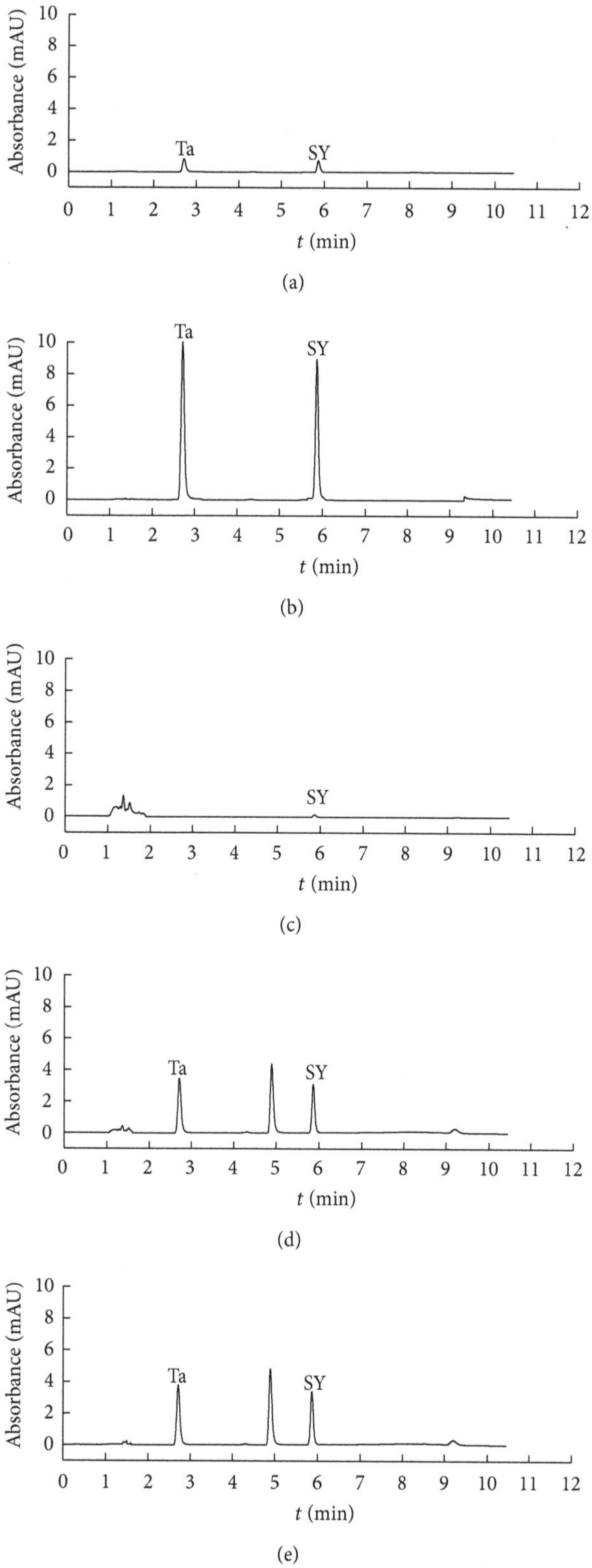

Figure 7: The chromatograms of standard solution and sample solution before and after extraction. (a) Standard solution of 0.25 μg/mL Ta and 0.25 μg/mL SY; (b) standard solution of 0.25 μg/mL Ta and 0.25 μg/mL SY after IL-salt ATPS extraction; (c) candy sample solution without pretreatment; (d) candy sample solution with polyamide adsorption method; (e) and candy sample solution with IL-salt ATPS (V_{IL} = 0.30 mL, $m_{K_2HPO_4}$ = 3.0 g).

were chromatograms of candy sample enriched by polyamide adsorption and IL-salt ATPS, respectively. Figure 7 showed that (1) the standard solution (curve (a)) and real sample (curve (c)) have been enriched (curves (b) and curve (d)) by IL-salt ATPS extraction; (2) the response signal of Ta and SY determined by IL-salt ATPS extraction-HPLC was consistent with GB/T 5009.35-2003 method (curves (d) and (e)).

The proposed procedure has been applied to the determination of Ta and SY in soft drink samples, candy samples, and instant powdered drink. The recoveries of Ta and SY in different food samples were in the range of 93.5–97.4% and 94.0–97.2%, respectively (Table 2). When compared with those achieved by the GB/T 5009.35-2003 method, as shown in Table 2, no significant difference was observed between the proposed method and GB/T 5009.35-2003 ($P > 0.05$).

4. Conclusion

A simple and rapid IL-based ATPS consisting of $[C_4MIM]Br$ and K_2HPO_4 coupled with HPLC-UV was developed for the sensitive and simultaneous determination of sunset yellow and tartrazine in soft drink samples and candy samples. In this research, direct injection of the $[C_4MIM]Br$ phase into HPLC system for the quantification of Ta and SY was proposed. This method proved to be efficient, simple and fast for separation/analysis of Ta and SY in soft drink samples, candy samples and instant powdered drink.

Disclosure

The research proposed in this paper does not contain any studies with human or animals subjects for the following reasons: (1) the materials used in this work are common chemical reagents without any animals and human involved. (2) The real samples studied in our work are local soft drinks, candy samples, and instant powder drink, which can be easily purchased and obtained from local markets. Therefore, there are no ethical issues with human or animal subjects in our study.

Conflict of Interests

The authors have financial relationships with the organizations that sponsored the research, listed as follows. (1) Professor Xiashi Zhu has received research grants from the National Natural Science Foundation of China (20875082, 2115500) and Foundation of the Priority Academic Program Development of Jiangsu Higher Education Institutions. (2) Mrs. Ou Sha has received the project funded by the Science and Technology Project of Lianyungang City (SH1209) and the Open Fund Project of Science and Technology of Jiangsu Marine Resources Development Research Institute (JSIMR201202). (3) Mr Weixing Ma has received the project funded by the Science and Technology Project of Lianyungang City (SH1209) and the Open Fund Project of Science and Technology of Jiangsu Marine Resources Development Research Institute (JSIMR201202).

TABLE 2: Determination of SY and Ta in food sample and recovery test ($n = 3$).

Sample	Ta[a] (μg/g)				SY[a] (μg/g)			
	Added	Found	Recovery	GB-found[b]	Added	Found	Recovery	GB-found[b]
Soft drink 1	—	6.64 ± 0.14	—	6.51 ± 0.21	—	NT	—	ND
	5.0	11.46 ± 0.16	96.4%		5.0	4.86 ± 0.16	97.2%	
	10.0	16.38 ± 0.14	97.4%		10.0	9.45 ± 011	94.5%	
Soft drink 2	—	1.42 ± 0.24	—	1.38 ± 0.15	—	17.67 ± 0.32	—	17.09 ± 0.84
	2.00	3.32 ± 0.10	95.0%		10.00	27.18 ± 0.25	95.1%	
	5.00	6.21 ± 0.18	95.8%		20.00	36.59 ± 0.16	94.6%	
Soft drink 3	—	ND	—	ND	—	22.76 ± 0.23	—	21.86 ± 0.58
	5.00	4.76 ± 0.12	95.2%		15.00	36.98 ± 0.21	94.8%	
	10.00	9.41 ± 0.26	94.1%		20.00	41.88 ± 0.15	95.6%	
Candy 1	—	0.11 ± 0.05	—	1.11 ± 0.13	—	1.26 ± 0.15	—	1.20 ± 0.09
	2.00	2.07 ± 0.14	98.0%		2.00	3.19 ± 0.21	95.5%	
	4.00	3.85 ± 0.18	93.5%		4.00	5.04 ± 0.15	94.5%	
Candy 2	—	ND	—	ND	—	0.39 ± 0.02	—	0.38 ± 0.09
	0.50	0.47 ± 0.12	94.0%		0.50	0.86 ± 0.05	94.0%	
	1.00	0.94 ± 0.13	94.0%		1.00	1.34 ± 0.07	95.0%	
Papaya powder	—	ND	—	ND	—	1.50 ± 0.17	—	1.64 ± 0.09
	2.00	1.89 ± 0.10	94.5%		3.00	4.34 ± 0.13	94.7%	
	5.00	4.82 ± 0.18	96.4%		6.00	7.21 ± 0.12	95.2%	

[a] $\overline{x} \pm ts/\sqrt{n}$ at 95% confidence ($n = 3$).
[b] GB/T 5009.35-2003.
ND: not detected.

Acknowledgments

The authors acknowledge the financial support from the National Natural Science Foundation of China (21375117; 21155001), Foundation of the Priority Academic Program Development of Jiangsu Higher Education Institutions and the Project funded by the Science and Technology Project of Lianyungang city (SH1209), and the Open Fund Project of Science and Technology of Jiangsu Marine Resources Development Research Institute (JSIMR201202).

References

[1] M. Kucharska and J. Grabka, "A review of chromatographic methods for determination of synthetic food dyes," *Talanta*, vol. 80, no. 3, pp. 1045–1051, 2010.

[2] M. M. Kamel and S. H. El-lethey, "The potential health hazard of tartrazine and levels of hyperactivity, anxiety-like symptoms, depression and anti-social behaviour in rats," *Journal of American Science*, vol. 7, pp. 1211–1218, 2011.

[3] K. S. Khera and I. C. Munro, "A review of the specifications and toxicity of synthetic food colors permitted in Canada," *CRC Critical Reviews in Toxicology*, vol. 6, no. 2, pp. 81–133, 1979.

[4] N. E. Llamas, M. Garrido, M. S. D. Nezio, and B. S. F. Band, "Second order advantage in the determination of amaranth, sunset yellow FCF and tartrazine by UV-vis and multivariate curve resolution-alternating least squares," *Analytica Chimica Acta*, vol. 655, no. 1-2, pp. 38–42, 2009.

[5] N. Dossi, R. Toniolo, A. Pizzariello, S. Susmel, F. Perennes, and G. Bontempelli, "A capillary electrophoresis microsystem for the rapid in-channel amperometric detection of synthetic dyes in food," *Journal of Electroanalytical Chemistry*, vol. 601, no. 1-2, pp. 1–7, 2007.

[6] S. Chanlon, L. Joly-Pottuz, M. Chatelut, O. Vittori, and J. L. Cretier, "Determination of Carmoisine, Allura red and Ponceau 4R in sweets and soft drinks by differential pulse polarography," *Journal of Food Composition and Analysis*, vol. 18, no. 6, pp. 503–515, 2005.

[7] S. P. Alves, D. M. Brum, É. C. Branco de Andrade, and A. D. Pereira Netto, "Determination of synthetic dyes in selected foodstuffs by high performance liquid chromatography with UV-DAD detection," *Food Chemistry*, vol. 107, no. 1, pp. 489–496, 2008.

[8] M. L. S. Silva, M. B. Q. Garcia, J. L. F. C. Lima, and E. Barrado, "Voltammetric determination of food colorants using a polyallylamine modified tubular electrode in a multicommutated flow system," *Talanta*, vol. 72, no. 1, pp. 282–288, 2007.

[9] N. Pourreza and M. Ghomi, "Simultaneous cloud point extraction and spectrophotometric determination of carmoisine and brilliant blue FCF in food samples," *Talanta*, vol. 84, no. 1, pp. 240–243, 2011.

[10] M. S. El-Shahawi, A. Hamza, A. A. Al-Sibaai, A. S. Bashammakh, and H. M. Al-Saidi, "A new method for analysis of sunset yellow in food samples based on cloud point extraction prior to spectrophotometric determination," *Journal of Industrial and Engineering Chemistry*, vol. 19, no. 2, pp. 529–535, 2013.

[11] Z. Zhang, Z. Zhang, Y. Fernández et al., "Adsorption isotherms and kinetics of methylene blue on a low-cost adsorbent recovered from a spent catalyst of vinyl acetate synthesis," *Applied Surface Science*, vol. 256, no. 8, pp. 2569–2576, 2010.

[12] Y. C. Sharma, "Optimization of parameters for adsorption of methylene blue on a low-cost activated carbon," *Journal of Chemical and Engineering Data*, vol. 55, no. 1, pp. 435–439, 2010.

[13] Determination of synthetic colour in foods, GB/T 5009.35-2003, China, 2003.

[14] F. Ruiz-Ruiz, J. Benavides, O. Aguilar, and M. Rito-Palomares, "Aqueous two-phase affinity partitioning systems: current applications and trends," *Journal of Chromatography A*, vol. 1244, pp. 1–13, 2012.

[15] R. Hamamoto, M. Kamihira, and S. Iijima, "Specific separation of animal cells using aqueous two-phase systems," *Journal of Fermentation and Bioengineering*, vol. 82, no. 1, pp. 73–76, 1996.

[16] H. D. Willauer, J. G. Huddleston, and R. D. Rogers, "Solute partitioning in aqueous biphasic systems composed of polyethylene glycol and salt: The partitioning of small neutral organic species," *Industrial and Engineering Chemistry Research*, vol. 41, no. 7, pp. 1892–1904, 2002.

[17] J. L. Anderson, J. Ding, T. Welton, and D. W. Armstrong, "Characterizing ionic liquids on the basis of multiple solvation interactions," *Journal of the American Chemical Society*, vol. 124, no. 47, pp. 14247–14254, 2002.

[18] K. E. Gutowski, G. A. Broker, H. D. Willauer et al., "Controlling the aqueous miscibility of ionic liquids: aqueous biphasic systems of water-miscible ionic liquids and water-structuring salts for recycle, metathesis, and separations," *Journal of the American Chemical Society*, vol. 125, no. 22, pp. 6632–6633, 2003.

[19] Y. Wang, J. Han, J. Liu, Y. Hu, C. Sheng, and Y. Wu, "Liquid-liquid equilibrium phase behavior of iminazolium-based ionic liquid aqueous two-phase systems composed of 1-alkyl-3-methyl imidazolium tetrafluoroborate and different electrolytes $ZnSO_4$, $MgSO_4$ and Li_2SO_4 at 298.15 K: experimental and correlation," *Thermochimica Acta*, vol. 557, pp. 68–76, 2013.

[20] Y. Li, M. Zhang, H. Su, Q. Liu, and W. Guan, "Liquid-liquid equilibria of aqueous two-phase systems of the ionic liquid brominated N-ethyl pyridine and sodium dihydrogen phosphate, sodium sulfate, ammonium citrate, and potassium tartrate at different temperatures: Experimental determination and correlation," *Fluid Phase Equilibria*, vol. 341, pp. 70–77, 2013.

[21] J. G. Huddleston, A. E. Visser, W. M. Reichert, H. D. Willauer, G. A. Broker, and R. D. Rogers, "Characterization and comparison of hydrophilic and hydrophobic room temperature ionic liquids incorporating the imidazolium cation," *Green Chemistry*, vol. 3, no. 4, pp. 156–164, 2001.

[22] P. A. Albertsson, "Inorganic compound," in *Partition of Cell Particles and Macromoleciles*, p. 23, Wiley, New York, NY, USA, 3rd edition, 1986.

[23] D. Vollhardt and V. B. Fainerman, "Characterisation of phase transition in adsorbed monolayers at the air/water interface," *Advances in Colloid and Interface Science*, vol. 154, no. 1-2, pp. 1–19, 2010.

[24] J. A. Deanm, *Lange's Handbook of Chemistry*, McGraw Hill, New York, NY, USA, 1985.

[25] Wuhan University, *Analytical Chemistry. Section 3. Inorganic Compound, 3.12*, Higher Education Press, Wuhan, China, 3rd edition, 1995.

Permissions

All chapters in this book were first published in JAMC, by Hindawi Publishing Corporation; hereby published with permission under the Creative Commons Attribution License or equivalent. Every chapter published in this book has been scrutinized by our experts. Their significance has been extensively debated. The topics covered herein carry significant findings which will fuel the growth of the discipline. They may even be implemented as practical applications or may be referred to as a beginning point for another development.

The contributors of this book come from diverse backgrounds, making this book a truly international effort. This book will bring forth new frontiers with its revolutionizing research information and detailed analysis of the nascent developments around the world.

We would like to thank all the contributing authors for lending their expertise to make the book truly unique. They have played a crucial role in the development of this book. Without their invaluable contributions this book wouldn't have been possible. They have made vital efforts to compile up to date information on the varied aspects of this subject to make this book a valuable addition to the collection of many professionals and students.

This book was conceptualized with the vision of imparting up-to-date information and advanced data in this field. To ensure the same, a matchless editorial board was set up. Every individual on the board went through rigorous rounds of assessment to prove their worth. After which they invested a large part of their time researching and compiling the most relevant data for our readers.

The editorial board has been involved in producing this book since its inception. They have spent rigorous hours researching and exploring the diverse topics which have resulted in the successful publishing of this book. They have passed on their knowledge of decades through this book. To expedite this challenging task, the publisher supported the team at every step. A small team of assistant editors was also appointed to further simplify the editing procedure and attain best results for the readers.

Apart from the editorial board, the designing team has also invested a significant amount of their time in understanding the subject and creating the most relevant covers. They scrutinized every image to scout for the most suitable representation of the subject and create an appropriate cover for the book.

The publishing team has been an ardent support to the editorial, designing and production team. Their endless efforts to recruit the best for this project, has resulted in the accomplishment of this book. They are a veteran in the field of academics and their pool of knowledge is as vast as their experience in printing. Their expertise and guidance has proved useful at every step. Their uncompromising quality standards have made this book an exceptional effort. Their encouragement from time to time has been an inspiration for everyone.

The publisher and the editorial board hope that this book will prove to be a valuable piece of knowledge for researchers, students, practitioners and scholars across the globe.

List of Contributors

Sara Di Salvo, Martina Casalini, Sara Marchionni, Teresa Adani, Maurizio Ulivi, Simone Tommasini and Paul P. A. Mazza
Dipartimento di Scienze della Terra, Universit`a degli Studi di Firenze, via Giorgio La Pira 4, 50121 Firenze, Italy

Riccardo Avanzinelli, Sandro Conticelli and Lorella Francalanci
Dipartimento di Scienze della Terra, Universit`a degli Studi di Firenze, via Giorgio La Pira 4, 50121 Firenze, Italy
C.N.R., Istituto Geoscienze e Georisorse, U.O. di Firenze, via Giorgio La Pira 4, 50121 Firenze, Italy

Eleonora Braschi and Andrea Orlando
C.N.R., Istituto Geoscienze e Georisorse, U.O. di Firenze, via Giorgio La Pira 4, 50121 Firenze, Italy

Feng Guan, Yu-Ting Jin, Jin Zhao, Ai-Chun Xu and Yuan-Yuan Luo
College of Life Sciences, China Jiliang University, Hangzhou 310018, China

Sabriye Yusan
Department of Chemistry, Faculty of Science, Universiti Teknologi Malaysia, 81310 Johor Bahru, Johor, Malaysia
Institute of Nuclear Science, Ege University, Bornova, 35100 Izmir, Turkey

Mokhlesur M. Rahman, Nasir Mohamad and Tengku M. Arrif
Institute for Community Development & Quality of Life (i-CODE), Universiti Sultan Zainal Abidin, 21300 Kuala Nerus, Terengganu, Malaysia

Ahmad Zubaidi A. Latif
Faculty of Medicine, Universiti Sultan Zainal Abidin, 21300 Kuala Nerus, Terengganu, Malaysia

Mohd Aznan M. A.
Faculty of Medicine, International Islamic University Malaysia, 25200 Kuantan, Malaysia

Wan Sani B. Wan Nik
School of Ocean Engineering, Universiti Malaysia Terengganu, 21300 Kuala Nerus, Terengganu, Malaysia

Peng Li
The Graduate School, Tianjin Medical University, 22 Qixiangtai Road, Heping District, Tianjin 300071, China

Yongcheng Hu
Department of Orthopedic Oncology, Tianjin Hospital, 406 Jiefang South Road, Tianjin 300210, China

Mes in Redi-Abshiro, Bhagwan Singh Chandravanshi and Estifanos Ele
Department of Chemistry, Addis Ababa University, Addis Ababa, Ethiopia

Ayalew Debebe
Department of Chemistry, Addis Ababa University, Addis Ababa, Ethiopia
Department of Chemistry, Haramaya University, Dire Dawa, Ethiopia

Shibru Temesgen
Department of Statistics, Addis Ababa University, Addis Ababa, Ethiopia

Qian Wang
Department of Animal Science and Veterinary Medicine, Tianjin Agricultural University, Tianjin 300384, China
College of Food Science and Engineering, Northwest A&F University, Yangling, Shaanxi 712100, China

Rui Zhao
Department of Food Science and Biological Engineering, Tianjin Agricultural University, Tianjin 300384, China

Yan Cui, Cuicui Liu, Liya Liang and Shijie Yan
Department of Food Science and Biological Engineering, Tianjin Agricultural University, Tianjin 300384, China
Tianjin Engineering and Technology Research Center of Agricultural Products Processing, Tianjin 300384, China

Rui Tan
College of Medcine, Southwest Jiaotong University, Chengdu 610031, China

Jiaji Ding
College of Medcine, Southwest Jiaotong University, Chengdu 610031, China
Institute of Medicinal Plant Development, Chinese Academy of Medical Sciences & Peking Union Medical College, Beijing 100193, China

Caimei Gu and Linfang Huang
Institute of Medicinal Plant Development, Chinese Academy of Medical Sciences & Peking Union Medical College, Beijing 100193, China

Xiaojie Sun, Haiyan Ding, Lihong Xing, Yuxiu Zhai and Zhaoxin Li
Key Laboratory of Testing and Evaluation for Aquatic Product Safety and Quality, Ministry of Agriculture and Rural Affairs, Yellow Sea Fisheries Research Institute, Chinese Academy of Fishery Sciences, Qingdao 266071, China

Jie Tan
Key Laboratory of Testing and Evaluation for Aquatic Product Safety and Quality, Ministry of Agriculture and Rural Affairs, Yellow Sea Fisheries Research Institute, Chinese Academy of Fishery Sciences, Qingdao 266071, China
Qingdao National Laboratory for Marine Science and Technology, Qingdao 266000, China

Xiaojie Tan
The Affiliated Hospital of Qingdao University, Qingdao 266000, China

Jun Xing
Key Laboratory of Analytical Chemistry for Biology and Medicine, Ministry of Education, College of Chemistry and Molecular Sciences, Wuhan University, Wuhan 430072, China

Roberta G. Machicote and Marcela A. Castillo
División Química Analítica, Departamento de Química, Facultad de Ciencias Exactas, Universidad Nacional de La Plata, Calle 47, Esq. 115, 1900 La Plata, Argentina

Maria E. Pacheco and Liliana Bruzzone
Laboratorio de Investigación y Desarrollo de Métodos Analíticos (LIDMA), Facultad de Ciencias Exactas, Universidad Nacional de La Plata, Calle 47, Esq. 115, 1900 La Plata, Argentina

Katarzyna Madej and Katarzyna Janiga
Department of Analytical Chemistry, Faculty of Chemistry, Jagiellonian University, Krakow 30-387, Poland

Wojciech Piekoszewski
Department of Analytical Chemistry, Faculty of Chemistry, Jagiellonian University, Krakow 30-387, Poland
Department of Food Science and Technology, School of Biomedicine, Far Eastern Federal University, Vladivostok, Russia

Yong Dae Park, Sungbeom Lee, Chul-Hong Park, Hyoung-Woo Bai and Seung Sik Lee
Research Division for Biotechnology, Advanced Radiation Technology Institute, Korea Atomic Energy Research Institute, Jeongeup 580-185, Republic of Korea

Jeum-Jong Kim
Technology Innovation Support Team, Korea Research Institute of Chemical Technology (KRICT), Deajeon 305-600, Republic of Korea

Jihye Lee and Yeonhee Lee
Advanced Analysis Center, Korea Institute of Science and Technology, Seoul 02792, Republic of Korea

Yun Jung Jang
Advanced Analysis Center, Korea Institute of Science and Technology, Seoul 02792, Republic of Korea
Department of Materials Science and Engineering, Korea University, Seoul 02841, Republic of Korea

Donghwan Kim
Department of Materials Science and Engineering, Korea University, Seoul 02841, Republic of Korea

Kang-Bong Lee
Green City Technology Institute, Korea Institute of Science and Technology, Seoul 02792, Republic of Korea

Shoujuan Wang, ZhongmingLiu, Xin Zhao, Guihua Yang and Fangong Kong
State Key Lab of Biobased Material and Green Papermaking, Key Laboratory of Pulp and Paper Science and Technology Ministry of Education, Qilu University of Technology (Shandong Academy of Sciences), Jinan, Shandong 250353, China

Zhen Xu and Yongyi Fang
Xuancheng Product Quality Supervision and Inspection Institute, Xuan Cheng, Anhui 242000, China

Da-Som Kim, Hoe-Sung Kim and Eui-Cheol Shin
Department of Food Science, Gyeongnam National University of Science and Technology, Jinju 52725, Republic of Korea

Kyoung-Tae Lee and Yong Bae Park
Southern Forest Resources Research Center, National Institute of Forest Science, Jinju 52817, Republic of Korea

Dong-Lee Hong, Sung-Rae Cho and Yang-Bong Lee
Department of Food Science and Technology, Pukyong National University of Science and Technology, Busan 48513, Republic of Korea

Jeong Hoon Pan and Jae Kyeom Kim
School of Human Environmental Sciences, University of Arkansas, Fayetteville, AR 72701, USA

Sang A Park, Soi Jeong, Young Ho Choe and Young-Min Hyun
Department of Anatomy and Brain Korea 21 PLUS Project for Medical Science, Yonsei University College of Medicine, Seoul 03722, Republic of Korea

Samuel Melaku, Zewdu Gebeyehu and Rajeev B. Dabke
Department of Chemistry, Columbus State University, Columbus, GA 31907, USA

Amani Taamalli and Mokhtar Zarrouk
Laboratoire de Biotechnologie de l'Olivier, Centre de Biotechnologie de Borj-Cedria, Hammam-Lif, Tunisia

David Arráez Román, Ana María Gómez Caravaca and Antonio Segura Carretero
Center of Research and Development of Functional Food, Health Science Technological Park, Avda. del Conocimiento s/n, 18100 Granada, Spain
Department of Analytical Chemistry, Faculty of Sciences, University of Granada, 18071 Granada, Spain

Hussain Al-Thaiban, Nada Al-Tamimi and Murad Helaleh
Anti Doping Lab-Qatar, Toxicology Multipurpose laboratory, Doha, Qatar

Ioannis K. Karabagias
Laboratory of Food Chemistry, Department of Chemistry, University of Ioannina, Ioannina 45110, Greece

Artemis P. Louppis and Chara Papastephanou
cp Foodlab Ltd., Polifonti 25, 2047 Strovolos, Nicosia, Cyprus

Stavros Kontakos
Department of Social Administration and Political Science, Democritus University of)race, Komotini 69100, Greece

Chryssoula Drouza
Department of Agricultural Sciences, Biotechnology and Food Science, Cyprus University of Technology, 3036 Limassol, Cyprus

Qianchun Zhang, Siqi Xie, Yanqun Yang, Yun Wu, Xingyi Wang, Jincheng Wu, Li Zhang, Junyu Chen and Yuan Wang
School of Biology and Chemistry, Xingyi Normal University for Nationalities, Xingyi 562400, China

Svetla Todinova
Institute of Biophysics and Biomedical Engineering, Bulgarian Academy of Sciences, Bl. 21, Acad. G. Bonchev Str., Sofia 1113, Bulgaria

Yuliana Raynova and Krassimira Idakieva
Institute of Organic Chemistry with Centre of Phytochemistry, Bulgarian Academy of Sciences, Bl. 9, Acad. G. Bonchev Str., Sofia 1113, Bulgaria

N. M. Dubova, G. B. Slepchenko and M. S. Ostapenko
Physical and Analytical Chemistry Department, National Research Tomsk Polytechnic University, Tomsk, Russia

I. A. Khlusov
Experimental Physics Department, National Research Tomsk Polytechnic University, Tomsk, Russia
Laboratory of Immunology and Cell Biotechnologies, Immanuel Kant Baltic Federal University, Kaliningrad, Russia
Morphology and General Pathology Department, Siberian State Medical University, Tomsk, Russia

E. A. Nesterov
Laboratory No. 31 of the Nuclear Reactor, National Research Tomsk Polytechnic University, Tomsk, Russia

Michaela Dufka and Bohumil Dočekal
Institute of Analytical Chemistry of the Czech Academy of Sciences, v.v.i., Veveří 97, 602 00 Brno, Czech Republic

Stefan F. Mühlebach
Clinical Pharmacy and Epidemiology, Hospital Pharmacy, University of Basel, Spitalstrasse 26, CH-4031 Basel, Switzerland

Raphael Hösli
Clinical Pharmacy and Epidemiology, Hospital Pharmacy, University of Basel, Spitalstrasse 26, CH-4031 Basel, Switzerland
Spitalzentrum Biel, Apotheke, Vogelsang 84, CH-2501 Biel-Bienne, Switzerland

Stefan König
Division of Forensic Medicine, University of Bern, Bühlstrasse 20, CH-3012 Bern, Switzerland

Index

www.ingramcontent.com/pod-product-compliance
Lightning Source LLC
LaVergne TN
LVHW081015210926
848485LV00004B/869
9781639890477